INTRODUCTION TO PROGRAMMING AND PROBLEM-SOLVING USING
SCALA

SECOND EDITION

CHAPMAN & HALL/CRC
TEXTBOOKS IN COMPUTING

Series Editors

John Impagliazzo
Professor Emeritus, Hofstra University

Andrew McGettrick
Department of Computer
and Information Sciences
University of Strathclyde

Aims and Scope

This series covers traditional areas of computing, as well as related technical areas, such as software engineering, artificial intelligence, computer engineering, information systems, and information technology. The series will accommodate textbooks for undergraduate and graduate students, generally adhering to worldwide curriculum standards from professional societies. The editors wish to encourage new and imaginative ideas and proposals, and are keen to help and encourage new authors. The editors welcome proposals that: provide groundbreaking and imaginative perspectives on aspects of computing; present topics in a new and exciting context; open up opportunities for emerging areas, such as multi-media, security, and mobile systems; capture new developments and applications in emerging fields of computing; and address topics that provide support for computing, such as mathematics, statistics, life and physical sciences, and business.

Published Titles

CHAPMAN & HALL/CRC
TEXTBOOKS IN COMPUTING

INTRODUCTION TO PROGRAMMING AND PROBLEM-SOLVING USING
SCALA

SECOND EDITION

Mark C. Lewis
Lisa L. Lacher

CRC Press
Taylor & Francis Group
Boca Raton London New York

CRC Press is an imprint of the
Taylor & Francis Group, an **informa** business

A CHAPMAN & HALL BOOK

CRC Press
Taylor & Francis Group
6000 Broken Sound Parkway NW, Suite 300
Boca Raton, FL 33487-2742

© 2017 by Taylor & Francis Group, LLC
CRC Press is an imprint of Taylor & Francis Group, an Informa business

No claim to original U.S. Government works

Printed on acid-free paper
Version Date: 20160607

International Standard Book Number-13: 978-1-4987-3095-2 (Pack - Book and Ebook)

Visit the Taylor & Francis Web site at
http://www.taylorandfrancis.com

and the CRC Press Web site at
http://www.crcpress.com

Printed and bound by CPI Group (UK) Ltd, Croydon, CR0 4YY

Contents

List of Figures

List of Tables

Preface

Welcome to "Introduction to Programming and Problem Solving Using Scala". This book is intended to be used in first semester college classrooms to teach students beginning programming. To accomplish this task, the Scala[1] programming language is used. The book was constructed with a focus on the topics students need to know. These topics were then woven into a format that was deemed the best way to communicate those ideas. Because there are two very different audiences who might be reading this, the rest of the preface is split into two sections.

To the Student

Welcome to the world of programming. You are about to embark on a field of study that will hopefully open your mind to new ways of thinking and impact the way you view everything in the world around you. For students who intend to major in Computer Science and make careers working with computer technology, this is your first step in learning how to communicate with the computer and to instruct the computer in how it should solve problems for you.

For those who are not planning to make a career in a computing field, a course in computer programming can still be remarkably beneficial. Computer programming is, fundamentally, about problem solving. It is about figuring out how to solve problems and express them in ways that the computer can understand. Your entire future life is, in one way or another, going to deal with how to solve problems. You will need to have approaches to solving these problems well formed in your own head and be able to communicate them to other people in non-ambiguous ways so that they understand what you want and can act on it. Learning to program computers will help you develop and hone this ability.

There are more direct benefits as well. The world you live in is already extremely dependent on computing. Many of the things you own that you do not call a computer include microchips and digital processing. Those devices/objects run programs to complete their tasks. Many of the tasks you do during an average day are also enabled by computer power and the ability to process and store information quickly. Activities like browsing the web for work or leisure explicitly involve this. Other activities such as financial transactions made with anything other than cash implicitly involve it. This ubiquitous presence of digital processing is only going to grow over time. So even if you never have to directly write programs as part of your future life, you can only benefit from having an understanding of what is going on under the hood. Even if you do not write the software yourself, there is a good chance that you will interact with those who do, perhaps as you or your company strives to gain a competitive advantage by improving the information technology you use to do your

---

[1]Code samples in this book were compiled and tested using Scala 2.12.x.

work. Even a basic knowledge of programming will help you understand what these people are doing and communicate more effectively with them.

Those who do not intend to make programming into a career might occasionally come across problems where they realize that they could save a lot of time if they could get a computer to do it for them. When that happens, even a basic knowledge of programming can make your life a whole lot better. Data is becoming a bigger and bigger part of the business world. The vast amounts of data created and collected by modern technology can not be processed by humans manually. To sift through this data to find meaning requires using machines, and that means using software. You can either pay someone else to write it or you can write it yourself. That latter is only an option if you know how to program.

While many might be tempted to view programming as an advanced concept, the reality is this book is going to take you back to your early education quite a few times. The reason for this is you spent a lot of time in those early years learning how to solve certain basic problems. The approaches became second nature to you and you no longer think about them, you just do them. In the context of programming you have to go back and examine how you solve problems, even ones you have been doing for a long time. The reason for this is that now you have to tell a computer how to do those same things, and the computer will need detailed instructions. The reality is, the programming presented in this book is not an advanced topic, it is basic logic and problem solving, done at a level you probably have not worked at since you were much younger.

We often refer to programming as an art. Programming shares many features with traditional arts like creative writing and painting or sculpture. The programmer first creates an image in his/her mind of what he/she wants to bring into existence. This is followed by a period of work bringing that mental image into a more real form. The end result is something new, born from the imagination of the creator, that can be experienced by others.[2] The digital medium might, in some sense, seem less real than paint, clay, or stone, but it is also remarkably dynamic. You can create things in a computer that you have no chance of creating in the real world by hand. You can also create things that are remarkably useful. All the software you interact with every day, whether to get things done or just for entertainment, was written by programmers. The exercises and projects in this book have been created with the goal of giving you the ability to express your creativity.

Programming has another characteristic in common with other creative arts, if you want to be good at it, you need to practice. There are many ways to program the solution to any problem. That introduces a lot of nuance. This book will strive to instruct you in regards to the strengths and weaknesses of different approaches, but to really understand how to write good code to solve a problem, you need to have the experience of solving similar ones. Imagine an art major who never draws or paints except for class projects, not even doodling in notes, or a creative writing major who never writes a story except the ones required for class. Hopefully those hypothetical people seem silly to you. A computer science major who never writes code beyond what is assigned for class is exactly the same. So explore the art of programming. Have fun with it and try to do something that interests you.

Some people might wonder why they should be using Scala to learn programming. Scala is a fairly new language that has gained a lot of momentum since it was adopted by Twitter in 2009, but there are definitely other languages that are used more for professional development. Scala is newer than languages like Java and Python, and this means that it has integrated lessons learned from those languages. It has more modern features that those languages are struggling to integrate. In particular, Scala has found significant use in developing web sites that have lots of users and doing big data analysis. However, the

---

[2]We particularly like the tag line of @muddlymon on Twitter during the time we were writing this book, which read "I make things out of ideas by wiggling my fingers."

real reasons for starting in Scala stem from the advantages it brings to beginning programmers. Scala has tools like Python and other scripting languages that make it easy to write small programs and to get started without dealing with too much overhead. Unlike Python and other scripting languages, it does robust error checking to help you figure out when you mess things up. Also, Scala has strong support for object-oriented programming and functional programming, so it is able to keep growing with you beyond your first course in programming. You can branch out to more advanced topics without having to learn a different language. In addition, Scala is powerful and expressive, so that you can do a lot of things easily. We consider this to be a significant benefit for the novice programmer as you want to write things that are interesting sooner rather than later.

Using this Book

The goal of this book is to help you in your efforts to learn how to program. In addition to the practice that was just mentioned, there are some suggestions for how to get the most out of this book. Clearly, working as many exercises and projects as you can counts as practice. In addition, you should really read descriptions in this book and make sure you understand the code samples shown in it. All too often, students treat the task of reading course material as sounding out all the words in your head. While that fits the denotative definition, the connotation is that you string the words together and understand what they mean. Read for understanding, not just so you can say you have gotten through a chapter. You should also "follow along" with the material by writing code. The best way to internalize the details of a language, and you are learning a new language, is to type things in, so that you have to pay attention to the details. Gaining understanding of early material is especially important in this topic because the material in any given chapter is typically highly dependent on the chapters before it. If you fail to understand one chapter, you are not likely to understand the ones that follow it either. As such, effort spent on the early chapters will pay off later on. Fortunately, there are resources that accompany this book that can help with this.

Students coming into this course who have a background in programming should consider taking a "fast path" to chapters 5 and 6. You do not want to skip them completely as there are aspects of Scala that will likely differ from what you have done before. One way to approach this is the go to the end of any given chapter and try to do the exercises. Refer back to the chapter as needed to complete them. Once you can do the exercises and a project or two, you should feel comfortable moving on.

Book Website

The authors have posted a number of different supplements to this book at `http://book.programmingusingscala.net` that can help you work through the material. Some of the material is code or prose that could not be put into the book due to length restrictions. The most useful things on the website are things that cannot be represented in static text. In addition, the code samples for the book have been posted at `https://github.com/MarkCLewis/ProblemSolvingUsingScala`.

There are video lectures posted on a YouTube channel (`https://www.youtube.com/channel/UCEvjiWkK2BoIH819T-buioQ`) that follow along with the material in this book. They are organized in playlists by chapter. Programming is a very non-linear process. This book typically shows completed pieces of code that do what we want. It is hard to demonstrate the process of building that code in the format of a book. The videos show construction of code from the ground up and include descriptions of the programmers thoughts as it is being done. This type of "live coding" is invaluable in learning to program as it lets you into the mind of a more experienced programmer where you can see the thought processes associated with the development of the code.

The web site also includes solutions of some exercises for you to reference along with sample implementations of certain projects. Some of the projects, especially in the second half of the book, can be challenging to describe in text. To give you a better idea of what is expected, the author has implemented sample solutions that you can run to see what they do. The site also includes some additional exercises and links to data files and remote sites that make the exercises more relevant.

To the Instructor

If you are reading this, it likely means that you are already aware of many of the features of Scala that make it a great programming language.[3] The flexibility of Scala means that things can be covered in many different ways. The approach taken in this book might be summarized as semi-functional/semi-imperative with objects later. It is worth describing exactly what is meant by that.

This book takes advantage of the aspects of Scala that support programming in the small. It is expected that students will operate mainly in the REPL and in a scripting environment through at least the first 10 chapters and possibly through the first 15. The benefit of this is that you can focus on logic instead of extra keywords and scoping rules. Students are easily overwhelmed early on and this approach helps to flatten the learning curve a bit.

Scala is purely object-oriented, so students will be using objects and calling methods on objects very early on. However, the construction of classes to build their own objects is postponed until later.[4] Fitting with this approach, programs are written as scripts, not as applications, until the last chapter. So the `object` keyword and things like `def main(args: Array[String]) : Unit` are postponed until students are ready to write programs that are big enough that they need the organization provided by doing proper OO across multiple files.

The approach is described as semi-functional because there is significant use of higher order functions with function literals/lambda expressions and immutable style is often preferred. However, this is not enforced in a zealotous way, and mutable state is introduced early on. Students are shown mutable and immutable approaches to a number of problems and the benefits and pitfalls of each approach are discussed.

While the ideal way to go through this book is in a linear fashion, particularly if you want to use the end-of-chapter projects as assignments, strict linearity is not required and you can choose to move some topics around so that concepts like class declarations are covered earlier. Though it would be a bit more challenging, it would be possible to use this book in a more purely functional manner by simply having students avoid var declarations, assignments into mutable collections like arrays, and the use of while loops.

---

[3]If you are currently unconvinced of the benefits of Scala for teaching CS1 and CS2, there is a complete discussion at http://book.programmingusingscala.net.

[4]That material is covered very briefly at the end of this volume, but most of the treatment for doing real OO is postponed to the second volume "Object-orientation, Abstraction, and Data Structure Using Scala"[1].

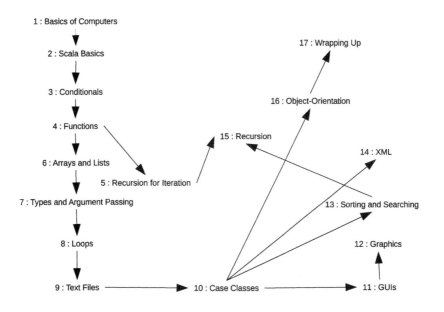

FIGURE 1: This figure shows the dependencies between different chapters in the book so that you can plan what you want to cover or to allow you to intelligently choose alternate paths through the book. Arrows point toward later chapters that use material from an earlier chapter.

Using this Book

This book is intended to cover CS1. When looking at the chapter list, you might feel that chapters 11 and 12 are outliers in certain ways. They definitely break from the straight algorithmic, problem solving nature of the other chapters. This break was placed at this point in the book at the request of students who often find visual interfaces more compelling than text based ones. Thanks to the syntax of Scala, it is possible to build GUIs without explicitly declaring classes or using keywords associated with it, such as **extends**. It is possible to skip these, but not advised as graphics play a role in many of the examples later in the book.

If you have a desire to cover full OO concepts earlier, chapter 16 can be moved to closely follow chapter 10. There are also a number of chapters through the book that contain material that can be skipped completely. For example, chapter 14 covers XML. Scala makes this fairly easy to do and it fits in well with the projects that are given to have students work on data stored in a formatted manner. However, if you do not want to spend time on that topic it can be skipped with little impact on later content other than projects and occasional usage in later chapters.

To help you decide what you can and cannot skip, figure 1 shows rough dependencies for different chapters in the book. Arrows point from one chapter to later ones that have direct dependencies. There will be occasional references to material that is not in the line of the dependency arrows, but the material should still be understandable.

The course web site, **http://book.programmingusingscala.net**, includes a number of different types of material that can be helpful for instructors. There are solutions to certain exercises so you should check there before using exercises for grading purposes. Additional exercises with recent links and data sets are also available on the web.

There are executable JAR files for some of the projects to help students, and instructors, understand what is being asked for from the student.

In addition, there are videos posted for every chapter of the text. These are generally "live coding" sessions. Instructors should feel free to use these as pre-lectures or to use with a flipped classroom format.

Chapter 1

Basics of Computers, Computing, and Programming

In all things it is good to understand some foundational material before going into the details. This helps to ground you and give you some context. Inevitably, you already have some experience with computers and that does give you a bit of context to work in. However, it is quite possible that your experience is limited to rather recent technology. One of the factors that shapes the world of computer science is that it is fast moving and ever changing. Knowing how we got to where we are can perhaps help us see where we will be going.

1.1 History

One might think that Computer Science is a field with a rather short history. After all, computers have not existed all that long and have been a standard fixture for even less time. However, the history of Computer Science has deep roots in math that extend far back in time. One could make a strong argument that the majority of Computer Science is not even really about computers. This is perhaps best exemplified in this quote by Edsger Dijkstra, "Computer science is no more about computers than astronomy is about telescopes."[2] Instead, Computer Science is about ALGORITHMs. An algorithm is a formal specification for stating a method to solve a problem. The term itself is a distortion of the name al-Khwārizmī. He was Persian mathematician who lived in the 11th century and wrote the *Treatise on Demonstration of Problems of Algebra*, the most significant treatise on algebra written before modern times. He also wrote *On the Calculation with Hindu Numerals*, which presented systematic methods of applying arithmetic to algebra.

One can go even further back in time depending on how flexibly we use the term computation. Devices for facilitating arithmetic could be considered. That would push things back to around 2400 BCE. Mechanical automata for use in astronomy have also existed for many centuries. However, we will focus our attention on more complete computational devices, those that can be programmed to perform a broad range of different types of computation. In that case, the first real mechanical computer design would have been the Analytical Engine which was designed by Charles Babbage and first described in 1837. Ada Lovelace is often referred to as the first programmer because her notes on the Analytic Engine included what would have been a program for the machine. For various reasons, this device was never built, and, as such, the first complete computers did not come into existence for another 100 years.

It was in the 1940s that computers in a form that we would recognize them today came into existence. This began with the Zuse Z3 which was built in Germany in 1941. By the end of the 1940s there were quite a few digital computers in operation around the world including the ENIAC, built in the US in 1946. The construction of these machines was influenced in large part by more theoretical work that had been done a decade earlier.

One could argue that the foundations of the theoretical aspects of Computer Science began in 1931 when Kurt Gödel published his incompleteness theorem. This theorem, which proved that in any formal system of sufficient complexity, including standard set theory of mathematics, would have statements in it that could not be proved or disproved. The nature of the proof itself brought in elements of computation as logical expressions were represented as numbers, and operations were transformations on those numbers. Five years later, Alan Turing and Alonzo Church created independent models of what we now consider to be computation. In many ways, the work they did in 1936 was the true birth of Computer Science as a field, and it enabled that first round of digital computers.

Turing created a model of computation called a Turing machine. The Turing machine is remarkably simple. It has an infinite tape of symbols and a head that can read or write on the tape. The machine keeps track of a current state, which is nothing more than a number. The instructions for the machine are kept in a table. There is one row in the table for each allowed state. There is one column for each allowed symbol. The entries in the table give a symbol to write to the tape, a direction to move the tape, and a new state for the machine to be in. The tape can only be moved one symbol over to the left or right or stay where it is at each step. Cells in the table can also say stop in which case the machine is supposed to stop running and the computation is terminated.

The way the machine works is that you look up the entry in the table for the current state of the machine and symbol on the tape under the head. You then write the symbol from the table onto the tape, replacing what had been there before, move the tape in the specified direction, and change the state to the specified state. This repeats until the stop state is reached.

At roughly the same time that Turing was working on the idea of the Turing machine, Alonzo Church developed the lambda calculus. This was a formal, math based way of expressing the ideas of computation. While it looks very different from the Turing machine, it was quickly proved that the two are equivalent. That is to say that any problem you can solve with a Turing machine can be solved with the lambda calculus and the other way around. This led to the so-called Church-Turing thesis stating that anything computable can be computed by a Turing machine or the lambda calculus, or any other system that can be shown to be equivalent to these.

1.2 Hardware

When we talk about computers it is typical to break the topic into two parts, hardware and software. Indeed, the split goes back as far as the work of Babbage and Lovelace. Babbage designed the hardware and focused on the basic computation abilities that it could do. Lovelace worked on putting together groups of instructions for the machine to make it do something interesting. Her notes indicate that she saw the further potential of such a device and how it could be used for more than just doing calculations.

To understand software, it is helpful to have at least some grasp of the hardware. If you continue to study Computer Science, hopefully you will, at some point, have a full course that focuses on the nature of hardware and the details of how it works. For now, our goal is much simpler. We want you to have a basic mental image of how the tangible elements of a computer work to make the instructions that we type in execute to give us the answers that we want.

The major hardware components of a computer include the central processing unit (CPU), memory, and input/output devices. Let's take a closer look at each of these in more detail.

1.2.1 Central Processing Unit

Modern computers work by regulating the flow of electricity through wires. Most of those wires are tiny elements that have been etched into silicon and are only tens of nanometers across. The voltage on the wires is used to indicate the state of a bit, a single element of storage with only two possible values, on or off. The wires connect up transistors that are laid out in a way that allows logical processing. While a modern computer processor will include literally hundreds of millions to billions of transistors, we can look at things at a much higher level and generally ignore that existence of those individual wires and transistors.

In general, modern computers are built on the von Neumann architecture with minor modifications. John von Neumann was another one of the fathers of computing. One of his ideas was that programs for a computer are nothing more than data and can be stored in the same place as all other data. This can be described quite well with the help of the basic diagram in figure 1.1. There is a single memory that stores both the programs and the data used by the program. It is connected to a Central Processing Unit (CPU) by a bus. The CPU, which can be more generally called a processor, has the ability to execute simple instructions, read from memory, and write to memory. When the computer is running, the CPU loads an instruction from memory, executes that instruction, then loads another. This happens repeatedly until the computer stops. This simple combination of a load and execute is called a cycle.

One of the things the CPU does is to keep track of the location in memory of the next instruction to be executed. We call a location in memory an ADDRESS. After the instruction is executed, it moves forward each time to get to the next instruction. Different types of computers can have different instructions. All computers will have instructions to read values from memory, store values to memory, do basic math operations, and change the value of the execution address to jump to a different part of the program.

The individual instructions that computers can do are typically very simple. Computers get their speed from the fact that they perform cycles very quickly. Most computers now operate at a few gigahertz. This means that they can run through a few billion instructions every second. There are a lot of complexities to real modern computers that are needed to

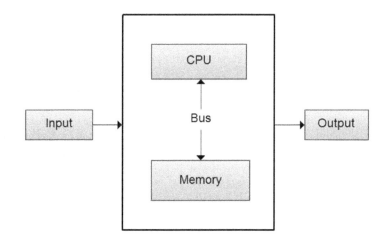

FIGURE 1.1: This is the most basic view of the von Neumann shared memory architecture which contains a CPU, memory, and input and output mechanisms. A CPU and the memory are connected by a bus. The memory stores both data and the programs themselves. The CPU can request information from memory or write to memory over the bus.

make that possible, which are not encompassed by this simple image of a computer. In many ways, these are details that we do not have to worry about too much at the beginning, but they are important in professional programming because they can have a profound impact on the performance of a program.

1.2.2 Memory

Memory on a computer is used to store information. There are two basic types of memory: primary memory (storage) which is known as RAM (Random Access Memory) and secondary memory (storage). Examples of secondary storage include hard drives, solid state drives, flash drives, and magnetic tape. RAM is also referred to as main memory which is directly connected to the CPU. All programs must be loaded into RAM before they can execute and all data must be loaded into RAM before it can be processed. Figure 1.2 gives you the basic idea of main memory. RAM is considered because when a computer gets turned off, everything in main memory is lost unless it has been saved to secondary storage first. Secondary storage is considered because it allows programs and data to be stored permanently, and these can be accessed even after the power has been cycled.

Typically speed goes inverse of size for memory, and this is true of disks and RAM. Disk drives are significantly slower to access than RAM, though what is written to them stays there even when the machine is turned off. RAM is faster than the disk, but it still is not fast enough to respond at the rate that a modern processor can use it. For this reason, processors generally have smaller amounts of memory on them called cache. The cache is significantly faster than the RAM when it comes to how fast the processor can read or write values. Even that is not enough anymore, and modern processors will include multiple levels of cache referred to as L1, L2, L3, etc. Each level up is generally bigger, but also further from the processor and slower.

Some applications have to concern themselves with these details because the program runs faster if it will fit inside of a certain cache. If a program uses a small enough amount of memory that it will fit inside of the L2 cache, it will run significantly faster than if it does

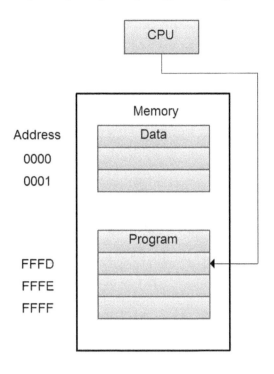

FIGURE 1.2: This is a basic view of the CPU and a program in memory.

not. We will not generally worry about this type of issue, but there are many professional developers who do.

A significant difference between our simple computer illustration and today's computers is that modern processors have multiple cores. What that means in our simple picture is that the CPU is not doing one instruction at a time, it is doing several. This is what we call parallel processing. When it happens inside of a single program it is referred to as multithreading. This is a significant issue for programmers because programs have not historically included multithreading and new computers require it in order to fully utilize their power. Unfortunately, making a program multithreaded can be difficult. Different languages offer different ways to approach this. This is one of the strengths of Scala that will be hit on briefly in this book and addressed in detail in *Object Orientation, Abstraction, and Data Structures Using Scala*[1] which is the second volume for CS2.

1.2.3 Input/Output Devices

For a computer to perform tasks, it needs to get input in the form of instructions and data. Devices that provide this capability are called input devices. Examples of input devices include a keyboard, mouse, and secondary storage. The keyboard is the STANDARD INPUT. Output devices enable the computer to display the results of the instructions and calculations. Some examples of output devices include the printer, terminal screen, and secondary storage. The terminal screen is the STANDARD OUTPUT.

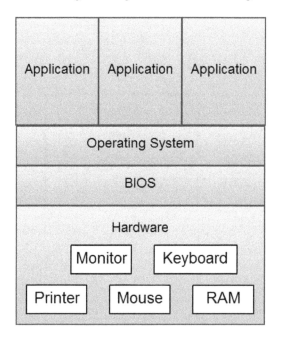

FIGURE 1.3: This is a basic view of the layers of software on top of a computer's hardware.

1.3   Software

The programs that are run on hardware are typically called software. The software is not a physical entity. However, without some type of software, the hardware is useless. It is the running of a program that makes hardware useful to people. As with the hardware, software can be seen as having multiple parts. It also has a layering or hierarchy to it as can be seen in figure 1.3. At the base is a layer that gives the computer the initial instructions for what to do when it is turned on. This is often called the BIOS (Basic Input/Output System). The BIOS is generally located on a chip in the machine instead of on the more normal forms of memory like the disk or RAM. Instructions stored in this way are often called firmware. This term implies that it is between the software and the hardware. The firmware really is instructions, just like any other software. In a sense it is less soft because it is stored in a way that might be impossible to write to or which is harder to write to.

The BIOS is responsible for getting all the basic functionality started up on the machine. Sitting on top of the BIOS is the Operating System (OS). The Operating System is responsible for controlling the operations of the machine and how it interacts with the user. The OS is also responsible for writing files to disk and reading files from disk. In addition, it has the job of loading other programs into memory and getting them started. Over time, the amount of functionality in operating systems has grown so that they are also expected to present nice interfaces and have all types of other "basic" functionality that really is not so basic.

At the top level are the application and utility programs that the user runs. When the user instructs the operating system to run a program, the operating system loads that program into memory and sets the execution address so that the computer will start running the program. The breadth of what programs can do is nearly unlimited.[1] Everything that runs on every digital device in the world is a program. You use them to type your papers and do your e-mail. They likely also run the fuel injection system in your car and control the lights at intersections. Programs regulate the flow of electricity through the power grid and the flow of water to your indoor plumbing. Programs do not just give you the apps on your phone, they are running when you talk on the phone to compress your speech into a digital form and send it out to a local tower where another program examines it and sends it on toward the destination. On the way, it likely passes through multiple locations and gets handled by one or more programs at each stop. At some point on the other end, another program takes the digital, compressed form and expands it back out to analog that can be sent to a speaker so the person you are talking to can hear it. Someone wrote each of those programs and over time more programs are being written that serve more and more different purposes in our lives.

In the last section we mentioned that newer processors have multiple cores on them. The availability of multiple cores (and perhaps multiple processors) is significant for software as well. First, they give the OS the ability to have multiple things happening at one time. All but the simplest of operating systems perform multitasking. Multitasking is a method that allows multiple tasks to be performed during the same time period and allows the OS to have multiple programs or processes running simultaneously. This can be done on a single core by giving each process a short bit of time and then switching between them. When there are multiple cores present, it allows the programs to truly run multiple processes all at once, which is known as multiprocessing.

Each process can also exploit its own parallelism by creating multiple threads. The OS is still responsible for scheduling what threads are active at any given time. This allows a single program to utilize more of the resources of a machine than what is present on a single core. While this does not matter for some specialized applications, the use of multiple cores has become more and more commonplace and the core count on large machines as well as smaller devices is currently climbing at an exponential rate. As a result, the need to multithread programs becomes ever more vital.

This increasing number of cores in machines has led to another interesting development in the area of servers. Servers are more powerful computers that are used by companies to store large amounts of data and do lots of processing on that data. Everything you do on the web is pulling data from servers. If you are at a University, odds are that they have at least one room full of servers that act as a control and data storage center for the campus.

The nature of the work that servers do is often quite different from a normal PC. A large fraction of their job is typically just passing information around, and the workload for that can be very unevenly distributed. The combination of this and multiple cores has led to an increase in the use of virtualization. A virtual machine is a program that acts like a computer. It has a BIOS and loads an OS. The OS can schedule and run programs. This whole process happens inside of a program running potentially on a different OS on a real machine. Using virtualization, you can start multiple instances of one or more operating systems running on a single machine. As long as the machine has enough cores and memory, it can support the work of all of these virtual machines. Doing this cuts down on the number of real machines that have to be bought and run, reducing costs in both materials and power consumption.

---

[1]There are real limitations to computing that are part of theoretical computer science. There are certain problems that are provably not solvable by any program.

1.4   Nature of Programming

Every piece of software, from the BIOS of each device to the OS, and the multitude of applications they run, is a program that was written by a programmer. So what is this thing we call programming, and how do we do it? How do we give a computer instructions that will make it do things for us? In one sense, programming is just the act of giving the computer instructions in a format that it can work with. At a fundamental level, computers do nothing more than work with numbers. Remember the model in figure 1.2. Each cycle the computer loads an instruction and executes it. There was a time when programming was done by writing the specific, low level instructions that the machine executes. We refer to the language of these instructions as machine language. While machine language is really the only language that the computer understands, it is not a very good language for humans to work in. The numbers of machine language do not hold inherent meaning for humans, and it is very easy to make mistakes. To understand this, consider the following code that calculates a wage using the math $wage = rate * hours$ using MIPS machine language.

```
100011 00000 00010 0000000000000000   # Load rate into register 2
100011 00001 00011 0000000000000000   # Load hours into register 3
000000 00010 00011 00100 00000 011000 # Multiply registers 2 and 3,
                                       #  and store in 4
101011 00100 00101 0000000000000000   # Store result from register 4
```

This shows the instructions in binary; we will discuss binary in detail in chapter 2. Spaces have been inserted to make it easier to see the parts of the instructions, and comments are put in to explain what each command, made up of 32 0s and 1s is doing. Real machine code does not have spaces, comments, or even separate lines. The commands are generally written/viewed in hexadecimal instead of binary as well. So this code would really look like the following.

```
8c0200008c230000432018ac850000
```

Even without knowing anything about programming, you can probably tell that this would be very hard to work with. For this reason, people have developed better ways to program computers than to write out machine language instructions.

The first step up from machine language is assembly language. Assembly language is basically the same as machine language in that there is an assembly instruction for each machine language instruction. However, the assembly instructions are entered as words, known as mnemonics, that describe what they do. The assembly language also helps to keep track of how things are laid out in memory so that programmers do not have to actively consider such issues the way they do with machine language. The machine language above can be written as the following assembly language instructions.

```
lw $s0, $s2, 0
lw $s1, $s3, 0
mult $s2, $s3, $s4
sw $s4, $s5, 0
```

The first group of six numbers in the binary machine code corresponds to an instruction. For example, you can see that 100011 becomes lw, which is short for "load word". To get the computer to understand assembly language, we employ a program that does a translation from assembly language to machine language. This program is called an assembler.

Even assembly language is less than ideal for expressing the ideas that we want to put into programs. For this reason, other languages have been created. These higher level languages use more complete words and allow a more complex organization of ideas so that more powerful programs can be written more easily. Our earlier pieces of code would be written as `wage = rate*hours` in most modern languages. The computer does not understand these languages either. As such, they either employ programs called compilers that translate the higher level languages into assembly then down to machine language or programs called interpreters that execute the instructions one at a time without ever turning them into machine language.

There are literally hundreds of different programming languages. Each one was created to address some deficiency that was seen in other languages or to address a specific need. This book uses the Scala programming language. It is hard to fully explain the benefits of Scala, or any other programming language, to someone who has not programmed before. We will just say that Scala is a very high-level language that allows you to communicate ideas to the computer in a concise way, and gives you access to a large number of existing libraries to help you write programs that are fun, interesting, and/or useful.

Early on in your process of learning how to program, you will likely struggle with figuring out how to express your ideas in a programming language instead of the natural language that you are used to. Part of this is because programming languages are fundamentally different than natural languages in that they do not allow ambiguity. In addition, they typically require you to express ideas at a lower level than you are used to with natural language. Both of these are a big part of the reason why everyone should learn how to program. The true benefits of programming are not seen in the ability to tell a computer how to do something. The real benefits come from learning how to break problems down.

At a very fundamental level, the computer is a stupid machine. It does not understand or analyze things.[2] The computer just does what it is told. This is both a blessing and a curse for the programmer. Ambiguity is fundamentally bad when you are describing how to do things, and it is particularly problematic when the receiver of the instructions does not have the ability to evaluate the different possible meanings and pick the one that makes the most sense. That means that programs have to be rigorous and take into account small details. On the other hand, you can tell computers to do things that humans would find incredibly tedious, and the computer will do it repeatedly for as long as you ask it to. This is a big part of what makes computers so useful. They can sift through huge amounts of data and do many calculations quickly without fatigue induced errors.

In the end, you will find that converting your thoughts into a language the computer can understand is the easy part. Yes, it will take time to learn your new language and to get used to the nuances of it, but the real challenge is in figuring out exactly what steps are required to solve a problem. As humans, we tend to overlook many of the details in the processes we go through when we are solving problems. We describe things at a very high level and ignore the lower levels, assuming that they will be implicitly understood. Programming forces us to clarify those implicit steps and, in doing so, forces us to think more clearly about how we solve problems. This skill becomes very important as problems get bigger and the things we might want to have implicitly assumed become sufficiently complex that they really need to be spelled out.

One of the main skills that you will develop when you learn how to program is the ability to break problems down into pieces. All problems can be broken into smaller pieces, and it is often helpful to do so until you get down to a level where the solution is truly obvious or

---

[2]At least they don't yet. This is the real goal of the field of Artificial Intelligence (AI). If the dreams of AI researchers are realized, the computers of the future will analyze the world around them and understand what is going on.

trivial. This approach to solving problems is called a top-down approach, because you start at the top with a large problem and break it down into smaller and smaller pieces until, at the bottom, you have elements that are simple to address. The solutions you get are then put back together to produce the total solution.

Another thing that you will learn from programming is that while there are many ways to break down almost any problem, not all of them are equally good. Some ways of breaking the problem down simply "make more sense". Granted, that is something of a judgement call, and might differ from one person to the next. A more quantifiable metric of the quality of how a problem is broken down is how much the pieces can be reused. If you have solved a particular problem once, you do not want to have to solve it again. You would rather use the solution you came up with before. Some ways of breaking up a problem will result in pieces that are very flexible and are likely to be useful in other contexts. Other ways will give you elements that are very specific to a given problem and will not be useful to you ever again.

There are many aspects of programming for which there are no hard and fast rules on how things should be done. In this respect, programming is much more an art than a science. Like any art, in order to really get good at it you have to practice. Programming has other similarities to the creative arts. Programming itself is a creative task. When you are programming, you are taking an idea that exists in your head and giving it a manifestation that is visible to others. It is not actually a physical manifestation. Programs are not tangible. Indeed, that is one of the philosophically interesting aspects of software. It is a creation that other people can experience, but they cannot touch. It is completely virtual. Being virtual has benefits. Physical media have limitations on them, imposed by the laws of physics. Whether you are painting, sculpting, or engineering a device, there are limitations to what can be created in physical space. Programming does not suffer from this. The ability of expression in programming is virtually boundless. If you go far enough in computer science you will learn where there are bounds, but even there the results are interesting because it is possible that the bounds on computation are bounds on human thinking as well.

This ability for near infinite expression is the root of the power, beauty, and joy of programming. It is also the root of the biggest challenge. Programs can become arbitrarily complex. They can become too complex for humans to understand what they are doing. For this reason, a major part of the field of Computer Science is trying to find ways to tame the complexity and make it so that large and complex ideas can be expressed in ways that are also easy for humans to follow and determine the correctness of.

1.5   Programming Paradigms

The fact that there are many ways to break up problems or work with problems has not only led to many different programming languages, it has led to whole families of different approaches that are called paradigms. There are four main paradigms of programming. It is possible others could come into existence in the future, but what appears to be happening now is that languages are merging the existing paradigms. Scala is one of the languages that is blurring the lines between paradigms. To help you understand this we will run through the different paradigms.

1.5.1 Imperative Programming

The original programming paradigm was the imperative paradigm. That is because this is the paradigm of machine language. So all the initial programs written in machine language were imperative. Imperative programming involves giving the computer a set of instructions that it is to perform. Those actions change the state of the machine. This is exactly what you get when you write machine language programs.

The two keys to imperative programming are that the programmer specifically states how to do things and that the values stored by the computer are readily altered during the computation. The converse of imperative programming would be declarative programming where the programmer states what is to be done, but generally is not specific about how it should be done. The Scala language allows imperative style programming and many of the elements of this book will talk about how to use the imperative programming style in solving problems.

1.5.2 Functional Programming

Functional programming was born out of the mathematical underpinnings of Computer Science. In the functional programming paradigm, the main program is written as a function that receives the program's input as its arguments and delivers the program's output as its result. The first functional languages were based very heavily on Alonzo Church's lambda calculus. In a way this is in contrast to imperative programming which bares a stronger resemblance to the ideas in the Turing machine. "Programming" a Turing machine is only loosely correlated to writing machine language, but the general ideas of mutable state and having commands that are taken one after the other are present on the Turing machine. Like the lambda calculus, functional languages are fundamentally based on the idea of functions in mathematics. We will see a lot more of the significance of mathematical functions on programmatic thinking in chapter 4.

Functional languages are typically more declarative than imperative languages. That is to say that you typically put more effort into describing what is to be done and a lot less in describing how it is to be done. For a language to be considered purely functional the key is that it not have mutable state, at least not at a level that the programmer notices. What does that mean? It means that you have functions that take values and return values, but do not change anything else along the way. The only thing they do is give you back the result. In an imperative language, little traces of what has been done can be dropped all over the place. Certainly, functional programs have to be able to change memory, but they always clean up after themselves so that there is nothing left behind to show what they did other than the final answer. Imperative programs can leave alterations wherever they want, and there is no stipulation that they set things back the way they found it. Supporters of functional programming often like to say that functional programs are cleaner. If you think of a program as being a person, this statement makes for an interesting analogy. Were a functional program to enter your room looking for something, you would never know, because it would leave no trace of its passing. At the end, everything would be as it had started except that the function would have the result it was looking for. An imperative program entering your room might not change anything, but more than likely it would move the books around and leave your pillow in a different location. The imperative program would change the "state" of your room. The functional one would not.

Scala supports a functional style of programming, but does not completely enforce it. A Scala program can come into your room, and it is easy to set it up so that it does not leave any trace behind, but if you want to leave traces behind you can. The purpose of this combination is to give you flexibility in how you do things. When a functional implementation is clean and easy you can feel free to use it. However, there are situations where the functional style has drawbacks, and in those cases you can use an imperative style.

1.5.3   Object-Oriented Programming

Object-oriented programming is a relative newcomer to the list of programming paradigms. The basic idea of object-oriented programming first appeared in the SIMULA67 programming language. As the name implies, this dates it back to the 1960s. However, it did not gain much momentum until the 1980s when the Smalltalk language took the ideas further. In the 1990s, object-orientation really hit it big and now virtually any new language that is created and expects to see wide use will include object-oriented features.

The basic idea of object-orientation is quite simple. It is that data and the functions that operate on the data should be bundled together into things called objects. This idea is called encapsulation, and we will discuss it briefly at the end of this book. This might seem like a really simple idea, but it enables a lot of significant extras that do not become apparent until you have worked with it a while.

Object-orientation is not really independent of the imperative and functional paradigms. Instead, object-oriented programs can be done in either a functional or an imperative style. The early object-oriented languages tended to be imperative and most of the ones in wide use today still are. However, there are a number of functional languages now that include object-oriented features.

Scala is a purely object-oriented language, a statement that is not true of many languages. What this means and the implications of it are discussed in detail in later chapters. For most of this book, the fact that Scala is object-oriented will not even be that significant to us, but it will always be there under the surface.

1.5.4   Logic Programming

The fourth programming paradigm is logic programming. The prime example language is Prolog. Logic programming is completely declarative. As the name implies, programs are written by writing logical statements. It is then up to the language/computer to figure out a solution. This is the least used of the paradigms. This is in large part because of significant performance problems. The main use is in artificial intelligence applications. There is some indication that logic programming could reappear in languages where it is combined with other paradigms, but these efforts are still in the early stages.

1.5.5   Nature of Scala

As was indicated in the sections above, Scala provides a mix of different paradigms. It is a truly hybrid language. This gives you, as the programmer, the ability to solve problems in the way that makes the most sense to you or that best suits the problem at hand. The Scala language directly includes imperative, functional, and object-oriented elements. The name Scala stands for Scalable Language; so, it is not completely out of the question for them to add a library in the future that could support logic programming to some extent if a benefit to doing so was found.

1.6 End of Chapter Material

1.6.1 Summary of Concepts

- Computing has a much longer history than one might expect with roots in mathematics.

- Hardware is the term used for the actual machinery of a computer.

- Software is the term used for the instructions that are run on hardware. Individual pieces of software are often called programs.

- The act of writing instructions for a computer to solve a problem in a language that the computer can understand or that can be translated to a form the computer understands is called programming.

- Programming is very much an art form. There are many different ways to solve any given problem, and they can be arbitrarily complex. Knowing how to design good programs takes practice.

- There are four different broad types or styles of programming called paradigms.

  - The imperative programming paradigm is defined by explicit instructions telling the machine what to do and mutable state where values are changed over time.

  - The functional paradigm is based on Church's lambda calculus and uses functions in a very mathematical sense. Mathematical functions do not involve mutable state.

  - Object-orientation is highlighted by combining data and functionality together into objects.

  - Logic programming is extremely declarative, meaning that you say what solution you want, but not how to do it.

  - Scala is purely object-oriented with support for both functional and imperative programming styles.

1.6.2 Exercises

1. Find out some of the details of your computer hardware and software.

   (a) Processor
       i. Who makes it?
       ii. What is the clock speed?
       iii. How many cores does it have?
       iv. How much cache does it have at different cache levels? (optional)

   (b) How much RAM does the machine have?

   (c) How much non-volatile, secondary storage (typically disk space) does it have?

   (d) What operating system is it running?

2. If you own a tablet computer, repeat exercise 1 for that device.

3. If you own a smartphone, repeat exercise 1 for that device.

4. Briefly describe the different programming paradigms and compare them.

5. List a few languages that use each of the different programming paradigms.

6. List a few languages that use compilers.

7. List a few languages that use interpreters.

8. What is the difference between computer multitasking and multiprocessing?

9. What is multithreading?

10. Compare and contrast an algorithm vs. a program.

11. Compare and contrast an interpreter vs. a compiler.

12. Do an Internet search for a genealogy of programming languages. Find and list the key languages that influences the development of Scala.

13. Go to a website where you can configure a computer (like `http://www.dell.com`). What is the maximum number of cores you put in a PC? What about a server?

14. Go to `http://www.top500.org`, a website that keeps track of the 500 fastest computers in the world. What are the specifications of the fastest computer in the world? What about the 500th fastest?

1.6.3 Projects

1. Search on the web to find a list of programming languages. Pick two or three and write descriptions of the basic properties of each including when they were created and possibly why they were created.

2. Compare and contrast the following activities: planning/building a bridge, growing a garden, painting a picture. Given what little you know of programming at this point, how do you see it comparing to these other activities?

3. Make a time line of a specific type of computer technology. This could be processors, memory, or whatever. Go from as far back as you can find to the modern day. What were the most significant jumps in technology, and when did they occur?

4. One of the more famous predictions in the field of computing hardware is Moore's law. This is the term used to describe a prediction that the number of components on an integrated circuit would double roughly every 18 months. The name comes from Intel® co-founder Gordon Moore who noted in 1965 that the number had doubled every year from 1958 to that point and predicted that would continue for at least a decade. This has now lead to exponential growth in computing power and capabilities for over five decades.[3] Write a short report on the impact this has had on the field and what it might mean moving forward.

5. Investigate and write a short report on some developments in future hardware technology. Make sure to include some things that could come into play when the physical limits of silicon are reached.

---

[3]You can see the exponential growth of supercomputer power on plots at Top500.org.

6. Investigate and write a short report on some developments in future software technology. What are some of the significant new application areas that are on the horizon? This should definitely include some look at artificial intelligence and robotics.

7. As computers run more and more of our lives, computer security and information assurance become more and more significant. Research some of the ways that computers are compromised and describe how this is significant for programmers writing applications and other pieces of software.

Additional exercises and projects can be found on the website.

Chapter 2

Scala Basics

It is time to begin our journey learning how to program with the Scala language. You can download Scala for free from `http://www.scala-lang.org` to run on Windows, Mac, or Linux (see the inset below for full instructions on how to install). In this book, we will use the command line to run Scala. If you do not have experience with the command line on your machine, you can refer to Appendix A for a brief introduction. Before looking at the language itself, we need to talk a bit about tools so that you can play along.

2.1 Scala Tools

After you have installed Scala on your machine there are several different programs that get installed in the `bin` directory under the Scala installation. To begin with, we will only concern ourselves with one of these: `scala`.[1] The `scala` command actually runs `scala` programs. There is a second command, `scalac`, that is used to compile `scala` text files into bytecode that is compatible with either the Java or .NET platform. We will only use `scalac` in the last chapter of this book, but we will begin using the `scala` command immediately.

Scala on your Machine

If you only use Scala on a machine in a computer lab, hopefully everything will have been set up for you so that you can simply type the name of a command and it will run. To run Scala on your own machine you can follow the instructions below.

Installation

Scala requires Java® to run so if you do not have Java installed you should go to `http://java.oracle.com` and download then install the most recent version of the Java SE JDK. When you install Java, you can go with the default install locations.

After you have Java installed you can install Scala. To download Scala go to `http://www.scala-lang.org`. On that site download the latest version of Scala. The code in this book was written to work with Scala 2.12.

Dealing with the `PATH`

If you are using Scala on your own machine, it is possible that entering `scala` or `scala.bat` on the command line could produce a message telling you that the command or program `scala` could not be found. This happens because the location of the installed programs are not in your default `PATH`.

The `PATH` is a set of directories that are checked whenever you run a command. The first match that is found for any executable file in a directory in the `PATH` will be run. If none of the programs in the `PATH` match what you entered, you get an error.

When you installed Scala, a lot of different stuff was put into the install directory. That included a subdirectory called "bin" with different files in it for the different executables. If you are on a Windows machine, odds are that you installed the program in `C:\Program Files (x86)\scala` so the `scala.bat` file that you want to run is in `C:\Program Files\scala\bin\scala.bat`. You can type in that full command or you can add the `bin` directory to your `PATH`. To do this go to Control Panel, System and Security, Advanced System Settings, Environment Variables, and edit the path to add `C:\Program Files\scala\bin` to the path.

Under Unix/Linux you can do this from the command line. Odds are that Scala was installed in a directory called `scala` in your user space. To add the `bin` directory to your path you can do the following:

```
export PATH=$PATH:/home/username/scala/bin
```

Replace "username" with your username. This syntax assumes you are using the bash shell. If it does not work for you, you can do a little searching on the web for how to add directories to your path in whatever shell you are running. To make it so that you do not have to do this every time you open a terminal, add that line to the appropriate configuration file in your home directory. If you are running the Bash shell on Linux this would be `.bashrc`.

---

[1] On a Windows system this commands should be followed by ".bat".

There are three ways in which the `scala` command can be used. If you just type in `scala` and press enter you will be dropped into the Scala REPL (Read-Execute-Print Loop). This is an environment where you can type in single Scala expressions and immediately see their values. This is how we will start off interacting with Scala, and it is something that we will come back to throughout the book because it allows us to easily experiment and play around with the language. The fact that it gives us immediate feedback is also quite helpful.

To see how this works, at the command prompt, type in `scala` and then press enter. It should print out some information for you, including telling you that you can get help by typing in `:help`. It will then give you a prompt of the form `scala>`. You are now in the Scala REPL. If you type in `:help` you will see a number of other commands you could give that begin with a colon. At this time the only one that is significant to us is `:quit` which we will use when we are done with the REPL and want to go back to the normal command prompt.

It is customary for the first program in a language to be Hello World. So as not to break with tradition, we can start by doing this now. Type the following after the `scala>` prompt.

```scala
println("Hello, World!");
```

If you do this you will see that the next line prints out "Hello, World!". This exercise is less exciting in the REPL because it always prints out the values of things, but it is a reasonable thing to start with. It is worth asking what this really did. `println` is a function in Scala that tells it to print something to standard output and follow that something with a newline character[2] to go to the next line. In this case, the thing that was printed was the string "Hello, World!". You can make it print other things if you wish. One of the advantages of the REPL is that it is easy to play around in. Go ahead and test printing some other things to see what happens.

The second usage of the `scala` command is to run small Scala programs as scripts. The term script is generally used to refer to short programs that perform specific tasks. There are languages that are designed to work well in this type of usage, and they are often called scripting languages. The design of Scala makes it quite usable as a scripting language. Unlike most scripting languages, however, Scala also has many features that make it ideal for developing large software projects as well. To use Scala for scripting, simply type in a little Scala program into a text file that ends with ".scala"[3] and run it by putting the file name after the `scala` command on the command line. So you could edit a file called `Hello.scala` and add the line of code from above to it. After you have saved the file, go to the command line and enter "scala Hello.scala" to see it run.

2.2 Expressions, Types, and Basic Math

All programming languages are built from certain fundamental parts. In English you put together words into phrases and then combine phrases into sentences. These sentences can be put together to make paragraphs. To help you understand programming, we will make analogies between standard English and programming languages. These analogies are

---

[2]You will find more information about the and other escape characters in section 2.8

[3]It is not technically required that your file ends with ".scala", but there are at least two good reasons you should do this. First, humans benefit from standard file extensions because they have meaning and make it easier to keep track of things. Second, some tools treat things differently based on extensions. For example, some text editors will color code differently based on the file extension.

not perfect. You cannot push them too far. However, they should help you to organize your thinking early in the process. Later on, when your understanding of programming is more mature, you can dispense with these analogies as you will be able to think about programming languages in their own terms.

The smallest piece of a programming language that has meaning is called a TOKEN. A token is like a word or punctuation mark in English. If you break up a token, you change the meaning of that piece, just like breaking up a word is likely to result in something that is no longer a word and does not have any meaning at all. Indeed, many of the tokens in Scala are words. Other tokens are symbols like punctuation. Let's consider the "Hello, World" example from the previous section.

```
println("Hello, World!");
```

This line contains a number of tokens: `println`, `(`, `"Hello, World!"`, and `)`.

When you think of putting words together, you probably think of building sentences with them. A sentence is a grouping of words that stands on its own in written English. The equivalent of a sentence in Scala, and most programming languages, is the STATEMENT. A statement is a complete and coherent instruction that we can give the computer. When you are entering "commands" into the REPL, they are processed as full statements. If you enter something that is not a complete statement in the REPL, instead of the normal prompt, you will get a vertical bar on the next line telling you that you need to continue the statement. The command listed above is a complete statement which is why it worked the way it did.

Note that this statement ends with a semicolon. In English you are used to ending sentences with a period, question mark, or exclamation point. Scala follows many other programming languages in that semicolons denote the end of a statement. Scala also does something called semicolon inference. Put simply, if a line ends in such a way that a semicolon makes sense, Scala will put one there for you. As a result of this, our print statement will work just as well without the semicolon.

```
println("Hello World!")
```

You should try entering this into the REPL to verify that it works. Thanks to the semicolon inference in Scala, we will very rarely have to put semicolons in our code. One of the few times they will really be needed is when we want to put two statements on a single line for formatting reasons.

While you probably think of building sentences from words in English, the reality is that you put words together into phrases and then join phrases into sentences. The equivalent of a phrase in Scala is the EXPRESSION. Expressions have a far more significant impact on programming languages than phrases have in English, or at the least programmers need to be more cognizant of expressions than English writers have to be of phrases. An expression is a group of tokens in the language that has a value and a TYPE.[4] For example, $2 + 2$ is an expression which will evaluate to 4 and has an Integer type.

Just like some phrases are made from a single word, some tokens represent things that have values on their own, and, as such, they are expressions themselves. The most basic of these are what are called LITERALS. Our sample line was not only a statement, it was also an expression. In Scala, any valid expression can be used as a statement, but some statements are not expressions. The `"Hello, World!"` part of our statement was also an expression. It is something called a string literal which we will learn more about in section 2.4.

Let us take a bit of time to explore these concepts in the REPL. Run the `scala` command

---

[4]Type is a construct that specifies a set of values and the operations that can be performed on them. Common types include numeric integer, floating-point, character, and boolean.

without any arguments. This will put you in the REPL with a prompt of scala>. In the last chapter we typed in a line that told Scala to print something. This was made from more than one token. We want to start simpler here. Type in a whole number, like 5, followed by a semicolon and hit enter. You should see something like this:

```
scala> 5;
res0: Int = 5
```

The first line is what you typed in at the prompt. The second line is what the Scala REPL printed out as a response. Recall that REPL stands for Read-Evaluate-Print Loop. When you type something in, the REPL reads what you typed, then evaluates it and prints the result. The term loop implies that this happens over and over. After printing the result, you should have been given a new prompt.

So what does this second line mean? The REPL evaluated the statement that you input. In this case, the statement is just an expression followed by a semicolon and the REPL was printing out the value of the expression you entered. As was mentioned above, the REPL needs you to type in full statements so that it can evaluate it. In this case, we typed in a very simple statement that has an expression called a NUMERIC LITERAL followed by a semicolon. This semicolon will be inferred if you do not add it in. We will take advantage of that and leave them out of statements below.

The end of the output line gives us the value of the expression which is, unsurprisingly, 5. What about the stuff before that? What does res0: Int mean? The res0 part is a name. It is short for "result0". When you type in an expression as a statement in the Scala REPL as we did here, it does not just evaluate it, it gives it a name so that you can refer back to it later. The name res0 is now associated with the value 5 in this run of the REPL. We will come back to this later. For now we want to focus on the other part of the line, :Int. Colons are used in Scala to separate things from their types. We will see a lot more of this through the book, but what matters most to us now is the type, Int. This is the type name that Scala uses for basic numeric integers. An integer can be either a positive or negative whole number. You can try typing in a few other integer values to see what happens with them. Most of the time the results will not be all that interesting, but if you push things far enough you might get a surprise or two.

What happens if you type in a number that is not an integer? For example, what if you type in 5.6? Try it, and you should get something like this:

```
scala> 5.6
res1: Double = 5.6
```

We have a different name now because this is a new result. We also get a different type. Instead of Int, Scala now tells us that the type is Double. In short, Double is the type that Scala uses by default for any non-integer numeric values. Even if a value technically is an integer, if it includes a decimal point, Scala will interpret it to be a Double. You can type in 5.0 to see this in action. Try typing in some other numeric values that should have a Double as the type. See what happens. Once again, the results should be fairly mundane. Double literals can also use scientific notation by putting the letter e between a number and the power of ten it is multiplied by. So 5e3 means $5 * 10^3$ or 5000.

So far, all of the expressions we have typed in have been single tokens. Now we will build some more complex expressions. We will begin by doing basic mathematical operations. Try typing in "5+6".

```
scala> 5+6
res2: Int = 11
```

This line involves three tokens. Each character in this case is a separate token. If you space things out, it will not change the result. However, if you use a number with multiple digits, all the digits together are a single token and inserting spaces does change the meaning.

There should not be anything too surprising about the result of 5+6. We get back a value of 11, and it has a type of Int. Try the other basic arithmetic operations of -, *, and /. You'll notice that you keep getting back values of type Int. This makes sense for addition, subtraction, and multiplication. However, the result of 5/2 might surprise you a little bit. You normally think of this expression as having the value of 2.5 which would be a Double. However, if you ask Scala for the result of 5/2 it will tell you the value is the Int 2. Why is this, and what happened to the 0.5? When both operands are of type Int, Scala keeps everything as Ints. In the case of division, the decimal answer you might expect is truncated and the fractional part is thrown away. Note that it is not rounded, but truncated. Why is this? It is because in integer arithmetic, the value of 5/2 is not 2.5. It is 2r1. That is to say that when you divide five by two, you get two groups of two with one remainder. At some point in your elementary education, when you first learned about division, this is probably how you were told to think about it. At that time you only had integers to work with so this is what made sense.

Scala is just doing what you did when you first learned division. It is giving you the whole number part of the quotient with the fractional part removed. This fractional part is normally expressed as a remainder. There is another operation called modulo that is represented by the percent sign that gives us the remainder after division. Here we can see it in action.

```
scala> 5%2
res3: Int = 1
```

The modulo operator is used quite a bit in computing because it is rather handy for expressing certain ideas. You should take some time to re-familiarize yourself with it. You might be tempted to say that this would be your first time dealing with it, but in reality, this is exactly how you did division yourself before you learned about decimal notation for fractions.

What if you really wanted 2.5 for the division? Well, 2.5 in Scala is a Double. We can get this by doing division on Doubles.

```
scala> 5.0/2.0
res4: Double = 2.5
```

All of our basic numeric operations work for Doubles as well. Play around with them some and see how they work. You can also build larger expressions. Put in multiple operators, and use some parentheses.

What happens when you combine a Double and an Int in an expression. Consider this example:

```
scala> 5.0/2
res5: Double = 2.5
```

Here we have a Double divided by an Int. The result is a Double. When you combine numeric values in expressions, Scala will change one to match the other. The choice of which one to change is fairly simple. It changes the one that is more restrictive to the one that is less restrictive. In this case, anything that is an Int is also a Double, but not all values that are Doubles are Ints. So the logical path is to make the Int into a Double and do the operation that way.

2.3   Objects and Methods

One of the features of the Scala language is that all the values in Scala are OBJECTS. The term object in reference to programming means something that combines data and the functionality on that data in a single entity. In Scala we refer to the things that an object knows how to do as METHODS. The normal syntax for calling a method on an object is to follow the object by a period (which we normally read as "dot") and the name of the method. Some methods need extra information, which we called arguments. If a method needs ARGUMENTS then those are put after the method name in parentheses.

In Scala, even the most basic literals are treated as objects in our program, and we can therefore call methods on them. An example of when we might do this is when we need to convert one type to another. In the sample below we convert the `Double` value `5.6` into an `Int` by calling the `toInt` method. In this simple context we would generally just use an `Int` literal, but there will be situations we encounter later on where we are given values that are `Doubles` and we need to convert them to `Ints`. We will be able to do that with the `toInt` method.

```
scala> 5.6.toInt
res6: Int = 5
```

One thing you should note about this example is that converting a `Double` to an `Int` does not round. Instead, this operation performs a truncation. Any fractional part of the number is cut off and only the whole integer is left.

We saw at the beginning of this chapter that Scala is flexible when it comes to the requirement of putting semicolons at the end of statements. Scala will infer a semicolon at the end of a line if one makes sense. This type of behavior makes code easier to write.

Methods that take one argument can be called using "infix" notation. This notation leaves off the dot and parentheses, and simply places the method between the object it is called on and the argument. If the method name uses letters, spaces will be required on either side of it. This type of flexibility makes certain parts of Scala more coherent and provides the programmer with significant flexibility. Though you did not realize it, you were using "infix" notation in the last section. To see this, go into Scala and type "5." then press tab. The Scala REPL has tab completion just like the command line, so what you see is a list of all the methods that could be called on the `Int`. It should look something like the following.

```
scala> 5.
%   +    >    >>>           isInstanceOf  toDouble  toLong    unary_+  |
&   -    >=   ^             toByte        toFloat   toShort   unary_-
*   /    >>   asInstanceOf  toChar        toInt     toString  unary_~
```

You have already seen and used some of these methods. We just finished using `toInt` on a `Double`. We can call `toDouble` on an `Int` as well. The things that might stand out though are the basic math operations that were used in the previous section. The +, -, *, /, and % we used above are nothing more than methods on the `Int` type. The expression 5+6 is really 5 .+ (6) to Scala. In fact, you can type this into Scala and see that you get the same result.

```
scala> 5 .+ (6)
res7: Int = 11
```

The space between the 5 and the . is required here because without it Scala thinks you want a `Double`. You could also make this clear using parentheses by entering `(5).+(6)`.

So when you type in 5+6, Scala sees a call to the method + on the object 5 with one argument of 6. We get to use the short form simply because Scala allows both the dot and the parentheses to be optional in cases like this.

2.4 Other Basic Types

Not everything in Scala is a number. There are other non-numeric types in Scala which also have literals. We will start simple and move up in complexity. Perhaps the simplest type in Scala is the `Boolean` type. Objects of the `Boolean` type are either `true` or `false`, and those are also valid literals for `Boolean`s.

```
scala> true
res8: Boolean = true

scala> false
res9: Boolean = false
```

We will see a lot more on `Boolean`s and what we can do with them in chapter 3 when we introduce Boolean logic.

Another type that is not explicitly numeric is the `Char` type. This type is used to represent single characters. We can make character literals by placing the character inside of single quotes like we see here.

```
scala> 'a'
res10: Char = a
```

The way that computers work, all character data is really numbers, and different numbers correspond to different characters. We can find out what numeric value is associated with a given character by using the `toInt` method. As you can see from the line below, the lowercase "a" has a numeric value of 97.

```
scala> 'a'.toInt
res11: Int = 97
```

Because characters have numeric values associated with them, we can also do math with them. When we do this, Scala will convert the character to its numeric value as an `Int` and then do the math with the `Int`. The result will be an `Int`, as seen in this example.

```
scala> 'a'+1
res12: Int = 98
```

In the last section you might have noticed that the `Int` type has a method called `toChar`. We can use that to get back from an integer value to a character. You can see from the following example that when you add 1 to `'a'` you get the logical result of `'b'`.

```
scala> ('a'+1).toChar
res13: Char = b
```

An object of the `Char` type can only be a single character. If you try to put more than one character inside of single quotes you will get an error. It is also an error to try to make a `Char` with empty single quotes. However, there are lots of situations when you want to be able to represent many characters, or even zero characters. This includes words, sentences, and many other things. For this there is a different type called a `String`. String literals are formed by putting zero or more characters inside of double quotes like we see in this example.

```scala
scala> "Scala is a programming language"
res14: String = Scala is a programming language
```

Notice that the type is listed as `String`.[5]

Certain operations that look like mathematical operations are supported for `Strings`. For example, when you use + with `Strings`, it does string concatenation. That is to say it gives back a new string that is the combined characters of the two that are being put together as shown here:

```scala
scala> "abc"+"def"
res15: java.lang.String = abcdef
```

This type of operation works with other types as well. The next example shows what happens when we concatenate a `String` with an `Int`. The `Int` is converted to a `String`, using the `toString` method, and normal string concatenation is performed.

```scala
scala> "abc"+123
res16: java.lang.String = abc123
```

This works whether the `String` is the first or second argument of the +.

```scala
scala> 123+"abc"
res17: java.lang.String = 123abc
```

In addition to concatenation, you can multiply a string by an integer, and you will get back a new string that has the original string repeated the specified number of times.

```scala
scala> "abc"*6
res18: String = abcabcabcabcabcabc
```

This can be helpful for things such as padding values with the proper number of spaces to make a string a specific length. You can do this by "multiplying" the string " " by the number of spaces you need.

The infix notation for calling a method was introduced earlier. We can show another example of this using the `String` type and the `substring` method. As the name implies, `substring` returns a portion of a `String`. There are two versions of it. One takes a single `Int` argument and returns everything from that index to the end of the `String`. Here you can see that version being called using both the regular notation and the infix notation.

```scala
scala> "abcd".substring(2)
res19: String = cd
```

```scala
scala> "abcd" substring 2
res20: String = cd
```

---

[5]The way you are running this, the real type is a `java.lang.String`. Scala integrates closely with Java and uses some of the Java library elements in standard code. This also allows you to freely call code from the Java libraries, a fact that has been significant in the adoption of Scala.

The indices in `String`s begin with zero, so 'a' is at index 0, 'b' is at index 1, and 'c' is at index 2. Calling `substring` with an argument of 2 gives back everything from the 'c' to the end.

The version of `substring` that takes two arguments allows us to demonstrate a different syntax where we just leave off the dot. In this case, the first argument is the first index to take and the second one is one after the last index to take. In math terms, the bounds are inclusive on the low end and exclusive on the high end. The fact that there are two arguments means that we have to have parentheses to group together the two arguments, However, we are not required to put the dot, and the method name can just be between the object and the arguments. Here are examples using the normal syntax and the version without the dot.

```
scala> "abcd".substring(1,3)
res21: String = bc

scala> "abcd" substring (1,3)
res22: String = bc
```

The space between the method and the parentheses is not required. Remember that in general Scala does not care about spaces as long as they do not break up a token. In this book, we will typically use the standard method calling notation, but you should be aware that these variations exist.

There are other types that are worth noting before we move on. One is the type `Unit`. The `Unit` type in Scala basically represents a value that carries no information.[6] There is a single object of type `Unit`. It is written in code and prints out as (). We have actually seen an example of code that uses `Unit`. The first program we saw in this chapter used a function called `println`. When we called `println` Scala did something (it directed the string to standard output), but did not give us back a value. This is what happens when we type in an expression that gives us back a value of `Unit` in the REPL.

Another significant type in Scala is the TUPLE. A tuple is a sequence of a specified number of specific types. Basically, a collection of values that is strict about how many and what type of values it has. We can make tuples in Scala by simply putting values in parentheses and separating them with commas as seen in the following examples.

```
scala> (5,6,7)
res23: (Int, Int, Int) = (5,6,7)

scala> ("book",200)
res24: (String, Int) = (book,200)

scala> (5.7,8,'f',"a string")
res25: (Double, Int, Char, String) = (5.7,8,f,a string)
```

The tuples in Scala provide a simple way of dealing with multiple values in a single package, and they will come up occasionally through the book. Note that the way we express a tuple type in Scala is to put the types of the values of the tuple in parentheses with commas between them, just like we do with the values to make a tuple object.

Tuples with only two elements can have special meanings in some parts of Scala. For that reason, there is an alternate syntax you can use to define these. If you put the token -> between two values, it will produce a 2-tuple with those values. Consider the following example.

---

[6]The equivalent in many other languages is called `void`.

```
scala> 3 -> "three"
res26: (Int, String) = (3,three)
```

The -> will only produce tuples with two elements though. If you try using it with more than two elements you can get interesting results.

```
scala> 4 -> 5 -> 6
res27: ((Int, Int), Int) = ((4,5),6)
```

So if you want tuples with more than two elements, stick with the parentheses and comma notation.

Once you have a tuple, there are two ways to get things out of them. The first is to use methods named _1, _2, _3, etc. So using `res21` from above we can do the following.

```
scala> res25._1
res28: Double = 5.7
```

```
scala> res25._3
res29: Char = f
```

The challenge with this method is that method names like _1 are not very informative and can make code difficult to read. We will see an alternative approach in section 2.7 that requires a bit more typing, but can produce more readable code.

2.5   Back to the Numbers

Depending on how much you played around with the topics in section 2.2 you might or might not have found some interesting surprises where things behaved in ways that you were not expecting. Consider the following:

```
scala> 1500000000+1500000000
res30: Int = -1294967296
```

Mathematicians would consider this to be the wrong answer. It is actually a reflection of the way that numbers are implemented on computers. The details of this implementation can impact how your programs work, so it is worth taking a bit of time to discuss it.

At a fundamental level, all information on computers is represented with numbers. We saw this with the characters being numbers. On modern computers all these numbers are represented in BINARY, or base two which represents numeric values using two different symbols: 0 (zero) and 1 (one). The electronics in the computer alternate between two states that represent 1 and 0 or on and off. Collections of these represent numbers. A single value of either a 0 or a 1 is called a BIT. It is a single digit in a binary number. The term BYTE refers to a grouping of 8 bits which can represent 256 different numbers. In Scala these will be between -128 and 127. To understand this, we need to do a little review of how binary numbers work.

You have likely spent your life working with decimal numbers, or base ten. In this system, there are ten possible values for each digit: 0, 1, 2, 3, 4, 5, 6, 7, 8, and 9. Digits in different positions represent different power of ten. So the number 365 is really $3*10^2+6*10^1+5*10^0$. There is nothing particularly unique about base ten other than perhaps it relates well to the number of digits on human hands. You can just as well use other bases, in which case

Value	Power of 2	Digit
296	256	1
40	128	0
40	64	0
40	32	1
8	16	0
8	8	1
0	4	0
0	2	0
0	1	0

FIGURE 2.1: Illustration of the conversion from decimal to binary using the subtraction method. This method works from the top down. To get the number in binary just read down the list of digits.

you need an appropriate number of symbols for each digit and each position represents a power of that base.

Binary uses a base of two. In binary we only need two different digits: 0 and 1. This is convenient on computers where the electronics can efficiently represent two states. The different positions represent powers of two: 1, 2, 4, 8, 16, 32, ... So the number $110101 = 1 * 32 + 1 * 16 + 0 * 8 + 1 * 4 + 0 * 2 + 1 * 1 = 53$. This example shows how you convert from binary to decimal. Simply add together the powers of two for which the bits have a value of one. A byte stores eight bits that would represent powers of two from 128 down to 1. The word "would" is used here because there is a significant nuance to this dealing with negative numbers that we will discuss shortly.

There are two basic approaches to converting from decimal to binary. One involves repeated subtraction of powers of two while the other involves repeated division by two. We will start with the first one and use the value 296 in decimal for the conversion. We start by finding the largest power of 2 that is smaller than our value. In this case it is $256 = 2^8$. So we will have a one in the 2^8 position or the 9th digit[7]. Now we subtract and get $296 - 256 = 40$ and repeat. The largest power of 2 smaller than 40 is $32 = 2^5$. So the digits for 2^7 and 2^6 are 0. Subtract again to get $40 - 32 = 8$. We now have $8 = 2^3$ so the final number in binary is 100101000. This procedure is written out the way you might actually do it in figure 2.1.

The other approach is a bit more algorithmic in nature and is probably less prone to error. It works based on the fact that in binary, multiplying and dividing by 2 moves the "binary point" the same way that multiplying or dividing by 10 moves the decimal point in the decimal number system. The way it works is you look at the number and if it is odd you write a 1. If it is even you write a 0. Then you divide the number by 2, throwing away any remainder or fractional part, and repeat with each new digit written to the left of those before it. Do this until you get to 0. You can also think of this as just dividing by two repeatedly and writing the remainder as a bit in the number with the quotient being what you keep working with.

The number 296 is even so we start off by writing a 0 and divide by 2 to get 148. That is also even so write another 0. Divide to get 74. This is also even so write another 0. Divide to get 37. This is odd so write a 1. Divide to get 18, which is even so you write a 0. Divide to get 9 and write a 1. Divide to get 4 and write a 0. Divide to get 2 and write a 0. Divide to get 1 and write that one. The next division gives you zero so you stop. This procedure is illustrated in figure 2.2.

---

[7]Remember that the first digit is $2^0 = 1$.

Value	Digit
1	1
2	0
4	0
9	1
18	0
37	1
74	0
148	0
296	0

FIGURE 2.2: Illustration of the conversion from decimal to binary using the repeated division method. This method works from the bottom up so you get the bits in the result starting with the smallest.

2.5.1 Binary Arithmetic

Now that you know how to go from binary to decimal and decimal to binary, let's take a minute to do a little arithmetic with binary numbers. It is certainly possible to do this by converting the binary to decimal, doing the arithmetic in decimal, then converting back to binary. However, this is quite inefficient and not worth it because it really is not hard to work in binary. If anything, it is easier to work in binary than in decimal. Let us begin with the operation of addition. Say we want to add the numbers 110101 and 101110. To do this you do exactly what you would do with long addition in decimal. As with decimal numbers, you start by adding the bits one column, at a time, from right to left. Just as you would do in decimal addition, when the sum in one column is a two-bit number, the least significant part is written down as part of the total and the most significant part is "carried" to the next left column. The biggest difference between decimal and binary addition is that in binary there is a lot more carrying. Here is a problem solved without showing the carries.

```
  110101
+ 101110
  -------
 1100011
```

Here is the same problem, but with numbers written above to show when there is a carry.

```
  1111
  110101
+ 101110
  -------
 1100011
```

Multiplication in binary can also be done just like in decimal, and you have a lot fewer multiplication facts to memorize. Zero times anything is zero and one times anything is that number. That is all we have to know. Let us do multiplication with the same numbers we just worked with. First we will get all the numbers that need to be added up.

```
    110101
  * 101110
  -----------
      1101010
```

```
  11010100
 110101000
11010100000
```

Adding these numbers is best done in pairs. The reason is that as soon as you add together 3 or more numbers in binary you have the capability to have to do something you are not accustomed to doing in decimal: carry a value up two digits. In decimal you would have to have a column sum up to one hundred or more for this to happen. However, in binary you only have to get to four (which is written as 100 in binary). That happens in this particular instance in the 6th digit. To reduce the odds of an error, it is better to add the values two at a time as we have shown here.

```
    1101010
  + 11010100
  -----------
  100111110
 + 110101000
  -----------
  1011100110
+11010100000
  -----------
100110000110
```

You can do division in the same way that you do long division with integers, but we will not cover that here.

2.5.2   Negative Numbers in Binary

We still have not addressed the question of how we represent negative numbers on a computer. The description that we have given so far only deals with positive values. Numbers that are interpreted this way are called UNSIGNED. All the numeric types in Scala are SIGNED, so we should figure out how that works.[8] To do this, there are two things that should be kept in mind. The first is that our values have limited precision. That is to say that they only store a certain number of bits. Anything beyond that is lost. The second is that negative numbers are defined as the additive inverses of their positive counterparts. In other words, $x + (-x) = 0$ for any x.

To demonstrate how we can get negative numbers, let's work with the number 110101 (53 in decimal). Unlike before, we will now limit ourselves to a single byte. So, we have 8 digits to work with, and the top digits are zeros. Our number stored in a byte is really 00110101. So the question of what should be the negative is answered by figuring out what value we would add to this in order to get zero.

```
  00110101
+ ????????
  --------
  00000000
```

Of course, there is nothing that we can put into the question marks to make this work. However, if we go back to our first fact (i.e. values have limited precision) we can see what we must do. Note that our total below has 9 digits. We do not need the total to be zero, we need eight digits of zero. So in reality, what we are looking for is the following.

---

[8]The `Char` is actually a 16-bit unsigned numeric value, but the normal numeric types are all signed.

Type	Bits	Min	Max
Byte	8	-128	127
Short	16	-32768	32767
Int	32	-2147483648	2147483647
Long	64	-9223372036854775808	9223372036854775807

TABLE 2.1: Integer types with their sizes and ranges.

```
  00110101
+ ????????
  --------
 100000000
```

This problem is solvable and the most significant 1, the one on the far left, will be thrown away because we can only store 8 bits in a byte. So the answer is given here.

```
  00110101
+ 11001011
  --------
 100000000
```

Note that the top bit is "on" in the negative value. The top bit is not exactly a sign bit, but if a number is signed, the top bit will tell us quickly whether the number is positive or negative. This style of making negatives is called TWO'S COMPLIMENT. In the early days of digital computing other options were tried, such as adding a sign-bit or a method called ones' compliment where the bits are simply flipped. However, two's compliment is used in machines today because it allows numeric operations to be done with negative numbers using the same circuitry as is used for positive numbers.

This process gives us the correct answer and is based on the proper definition of what a negative number is. Finding negatives using the definition of what a negative value is works and can be a fallback, but there is a simpler method. To get the two's compliment negative of a binary number of any size, simply flip all the bits and add one. You can verify that this approach works for our example above. It is left as an exercise for the student to figure out why this works.

2.5.3   Other Integer Types

There are larger groups of bits beyond the 8-bit bytes that have meaning in Scala. In fact, if you go back to section 2.3 and you look at the different methods on an `Int`, you will see that `toDouble` and `toChar` are not the only conversions we can do. Scala has other integer types called `Byte`, `Short`, and `Long`. A `Byte` in Scala is an 8-bit number. A `Short` is a 16-bit number. The `Int` that we have been using is a 32-bit number. The `Long` type is a 64-bit number. The reason for the odd behavior that was demonstrated at the beginning of section 2.5 is that we added two numbers together whose sum is bigger than what can be stored in the lower 31 bits of an `Int` and the OVERFLOW, as it is called, wrapped it around to a negative value. Table 2.1 shows the minimum and maximum values for each of the different integer types.

Occasionally you will need to use literals that are bigger than what an `Int` can store. You can do this with a `Long`. Making a numeric literal into a `Long` is done by simply adding an L to the end. You can see this here.

```
scala> 5000000000L
```

```
res31: Long = 5000000000
```

The value five billion is not a valid `Int`. If you leave off the `L` here you get an error. The `L` can be lower case, but then it looks a lot like the number one so it is better to use the upper case.

We talked about binary above, and Scala has a method that will let you see the binary form of a number. This method works on the four normal numeric types and `Char`. Here we use it to see the binary representation for 83 and -83 for the values as both `Int` and `Long` types.

```
scala> 83.toBinaryString
res32: String = 1010011
```

```
scala> -83.toBinaryString
res33: String = 11111111111111111111111110101101
```

```
scala> 83L.toBinaryString
res34: String = 1010011
```

```
scala> -83L.toBinaryString
res35: String =
  1111111111111111111111111111111111111111111111111111111110101101
```

The `toBinaryString` method does not display leading zeros, so the positive values only show seven digits in both formats. However, the negative form has many leading ones and all of these are printed.

2.5.4 Octal and Hexadecimal

Binary is what the machine uses, but it really is not that useful to humans. This is in large part due to the fact that the number of digits in a binary number is often large, even if the number itself is not what we consider large. There are two other bases that are commonly seen in programming and dealing with computers. They are base 8, OCTAL, and base 16, HEXADECIMAL or HEX. Like decimal, these bases allow you to represent fairly large numbers with relatively few digits. Unlike decimal, converting from octal or hex to binary and back is trivial. The reason for this is that 8 and 16 are powers of two.

To see this, let us start with octal. When working in base 8, the digits can be between 0 and 7 with each subsequent digit being a higher power of 8. The ease of converting to and from binary comes from the fact that 8 is 2^3. In binary the values 0 to 7 are represented with three bits between 000 and 111. The fourth and subsequent bits represent values that are multiples of eight. Because of this, we can convert a binary number to an octal number by grouping the bits into groups of three, starting with the least significant bit, and converting those groups.[9] So the binary number, 1010011 is 123 in octal. The lowest three bits, 011, convert to 3, the next three, 010, convert to 2, and the top bit is just a 1. We can use the `toOctalString` method to confirm this.

```
scala> 83.toOctalString
res36: String = 123
```

To go the other way, from octal to binary, we simply convert the octal digits to three digit

---

[9]It is very important to start grouping with the ones bit. Starting at the other end will give you the wrong answer if the last group has fewer than three bits.

binary numbers. So the octal value, 3726 converts to 011111010110. We can emphasize the groupings of bits by spacing them out: 011 111 010 110. This is 2006 in decimal.

Moving between hexadecimal and binary is similar. The catch is that now a single digit needs to have 16 possible values. So the 0-9 that we are used to will not suffice. It is typical to augment the normal digits with the letters A-F where A is 10 and F is 15. Because $16 = 2^4$, we use groups of 4 bits when converting between hexadecimal and binary. Once again, you start the process with the lower bits and work up. So 1010011 is grouped as 0101 0011 and becomes 53. We saw that 2006 in decimal is 011111010110. This groups as 0111 1101 0110 and becomes 7D6 in hex. Again, there is a method called toHexString that can be used on the numeric types to quickly get the hexadecimal representation of a number.

While toHexString give us hexadecimal representations of numeric values that we have in decimal, it is sometimes helpful to be able to enter values into programs using hexadecimal in a program. This can be done by prefixing a numeric literal with 0x. The following uses of this confirms the conversion we did for the numbers above.

```
scala> 0x53.toBinaryString
res37: String = 1010011

scala> 0x7D6.toBinaryString
res38: String = 11111010110
```

2.5.5 Non-Integer Numbers

We saw previously that if we type in a numeric value that includes a decimal point Scala tells us that it has type Double. The Double literal format is more powerful than just including decimal points. It also allows you to use scientific notation to enter very large or very small numbers. Simply follow a number by an e and the power of ten it should be multiplied by. So 15000.0 can also be written as 1.5e4.

The name Double is short for double precision floating point number. The full name includes information about the way that these numbers are stored in the memory of a computer. Like all values in a computer, the Double is stored as a collection of bits. To be specific, a Double uses 64-bits. This size is related to the double precision part. There is another type called Float that is a single precision floating point number and only uses 32-bits. In both cases, the internal representation uses floating point format. This is similar to scientific notation, but in binary instead of decimal. The bits in a floating point number are grouped into three different parts. We will call them s, e, and m and the value of the number is given by $(-1)^s * (1 + m) * 2^{(e-bias)}$. The first bit in the number is the sign bit, s. When that bit is on, the number is negative and when it is off it is positive. After the sign bit is a group of bits for the EXPONENT, e. Instead of using two's compliment for determining if the exponent is negative, the exponent is biased by a value that is picked to match with the number of bits in the exponent. Using a bias instead of two's compliment means that comparisons between floating point values can be done with the same logic used for integer values with the same number of bits. All remaining bits are used to store a MANTISSA, m. The stored mantissa is the fractional part of the number in normalized binary. So the highest value bit is $\frac{1}{2}$, the next is $\frac{1}{4}$, and so on. Table 2.2 below gives the number of bits used for e and m, the *bias*, and the range of numbers they can represent in the Double and Float types. The E notation is short for multiplication by 10 to that power.

As we have seen, floating point literals are considered to be of type Double by default. If you specifically need a Float you can append an f to the end of the literal. There are many other details associated with floating point values, but there is only one main point that will be stressed here. That is the fact that floating point values, whether Double or Float, are

Type	e Bits	m Bits	bias	Min	Max
Float	8	23	127	-3.4028235E38	3.4028235E38
Double	11	52	1023	-1.7976931348623157E308	1.7976931348623157E308

TABLE 2.2: Floating point types with sizes and ranges.

not Real numbers in the sense you are used to in math with arbitrary precision. Floating point numbers have limited precision. Like the integers, they can be overflowed. Unlike the integers, they are fundamentally imprecise because they represent fractional values with a finite number of bits. The real implications of this are seen in the following example.

```scala
scala> 1.0-0.9-0.1
res39: Double = -2.7755575615628914E-17
```

To understand why this happens, consider the simple fraction, $\frac{1}{3}$, the decimal representation of which 0.33333... In order to write this fraction accurately in decimal, you need an infinite number of digits. In math we can denote things like this by putting in three dots or putting a line over the digits that are repeated. For floating point values, the digits simply cut off when you get to the end of the mantissa. As such, they are not exact and the circuitry in the computer employs a rounding scheme to deal with this. This imprecision is not visible most of the time, but one immediate implication of it is that you cannot trust two floating point numbers to be equal if they were calculated using arithmetic. It also means that you should not use floating point numbers for programs that involve money. The decimal value 0.1 is a repeating fraction in binary, hence the problem in the example above, and as such, is not perfectly represented. Instead you should use an integer type and store cents instead of dollars.

2.6   The math Object

While on the topic of numbers, there are quite a few standard functions that you might want to do with numbers beyond addition, subtraction, multiplication, and division. There are a few other things you can get from operators that we will discuss later. Things like square root, logarithms, and trigonometric functions are not operators. They are found as methods in the math object. You can use tab completion in the REPL to see all the different methods that you can call on math and values stored in it.

```scala
scala> math.
BigDecimal                      ScalaNumericConversions max
BigInt                          abs                     min
E                               acos                    package
Equiv                           asin                    pow
Fractional                      atan                    random
IEEEremainder                   atan2                   rint
Integral                        cbrt                    round
LowPriorityEquiv                ceil                    signum
LowPriorityOrderingImplicits cos                        sin
Numeric                         cosh                    sinh
Ordered                         exp                     sqrt
Ordering                        expm1                   tan
```

PartialOrdering	floor	tanh
PartiallyOrdered	hypot	toDegrees
Pi	log	toRadians
ScalaNumber	log10	ulp
ScalaNumericAnyConversions	log1p	

Many of these probably do not make sense right now, and you should not worry about them. However, many of them should be identifiable by the name. So if we wanted to take a square root of a number, we could do the following.

```scala
scala> math.sqrt(9)
res40: Double = 3.0
```

You would use a similar syntax for taking cosines and sines. The functions provided in the math object should be sufficient for the needs of most people. Only two of the contents of math that start with capital letters are worth noting at this point. Pi and E are numeric constants for π and e.

```scala
scala> math.Pi
res41: Double = 3.141592653589793
```

Syntax versus Semantics

The terms SYNTAX and SEMANTICS are used very commonly when discussing programming languages. For natural languages, syntax can be defined as "the arrangement of words and phrases to create well-formed sentences in a language". This is a pretty good definition for programming languages other than we are not building sentences, we are building programs. The syntax of a programming language specifies the format or tokens and how tokens have to be put together to form proper expressions and statements as well as how statements must be combined to make proper programs.

As you will see, assuming that you have not yet, programming languages are much more picky about their syntax than natural languages. Indeed, the syntax of programming languages are specified in formal grammars. You do not really get the same type of artistic license in a programming language that you do in a natural language, as deviating from the syntax makes things incorrect and meaningless. Don't worry though, in expressive and flexible languages like Scala, you still have a remarkable amount of freedom in how you express things, and with experience, you can create beautiful solutions that follow the syntax of the language.

The semantics of a program deals with the meaning. Syntax does not have to be attached to meaning. It is just formal rules that are part of a formal system that specify if a program is well formed. It is the semantics of a language that tell us what something that follows the syntax actually means.

2.7   Naming Values and Variables

We have seen enough that you can solve some simple problems. For example, if you were given a number of different grades and asked to find the average, you could type in an expression to add them all up and divide by the number of them to get the average. We basically have the ability to use Scala now to solve anything we could solve with a calculator as well as doing some fairly simple string manipulation. We will develop a lot more over time, but we have to start somewhere. As it stands we are not just limited to solving problems we could do with a calculator, we are solving them the way we would with a calculator. We type in mathematical expressions the way we would write them on paper and get back an answer. Real programming involves tying together multiple lines of instructions to solve larger problems. In order to do this, we need to have a way to give names to values so we can use those values later.

There are two keywords in Scala that give names to values: `val` and `var`. To begin with, let us look at the full syntax of `val` and `var` in two samples. Then we can pull them apart, talk about what they do, see how they are different, and discuss what parts of them are optional.

```
scala> val age:Int = 2015-1996
age: Int = 19

scala> var average:Int = (2+3+4+5)/4
average: Int = 3
```

Syntactically the only difference between these two is that one says `val` and the other says `var`. That is followed by a name with a colon and a type after it. The rules for names in Scala are that they need to start with a letter or an underscore followed by zero or more letters, numbers, and underscores.[10] So `abc`, `abc123_def`, and `_Team2150` are all valid Scala names while `2points` is not. You also cannot use keywords as variable names. The only keywords that have been introduced so far are `val` and `var`, but there will be others, and you cannot use those as names for things as they are reserved by the language.

Scala is also case sensitive. So the names `AGE`, `age`, `Age`, and `agE` are all different. In general, it is considered very poor style to use names that differ only in capitalization as it can quickly lead to confusion. Most names will not involve underscores either, and numbers only appear where they make sense. Scala borrows a standard naming convention from Java called camel case. The names of values begin with a lower case letter and the first letter of subsequent words are capitalized. For example, `theClassAverage` is a name that follows this convention. Type names use the same convention except that the first letter is capitalized. This is called camel case because the capital letters look like humps.

The types in both of these examples are followed by an equal sign and an expression. Unlike many other programming languages, this is not optional in Scala. In Scala, when you declare a `val` or `var`, you must give it an initial value.[11]

While the initial value is not optional, the type generally is. Scala is able to figure out

---

[10]Scala also allows names that are either made entirely of operator symbols or have a standard name followed by an underscore and then operator symbols. Symbolic names should only be used in special situations, and using them improperly makes code difficult to read. For this reason, we will ignore these types of names for now.

[11]There are very good reasons for requiring initialization of variables. Even in languages that do not require it, a programmer can make his/her life a lot easier by initializing all variables at creation. The declaration and initialization should ideally happen at the point where you have a real value to put into the variable. This prevents many errors and as a result, can save you a lot of time in your programming.

var a = 5

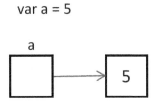

FIGURE 2.3: This figure shows how you should think about value and variable declarations in Scala. The variable itself stores a reference to an object. The difference between val and var is whether or not you can change what is referred to, not whether that object can be modified.

the types of things for us in many situations. If we leave off the colon and the type, Scala will simply use whatever type it infers is appropriate for the expression in the initial value. Most of the time, the type that it gives us will be exactly what we want. Using this we could instead have the following shorter forms of these declarations.

```scala
scala> val age = 2015-1996
age: Int = 19

scala> var average = (2+3+4+5)/4
average: Int = 3
```

The reason for using a val or var declaration is that they give a name to the value that we can refer back to later. For example, we could now type in age+10 and Scala would give us 29. The names serve two purposes. They prevent us from typing in expressions over and over again. They also help give meaning to what we are doing. You should try to pick names that help you or other readers figure out what is going on with a piece of code.

So far we have discussed all the similarities between val and var and you might be wondering in what way they are different. The declarations themselves are basically identical. The difference is not in the syntax, but in the meaning, or semantics. A val declaration gives a name to a reference to a value. That reference cannot be changed. It will refer to the thing it was originally set to forever. In the REPL, you can declare another val with the same name, but it does not do anything to change the original. A var declaration, on the other hand, allows the reference to change. In both cases we are not naming the value, we are naming a box that stores a reference the value. The significance of this will be seen in section 7.7. Figure 2.3 shows a visual representation of how you should picture what a val or var declaration does in Scala.

The act of changing the reference stored in one of these boxes we call variables is referred to as an ASSIGNMENT. Assignment in Scala is done with the same equal sign that was used to set the initial value. In an assignment though there is no val or var keyword. If you accidentally include either var or val you will be making a new variable, not changing the old one.

```scala
scala> average = 8
average: Int = 8
```

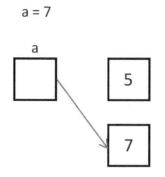

FIGURE 2.4: This figure shows what you might imagine happening with each of the lines assigning new values to the variable `average`.

```
scala> average = 2*average
average: Int = 16
```

The first assignment causes the box named average to change from referring to the object 3 to the object 8. The second one uses the previously referred to value and multiplies it by two, then stores a reference to that new value back into the variable. The effects of these lines are illustrated in figure 2.4.

As a general rule, you should prefer `val` declarations over `var` declarations. Try to make everything a `val`, and only convert it to a `var` if you find that you truly need to do so. The reason for this is that it simplifies the logic of your program and makes it less likely that you will mess things up. Things that change are harder to reason about than things that stay the same.

2.7.1   Patterns in Declarations

There is a bit more to the initialization of `val` and `var` declarations than was mentioned above. Technically, the initialization is able to do something called PATTERN MATCHING that we will get to in detail in chapter 5. For now, the only aspect we will care about is that we can put tuples on the left hand side of the equals sign where we would normally put just a variable name. First, let us see what happens if we do a `val` declaration with a tuple on the right hand side.

```
scala> val t = (100,5.7)
t: (Int, Double) = (100,5.7)
```

Note that `t` refers to the tuple and has a type (`Int`,`Double`). This is exactly what we would expect. The power that pattern matching provides is that if you put multiple names inside of parentheses on the left of the equals, much like a tuple, all the names will be bound. That type of behavior is shown here.

```
scala> val (price,weight) = t
price: Int = 100
weight: Double = 5.7
```

The same can be done with a `var` and then all the names will have the ability to change what they refer to. This is the second way of getting values out of tuples. It is more readable because we can pick meaningful names for the variables. After doing the example above, you could use `price` and `weight` instead of `t._1` and `t._2`.

2.7.2 Using Variables

Let us use the ability to name values to do a little problem solving. We are given a total time in seconds, and we want to know what that is in hours, minutes, and seconds. We then want to print that out in a reasonable format of "hh:mm:ss". The first step in solving this problem is to figure out how to go from just seconds to hours, minutes, and seconds. Once we have that, we can worry about formatting it to get the right string value.

How do we get from seconds to hours, minutes, and seconds? First, how do you get from seconds to minutes? That is fairly easy, you simply divide by 60. Thanks to the fact that integer division truncates, you will get the proper number of whole minutes. Here are two lines that define a number of total seconds as well as a number of total minutes.

```scala
scala> val totalSeconds = 123456
totalSeconds: Int = 123456

scala> val totalMinutes = totalSeconds/60
totalMinutes: Int = 2057
```

That number of minutes is not exactly the amount of time we want though. There are seconds left over. How do we figure out how many seconds we should display? We could do `totalSeconds-(60*totalMinutes)`, but a simpler expression is used here.

```scala
scala> val displaySeconds = totalSeconds%60
displaySeconds: Int = 36
```

The modulo gives us the remainder after we have gotten all the full groups of 60. That is exactly what we want. Now how do we get the number of hours and the number of minutes to display? The math is the same because there are 60 minutes in each hour.

```scala
scala> val displayMinutes = totalMinutes%60
displayMinutes: Int = 17

scala> val displayHours = totalMinutes/60
displayHours: Int = 34
```

What we see from this is that 123456 seconds is 34 hours, 17 minutes, and 36 seconds. We could repeat this same process for a different number of seconds if we used a different value for `totalSeconds`.

Now that we have these values, we want to figure out how to get them into a string with the format "hh:mm:ss". A first attempt at that might look like the following.

```scala
scala> val finalString = displayHours+":"+displayMinutes+":"+displaySeconds
finalString: String = 34:17:36
```

For this particular number of seconds, this works just fine. However, if you play around with this at all, you will find that it has a significant shortcoming. If the number of minutes or seconds is less than 10, only one digit is displayed when we want two. So we need to come up with a way to get a leading zero on numbers that only have one digit. To do this, we will break the problem into two steps.

The first step will be to get the number of minutes and seconds as `String`s.

```scala
scala> val min=displayMinutes.toString
min: String = 17

scala> val sec=displaySeconds.toString
sec: String = 36
```

This might seem odd, but the string version has something that the number itself does not, an easy way to tell how many digits/characters are in it. When there is only one digit, we want to add an extra zero. When there is not, we leave it as is. We can get this effect by using the ∗ method on the `String` and a little math. The short names were selected to keep our expression shorter for formatting, but that is not required.

```scala
scala> val finalString=displayHours+":"+("0"*(2-min.length))+min+":"+(
 | "0"*(2-sec.length))+sec
finalString: String = 34:17:36
```

The result for these values is the same, but we could force some different value into `min` and `sec` to see that this does what we want.

```scala
scala> val min="5"
min: String = 5

scala> val sec="2"
sec: String = 2

scala> val finalString=displayHours+":"+("0"*(2-min.length))+min+":"+(
 | "0"*(2-sec.length))+sec
finalString: String = 34:05:02
```

2.8   Details of `Char` and `String`

There is a lot more to `Char` and `String` than we covered in section 2.4. Some of it you really should know before we go further. We saw how we can make character literals or string literals that contain keys that appear on the keyboard and that go nicely into a text file. What about things that we cannot type as nicely or that have other meanings? For example, how do you put double quotes in a `String`? Typing the double quote closes off the string literal instead of putting one in. You are not allowed to have a normal string break across lines, so how do you get a newline in a String?

2.8.1   Escape Characters

We can do all of these things and more with ESCAPE CHARACTERS. These are denoted by a backslash in front of one or more characters. For example, if you want to put a double quote in a string, simply put a backslash in front of the double quote. You can insert a newline with a \n. If you want to insert a backslash simply put in two backslashes. Table 2.3 shows some commonly used escape characters.

In addition to escape characters, the backslash can be used to put any type of special

Literal	Meaning	Unicode Hex Encoding
\b	backspace	\u0008
\f	form feed	\u000C
\n	line feed	\u000A
\r	carriage return	\u000D
\t	tab	\u0009
\"	double quote	\u0022
\'	single quote	\u0027
\\	backslash	\u005C

TABLE 2.3: Table of special character escape sequences in Scala.

character into a string. If you know the UNICODE value for a special character, you can put \u followed by four hexadecimal digits in a string to specify that character.[12]

2.8.2 Raw Strings

There are some times when using the escape characters becomes a pain. For example, there are times when you need to build strings that have a number of backslashes. Each one you want requires you to put in two. This can get unwieldy. In addition, if you have a long, multi-line string, it can be difficult to format the string the way you want. For these types of situations, Scala includes a special form of string that begins and ends with three double quotes. Anything you put between the set of three double quotes is taken to be part of the string without alteration. These types of strings are called RAW STRINGS. The following shows an example of using this to enter a long string in the REPL.

```
scala> """This is a long string.
| It spans multiple lines.
| If I put in \n and \\ or \" they are taken literally."""
res42: String =
This is a long string.
It spans multiple lines.
If I put in \n and \\ or \" they are taken literally.
```

2.8.3 String Interpolation

In section 2.7.2, there were a number of expressions that put together strings using plus signs for concatenation. This approach can be challenging to write and read in code.[13] For that reason, there is an alternate approach to building strings that include values called STRING INTERPOLATION. The syntax for doing string interpolation is to put a "s" or a "f" in front of the string,[14] then put expressions in the string that begin with a dollar sign if they are to be evaluated and their values inserted.

The earlier example originally put together the string for the time using the expression

```
displayHours+":"+displayMinutes+":"+displaySeconds
```

---

[12]The topic of Unicode characters is beyond the scope of this book, but a simple web search will lead you to descriptions and tables of the different options.

[13]Using + to build long strings is also inefficient.

[14]The string interpolation mechanism in Scala is extensible, and programmers can add other options. The "s" and "f" forms are the main ones supported by the standard libraries.

Using string interpolation, this could be written as

```
s"$displayHours:$displayMinutes:$displaySeconds"
```

More complex expressions can be inserted into the string by enclosing the expression in curly braces after the dollar sign.

```
scala> val age = 2015-1996
age: Int = 19
scala> s"$age+10 = ${age+10}"
res43: String = 19+10 = 29
```

Here `$age` is nested inside an `s` processed string. The `s` interpolator knows to insert the value of the variable `age` at this location(s) in the string. There is no set rule for when you should use string interpolation instead of concatenation. You should pick whichever option you find easiest to read and understand.

The "f" interpolation requires that you place a format specifier after the expression. Coverage of these format specifiers is beyond the scope of this book. The interested reader is encouraged to look up details on his/her own. It should be noted that the format strings used by Scala are heavily based on those used in the printf function for the C programming language, and they appear in many libraries across different languages.

2.8.4   String Methods

There are many methods that you can call on the String type. Tab completion shows you some of them.

```
scala> "hi".
+                concat          isInstanceOf       startsWith
asInstanceOf     contains        lastIndexOf        subSequence
charAt           contentEquals   length             substring
chars            endsWith        matches            toCharArray
codePointAt      equalsIgnoreCase offsetByCodePoints toLowerCase
codePointBefore  getBytes        regionMatches      toString
codePointCount   getChars        replace            toUpperCase
codePoints       indexOf         replaceAll         trim
compareTo        intern          replaceFirst
compareToIgnoreCase isEmpty      split
```

These are the methods that come from the Java String type, and they provide a lot of the basic functionality that one needs when working with strings. Through a language feature in Scala called implicit conversions, there are others that are also available. The listing below shows those. You can see that it includes multiplication, as introduced earlier. It also includes methods like `toInt` and `toDouble`, which will convert strings with the proper values to those types.

```
*       foldLeft      mkString       stripLineEnd
++      foldRight     nonEmpty       stripMargin
++:     forall        padTo          stripPrefix
+:      foreach       par            stripSuffix
/:      format        partition      sum
:+      formatLocal   patch          tail
:\      groupBy       permutations   tails
>       grouped       prefixLength   take
>=      hasDefiniteSize product       takeRight
```

addString	head	r	takeWhile
aggregate	headOption	reduce	to
apply	indexOf	reduceLeft	toArray
asInstanceOf	indexOfSlice	reduceLeftOption	toBoolean
canEqual	indexWhere	reduceOption	toBuffer
capitalize	indices	reduceRight	toByte
collect	init	reduceRightOption	toDouble
collectFirst	inits	replaceAllLiterally	toFloat
combinations	intersect	repr	toIndexedSeq
compare	isDefinedAt	reverse	toInt
compareTo	isEmpty	reverseIterator	toIterable
contains	isInstanceOf	reverseMap	toIterator
containsSlice	isTraversableAgain	sameElements	toList
copyToArray	iterator	scan	toLong
copyToBuffer	last	scanLeft	toMap
corresponds	lastIndexOf	scanRight	toSeq
count	lastIndexOfSlice	segmentLength	toSet
diff	lastIndexWhere	seq	toShort
distinct	lastOption	size	toStream
drop	length	slice	toString
dropRight	lengthCompare	sliding	toTraversable
dropWhile	lines	sortBy	toVector
endsWith	linesIterator	sortWith	union
exists	linesWithSeparators	sorted	updated
filter	map	span	view
filterNot	max	split	withFilter
find	maxBy	splitAt	zip
flatMap	min	startsWith	zipAll
fold	minBy	stringPrefix	zipWithIndex

Going through all these methods is well beyond the scope of this chapter, but it is beneficial to see examples that use some of them. To do this, let us consider a situation where we have a person's name written as "*first last*". We wish to build a new string that has the same name in the format of "*last, first*". In order to do this, we must first find where the space is, then get the parts of the original string before and after the space. Once we have done that, we can simply put the pieces back together in the reverse order with a comma between them.

To find the location of the space, we will use the `indexOf` method. This method gives us a numeric index for the first occurrence of a particular character or substring in a string.

```scala
scala> val name = "Mark Lewis"
name: String = Mark Lewis

scala> val spaceIndex = name.indexOf(" ")
spaceIndex: Int = 4
```

The index of the space is 4, not 5, because the indexes in strings, and everything else except tuples, start counting at 0. So the 'M' is at index 0, the 'a' is at index 1, etc.

Now that we know where the space is, we need to get the parts of the string before and after it. That can be accomplished using the substring method.

```scala
scala> val first = name.substring(0,spaceIndex)
first: String = Mark

scala> val last = name.substring(spaceIndex+1)
```

```
last: String = Lewis
```

The first usage passes two arguments to substring. The first is the index of the first character to grab, in this case, it is 0. The second is the index after the last character to grab. In this case, it is the index of the space. The fact that the second bound is exclusive is significant, and it is a standard approach for methods of this nature in many languages. The second form takes a single argument, and it returns the substring from that index to the end of the string, making it ideal for getting the last name.

The two strings can now be put back together using concatenation or string interpolation. The following shows how to do it with string interpolation.

```
scala> val name2 = s"$last, $first"
name2: String = Lewis, Mark
```

One could also pull out the names using the splitAt method.

```
scala> val (first,last) = name.splitAt(spaceIndex)
first: String = Mark
last: String = " Lewis"

scala> val name2 = s"${last.trim}, $first"
name2: String = Lewis, Mark
```

The splitAt method returns a tuple, and we use a pattern here to pull out the two elements. Note that the space itself was included in the second element of the tuple. To get rid of that, we use the trim method. This method gives us back a new string with all leading and trailing whitespace removed.

If you only want a single character from a string, you can get it by INDEXING into the string with parentheses. Simply specify the index of the character you want in parentheses after the name of the string. So we could get the last initial from our original name string like this.

```
scala> name(spaceIndex+1)
res44: Char = L
```

2.8.5   Immutability of Strings

When looking through the list of methods on the String type, you might have noticed methods called toLowerCase and toUpperCase. These methods illustrate a significant feature of strings, the fact that they are IMMUTABLE. This means that once a string object has been created, it can not be changed. The toLowerCase method might sound like it changes the string, but it does not. Instead, it makes a new string where all the letters are lower case, and gives that back to us. This is illustrated by the following.

```
scala> val lowerName = name.toLowerCase
lowerName: String = mark lewis

scala> name
res45: String = Mark Lewis
```

The lowerName variable refers to a string that is all lower case, but when we check on the value of the original name variable, it has not been changed. None of the methods of String change the value. Any that look like they might simply give back modified values. This is what makes the String type immutable. The `trim` method used above demonstrates this same behavior. Most of the types we will deal with are immutable.[15]

2.9 Sequential Execution

Sequential execution is used when we write a program and want the instructions to execute in the same order that they appear in the program, without repeating or skipping any instructions from the sequence. So far, all of the program instructions we have written have been executed one after another in the same order that we have typed them in. The instructions have been executed sequentially.

Working in the REPL is great for certain tasks, but what if you have a sequence of things you want to do, and you want to do it multiple times. Having to type in the same set of instructions repeatedly is not a very good option. The time conversion above is a perfect example of that. If we want to do this for a different number of seconds, we have to repeat all the commands we just performed. Indeed, you cannot really say that you have programmed until you put in a fixed set of instructions for solving a problem that you can easily run multiple times. That is what a program really is. So now it is time to write our first program of any significance.

We have used the REPL to enter commands one at a time. This is a great way to test things out in Scala and see what a few commands do. A second way of giving commands to Scala is to write little programs as SCRIPTs. The term script is used to describe small programs that perform specific tasks. There are languages, called scripting languages, that have been created specifically to make the task of writing such small programs easier. Scala is not technically a scripting language, but it can be used in that way. The syntax was created to mirror a lot of the things that are commonly put into scripting languages, and if you run the `scala` command and give it the name of a file that contains Scala code, that file will be run as a script. The statements in it are executed in order.[16] The script for our time conversion looks like this.

Listing 2.1: TimeConvert.scala

```
val totalSeconds = 123456
val displaySeconds = totalSeconds%60
val totalMinutes = totalSeconds/60
val displayMinutes = totalMinutes%60
val displayHours = totalMinutes/60
val sec = displaySeconds.toString
val min = displayMinutes.toString
val finalString = displayHours+":"+("0"*(2-min.length))+min+
    ":"+("0"*(2-sec.length))+sec
println(finalString)
```

If you put this into a file called `TimeScript.scala` and then run `scala TimeScript.scala`, you will get the output `34:17:36`. The `println` statement is required for the script because unlike the REPL, the script does not print out values of all statements. You can run through

---

[15]The first mutable type we will encounter will be the Array type in chapter 6. That chapter will go further in demonstrating the significance of this distinction.

[16]We will see later that the statements are not always executed in order because there are statements that alter the flow of control through the program. Since we have not gotten to those yet though, execution is completely sequential at this point.

this code in the REPL using the `:load` command. If you do "`:load TimeScript.scala`" you will see it print out all of the intermediate values as well as the result of the `println`.

This script allows us to run the commands repeatedly without retyping. By editing the value of `totalSeconds`, we can test other total times fairly quickly. However, a better solution would be to allow a user to tell us how many seconds to use every time we run the script. We can easily get this behavior by replacing the top line of the script we had with these three lines.

Listing 2.2: TimeConvert2.scala

```scala
import io.StdIn._
print("Enter the number of seconds. ")
val totalSeconds = readInt()
```

The second line prints a prompt to let the user know that we are waiting for something to be input. After that we have altered the initialization of `totalSeconds` so that instead of giving it a fixed number, it is initialized to the value returned by `readInt`. This calls a function that reads in a single integer from the user. The first line is there because the full name of `readInt` is `io.StdIn.readInt`. The `import` statement allows us to use a shortened name whenever we want to read a value. The underscore in the import causes it to bring in other functions such as `readLine` and `readDouble` which allow us to read in strings and double values respectively. If you make this change and run the script, you will be able to enter any number of seconds, assuming it is a valid `Int`, and see it converted to hours, minutes, and seconds.

The following code shows the usage of `readLine` and `readDouble`.

```scala
import io.StdIn._
val name = readLine()
val number = readDouble()
```

Note that all of these functions read a full line from the user and expect it to match the desired type. If you want the user to enter multiple numbers on one line, you cannot use `readInt` or `readDouble`. For that you would have to read a `String` with `readLine`, then break it apart and get the numeric values.

2.9.1   Comments

When writing programs in files, not in the REPL, it is often useful to include plain English descriptions of parts of the code. This is done by writing comments. If you are writing code for a course, you likely need to have your name in the code. Your name is likely not valid Scala, so it should go in a comment. Different instructors and companies will have different commenting standards that you should follow. In a professional setting, comments are used primarily for two reasons. The first is to indicate what is going on in the code, particularly in parts of the code that might be difficult for readers to understand. The second is for documentation purposes using tools that generate documentation from code.

There are two basic comment types in Scala, single line comments and multiline comments. Single line comments are made by putting `//` in the code. Everything after that in the line will be a comment and will be ignored when the program is compiled and run. Multiline comments begin with `/*` and end with `*/`. You can put anything you want between those, and they can be spaced out across many lines. Code shown in this book will have limited commenting as descriptions of the code appear in the text of the book, and there is little point in duplicating that content.

2.10 A Tip for Learning to Program

In many ways, learning to program, whether in Scala or any other programming language, is very much like learning a new natural language. The best way to learn a natural language is through immersion. You need to practice it and be surrounded by it. The key is to not simply memorize the rules and vocabulary, but to put them into use and learn them through regular usage. You should strongly consider approaching programming in the same way.

So what does it mean to immerse yourself in a programming language? Clearly you are not going to have conversations in it or enjoy television or radio broadcasts in it. The way to immerse yourself in a programming language is to take a few minutes every day to write in it. You should consider trying to spend 15-30 minutes each day writing code. The REPL in Scala is an excellent tool for you to enter in statements to see what they do. Try to play around with the language. Instead of approaching it as memorizing keywords and rules, try to put things into use. The things that you use frequently will stick and become natural. Those things that you do not use regularly you will have to look up, but that is normal. Programmers, even professional programmers with many years of experience in a language, still keep references handy.

Over time, the number of lines of code that you write in these short time intervals each day will grow as the basics become second nature and you begin to practice more advanced concepts. By the end of this book you might find yourself writing a hundred lines or so of functional code on certain days during that time span. By that time you will hopefully also pick up a "pet project", something that you are truly interested in programming and that you will think about the structure and logic of much more frequently.

Especially early on, you might find it hard to think of anything that you can write. To help you with this, many of the chapters in this book contain a "Self-Directed Study" section, like the one below. Use these as a jumping off point for the material in each chapter. After that will come a set of exercises and often a set of larger problems called projects. Remember that one of the significant goals of learning to program is improving your problem solving skills. While the Self-Directed Study section will help you to familiarize yourself with the details presented in a chapter, the exercises and projects are actual problems that you are supposed to solve in a formal way using Scala. You should use these to help provide you with the immersion you need to learn the language.

2.11 End of Chapter Material

2.11.1 Problem Solving Approach

Many students who are new to programming struggle with putting the English descriptions for solving a problem that they have in their head into whatever programming language they happen to be learning. The reality is that for any given line of code, there are a fairly small number of "productive" things that you could write. In the REPL you can test out any statement that you want, but in a script, an expression like 4+5 does not do much when used alone as a statement. Sections like this one will appear at the end of a number of chapters as we introduce new concepts that might stand alone as statements, or which alter statements we have talked about previously in a significant way. The goal of these sections is to help focus your thinking so you can narrow down the list of possibilities any time that you are trying to decide what to put into the next line of code.

Given what we have just learned, there are only three types of statements that you would put in a script that stand alone:

1. A call to `print` or `println` to display information to the user. The function name should be followed with parentheses that contain the *expression* you want to print.

2. A variable declaration using `val` or `var`. A `val` declaration would look like `val` *name* = *expression*. The *name* could be followed with a colon and a type, though most of the time those will be left off.

3. An assignment into a previously declared `var` of the form *name* = *expression*. The *expression* must have a type that agrees with the type the variable was created with.

If you want to read information using a function like `readLine()`, `readInt()`, or `readDouble`, that should appear as part of an *expression* in one of the above statements. Remember to include the `import io.StdIn._` statement at the top of your file if you are going to be reading user input.

2.11.2   Summary of Concepts

- When you install Scala on your computer you get a number of different executable commands.

  - The `scala` command can run scripts or applications. If no argument is given it opens up the REPL for you to type in individual statements.
  - The `scalac` command is used to compile Scala source code to bytecode.

- Programming languages have relatively simple rules that they always follow with no ambiguity.

  - Tokens are the smallest piece with meaning. They are like words in English.
  - Expressions are combinations of tokens that have a value and a type.
  - Statements are complete instructions to the language. In Scala, any expression is a valid statement.
  - The simplest expressions are literals.
    * `Int` literals are just numbers with no decimal points like 42 or 365.
    * Adding an `L` to the end of an integer number makes a `Long` literal.
    * Numbers that include decimal points or scientific notation using `e` syntax are of the type `Double`.
    * Adding an `f` to the end of a number makes it a `Float`.
    * `Char` literals are single characters between single quotes.
    * `String` literals can have multiple characters between double quotes. Raw strings start and end with three double quotes and allow newlines.

- An object is a combination of information and functionality that operates on that information.

  - The information is called data members, fields, or properties.
  - The functionality is called methods.

- All values in Scala are objects.
- Methods are normally invoked using the "dot" notation. Arguments go in parentheses after the method name.
  * Scala allows the . to be left out if there is at least one argument to the method.
  * Parentheses are also optional for argument lists of length zero or one.
  * Operators are really method calls. So 4+5 is really (4).+(5).

- Numbers in computers are not exactly like numbers in math, and you need to know some of the differences so you will understand when they lead to unexpected behavior.

  - All values stored in a computer are stored in binary, base 2, numbers. Each digit is called a bit. Different types use different numbers of bits for storage. The finite number of bits means that there are minimum and maximum values that can be stored in each type.
  - Negative integer values are stored using two's compliment numbers.
  - Binary numbers require a large number of digits, though they are all either 0 or 1, and converting to and from decimal is non-trivial. For this reason, computer applications frequently use base 8, octal, and base 16, hex. You can make a hexadecimal literal by putting a leading 0x on an integer.
  - Non-integer numeric values are stored in floating point notation. This is like scientific notation in binary. These use the types Float and Double. Due to finite precision, not all decimal numbers can be represented perfectly in these numbers, and there are small rounding errors for arithmetic.

- Additional mathematical functions, like trigonometric functions and square root are methods of the math object.

- You can declare variables using the keywords val and var. The name of a variable should start with a letter or underscore and can be followed by letters, underscores, or numbers. A var declaration can be reassigned to reference a different value.

- String interpolation allows you to easily put values into strings without using + to concatenate them.

- There are many methods you can call on strings that allow you to do basic operations on them.

- Strings are immutable. That means that once a string is created, it is never changed. Methods that look like they change a string actually make a new string with the proper alterations.

- Instructions can be written together in scripts. The default behavior of a script is for lines of code to execute sequentially. Script files should have names that end with .scala. You run a script by passing the filename as a command-line argument to the scala command.

- Learning a programming language is much like learning a natural language. Do not try to memorize everything. Instead, immerse yourself in it and the things you use frequently will become second nature. Immersion in a programming language means taking a few minutes each day to write code.

2.11.3   Self-Directed Study

Enter the following statements into the REPL and see what they do. Try some variations to make sure you understand what is going on. Not all of these will be valid. You should try to figure out why.

```scala
scala> val a=5
scala> val b=7+8
scala> var c=b-a
scala> a=b+c
scala> c=c*c
scala> b=a/c
scala> b%2
scala> b%4
scala> b%a
scala> 0.3*b
scala> val name = "Your name here."
scala> name.length
scala> name+a
scala> println("Hi there "+name)
scala> println("\n\n\n")
scala> println("""\n\n\n""")
scala> 'a'+5
scala> ('a'+5).toChar
scala> math.Pi/2
scala> math.sqrt(64)-4.0
scala> math.sqrt(1e100)
scala> math.cos(math.Pi)
scala> 3000000000
scala> 3000000000L
scala> 3000000000.0
scala> 3e9
scala> 1/0
scala> 1.0/0.0
scala> 0.0/0.0
```

2.11.4   Exercises

1. What are the types of the following expressions?

   (a) 1

   (b) 1.7

   (c) 1.0

   (d) 'h'

   (e) "hi"

   (f) 5/8

   (g) 1+0.5

   (h) 7*0.5

   (i) "hi".length

2. Which of the following are valid Scala variable names?

    (a) 1stName

    (b) exam_1

    (c) Four

    (d) H

    (e) 4July

    (f) _MIXUP

    (g) GO!

    (h) last name

    (i) May4Meeting

    (j) sEcTiOn1

    (k) version.2

3. Do the following 8-bit binary arithmetic by hand.

    (a) $10101101_2 + 11010100_2$

    (b) $00111110_2 + 00111011_2$

    (c) $01001010_2 * 00110010_2$

4. Convert the following decimal values to binary (8-bit), hex (2-digit), and octal (3-digit) by hand.

    (a) 7

    (b) 18

    (c) 57

    (d) 93

    (e) 196

5. Convert the following hex values to binary and decimal by hand.

    (a) 0x35

    (b) 0x96

    (c) 0xA8

    (d) 0x7F

6. Convert the following decimal values to binary (8-bit) and hex (2-digit) by hand.

    (a) -87

    (b) -32

    (c) -105

    (d) -1

7. Write a script that will calculate the cost of a fast food order for a burger stand that sells hamburgers, french fries, and milkshakes. Hamburgers cost $2.00, french fries cost $1.00, and milkshakes cost $3.00. The tax rate is 8%. Ask the customer how many of each item they would like, then display a receipt that shows the price of each item, how many of each item was ordered, the extension (i.e. price * quantity), the subtotal, the amount of tax, and then the total due.

8. Kyle is really hungry for pizza and wants to get the best pizza deal. He is not sure if it is a better deal to get two cheaper medium sized pizzas or one large pizza. Write a script that will help him figure out which to order. He can get two 12 inch pizzas for $12.00 or one 14 inch pizza for $9.00. Hint: Use the formula "ÏĂ x r squared" to find the area of a pizza in inches.

9. Bryn just started a new job, but is already thinking of retirement. She wants to retire as a millionaire. She plans on saving $700 per month and expects to receive an annual return of 7%. Will she be a millionaire if she retires in 30 years? Write a script to figure this out.

10. A town administrator in west Texas is trying to decide if she should build a larger water tower. The town has a water tower that contains 20,000 gallons of water. If there is no rain, write a script that will calculate the number of weeks the water will last based on the town's usage (provided by the town administrator). The weekly usage does not exceed 20,000 gallons.

11. Write a script that will calculate how far a projectile will go given a launch speed and an angle ignoring friction. Assume that the projectile is launched from ground level with a certain speed in m/s and at a certain angle in radians. Use the fact that acceleration due to gravity is $9.8m/s^2$. The steps in doing this would be to calculate the speed parallel and perpendicular to the ground with math.sin and math.cos, then figure out how long it takes for the projectile to slow to a vertical speed of zero $(v = v_0 - at)$, and use double that time as how long it stays in the air.

12. Quinn is planning on seeding a new lawn of about 1500 square feet in her backyard this spring and wants to know how much top soil and grass seed she should buy. She lives in a new development and it is pretty barren and rocky in the backyard; so, she plans to put down 6 inches of top soil. Most grass seed mixes recommend 4 to 5 lbs. of grass seed for each 1,000 square feet of land for calculating grass seed. How many cubic yards of top soil and how pounds of grass seed does she need to buy? Write a script to help her find the answer.

13. Using Scala as a calculator, figure out how much you have to make each year to bring home $100,000 assuming a 27% tax rate.

14. In the REPL, declare a variable with the type **String** that has the name **str**. Give it whatever value of string you want. On the next line, type **str.** then hit tab to see the methods for **String**. By playing around with the REPL try to figure out what the following methods do.

    - **toUpperCase**
    - **trim**
    - **substring** – This method takes two **Int** arguments.
    - **replace** – This method can be called with two **Char** arguments or two **String** arguments.

15. Kepler's third law of planetary motion says that $P^2 \propto a^3$, where P is the orbital period, and a is the semi-major axis of the orbit. For our Sun, if you measure P in years and a in Astronomical Units (AU), the proportionality becomes equality. Look up the semi-major axis values for three bodies in our solar system other than the Earth and use Scala as a calculator to find the period according to Kepler's third law.

16. In this option you will write a little script that does part of a 1040-EZ. We have not covered enough for you to do the whole thing, but you can write enough to ask questions and do most of the calculations for lines 1-13. Do what you can and remember this is a learning exercise.

17. Your goal for this exercise is to write a script to calculate the cost of a simple three-ingredient recipe. You start by asking the user for the names and amounts of the three ingredients. Then prompt them for the cost per unit of each ingredient. Output the total cost of each ingredient and for the whole recipe. To make things simple, feel free to use the `Double` type.

Chapter 3

Conditionals

We solved some basic problems in the last chapter, but the techniques that we have access to are fundamentally limited at this point. The real problem is that every line in a script executes in the same order from top to bottom every time the script is run. In most real problems, we need to be able to do what is called conditional execution, where choices are made and some things happen only in certain situations. In this chapter we will learn the most fundamental method for doing conditional execution in Scala and see how we can use it to solve some problems.

In order to do this properly, we need to develop a formalized way to express logic and put it into our programs. This system, called Boolean logic, will allow us to state the conditions under which various parts of our program will or will not happen.

3.1 Motivating Example

You have been asked to write a program that will calculate charges for people visiting a local amusement park. There are different charges for adult vs. child, whether they are bringing in a cooler, and whether they want to also get into the water park. We need to write code that will tell us how much the person pays. We will have the user input the needed information such as an Int for the persons age, a Boolean for whether they have a cooler, and another Boolean for whether they want to get into the water park.

This is something that we could not do last chapter because we did not have a way of performing logic and making decisions. We could not say that we wanted to do something only in a particular situation. This ability to do different things in different situations is called conditional execution, and it is a very important concept in programming. It is also critical for problem solving in general. Conditional execution gives you the ability to express logic and to solve much more complex problems than you could do without it.

3.2   The `if` Expression

Virtually all programming languages have a construct in them called `if`. For this construct, you have a condition where one thing should happen if the condition is true. If the condition is false, then either nothing or something different will happen. In non-functional languages, the `if` construct is a statement. It has no value and simply determines what code will be executed. In Scala and other functional languages, the `if` is an expression which gives back a value. Scala allows you to use it in either style. The syntax of an `if` is: `if` (*condition*) *trueExpression* `else` *falseExpression*. The `else` clause is optional.

Let us start with an example, then we can broaden it to the more general syntax. Take the ticket price example and consider just the person's age. Say that we want to consider whether a person should pay the $20 children's rate or the $35 adult rate. For our purposes, a child is anyone under the age of 13. We could make a variable with the correct value with the following declaration using an `if` expression.

```
val cost = if (age<13) 20 else 35
```

This assumes that `age` is an `Int` variable that has been defined prior to this point. The first part of this line is a basic variable declaration as discussed in chapter 2. The next part contains an `if` expression which checks the age and gives back one of two values depending on whether `age<13` is true or false.

When using `if` as an *expression* we always need to include an `else` because there has to be a value that is given to the variable if the condition is false. This same type of behavior can be also accomplished with a `var` using the `if` as a *statement*.

```
var cost=20
if (age>=13) cost=35
```

This code creates a `var` and gives it the initial value of 20. It then checks if the age is greater than or equal to 13 and if so, it changes the value to 35. Note that with this usage, the `else` clause is not required. Here we only want to do something if the condition is true. In Scala you should generally prefer the first approach. It is shorter, cleaner, and leaves you with a `val` which you can be certain will not be changed after it is created. The first approach is a functional approach while the second is an imperative approach.

In general, the format of the `if` is as follows:

```
if (condition) expr1 [else expr2]
```

The square brackets here denote that the `else` and the second expression are optional. The condition must always be surrounded by parentheses and can be replaced by any expression of type `Boolean`. The expressions `expr1` and `expr2` can be any expression you want, though they should typically have the same type.

Often, one or both of the expressions will need to be more complex. Consider the time conversion script from section 2.9. There are a few places where we could have used conditionals in that example, but a major one would have been to check if the user input was a positive value, as what was done does not make much sense for negative values. There are a number of lines that should only be executed if the value is positive. The example below uses a code block to group those all together on the true branch of the if statement.

Listing 3.1: TimeConvertIf.scala

```scala
import io.StdIn._
print("Enter the number of seconds. ")
val totalSeconds = readInt()
if (totalSeconds > 0) {
  val displaySeconds = totalSeconds%60
  val totalMinutes = totalSeconds/60
  val displayMinutes = totalMinutes%60
  val displayHours = totalMinutes/60
  val sec = displaySeconds.toString
  val min = displayMinutes.toString
  val finalString = displayHours+":"+("0"*(2-min.length))+min+
    ":"+("0"*(2-sec.length))+sec
  println(finalString)
} else {
  println("This only works for a positive number of seconds.")
}
```

A code block is just code enclosed in curly braces. If you want multiple lines of code to be executed after an if or an else you should group those lines with curly braces. In the above example, the else branch did not need to have curly braces because the code following the else is a single line, but they were added to make the code consistent. It is considered good practice to use curly braces on if statements unless they are extremely simple and fit on one line.

The other thing to note about this example is that the code inside of the curly braces is uniformly indented. Scala does not care about indentation, but human programmers do, and this is one style rule that virtually all programmers feel very strongly about: nested code needs to be indented. How far you indent varies between programmers, languages, and tools. In Scala, it is customary to indent two spaces, and that is the style that is followed in this book.

The placement of brackets is something else to pay attention to. In this book, we use the style shown here, where the open bracket goes at the end of the line it is opening, and the close bracket is at the end and not indented. The most common alternate format has a newline before the open curly brace so that it appears lined up below the "i" in if. It really does not matter what style you use, as long as you are consistent in your usage.[1]

The previous example uses the if as a statement, and the block of code is a statement. A block of code can be used as an expression. The value of the expression is the value of the last thing in the block.

```scala
scala> {
     | println("First line")
     | 4+5
     | }
First line
res0: Int = 9
```

The time formatting example can be done using an if expression in the following way.

---

[1] Many employers will have style guides for these types of things. In that case, you will simply use whatever style they prefer.

Listing 3.2: TimeConvertIfExpr.scala

```scala
import io.StdIn._
print("Enter the number of seconds. ")
val totalSeconds = readInt()
val response = if (totalSeconds > 0) {
  val displaySeconds = totalSeconds%60
  val totalMinutes = totalSeconds/60
  val displayMinutes = totalMinutes%60
  val displayHours = totalMinutes/60
  val sec = displaySeconds.toString
  val min = displayMinutes.toString
  displayHours+":"+("0"*(2-min.length))+min+":"+("0"*(2-sec.length))+sec
} else {
  "This only works for a positive number of seconds."
}
println(response)
```

Common Bug

One common error that novice programmers encounter with the `if` statement occurs when they put more than one expression after the `if` or the `else` without using curly braces. Here is an example of how that can happen. We start with the following code that calculates an area.

```scala
val area = if (shape=="circle")
  math.Pi*radius*radius
else
  length*width
```

Here the `if` is broken across multiple lines so there is not one long line of code. This can make it easier to read. (Note that this code does not work in the REPL as it ends the expression when you hit enter before the else.) There are no curly braces, but they are not needed because each part of the `if` has only a single expression.

Now consider the possibility that the program this is part of does not work. To help figure out what is going on, you put print statements in so that you know when it uses the circle case and when it uses the rectangle case. The error comes when the code is changed to something like this.

```scala
val area = if (shape=="circle")
  println("Circle")
  math.Pi*radius*radius
else
  println(Rectangle)
  length*width
```

Now both branches have two expressions, and we have a problem. This code needs curly braces, but they are easy to forget because the indentation makes the code look fine. Scala does not care about indentation. That is only for the benefit of humans. In reality, this code should look something like the following.

```scala
val area = if (shape=="circle") {
```

```
   println("Circle")
   math.Pi*radius*radius
} else {
   println("Rectangle")
   length*width
}
```

Returning to the theme park example, what about the other parts of our admission park entrance cost? We also wanted to check if the person had a cooler or if they wanted to get into the water park. These should both be variables of type `Boolean`. We might call them `cooler` and `waterPark`. Let us say it costs an additional $5 to bring in a cooler and $10 to go to the water park. If we used the `if` as an expression, we can type in the following:

```
val cost = (if (age<13) 20 else 35)+(if (cooler) 5 else 0)+(if (waterPark) 10 else
   0)
```

Here we are adding together the results of three different `if` expressions. This format is somewhat specific to functional languages. It would be more common in most languages to see this instead:

```
var cost = 20
val cooler = false
val waterPark = true
if (age >= 13) cost = 35
if (cooler) cost = cost+5
if (waterPark) cost = cost+10
```

In this second form, we use `if` as a statement instead of an expression and have the body of the `if` change, or mutate, the value of the variable cost.

Note on Style

While there are differences between the functional and imperative versions of code in different applications in regard to things like performance and the likelihood of errors, at this stage those should not be your top concerns. You should pick the style that makes the most sense to you. Later you can evaluate the differences and pick the approach that best fits the task.

In the second and third `if` statements, the name `cost` is repeated. This type of repetition is often avoided in programming. Many languages, including Scala, include operations that allow us to avoid it. When the duplication is like this, with an assignment and a variable appearing on both sides of the equals sign, it is possible to use an abbreviated syntax where the operator is placed in front of the equals sign in the following manner:

```
var cost = 20
val cooler = false
val waterPark = true
if (age >= 13) cost = 35
if (cooler) cost += 5
if (waterPark) cost += 10
```

Note that the += is a token, so you cannot put a space between the symbols. Doing so will cause an error.

3.3 Comparisons

The first if statement shown in the previous section uses >= to do a comparison between two values. You likely had no problem figuring out that this can be read as greater than or equal to. Your keyboard does not have a ≥ key, so instead we use two characters in a row. All of the normal comparisons that you are used to exist in Scala, but some, like the greater than or equal to, differ from how you are used to writing them on paper.

The simplest comparison you can do is to check if two things are the same or different. You read this as saying that two things are equal or not equal to one another. In Scala we represent these with == and != respectively. Note that the check for equality uses two equal signs. A single equal sign in Scala stands for assignment, which we have already seen stores a value into a variable. The double equal sign checks if two expressions have the same value and produces a Boolean value with the result of the comparison. Here are a few examples of the use of this in the REPL.

```
scala> 2 == 3
res1: Boolean = false

scala> 7 == 7
res2: Boolean = true

scala> 'a' == 'a'
res3: Boolean = true

scala> "hi" == "there"
res4: Boolean = false

scala> "mom".length == "dad".length
res5: Boolean = true
```

The != operator is basically the opposite of ==. It tells us if two things are different and should be read as not equal. As we will see, the exclamation point, pronounced "bang" in many computing contexts, means "not" in Scala. Any of the examples above could use != instead of ==, and the result would have been the opposite of what is shown.

In addition to equality and inequality, there are also comparisons of magnitude like the age >= 13 that we used above. The comparisons of magnitude in Scala are done with <, >, <=, and >=. These also give us back a value of the Boolean type so it will be either true or false. These comparison operators are also known as relational operators. The order of the characters in <= and >= is significant. They are in the same order that you say them, "less than or equal to" and "greater than or equal to"; so, it will not be hard to remember. If you reverse the order, Scala will not be able to figure out what you mean and will return an error.

```
scala> 5=<9
<console>:6: error: value =< is not a member of Int
       5=<9
         ^
```

You can use == or != on any of the different types in Scala, both those we have talked about and everything that we have not yet talked about. This is not the case for the magnitude comparisons. While you can use <, >, <=, and >= for many of the types that we have seen so far, not every type has an order where these comparisons makes sense. For example, they do not work on tuples. Later in the book we will have types for things like colors, shapes, and fruits. Magnitude comparisons will not make sense with these either, and they will not work if you try to use them.

Equality vs. Identity (Advanced)

If you have experience with Java, you might find the behavior of == confusing. This is because in Scala, == does what most people expect, it checks for equality between values. For anyone who has not programmed in Java he/she might wonder what other options there are. We will see later that objects with the same value are not necessarily the same objects. If you go to a store and pick up two boxes of the same type of item, they are basically equal as far as you can tell, but they are not the same object. Each one has its own identity.

There are times in programming when you do not want to just check if two things are equal, you want to actually know if they are the same thing. This requires doing a check for identity. In Scala we use eq to check if two things are the same thing. Here is an example of where eq and == return different results.

```
scala> "sssss" eq "s"*5
res8: Boolean = false

scala> "sssss" == "s"*5
res9: Boolean = true
```

3.4 Boolean Logic

Imagine if the theme park had a policy where seniors are charged the same amount as children. So now anyone over 65 or anyone under 13 should pay the reduced rate. We could accomplish this by having separate if statements in either the functional or the imperative manner.

```
val cost = if (age < 13) 20 else if (age > 65) 20 else 35
```

or

```
var cost = 20
if (age >= 13) cost = 35
if (age > 65) cost = 20
```

Both of these are verbose, potentially inefficient, and prone to errors. The errors occur because in both we had to enter the number 20 in two places. What we would really like to say is exactly what we said in English. We want to say that we use the lower rate if the

Description	Usage	Meaning
and	a && b	True if both a and b are true.
or	a || b	True if a or b is true. Allows both being true.
exclusive or (xor)	a ^ b	True if either a or b is true, but not both.
not	!a	True if a is false and false if a is true.

TABLE 3.1: This table shows the Boolean operators for Scala and what each one does.

a && b	a=true	a=false
b=true	true	false
b=false	false	false

a || b	a=true	a=false
b=true	true	true
b=false	true	false

a ^ b	a=true	a=false
b=true	false	true
b=false	true	false

	!a
a=true	false
a=false	true

TABLE 3.2: Truth tables for the different Boolean operators.

person is under 13 or over 65 and use the higher rate otherwise. Boolean logic gives us the ability to say this.

There are four different BOOLEAN OPERATORS we can use to build complex Boolean expressions from simple ones. These are shown in table 3.1.

We can use the || operator just like we used "or" in our English description of what we wanted above. If we do this, our functional approach would simplify to this.

```scala
val cost = if (age < 13 || age > 65) 20 else 35
```

We can use && to say "and".

```scala
var cost = 20
if (age >= 13 && age <= 65) cost = 35
```

The first one reproduces the English description and uses an "or" to give a single **Boolean** expression for when the lower rate is charged. The second one is the converse and uses an "and" to determine when the higher rate should be charged. The second expression could be written instead with an || and a ! to make it more explicit that it is the converse of the first one.

```scala
var cost = 20
if (!(age < 13 || age > 65)) cost = 35
```

The extra set of parentheses is required here so that the not is done for the whole expression.

It is worth noting that || is not the "or" of normal English. In normal usage, when you say "or" you mean the logical exclusive or, ^. For example, when a parent offers their child the option of "cake or cookies", the parent is not intending for the child to take both. The inclusive or, | |, allows both. The exclusive or, ^, does not.

Another example would be code that tells us if two rectangles intersect. Each square is defined by its top left corner as x and y coordinates along with the edge lengths of the two rectangles. We want a result that is a `Boolean` telling us whether or not the rectangles intersect. Before we can write a program to do this, we need to figure out how we would do this independent of a program.

Your first inclination might be to say that given two rectangles it is obvious whether or not they intersect. Indeed, novice programmers are tempted to give that type of description to many problems. This is not because the solution really is obvious, it is because novice programmers have not developed their skills at analyzing what they do when they solve a problem. This is something that one develops over time, and it is a requirement for any real programmer.

So what are you missing when you say that a problem like this is obvious? Given a set of four numbers as is the case here, most people would not find the solution obvious. To make it obvious they would draw the squares and look at them. This use of your visual processing is basically cheating and implicitly brings into play a large amount of processing that your brain does automatically. For this reason, we will avoid any type of solution where you would be tempted to say you would "look at" something.

Even if we did take the approach of drawing the squares, that is not as straightforward as you might picture. When you picture drawing squares, you likely picture squares of nearly the same size that are in easy to draw coordinates. Plotting gets a lot harder if one of the squares is millions of times larger than the other. If your program is trying to see if an ant has entered a building, that is not at all out of the question. So we cannot settle for "just look at it" or "it is obvious". That means we have to figure out what it really means for two squares to be intersecting using just the numbers.

While looking at it is not allowed as a solution, it can be helpful to figure out what we really mean. Draw a number of different squares on paper. Label them with the numbers 1 and 2. What are different possible relative positions for the squares? What are cases when they do not intersect? What has to happen for them to intersect? These are not rhetorical questions. Go through this exercise and come up with something before you read on. The ability to break down a problem that is "obvious" into the real steps that humans go through on a subconscious level is a cornerstone of programming. It is also useful for those who do not intend to be programmers as it can help you to understand your thinking on all manner of topics.

When you went through the process of drawing the squares, one of the things you might have found was that squares can overlap in either x or y direction, but they only intersect if they overlap in both x and y. That gives us something to work with. Our answer could say something like, `overlapX && overlapY`. All we have to do now is figure out what it means to overlap in a given direction. Even this has lots of different possibilities, but if you play with them you will see that any situation where there is an overlap satisfies the following: the minimum of the first range has to be less than the maximum of the second range and the maximum of the first range has to be greater than the minimum of the second range. Go back to your drawings and verify that this is true.

At this point we have the ability to say what we want. There are many ways that we could do so. We are going to pick an approach which breaks the problem down into smaller pieces. This will be easier for people to read. We said above that the squares overlap if their ranges overlap in both x and y. So we can write code that checks to see if two ranges overlap in one dimension, then the other. We can then combine these to see if the rectangles overlap.

```
val overlapX = x1 < x2+size2 && x1+size1 > x2
val overlapY = y1 < y2+size2 && y1+size1 > y2
val squareOverlap = overlapX && overlapY
```

The `Boolean` value, `squareOverlap`, tells us the answer to the question.

Short Circuit Operators

One other significant factor about the `Boolean` `&&` and `||` operators is that they are short circuit operators. This means that if the value they will give is known after the first argument is evaluated, the second argument will not be evaluated. For `&&`, this happens if the first argument is `false` because no matter what the second argument is, the final value will be `false`. Similarly, if the first argument of `||` is `true`, the final value will be `true` so there is no point spending time to evaluate the second argument. This will be significant to us later on when we get to expressions that can cause errors. The only thing we could do now that would cause an error during a run is to divide by zero. We can use that to demonstrate how short circuiting can prevent an error and that the `^` operator is not short circuited.

```
scala> val n = 0
n: Int = 0

scala> 4/n
java.lang.ArithmeticException: / by zero
      at .<init>(<console>:7)
      at .<clinit>(<console>)
      at RequestResult$.<init>(<console>:9)
      at RequestResult$.<clinit>(<console>)
      at RequestResult$scala_repl_result(<console>)
      at sun.reflect.NativeMethodAccessorImpl.invoke0(Native Method)
      at sun...

scala> n != 0 && 4/n == 6
res3: Boolean = false

scala> n == 0 || 4/n == 6
res4: Boolean = true

scala> n == 0 ^ 4/n == 6
java.lang.ArithmeticException: / by zero
      at .<init>(<console>:7)
      at .<clinit>(<console>)
      at RequestResult$.<init>(<console>:9)
      at RequestResult$.<clinit>(<console>)
      at RequestResult$scala_repl_result(<console>)
      at sun.reflect.NativeMethodAccessorImpl.invoke0(Native Method)
      at sun...
```

First Character
(other special characters)
* / %
+ -
:
= !
< >
&
^
|
(all letters)
(all assignment operators)

TABLE 3.3: Table of operator precedence in Scala. The precedence of an operator is determined by the first character.

3.5   Precedence

So far all of our expressions have been fairly small. Other factors can come into play when they get large. One factor that becomes significant is precedence. This is the order in which operations are done. You know from math that multiplication happens before addition, and Scala follows that same order of operation. Now you need to know where these new operators fit into the order of precedence list. All of the comparison and Boolean operators perform *after* all mathematical operators. This allows expressions like `a+5 < b*2` without using parentheses. Similarly, comparisons have higher precedence than the Boolean operators. The expressions for `overlapX` and `overlapY` used both of these facts as they combined addition, comparison, and `&&`.

As was mentioned earlier, operators are really just methods in Scala. The ones we have talked about so far are simply methods that are defined in the Scala libraries on the appropriate types. So `&&` and `||` are defined on the `Boolean` type. The comparison operators are defined on numeric types, etc. When you use operator notation with a method, the precedence is determined by the *first character* in the operator. Table 3.3 shows the order.

3.6   Nesting `ifs`

There is another interesting point that we have used implicitly, but is worth noting explicitly. This is that `if` expressions can be nested inside of one another. We saw this when we first tried to add senior citizens at a lower cost.

```scala
val cost = if (age<13) 20 else if (age>65) 20 else 35
```

The contents of the `else` on the first `if` is itself an `if`. This is a general property of most programming languages. The `if` in Scala needs some form of expression inside of it for the `true` and `false` possibilities and if it is an expression in Scala. As such, the `if` itself makes

Item	S	M	L
Drink	$0.99	$1.29	$1.39
Side	$1.29	$1.49	$1.59
Main	$1.99	$2.59	$2.99
Combo	$4.09	$4.99	$5.69

TABLE 3.4: Theme park food item costs.

a perfectly valid expression to nest. So you can nest `ifs` inside of one another as much as it makes sense for the problem you are solving.

To make this more explicit, let us go back to our theme park and this time consider concessions. The menu is not broad and is standard for fast food. They have drinks, fries, and various main course items like hamburgers and hot dogs. You can also get a combo which has one of each. For any of these you can specify a size. The cost is specified by the simple matrix shown in table 3.4.

We need to convert this table into code. We will start by reading in two strings. The first is the item type and the second is the size, both as `Strings`. In the end we want the variable `cost` as a `Double`.

Listing 3.3: ConcessionPrices.scala

```scala
import io.StdIn._
println("What item are you ordering?")
val item = readLine()
println("What size do you want?")
val size = readLine()
val cost = if (item == "Drink") {
    if (size == "S") 0.99
    else if (size == "M") 1.29
    else 1.39
  } else if (item == "Side") {
    if (size == "S") 1.29
    else if (size == "M") 1.49
    else 1.59
  } else if (item == "Main") {
    if (size == "S") 1.99
    else if (size == "M") 2.59
    else 2.99
  } else {
    if (size == "S") 4.09
    else if (size == "M") 4.99
    else 5.69
  }
println(s"That will cost $$$cost.")
```

This code has a top level set of `ifs` that pick the item type. Inside of each is an `if` statement that picks from the different sizes. The way this was written, it will default to a combo if the item is not recognized and to a large if the size is not recognized. There are better ways to deal with this, but this will work for now.

This code demonstrates a standard formatting style used with this type of structure where the only thing in the `else` is another `if`. Instead of putting another set of curly braces after the `else` and indenting everything, leave off the curly braces and just put the `if` there. This prevents the indentation from getting too deep.

3.7 Bit-Wise Arithmetic

Generally, programmers do not need to worry about operations at the bit level. However, there are situations when a programmer may like to be able to go to the level of an individual bit. Data compression and encryption are a couple of examples of where you might want to use bit-wise operations. The bits in a binary number are just like `Boolean` values. We can perform Boolean logic on the bits in integer values the way we would on standard `true` and `false` values. To do this we use slightly different operators. We use & and | instead of && and ||. The & is a bit-wise and. The | is a bit-wise or. The versions with a single character do not short circuit. The motivation for short circuiting does not make sense for this type of operation because a bit cannot be a complex operation to evaluate. The other difference is that we use ~ instead of ! for the bit-wise negation.

If you think back to chapter 2 you will remember that every number on a computer is represented in binary. We store values of one or zero to represent different powers of two. When we do bit-wise operations we simply take the bits in the same position and do the specified operations on them to get the bits in the result. To see how this works let us run through a set of examples on the binary operators where we use the four bit numbers 1100 and 1010.

```
  1100
& 1010
------
  1000

  1100
| 1010
------
  1110

  1100
^ 1010
------
  0110
```

Negation is pretty straightforward: ~1001=0110. Of course, these bit combinations all have numeric decimal values. We can put commands into Scala to do these same things. Our first value is 12 and our second is 10. The value we took the negation of is 9. Here are the operations performed in Scala. Check that they match the answers we got.

```
scala> 12 & 10
res3: Int = 8

scala> 12 | 10
res4: Int = 14

scala> 12 ^ 10
res5: Int = 6

scala> ~9
res6: Int = -10
```

The last one is interesting. When you check if it makes sense, remember that negative values are represented using two's compliment and an `Int` has 32 bits.

There are two other bit-wise operators that are worth mentioning. They are left-shift and right-shift and are written in Scala as `<<` and `>>`. These operators shift the bits in an integer value the specified number of positions. If you think about the meaning of bits, this is like multiplying or dividing by powers of two in the same way that adding zeros to the end of a decimal number is the same as multiplying by powers of ten. Some simple examples show you how we can use this to get powers of two.

```
scala> 1 << 0
res7: Int = 1

scala> 1 << 1
res8: Int = 2

scala> 1 << 2
res9: Int = 4

scala> 1 << 3
res10: Int = 8
```

Let's look at some of these a little closer. The decimal 1 is represented as 00000001 in binary and `<<` means shift to the left. So `1 << 0` is shifting zero bits to the left, thus the result unchanged and still decimal 1. However, `1 << 1` shifts 1 bit to the left. Shifting 1 bit to left changes 00000001 to 00000010 which is decimal 2. If we start with the decimal 1 and perform `1 << 2`, this shifts 2 bits to the left. Shifting 2 bits to left changes 00000001 to 00000100 which is decimal 4. There is a second version of right-shift written as `>>>`. The normal version does not move the sign bit so that signs are preserved. The second version does not do this and will shift the sign bit down along with everything else.

You might wonder why you would want to do these things. Using a single integer value as a collection of `Boolean` values is common in libraries based on the C language and frequently appears in operating system code. There is another usage that could will into play in projects later in this book. If you have adjusted display settings on a computer you have probably seen that colors can be represented as 8-bit, 16-bit, 24-bit, or 32-bit values. Given the abundance of memory on modern computers, most of the time people will use 32-bit color. Have you ever wondered what that means? If you have written a web page or looked at a web page you have seen colors represented as six hexadecimal digits. The first two digits specify how much red, the next two specify how much green, and the last two specify how much blue. This is called RGB for obvious reasons and is exactly what 24-bit color gives you. 32-bit color uses an extra 8-bits because computers can operate more quickly with memory in 32-bit chunks because they are typically wired to deal with a 32-bit integer. The additional 8-bits stores the alpha channel which can be used for transparency. It does not matter much for your screen, but it is something that can be used to nice effect in 2-D graphics which we will discuss in chapter 12.

32-bit color is often called ARGB because it has alpha, red, green, and blue values all packed into 32 bits. Each gets 8 bits or a byte. This is where bit-wise operations come into play. You might be given four different values for alpha, red, green, and blue for a color and need to pack them together into a single 32-bit integer. Alternately, you might be given a single ARGB value as a 32-bit integer and have to figure out the individual components. Indeed, both of these appear as exercises below.

3.8 End of Chapter Material

3.8.1 Problem Solving Approach

The `if` can be used as a statement. This is a 4th option for statements that you can put into a script. So when you are considering what your next line of code is you should be thinking of one of these four possibilities.

1. Call `print` or `println`.

2. Declare a variable with `val` or `var`. (Prefer `val`.)

3. Assign a value to a variable. (Only works with `var` declarations that you are avoiding.)

4. Write an `if` statement. Note that to make sense as a statement, the expressions for the true and false possibilities should include one of more statements from this list. You should use an `if` when the code needs to do different things in different situations. When the word "if" fits in your English description of the solution to the problem, odds are good that it fits into the code as well.

3.8.2 Summary of Concepts

- Constructs that allow different pieces of code to be executed depending on different conditions in the code are called conditionals.

- The most basic conditional is `if`.

  - The syntax is `if` (*condition*) *trueExpression* `else` *falseExpression*. The `else` clause is optional.

  - The *condition* needs to be an expression with the `Boolean` type.

  - In Scala it can be used as an expression with a value or as a statement. As an expression, you need to have an `else` clause.

  - Curly braces can define blocks of code that function as a large expression. The value of the expression is the value of the last statement in the block.

- The `Boolean` expressions often involve comparisons of values.

  - Any values can be checked for equality or inequality using == or != respectively.

  - Values that have a natural ordering can also be compared using <, <=, >, or >=.

  - These comparison operators are also known as relational operators.

- More complex Boolean expressions can be built by combining simple expressions using Boolean logic.

  - The || operator is an inclusive or. This means that it is `true` if either of the two arguments are `true` as well as when both are `true`.

  - The && operator represents a logical and. It is only true when both arguments are true.

  - The ^ operator represents exclusive or. This is "or" as used in normal English where the result is `true` if one argument or the other is `true`, but not both.

- The ! operator is logical negation.

- The || and && operators are short-circuit operators.

• When building large expressions, the order in which operators are applied is significant. This is called precedence. The precedence of operators in Scala depends on the first character in the operator. Table 3.3 gives a full list of precedence in Scala.

• Bits can be viewed as `Boolean` values. Bit-wise arithmetic is operations that work on numeric numbers as collections of bits instead of normal numbers.

- | is bit-wise or.

- & is bit-wise and.

- ^ is bit-wise xor.

- ~ is bit-wise negation.

3.8.3   Self-Directed Study

Enter the following statements into the REPL and see what they do. Try some variations to make sure you understand what is going on. Note that some lines read values so the REPL will pause until you enter those values. The outcome of other lines will depend on what you enter.

```scala
scala> val a = readInt()
scala> val b = readInt()
scala> val minimum = if (a<b) a else b
scala> if (minimum != (a min b)) {
  println("Oops, something went wrong.")
} else {
  println("That's good.")
}
scala> true && true
scala> true && false
scala> false && true
scala> false && false
scala> true || true
scala> true || false
scala> false || true
scala> false || false
scala> !true
scala> !false
scala> true ^ true
scala> true ^ false
scala> false ^ true
scala> false ^ false
scala> a < b || { println("a>=b"); a >= b }
scala> a < b && { println("a>=b"); a >= b }
scala> a match {
  case 7 => "That is a lucky number."
  case 13 => "That is an unlucky number."
  case _ => "I'm not certain about that number."
}
scala> 13 & 5
scala> a.toBinaryString
```

```
scala> a & 0xff
scala> (a & 0xff).toBinaryString
scala> a ^ 0xff
scala> (a ^ 0xff).toBinaryString
scala> (a << 3).toBinaryString
scala> ((a >> 8) && 0xff).toBinaryString
```

3.8.4 Exercises

1. Write Boolean expressions for the following:

   (a) Assume you have a variable called **age**. Tell if the person is old enough to legally drink.

   (b) Given a **height** in inches, tell if a person can ride an amusement park ride that requires riders to be between 48" and 74".

2. Determine if the following expressions are **true** or **false**. Assume the following, a=1, b=2, c=3, d=true, e=false.

   (a) a==1

   (b) c<b || b>c

   (c) a<=c && d==e

   (d) 1+2==c

   (e) d

   (f) !e

   (g) d || e

   (h) 6-(c-a)==b && (e || d)

   (i) c>b && b>a

   (j) a+b!=c || (c*a-b==a && c-a>b)

3. Determine if the following expressions are true or false. Assume the following, a=1, b=10, c=100, x=true, y=false.

   (a) x

   (b) x && y

   (c) a==b-9

   (d) a<b || b>a

   (e) !y && !x

   (f) (c/b)/b==b/b

   (g) (c+b+a==b*b+c/c) || y

   (h) a <= b && b <= c && c >= a

   (i) c/(b/b)==b/b

   (j) !(x || y)

4. The **reverse** method can be called on a **String**. Use this to write a script where the user inputs a word (use **readLine**) and you tell them whether or not it is a palindrome.

5. A year is a leap year if it is perfectly divisible by 4, except for years which are both a century year and not divisible by 400 (e.g. 1800 is not a leap year, while 2000 is a leap year). There were no leap years before 1752. Write a script that inputs a year and calculates whether or not it is a leap year.

6. Write a script that prompts the user to enter the coordinates of a point in a Cartesian plane and tells the user indicating whether the point is the origin, is located on the x-axis, is located on the y-axis, or appears in a particular quadrant. For example: (0,0) is the origin and (3, -4) is in the forth quadrant.

7. It is customary to express colors on a computer as a combination of red, green, and blue along with another value called alpha that indicates transparency. A single Int has 32 bits or 4 bytes. The four different color values are often packed into a single Int as an ARGB value. The highest byte is the alpha and below that you have red, green, and blue in order. Each byte can store values between 0 and 255. For alpha, 0 is completely transparent and 255 is completely opaque.

   Write code that reads four `Int` values for alpha, red, green, and blue and calculates an `Int` with the combined ARGB value. If one of the numbers passed in is outside the 0 to 255 range, use 0 or 255, whichever it is closer to. Note that bitwise operations are appropriate for this exercise. You might find hexadecimal representation of numbers to be useful as well.

8. Repeat the previous exercise, but this time the input should be `Doubles` between 0.0 and 1.0 that you convert to `Ints` in the proper range.

9. Write code that does the opposite of what you did for exercise 7. It should take an `Int` with an ARGB value and calculate the four `Int` values with the component values between 0 and 255.

10. Write code that does the opposite of what you did for exercise 8. It should take an `Int` with an ARGB value and calculate four `Double` values with the component values between 0.0 and 1.0.

11. Write a script that has the user input a location as x and y coordinate values which tells whether the point specified is in the unit circle.[2]

12. Write a script that accepts the month, day, and year of a date and outputs the number of that day within its year (i.e. Jan 1st is always 1, Dec 31st is either 365 or 366).

3.8.5   Projects

1. Write a script that asks the user for the coefficients of a quadratic equation (a, b, and c in $ax^2 + bx + c$). It should print the roots for those coefficients. It needs to handle different options for the roots and print one or two solutions with real or complex values as appropriate.

2. Write a script that tells you whether or not a rectangle overlaps with a circle. It needs to prompt the user for the required information for the two shapes that includes positions and sizes, and it prints an appropriate message based on the Boolean result.

---

[2]The unit circle is a circle centered on the origin with radius 1.

3. You have built a simple robot that is supposed to move along a straight line from wherever it is placed to a flag. The robot can move in one of two ways. It can roll any distance at a speed of one inch per second. It can jump and each jump takes one second and moves a predetermined amount. When it jumps it always jumps that amount, not a fraction of it. Write a script that reads the distance you start from the flag and the distance covered in each jump and prints how many seconds it will take to get to the flag using an optimal approach. (The trick with this problem is to make sure you consider all the possibilities.)

4. This problem starts a track of options that you can work on that build up to having a functioning ray tracer at the end of the semester that can render images in 3-D. For this first step you are to write a script that determines if and where a ray intersects either a plane or a sphere.

   The script should start by asking for information on the ray. That includes the start point of the ray, $\vec{r}_0$, and the direction vector for the ray, $\vec{r}$. It should then ask if you want to consider a plane or a sphere. The code will have an `if` that splits into two possibilities at that point. One will ask for information on the sphere, a center point and radius, and do the proper calculation for a sphere. The second will ask for information on a plane, a normal direction and distance, and do the proper calculation for the plane. The script finishes by printing out information about the intersection.

   A ray can be defined as a start point and a direction vector using the parametric equation $\vec{r}(t) = \vec{r}_0 + t * \vec{r}$, for $t \geq 0$. You find intersections by solving for the t that satisfies certain conditions like coming a certain distance from a point (for a sphere) or satisfying the equation $\vec{r}(t) \cdot \vec{n} = d$ (for a plane with normal $\vec{n}$ that is d units from the origin). You can view coming within a certain distance from a point as solving the equation $(\vec{r}(t) - \vec{n}) \cdot (\vec{r}(t) - \vec{n}) = radius^2$. With a little algebra you can reduce this to a quadratic equation. You only care about the solution with the smaller value of t. This option is a bit more conceptually challenging, but if you do all the ray tracing options the results can be impressive. The web site at `http://book.programmingusingscala.net` includes a number of images made using a ray tracer the authors wrote in Scala.

5. You have been going through your book of *Simple 2-Ingredient Recipes* looking for something to cook. The problem is, you are really running low on ingredients. You only have four items, and they are in short supply. For this project you will write a script that will take the four items you have and how much of each you have. After that it will ask for two ingredients and how much of each is needed. It should output whether or not you have enough stuff to make that recipe.

   For the input, items will be strings and you do not care about units on the amount. Item names have to match exactly to be the same and assume the same units are used for any given item.

6. Imagine that you were considering taking a number of different majors or minors for college and you want to write a program to do a little math and help you see what different approaches would require. You have four major/minor interests. For each one the user should input a name and the number of hours it requires. You also have a number of hours for the core curriculum, a minimum number of hours for graduation, and how many hours you bring in from outside (AP/IB/Transfer). Prompt the user to enter those 11 values. Then run through the four major/minor interests and let them enter "yes" or "no" for whether they will do it to test an option. After that, print out

the average number of hours that need to be taken each of the 8-semesters of college to complete that combination.

7. You are trying to determine how many and what type of calories you need in a day. However, that depends on several factors including your age, height, gender, activity level, and goals (e.g. lose weight, stay the same, or gain weight). In order to solve this problem, you will need to calculate your Basal Metabolic Rate (BMR), your Total Energy Expended (TEE), and how many additional or fewer calories you might want to consume. The formula for the BMR for Men = 10 x weight(kg) + 6.25 * height(cm) - 5 x age(years) + 5. The formula for the BMR for Women = 10 x weight(kg) + 6.25 * height(cm) - 5 x age(years) - 161. To calculate your TEE you need to decide your activity level: sedentary = BMR * 1.2, lightly active = BMR * 1.375, moderately active = BMR * 1.55, very active = BMR * 1.725, and extra active = BMR * 1.9. Next you need to consider your goals. If you want to lose 1 pound, you need to consume 3500 less calories. The inverse is true if you want to gain weight (e.g. body building). The Centers for Disease Control and Prevention suggest that if you want to lose weight, aim to lose about 1 to 2 pounds each week. After you have determined your weight goals and included them in your calculations, then you calculate the amount of carbohydrates, fat and protein you need to stay active and healthy. Men and women of all ages should get 45 to 65 percent of their total daily calories from carbohydrates, 10 to 30 percent of their total daily calories from protein, and 20 to 35 percent of their total daily calories from fats. Write a script to determine how much carbohydrates, fat, and protein you should eat daily to meet your goals.

8. Write a script that will help someone decide whether it is more cost effective to walk, ride a bike, or drive a car to work. Assume that the more time that they spend commuting to work, then the fewer hours they will work. Time is money! Thus, although your transportation costs for walking are zero, you still will experience a lost opportunity cost by not being able to work during that time spent walking and this needs to be added to the overall travel cost. Assume a 17 minute mile walking pace, a 5 minute per mile biking pace, and a 3 minute per mile driving pace. Assume that bike maintenance is 1 cent per mile. Also assume that your car gets 25 mpg, gas costs $3 per gallon and car maintenance costs 6 cents per mile. Ask the user the distance they have to travel to work and what mode of transportation they wish to use in order to calculate their total travel costs per day.

9. Internet telephony service providers offer calling services between computers and regular phones not connected to the Internet. Write a script to help people determine what their calling costs would be based on how many minutes they expect to talk in a month, whether they will be using a monthly subscription or pay-as-you-go service, what country they will be making those calls to and whether they will be making calls to both mobile and landline phones or just to a landline phone. Use the following information to help you with your task:

	Monthly Subscription				Pay-As-You-Go (cents/min)	
	120 mins	400 mins	800 mins	Unlimited	Mobile/Landline	Landline
China	$1.19	$3.89	$7.79	$13.99	2	2
India	$1.42		$8.99	$19.99	1.5	1.5
Mexico				$1.79	3.5	1
UK				$2.09 for mobile/landline $1.19 landline	10	2.3
USA				$2.99	2.3	2.3

Chapter 4

Functions

Earlier in the book we made the statement that programming was all about problem solving. It is basically the art of giving a computer a set of instructions to solve a problem in terms that the computer can understand. One of the key foundational elements of good problem solving is problem decomposition. The idea of problem decomposition is that good solutions of large, complex problems are built out of solutions to smaller, more manageable problems. This is especially important in programming because the problems can be arbitrarily complex and, at the base, we have to break them down to the level where we can communicate them to the computer, which is fundamentally quite simplistic.

One of the key elements of problem decomposition in programming is the FUNCTION. A function is a collection of statements to solve a specific problem that we can call as needed. You can think of it as a little, self-contained program that performs a task that we can reuse over and over again. The scripts we have built to this point share some characteristics with functions. They were collections of statements we put together to do something repeatedly without retyping it. We have also looked at a few built-in methods that can be viewed as functions. Do you remember some of the math methods such as min and max? You know enough about programming in Scala to be able to write code to find the minimum or maximum of two numbers, but it is certainly useful to not have to type that code over and over every time we would want to find an answer. Functions are very beneficial to a programmer.

4.1   Motivating Example

Imagine that you needed to write a script for a resort where you had to calculate the taxes on a hotel room, gift shop purchases, and the restaurant purchases made by a guest.

You could accomplish this by repeating the code to do the calculation in multiple places, but there are many problems with this approach. The ability to cut and paste means that it will not take all that much extra effort, but it will make your code a lot longer and harder to manage. The primary problem arises if you realize that you have to change something in the calculation. If you have multiple copies of the code that do the calculation you have to find all of them and change them. This is an activity that is not only tedious, it is error prone because it is very easy to miss one or more of the copies.

The way to get around this is to use functions. A function lets you group together all the commands that are used in a particular calculation so that they can be called on from other parts of the code. You can specify what information needs to be provided for the function to do its work. In this way, functions provide flexibility. Instead of doing the copy and paste, you simply call that one function from different parts of the code, passing in the information that is needed at that point. If you realize something needs to change in the calculation, you simply change the function, and all the calls to it will use the modified version.

4.2   Function Refresher

Functions are things that you are familiar with from many different math classes, going back to algebra. The simplest example of a function from math might be something like $f(x) = x^2$. This says that f is a function that takes a number, given the name x, and it has the value of that number squared. Note that when the function is written, the FORMAL PARAMETER, x, has no value. For a function like this in math, it was usually understood that x was a real number, but that does not have to be the case. Complex numbers work as well. For other functions you might want to limit yourself to integers.

The value of x is specified when we *use* the function. For example, if we say $f(3)$ then 3 is the ARGUMENT to the function and x takes on the value of 3 when we figure out the function's value. When we do this we find that $f(3) = 9$.

That was an example of a function of one variable. We can also have functions of two or more variables. For example, you might have $g(x, y) = x + y^2$. When this function is called, we provide two values as arguments and they are used as the value of x and y. So $g(3, 5) = 3 + 25 = 28$. The value of x is 3, and the value of y is 5. You know this because the order of the arguments matches the order of the formal parameters.

In a general sense, a function is a mapping that associates input values with output values. In your math classes you have likely focused on functions that map from numbers into numbers. We have seen that our programs can work with numbers, but there are other types that we can work with as well. Some simple examples of this include the following:

- A function that takes a String and returns how many vowels are in it.

- A function that takes a String and returns the first uppercase letter in it.

- A function that takes two points as (x, y) pairs and returns the distance between them.

All of these have parameters that are not simple numbers. In terms you used in algebra, we would say that their domain is not just the real numbers. They also give us back values from different types. In algebra this would be the range. The first example can only produce a non-negative integer. The second produces a character. The third is a real number.

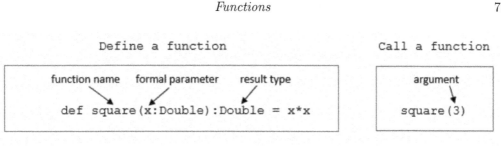

FIGURE 4.1: This figure shows examples of a function definition and a function call.

4.3 Making and Using Functions

There are a number of ways to make functions in Scala. We will begin with the standard definition format. In chapter 2 we saw that we can use the keywords `val` and `var` to declare variables. In a similar way, we use the keyword `def` to define a function. We will start with the simple mathematical examples that were used in the previous section and create Scala functions that match them in the REPL.

```scala
scala> def square(x:Double):Double = x*x
square: (x: Double)Double
```

In math, functions are given names like f and g. This works fine for a limited context, but it does not scale up well if you are using a lot of different functions. We use the name `square` here because it describes what the function does. The name is written after the `def` and after that we have a set of parentheses that contain a name and a type separated by a colon. This is like a `val` declaration except that it does not include an initial value[1] and the type is required. The value is specified when this function is called, just like for the math function. The contents of the parentheses are the formal parameters of the function.

After the parentheses is another colon and the type that this function gives us back, or returns. In Scala we refer to this as the RESULT TYPE. In this case, because we are mirroring the mathematical function, the type is `Double`. After the result type is an equal sign and the expression `x*x`. Once this has been entered, the REPL tells us that we have created something called `square` with the type `(x: Double)Double`. This is Scala's way of telling us that we have a function that takes a `Double` and results in a `Double`.

Now that we have defined square we can use it. In the last section we used our function f with $f(3)$. The syntax in Scala is the same.

```scala
scala> square(3)
res0: Double = 9.0
```

We give the name of the function followed by parentheses with the value we want to use for `x`. When we do this with the value 3, Scala gives us back a value of 9.0 as a `Double`. This particular value could also be represented as an `Int`, but the function says it works with type `Double` and so the `Int` that we passed in was converted to a `Double` and the result is also a `Double`.

---

[1]Scala allows you to provide a value here, called a default value. That is something you will not do often; so, it is not covered until chapter 10.

Let us now look at the second example math function. In this case we will use the short name g and make it work with the Int type instead of Double.

```
scala> def g(x:Int,y:Int):Int = x+y*y
g: (x: Int,y: Int)Int
```

Note that both of the parameters are put in the parentheses and separated by a comma. In general, you can have an arbitrary number of parameters. They can all have different types, and they will be separated by commas. The function is called by giving the function name with values separated by commas.

```
scala> g(3,5)
res1: Int = 28
```

In this example we used simple literals for the arguments to the function, but the arguments can be any expression with the proper type. For example, we could do the following:

```
scala> val a = 5
a: Int = 5

scala> g(a+2,square(3).toInt)
res2: Int = 88
```

Here we show that the arguments to g can include a reference to a val or a call to the square function. The call to square needs to be followed by a conversion to an Int because g is declared to take type Int and the result type of square is Double. If you leave that out, you will get an error like this:

```
scala> g(a+2,square(3))
<console>:9: error: type mismatch;
 found : Double
 required: Int
        g(a+2,square(3))
              ^
```

With the val and var declarations, we saw that we could typically leave off the types and Scala would figure them out for us. With functions, the types on the parameters are required. The result type is not required as it can often be inferred, but it is considered better style to specify it.[2] The reason this is preferred is that if you tell Scala a type and make a mistake in the function that causes it to result in a different type, Scala can tell you earlier that something has gone wrong.

These functions are very short. If we want to write longer functions, we need to create code blocks using curly braces. A block is composed of statements. The block, itself, is also either a statement or an expression. When used as an expression, the value you get from it is the value of the last expression in the block. To see how this works, we could take the statements that we had in our script at the end of chapter 2 and put them in a function. The function might look something like this if we type it into the REPL.

```
scala> def secondsToTimeString(totalSeconds:Int):String = {
     | val displaySeconds = totalSeconds%60
     | val totalMinutes = totalSeconds/60
     | val displayMinutes = totalMinutes%60
```

---

[2]The exception to this is recursive functions where the result type is required. More on that in chapter 5.

```
| val displayHours = totalMinutes/60
| val sec = displaySeconds.toString
| val min = displayMinutes.toString
| displayHours+":"+("0"*(2-min.length))+min+":"+("0"*(2-sec.length))+sec
| }
secondsToTimeString: (totalSeconds: Int)String
```

This function takes a single `Int` of the number of seconds. It results in a `String`. After the equal sign is an open curly brace that opens a block of code. This block is closed at the end of the function with a closing curly brace placed after an expression that gives us the String we want to return.

The advantage of this approach is we can easily call the function using different numbers of seconds, even in the REPL.

```
scala> secondsToTimeString(123456)
res0: String = 34:17:36

scala> secondsToTimeString(654321)
res1: String = 181:45:21
```

In this way, functions allow us to organize code and give useful names to things that we might want to do frequently.

This code also does something else that we have never seen before, it declares variables inside of a function. When we do this the variables are called LOCAL VARIABLES, and they can only be accessed inside of the function. If, after typing in this function, you try to use `sec`, or any of the other local variables, outside the function, you will get an error message.

```
scala> sec
<console>:6: error: not found: value sec
       sec
       ^
```

The part of the code over which a name can be accessed is called the SCOPE of that name. When you declare a variable or a function in a script outside any block of code, it has a scope that goes through the rest of the script. If you declare it in a block of code, it has a scope that goes from the point of the declaration to the end of the block it is declared in. A block is denoted with curly braces, so you can use the name until you get to the right curly brace that closes off the block the name is declared in. As a general rule, you want to minimize the scope of variables so they can only be accessed where they are needed. This reduces the amount of code you have to look through if something goes wrong with that variable. This is particularly true of `var` declarations.

> **Trap:** New programmers are often tempted to have functions use a `var` that was declared outside of the function. The majority of the time, this is a bad idea as it is very error prone, and makes the functions harder to use or less reusable.

These examples utilize the mathematical concept of a function as something that takes input values and maps them to output values. In programming it is possible for functions to do things other than give back a value. We have already seen an example of such a function. Our first program called `println`. You might notice that if you use `println` in the REPL, a value is printed, but there is not a result value shown. This is because `println` gives us back a value of type `Unit`, which was discussed in chapter 2. Here is a sample function that simply prints, so the return type is `Unit`.[3]

```scala
scala> def printMultiple(s:String,howMany:Int):Unit = {
     | println(s*howMany)
     | }
printMultiple: (s: String,howMany: Int)Unit

scala> printMultiple("hi ",3)
hi hi hi
```

Tip: The best functions result in values and do not print or read input from a user. The follow list gives reasons for this.

- Result values can be used for further computations, printed values cannot be.

- Printing ties a function to working with standard output.

- Reading input ties the function to working with standard input.

Right now our only output and input are the standard ones, so having a function that only does printing and reading tasks is not so limiting. However, mixing calculations or other computations into a function that does input and output still results in functions that are generally less useful. It is better to have the extra computation done in a separate function that has the proper result type. The function doing the input and output can use that separate function, as can other parts of the code.

Before moving on, we will do another example function that is more interesting, calculating an average in a course. The course average is calculated by combining the average of a number of different grades. For example, you might have tests, assignments, and quizzes. Each of these might contribute a different fraction of the total grade. The tests might be worth 40%, assignments worth 40%, and quizzes worth 20%. You will also have a different number of each of these grades. For example, you might have 2 tests, 3 assignments, and 4 quizzes. In many courses, part of the course average is computed by dropping a lowest grade and taking the average of what remains. In this case, let us assume that the lowest quiz grade is dropped.

What we want to do is write a function that takes all the grades, and returns the average for the course. To start with, it might look something like this:

```scala
def courseAverage(test1:Double,test2:Double,assn1:Double,
    assn2:Double,assn3:Double,quiz1:Double,quiz2:Double,
    quiz3:Double,quiz4:Double):Double = {
  ???
}
```

This function takes in nine `Double`s and produces an average for the full class as a `Double`. We want to calculate the averages for each of the different parts of the grade separately, then combine them with the proper percentages. Filling in a bit more gives the following.

```scala
def courseAverage(test1:Double,test2:Double,assn1:Double,
```

---

[3]As of the time of this edition, Scala supports a shortened syntax for functions that return `Unit`. This syntax is considered poor form, and is to be deprecated in a future release of the language, so it is not introduced here.

```scala
      assn2:Double,assn3:Double,quiz1:Double,quiz2:Double,
      quiz3:Double,quiz4:Double):Double = {
  val testAve = (test1+test2)/2
  val assnAve = (assn1+assn2+assn3)/3
  val minQuiz = ???
  val quizAve = (quiz1+quiz2+quiz4-minQuiz)/3
  testAve*0.4+assnAve*0.4+quizAve*0.2
}
```

All that is left to do is to figure out the minimum quiz grade. To accomplish this, we will use a method called `min` that is defined on the `Int` type. If you have two `Int` values, a and b, then the expression a `min` b will give you the smaller of the two values. This is a call to the `min` method on `Int` and could be written as `a.min(b)`. However, the operator syntax, where the dot and parentheses are left off, is superior when we have multiple values because we can put the values in a row to get the smallest of all of them. This gives us a complete version of the function which looks like the following.

Listing 4.1: CourseAverage.scala

```scala
def courseAverage(test1:Double,test2:Double,assn1:Double,
      assn2:Double,assn3:Double,quiz1:Double,quiz2:Double,
      quiz3:Double,quiz4:Double):Double = {
  val testAve = (test1+test2)/2
  val assnAve = (assn1+assn2+assn3)/3
  val minQuiz = quiz1 min quiz2 min quiz3 min quiz4
  val quizAve = (quiz1+quiz2+quiz3+quiz4-minQuiz)/3
  testAve*0.4+assnAve*0.4+quizAve*0.2
}
```

If you put this into the REPL, you can call it and get output as shown here.

```scala
scala> courseAverage(90,80,95,76,84,50,70,36,89)
res4: Double = 81.93333333333334
```

Use of ??? (Aside)

The code examples above used ??? in parts of the code that had not yet been filled in. The three question marks is actually a construct defined in the Scala libraries, and you can put the first one in a script and run the script just fine as long as it never calls that function. If the function were called, the program would crash with a `NotImplementedError` when the line with the ??? was executed. This can be very handy when you are writing larger programs and want to test some parts of the code when you have not gotten around to completing everything.

The second example does not work because ??? results in a type called `Nothing`, and you cannot do math operations with `Nothing`. For this reason, the ??? is generally not as helpful as an expression that is used in some way, but it is still a useful feature that you should take advantage of.

4.4 Problem Decomposition

As we have said, programming is about problem solving. One of the major aspects of problem solving is breaking hard problems into smaller pieces. There are a number of reasons to do this. The most obvious reason is that big problems are hard to solve. If you can break a big problem into smaller pieces, it is often easier to solve the pieces and then build an answer to the full problem from the answers to the pieces. This can be repeated on each sub-problem until you get to a level where the problems are simple to solve.

A second advantage to breaking problems up is that the solutions to the pieces might be useful in themselves. This allows you to potentially reuse pieces of code. The ability to reuse existing code is critical in programming. There are many different ways that you can break up a problem. A good way to tell if one is better than another is if one gives you functions that you are more likely to be able to reuse.

A third advantage to decomposing a problem is that the resulting code can be easier to understand and modify. If you give the functions good names, then people can easily read and understand top level functions and have a better idea of what is going on in the low level functions. In addition, if you decide at some point that you need to modify how something is done, it is easier to determine where the change needs to be made and how to make it if the problem has been broken down in a logical way.

A fourth advantage is that they are easier to debug. When you mess something up in your code it is called a BUG.[4] The act of removing these flaws is debugging. It is a lot easier to find and fix bugs in small pieces of code than in large pieces of code. Functions let you make small pieces of code operate independently, so you can verify each function works on its own, independent of others.

We looked at two different problems and saw how they could be programmed as a set of instructions that we can put inside of a function. The question now is how could we break these things up? If you go back to the original discussions, the descriptions of how to solve the problems was done piece by piece in a way that worked well for breaking them apart.

Let us begin with the grading program. There are four basic parts to this problem. We find the average of the tests, the assignments, and the quizzes. Once we have those, we combine them using the proper percentages. Each could be broken out into a separate function. If we write this into a file it might look like the following.

Listing 4.2: CourseAverageParts.scala

```scala
def testAve(test1:Double,test2:Double):Double = (test1+test2)/2

def assnAve(assn1:Double,assn2:Double,assn3:Double):Double =
  (assn1+assn2+assn3)/3

def quizAve(quiz1:Double,quiz2:Double,quiz3:Double,quiz4:Double):Double = {
  val minQuiz = quiz1 min quiz2 min quiz3 min quiz4
  (quiz1+quiz2+quiz3+quiz4-minQuiz)/3
}

def fullAve(test:Double,assn:Double,quiz:Double):Double =
  test*0.4+assn*0.4+quiz*0.2
```

---

[4]The name "bug" has a long history in engineering, going back to at least Thomas Edison. It was pulled into the field of electronic computers when an actual moth was found in a relay in the Mark II in 1947. The name implies that these things accidentally crawl into software without any action for programmers. In reality, you create the bugs in your software when you do things that are incorrect.

```
def courseAverage(test1:Double,test2:Double,assn1:Double,
    assn2:Double,assn3:Double,quiz1:Double,quiz2:Double,
    quiz3:Double,quiz4:Double):Double = {
  val test=testAve(test1,test2)
  val assn=assnAve(assn1,assn2,assn3)
  val quiz=quizAve(quiz1,quiz2,quiz3,quiz4)
  fullAve(test,assn,quiz)
}
```

Once you have this in a file you can either use this as part of a script or use the `:load` command in the REPL to load it in and call the functions directly. Note that this version requires a lot more typing than what we had before. That will not always be the case, but it is for this example and many small examples where we cannot reuse functions. Even though this version requires more typing, it has the advantages of being easier to understand and alter because the functionality is broken out into pieces.

This example is one where we could potentially have some reuse if we just knew a bit more Scala. Both `testAve` and `assnAve` do nothing more than take the average of numbers. Because one averages two numbers while the other averages three, we do not yet know how to write a single function to handle both cases. We will fix that in chapter 6.

Our other example was for converting a number of seconds into a properly formatted time. When we originally discussed this, we broke it into two problems. First, we had to figure out how many seconds, minutes, and hours a given number of seconds was. After that, we had to format the string properly. We will maintain that separation of work with functions. If we write this in a file, it might look like the following.

Listing 4.3: TimeConvertParts.scala

```
def calcHMS(totalSeconds:Int):(Int,Int,Int) = {
  val displaySeconds = totalSeconds%60
  val totalMinutes = totalSeconds/60
  val displayMinutes = totalMinutes%60
  val displayHours = totalMinutes/60
  (displayHours,displayMinutes,displaySeconds)
}

def formatHMS(numHours:Int,numMinutes:Int,numSeconds:Int):String = {
  val sec = numSeconds.toString
  val min = numMinutes.toString
  numHours+":"+("0"*(2-min.length))+min+":"+ ("0"*(2-sec.length))+sec
}

def secondsToTimeString(totalSeconds:Int):String = {
  val (h,m,s) = calcHMS(totalSeconds)
  formatHMS(h,m,s)
}
```

This code does something that we have not seen a function do yet. The `calcHMS` function returns a tuple. This function needs to give us back three values for the hours, minutes, and seconds. This type of situation is common in programming. How do you get functions to return multiple values? Different languages have different solutions to this problem. In Scala, the most direct solution is to have the function return a tuple as we see here.

This approach of taking a bigger problem and breaking it down into pieces is often called a top-down approach. The mental image is that you have the full problem at the top

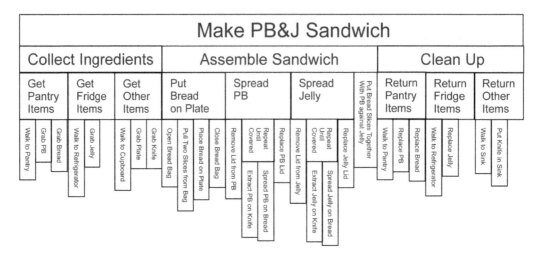

FIGURE 4.2: This figure shows a graphical breakdown of making a peanut butter and jelly sandwich.

and in each level below it you have smaller pieces that are combined to make a solution to the level above them. This structure stops at the bottom when you get to something that can be solved relatively easily. In contrast to top-down, it is also possible to approach problems from the bottom-up. This approach works when you have a certain familiarity with solving a particular type of problem and you know what types of pieces you will need or when your top level problem is not completely well defined and you have to try some different approaches. This can happen when you are building software for a customer, as the customer may have only a vague idea of what they want their program to look like. In that situation, you can build pieces you know will be useful and try putting them together in different ways until you find something that the customer is happy with.

To help you understand this, consider the common example of making a peanut butter and jelly sandwich shown in figure 4.2. Often this example is used to help illustrate the explicitness required for programming. Here the main focus is on how the problem can be broken down and solved piece-by-piece. At the top is the full problem. That is broken into three subproblems, which are each broken down further. In order to be truly explicit, the steps would have to be broken down much further. Imagine if the instructions were going to a robot instead of a human. To have it "Grab Jelly" you would have to instruct it on moving its arm to the proper location, grasping, etc.

This example also shows that the problem probably contains some reusable parts. There are several boxes at the bottom that start with "Grab" or "Walk to". Ideally we would find a way to write a function for walking that could be passed information on where to go.

Instead of viewing this in the graphical way shown in the figure, it can also be viewed as an outline like the one shown below. Nesting shows how one action is broken down into others.

- Make PB&J Sandwich

1. Collect Ingredients

    (a) Get Pantry Items
        i. Walk to Pantry
        ii. Grab Peanut Butter

        iii. Grab Bread

   (b) Get Fridge Items

        i. Walk to Fridge

        ii. Grab Jelly

   (c) Get Other Items

        i. Walk to Cupboard

        ii. Grab Plate

        iii. Grab Knife

2. Assemble Sandwich

   (a) Put Bread on Plate

        i. Open Bread Bag

        ii. Pull Two Slices from Bag

        iii. Place Bread on Plate

        iv. Close Bread Bag

   (b) Spread Peanut Butter

        i. Remove Lid from Peanut Butter

        ii. Repeat Until Bread is Covered

           A. Extract Peanut Butter on Knife

           B. Spread Peanut Butter on Bread

        iii. Replace Peanut Butter Lid

   (c) Spread Jelly

        i. Remove Lid from Jelly

        ii. Repeat Until Bread is Covered

           A. Extract Jelly on Knife

           B. Spread Jelly on Bread

        iii. Replace Jelly Lid

   (d) Put Bread Slices Together with Peanut Butter Against Jelly

3. Clean Up

   (a) Return Pantry Items

        i. Walk to Pantry

        ii. Replace Peanut Butter

        iii. Replace Bread

   (b) Return Fridge Items

        i. Walk to Refrigerator

        ii. Replace Jelly

   (c) Return Other Items

        i. Walk to Sink

        ii. Put Knife in Sink

This outline format and the indentation that comes with it look more like the code that we will write in Scala.

The problems that we solved above in Scala had solutions such that the decomposed solution required more text than the original, monolithic solution. We have expressed that this is not such a bad thing as the decomposed solutions have certain advantages. There is another factor that you should keep in mind when decomposing problems. The discussion of top-down approach requires that you keep breaking the problem down until you get to something that is fairly simple to solve. This is not just practical for the problem solving, it turns out that there are good reasons to keep functions short. Long functions are much more likely to include errors than small functions are. For this reason, it is generally advised that programmers keep their functions relatively short. How short you make your functions can be a matter of style and personal preference, but there is a good rule you can follow that is backed up by research. The rule is that functions should be kept short enough that they fit completely on your screen at one time. When a function gets long enough that it does not all fit on the screen, the programmer begins having to rely upon his/her memory of what is in the parts that are off the screen. Human memory is not a great thing to rely upon, and as a result, the rate of errors goes up significantly when there are parts of the function that you cannot see on the screen.

We can also employ the `if` construct that we learned in chapter 3 inside of functions. Consider the code we wrote to calculate the cost of an item purchased at the theme park. This can be converted to the following function.[5]

Listing 4.4: ItemCost.scala

```scala
def itemCost(item:String, size:String):Double = {
  if (item=="Drink") {
    if (size=="S") 0.99
    else if (size=="M") 1.29
    else 1.39
  } else if (item=="Side") {
    if (size=="S") 1.29
    else if (size=="M") 1.49
    else 1.59
  } else if (item=="Main") {
    if (size=="S") 1.99
    else if (size=="M") 2.59
    else 2.99
  } else {
    if (size=="S") 4.09
    else if (size=="M") 4.99
    else 5.69
  }
}
```

This function does not include reading the input from the user. Instead, it is assumed that the code that calls it will read the values and pass them in. In general, passing values into functions and returning values back is more flexible than reading them and printing results to output in the function. A function that does reading and/or printing internally cannot be reused if the values come from something other than standard input or if you want to

---

[5]The `Double` type really should not be used for money because it rounds. Remember that numbers in the computer are represented in binary. The value 0.1 in decimal is an infinite repeating binary number. That number gets truncated after a certain number of bits so the value 0.1 cannot be represented perfectly in a `Double`. That is fine for most applications, including scientific applications. It is not desirable for an application that deals with real money.

format the output in a different way. Alternately, you might want to add up all the costs for a day. You cannot add things that are printed. That requires having the values returned.

To understand, consider the simple example of wanting to know how much should be paid for two items.

```scala
import io.StdIn._
println("What is the first item?")
val item1 = readLine()
println("What size?")
val size1 = readLine()
println("What is the second item?")
val item2 = readLine()
println("What size?")
val size2 = readLine()
val totalCost = itemCost(item1,size1)+itemCost(item2,size2)
println("The total cost is "+totalCost)
```

The line where we calculate `totalCost` shows the real benefit of putting our code in a function that returns a value.

4.5   Function Literals/Lambda Expressions/Closure

In chapter 2 we saw that the simplest form of expressions for `Int`, `Double`, `String`, and some other types was the literal form. For example, just the number 5 is a literal of type `Int`. Literals allow you to have an expression of a type that you can write in a short format without declaring a name for it. Scala, because it includes elements of functional programming, allows you to express functions as literals too. This construct is also referred to as a LAMBDA EXPRESSION or a CLOSURE.

The syntax for a function literal starts with a set of parentheses that have a comma separated list of arguments in them followed by an arrow made from an equals sign and a greater than sign with the body of the function after that. This type of arrow is often read as "rocket". So if we go back to our first two functions we might write them in this way.

```scala
scala> (x:Double)=>x*x
res9: Double => Double = <function1>

scala> (x:Int,y:Int)=>x+y*y
res10: (Int, Int) => Int = <function2>
```

In this type of usage, the type of the arguments is required. As we will see, there are many usages where it is not required. The basic rule is that if Scala can figure out the types from where the function literal is used, it will do so. If the function were longer, one could use curly braces to hold multiple statements. In theory one can use function literals for functions of any length. In practice this is a style issue, and you will not want to use function literals that are too long. The function literals will be embedded in other functions, and you do not want those functions to exceed a screen size in length; so, the literals should not be more than a few lines at most. The majority of function literals will be like these and fit on a single line.

Scala has an even shorter form for some special cases of function literals. This form uses underscores to represent parameters and skips the parameter list and rocket completely. It can only be used when each parameter is used only once and in the same order they are passed in. In this form, the function literal `(x:Int,y:Int)=>x+y` could be represented as `((_:Int)+(_:Int))`. When Scala can figure out the types we can use the even shorter form of `(_+_)`. Note that this cannot be used for x^2 written as `x*x` because we use the x twice. It has significant limitations, but there are many situations where it is useful and convenient to use. For that reason, this format will appear frequently in later chapters.

4.6   Side Effects

In purely functional programming, a function takes inputs and calculates a value. The value it returns depends only on the arguments and the function does not do anything but give back that value. Often in programming, functions will do other things as well, or might return different values for the same inputs. These types of functions are not technically "functional". When they do other things we refer to that as SIDE EFFECTS. This terminology is like that for medications. You take a medication because it has a certain desired effect. However, it might have certain other effects as well called the side effects. While the side effects of a medication typically are not desired, there are times when side effects of functions are desired. The simplest example of this is printing. The print statement does not return anything. Instead it sends information to output. That is a side effect.

It is possible to have functions that return a value and have side effects. However, quite frequently you will find that functions with side effects do not return any data. To Scala that means their result type is `Unit`. The type `Unit` is a type that represents an object that carries no information. The `Boolean` type is the simplest type that carries information as it could be either **true** or **false**. The `Unit` type does not carry any information because there is only one instance of it, which is written as `()`.

Here is a simple example of a function that only includes side effects and results in `Unit`.

```scala
scala> def introduce(name:String):Unit = {
     | println("Hi, my name is "+name+".")
     | }
introduce: (name: String)Unit

scala> introduce("Lisa")
Hi, my name is Lisa.
```

This is common in non-functional programming. Indeed, some languages include a special construct, the PROCEDURE, for this.

The other significant form of side effect is an assignment. At this point that means an assignment to a **var**. One could write a function that has the primary purpose of changing the value of some **var** that is declared before it outside the function. When a **var** is declared outside of any block of code, it is said to have GLOBAL SCOPE. This means that it can be accessed anywhere in the file after to point of declaration.[6] As was already mentioned, this is often a poor way to do things as it is bug prone, and makes it harder to reuse the function. As a general rule, global scope should be avoided for anything that can be changed. Still,

---

[6]Due to the way that Scala scripts are compiled, global **vars** are actually in scope everywhere, but they are not initialized until the point of declaration. For this reason, you should not use them until after the point of declaration.

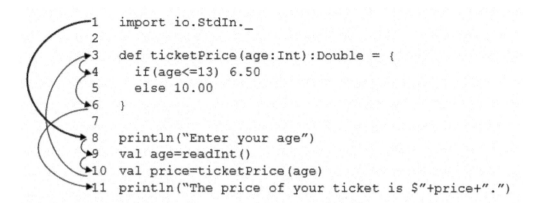

```
 1   import io.StdIn._
 2
 3   def ticketPrice(age:Int):Double = {
 4     if(age<=13)  6.50
 5     else 10.00
 6   }
 7
 8   println("Enter your age")
 9   val age=readInt()
10   val price=ticketPrice(age)
11   println("The price of your ticket is $"+price+".")
```

FIGURE 4.3: Here you can see how control flow moves through a short program that involves a function. Assume the user enters an age of 12. The code in the function is not executed until the point where it is invoked. Once it has completed, control returns to where it had been and then continues on to execute the next line of code after the call.

there are times when we will find it to be useful. The following code illustrates the type of use we might see for this later in the book.

```
var x = 0
def moveLeft():Unit = x -= 1
def moveRight():Unit = x += 1
```

4.7   Thinking about Function Execution

It is not that hard to understand the sequential model of program execution. As the program executes, you simply move down from one line to the next, doing whatever that line says to do. All that conditionals added was the possibility that certain parts of the code might get skipped. Functions can be a bit more complex. When you see a function defined in code, you would simply skip over it, and remember it is there. The code does not get executed until something calls it. When a function is called, control jumps to that function, and continues sequentially inside of the function. When the function is done, control jumps back to where it had been called from. This can be seen in figure 4.3.

In the presence of side effects, this is pretty much the only way you can think about program execution and functions, because the order in which control passes over those side effects is significant. If you have code that does not have side effects, you can think about it in terms of substitution. Without side effects, everything is either a declaration or an expression. Any expression can be substituted by something that is equivalent to it. The goal is to substitute for something that moves you closer to having a simple value. Here are the rules for substitution with different expressions.

- Expressions that contain only literals can be replaced with what they evaluate to (e.g. $4+5 \rightarrow 9$).

- Variables can be replaced with their values.

- `if`-expressions are evaluated based on the value of the condition.

    − `if (true)` *trueExpr* `else` *falseExpr* → *trueExpr*
    − `if (false)` *trueExpr* `else` *falseExpr* → *falseExpr*

- A function call is replaced by the body of the function where all parameters have been substituted by the arguments in the call.

To help understand these rules, we will run through a little example. One of the authors teaches a roller skating class[7] where most of the grade comes from various skills and an endurance test. The following code can be used to calculate how many points a particular student got from these components.

Listing 4.5: SkatingPoints.scala

```scala
def skillPoints(numSkills:Int):Int = 5*numSkills

def endurancePoints(numLaps:Int):Int = {
  if (numLaps < 20) 0
  else if (numLaps > 40) 20
  else numLaps-20
}

val skills = 9
val laps = 36
val points = skillPoints(skills) + endurancePoints(laps)
```

It has been broken up to use two small functions, one for each component. This example also hard codes the values instead of using `readInt` because that would be a side effect which would complicate the use of substitution.

What we want to do is find the value of `points` using substitution. The expression we want to evaluate is

```scala
skillPoints(skills) + endurancePoints(laps)
```

The first thing we will do is substitute in the values of the two `val`s. This gives us the following.

```scala
skillPoints(9) + endurancePoints(36)
```

Note that this is technically doing two things at once. That is to keep the discussion shorter. You should do the substitutions one at a time. If there is more than one available, you can pick which one to do first, because there are no side effects, the order it does not matter. Next we will expand out the call to `skillPoints`.

```scala
5*9 + endurancePoints(36)
```

Remember that this is done by replacing the call with the body of the function where parameters have been replaced by arguments. So the `numSkills` in `skillPoints` was replaced by the value 9. We can then evaluate 5*9.

```scala
45 + endurancePoints(36)
```

---

[7]Yes, it counts for college credit.

Now we substitute for the `endurancePoints` function call using the value 36 for `numLaps`.

```
45 + {
  if (36 < 20) 0
  else if (36 > 40) 20
  else 36-20
}
```

The first condition is false, so that `if` is replaced by the false expression, after the first else.

```
45 + {
  if (36 > 40) 20
  else 36-20
}
```

The next condition is also false, so that also simplifies to the false expression.

```
45 + {
  36-20
}
```

At this point, all that is left is doing some math with literals, so we get `45 + 16`, which simplifies to our final answer of `61`.

Higher Order Functions

Function literals would be nothing more than a novelty in Scala or any other programming language if it were not for higher order functions. A higher order function is a function that operates on other functions. This means that either we pass other functions into it, or it results in a function. Higher order functions are a functional programming concept, and they can be challenging to get ones head around, but we will see how useful they can be in the coming chapters.

As an example of a higher order function here, we will use the idea of composition. If we have two functions, $f(x)$ and $g(x)$, then the composition of f with g is $f(g(x))$. We can write a Scala function that does this as follows.

```
scala> def compose(f:Double => Double,g:Double => Double):Double => Double =
     | x => f(g(x))
compose: (f: Double => Double, g: Double => Double)Double => Double
```

Note that we write the function types themselves using arrows, "rocket". We could call this with either functions we build using `def` or function literals. We will do it first with the following two functions that have been defined using `def`.

```
scala> def plus5(x:Double):Double = x+5
plus5: (x: Double)Double
```

```
scala> def square(x:Double):Double = x*x
square: (x: Double)Double
```

Now we want to use compose to build new functions from these two using the compose function and then see our new functions working.

```
scala> val h=compose(plus5,square)
h: Double => Double = <function1>

scala> val j=compose(square,plus5)
j: Double => Double = <function1>

scala> h(3)
res0: Double = 14.0

scala> j(3)
res1: Double = 64.0
```

The function $h(x) = x^2 + 5$ while $j(x) = (x + 5)^2$. You can see that when we call these with an argument of 3, we get values of 14 and 64 respectively, just as would be expected.

We could define the same functions h and j using function literals as well.

```
scala> val h = compose(_+5, x => x*x)
h: Double => Double = <function1>

scala> val j = compose(x => x*x, _+5)
j: Double => Double = <function1>

scala> h(3)
res5: Double = 14.0

scala> j(3)
res6: Double = 64.0
```

If you had no real need for the functions **plus5** or **square**, you would likely use this latter format. They would be defined with **def** if you had a need to refer to them multiple times.

4.8   type Declarations

Scala provides a mechanism that allows you to provide alternate names for types. This can provide significant power when combined with concepts like abstraction that are beyond the scope of a first-semester course. However, for now, it can be used to help give you short meaningful names for tuple types that you use commonly.

The syntax of a **type** declaration begins with the keyword **type**. This is followed by the name you want to give the type. As with other names, you should pick something that is meaningful to you so that when you are reading through code, having this name helps it to make more sense. Unlike **var**, **val**, and **def**, it is the general style for type names to being with capital letters. You have probably noticed that the names of all the types we have encountered so far, like **Int** and **Double**, start with capital letters. This is a style choice, not something that Scala enforces, but it is highly recommended that you follow it. After the name is a equal sign followed by the type you are giving a new name to.[8]

As an example of this, you might put something like the following at the top of a script if you were going to write a lot of functions dealing with vectors in 3D.

```scala
type Vect = (Double, Double, Double)
```

In the rest of your code, you could refer to the type `Vect` instead of having to type out `(Double, Double, Double)`. This does not change how you interact with the tuple. You would still use the methods `_1`, `_2`, and `_3` or a `val` with a tuple pattern to get the values out of the `Vect`.

There can even be a value to using a `type` declaration to give a different name to a standard type like `Int` like this.

```scala
type Counter = Int
```

If you carefully use the `Counter` type through your program in places that call for it, you can easily change the size of the integer type that is being used. If you find that you need your counters to be able to go over `Int.MaxValue`, you can switch to using a `Long`. On the other hand, if your counters are always small and you have a need to save memory, you could consider going down to a `Short`. This type of usage is not common is Scala, but it is used significantly in C using a similar construct called **typedef** when writing libraries so that it is easy to modify types to fit the platform you are compiling for.

4.9 Putting It Together

In the last chapter we also wrote conditional code for calculating the admission cost to the theme park. This makes a good example of a function as well. Here is that converted to a function.

Listing 4.6: EntryCost.scala
```scala
def entryCost(age:Int, cooler:Boolean, waterPark:Boolean): Double = {
  (if (age<13 || age>65) 20 else 35) +
  (if (cooler) 5 else 0) +
  (if (waterPark) 10 else 0)
}
```

The information needed for the calculation is passed in and this function only does a calculation with the data, giving back a number.

Now we can put things together and write a script that will calculate the cost of a group of people with up to four members including the purchase of meals. We will assume that the whole group either brings a cooler or does not, that all would either go to the water park or not, and that they all order some size of combo meal. The following code could be put in a script with the `entryCost` and `itemCost` functions from earlier in the chapter.

Listing 4.7: ThemeParkBook.scala
```scala
def individualCost(cooler:Boolean, waterPark:Boolean): Double = {
  println("What is the person's age?")
```

---

[8]The old name continues to work. You just get to use this one as well to refer to the same thing.

```
  val age = readInt()
  println("What size combo are they ordering?")
  val size = readLine()
  entryCost(age,cooler,waterPark)+itemCost("Combo",size)
}


println("How many people are in your group? (1-4)")
val numPeople = readInt()
if (numPeople<1 || numPeople>4) {
  println("This script can not handle "+numPeople+" people.")
} else {
  println("Is the group bringing a cooler? (Y/N)")
  val cooler = readLine()=="Y"
  println("Will they be going to the water park? (Y/N)")
  val waterPark = readLine()=="Y"
  val totalCost = individualCost(cooler,waterPark) +
    (if (numPeople>1) individualCost(cooler,waterPark) else 0) +
    (if (numPeople>2) individualCost(cooler,waterPark) else 0) +
    (if (numPeople>3) individualCost(cooler,waterPark) else 0)
  println("The group cost is $"+totalCost+".")
}
```

Note that this code is not in a function. When you write a script, function declarations do exactly that, they declare a function. You need to have code outside of the functions that makes something happen, typically by calling one or more of the declared functions.

This script does have one significant limitation. It only works for 1-4 people. We do not yet have the ability to make a piece of code happen an arbitrary number of times. That is something we will learn how to do in the next chapter.

Ways to Run a Scala Program (Aside)

One thing that students occasionally struggle with is figuring out how they should run their Scala programs. There are three different ways that you can run Scala. Two of these you have seen: the REPL and scripts. A third, compiling to applications, will be dealt with briefly at the end of this book. To use the REPL you enter **scala** on the command line with no file to execute. To run a script, you enter **scala** on the command line and give it a file name that ends with .scala. At first it might seem that these two approaches are completely distinct. However, the REPL has a :load command that you can use to load in code from a Scala file. Using this, it is possible to load a script file into the REPL. There are times when this can be handy, but you need to understand the difference between these two approaches.

When you run a file as a script, the only thing that the user sees is the values that are printed out. If your file does not contain a print statement, nothing will be shown on screen and it will appear to the user that it did not work. If all the file has in it are declarations, statements beginning with **val**, **var**, or **def** (there are other types of declarations that we will learn about later, but these three cover what we know at this point), the script does not really do anything. You have to have statements that use the declarations for them to matter in the script.

This is not the case when you load a file into the REPL. In that usage, feedback is shown for every declaration in the form of a line that is printed with the type and value of the declaration. If you declared functions in the file, you can call them in the REPL after the file has been loaded. For this reason, files you load into the REPL do not always need statements at the end that call functions, and they often do not include print statements to display output.

Basically, the REPL gives you the ability to "play around" with the code. Using a script simply runs it and does what it says. You should ask your instructor about this, but odds are good that any code you submit for grading should be in the form of a script that includes calls that demonstrate the functionality of the script and prompt users for input. In this case, you can use the `:load` option in the REPL when you are working on the program, but before you turn it in, make sure it runs properly and does what you want when used as a script.

4.10   End of Chapter Material

4.10.1   Problem Solving Approach

The material covered in this chapter adds significant flexibility to our approach to problem solving. It also adds a number of choices you can consider when you think about what you might write for any particular line of code. You can now write functions. Technically, function declarations can go anywhere, including inside of other functions. It is best that they have some logical organization to them. In the text we will tend to group top level functions together at the top of scripts.

We also learned in this chapter about functions that are called only for their side effects. Technically this was part of our list previously in the form of calls to `print` and `println`. We now know that these are special cases of a general class of options. Given what we have learned, here is a revised list of what any productive line of code might be doing.

1. Call a function just for the side effects. Previously `print` or `println` were the only examples we had of this, but we can now be more general.

2. Declare something:

   - A variable with `val` or `var`.

   - A function with `def`. Inside of the function will be statements that can pull from any of these rules. The last statement of the function should be an expression that gives the result value.

   - A type declaration with `type` to give something a more meaningful name.

3. Assign a value to a variable.

4. Write an `if` statement.

4.10.2   Summary of Concepts

- Functions are used to break problems up into smaller pieces.

  - Help solve problems.

- Informative names make code more understandable.

- Smaller functions are easier to work with and easier to debug.

- Functions help code become more reusable.

- The functions in programming are similar to those from math. Information is passed in through formal parameters. The value of the parameters is determined at the time the function is called when arguments are passed in.

- Functions are declared in Scala using `def`. This is followed by the function name, an argument list, a result type, and then an equal sign with the expression for the function. Result type can be left off and inferred, but it is recommended you include it anyway.

- Functions can also be written as literals in Scala.

  - The rocket notation has the argument list and body separated by =>, which we read as "rocket".

  - Shorter notation uses underscores for arguments. Only works for certain functions.

- Functions that are called only for their side effects and do not need to return a value, return `Unit`.

- If you have functions that often take or return tuples, it can be useful to use a `type` declaration to give shorter, meaningful names to the tuples.

4.10.3   Self-Directed Study

Enter the following statements into the REPL and see what they do. Some will produce errors. You should figure out why. Try some variations to make sure you understand what is going on.

```
scala> def succ(n:Int):Int = n+1
scala> succ(1)
scala> succ(succ(1))
scala> var cnt = 0
scala> def inc { cnt=cnt+1 }
scala> cnt
scala> inc
scala> cnt
scala> inc
scala> inc
scala> cnt
scala> val f = (x:Double,y:Double)=>math.sqrt(x*x+y*y)
scala> f(3,4)
scala> def doThreeTimes(g:(Double)=>Double,x:Double) = g(g(g(x)))
scala> doThreeTimes(y=>y+1,1)
scala> doThreeTimes(y=>y*2,1)
scala> doThreeTimes(a=>a-5,100)
scala> doThreeTimes(_+1,1)
scala> doThreeTimes(_*2,1)
scala> def incChar(c:Char,offset:Int) = (c+offset).toChar
scala> incChar('a',1)
```

```
scala> incChar('a',2)
scala> incChar('z',-3)
scala> def tnp1(n:Int):(Int,Int) = {
    val odd = n%2
    (n/2*(1-odd),(3*n-1)*odd)
}
scala> odd
scala> var name = ""
scala> def introduction {
    println("What is your name?")
    name = readLine()
    println("Hello "+name)
}
scala> name
```

4.10.4 Exercises

1. What would be the types of the parameters and result of functions that do the following tasks. (Note that you are not supposed to write the functions, just say what their types are.)

   - Tell you how many words are in a person's name.
   - Take a person's name as normally written and give it back in "last, first" format.
   - Take three words and return the TLA (Three Letter Acronym) for them.
   - Take ten points in 2-space (x-y coordinates) and return how many are in the unit circle.
   - Take a number and tell how many distinct prime factors it has.
   - Take a number and tell how many positive factors it has.
   - Take a value in Fahrenheit and give back the equivalent in Celsius.
   - Take a value in Fahrenheit and give back the equivalent in Celsius and Kelvin.

2. Write functions that return a `Boolean` for the different parts of exercise 3.1.

3. Write a function that takes a `String` parameter and returns a `Boolean` telling if the argument is a palindrome.

4. Write a function for the solution to exercise 3.7.

5. Write a function for the solution to exercise 3.8.

6. Write a function for the solution to exercise 3.9.

7. Write a function for the solution to exercise 3.10.

8. Convert the code you wrote for exercise 3.11 into a function that takes the x and y values are arguments and returns a `Boolean` for whether a point is in the unit circle.

9. Write a function to convert a temperature from Fahrenheit to Celsius.

10. Write a function to convert from miles to kilometers.

11. Write a function to convert from seconds to years.

12. Write a function to convert from AU (Astronomical Units) to miles.

13. Write two functions that takes four numbers and returns the smallest. The first can use `min`, but the second cannot.

14. Write a function that takes three numbers and returns the median.

15. Write a function that sums all of the digits in an integer number entered by the user. For example, if the user enters 723, the function would return 12 which is equal to 7+2+3.

16. Write a function that receives the three angles of a triangle and determines if the input is valid. Then write another function that returns the type of triangle that is represented by the three angles: isosceles, equilateral, or scalene.

17. Write a function that computes the number of days in a year. Test your program for all the years from 2010 until 2020.

4.10.5 Projects

1. Write a set of functions to do the following operations on a 2-tuples of `Int` as if they were the numerator and denominator of rational numbers.

    (a) Addition
    (b) Subtraction
    (c) Multiplication
    (d) Division

   For example, the function for addition might start off as `def add(n1:(Int, Int), n2:(Int, Int)):(Int, Int)`.

2. Write a set of functions to do the following operations on 3-tuples of `Doubles` as if they were vectors.

    (a) Addition
    (b) Subtraction
    (c) Dot product
    (d) Cross product

   For example, the function for addition might start off as `def add(v1:(Double, Double, Double), v2:(Double, Double, Double)):(Double, Double, Double)`.

3. Write a set of functions to do the following operations on a 2-tuples of `Doubles` as if they were complex numbers.

    (a) Addition
    (b) Subtraction
    (c) Multiplication
    (d) Division
    (e) Magnitude

For example, the function for addition might start off as `def add(c1:(Double, Double), c2:(Double, Double)):(Double, Double)`.

4. Use your solution to exercise 3 to make a function that takes two complex numbers, z and c and returns the value $z^2 + c$.

5. Write a function to solve the quadratic equation for real values. Your solution should return a `(Double, Double)` of the two roots.

6. Enhance the solution to exercise 5 so that it returns a `((Double, Double),(Double, Double))`. This is a 2-tuple of 2-tuples. This represents the two roots and each root can be a complex number expressed as a `(Double, Double)`.

7. The hyperbolic trigonometric function, sinh, cosh, and tanh, are all defined in terms of power of e, the base of the natural logarithms. Here are basic definitions of each:

$$\sinh(x) = \tfrac{1}{2}(e^x - e^{-x})$$
$$\cosh(x) = \tfrac{1}{2}(e^x + e^{-x})$$
$$\tanh(x) = \tfrac{\sinh(x)}{\cosh(x)}$$

Write functions for each of these. You can use math.exp(x) to represent e^x.

8. Write two functions for converting between Cartesian and polar coordinate systems. The first takes `x:Double` and `y:Double` and returns a `(Double, Double)` with r and θ. The second takes `r:Double` and `theta:Double` and returns `(Double, Double)` with x and y. You can use the `Math.atan2(y:Double, x:Double)` function to get an angle. This avoids the problem of using division when x is zero.

9. This option has you doing a scientific calculation. We cannot do that much yet, but we will work our way up. We are going to play with calculating the non-greenhouse temperatures for planets with moderate to fast spin rates. This might seem like a complex thing to do, but it is not difficult. You need two pieces of information and a little algebra. The first piece of information is the Stefan-Boltzmann Law ($j^* = \sigma T^4$, $\sigma = 5.670400*10^{-8}\ [\frac{J}{s*m^2*K^4}]$) for the amount of energy given off in thermal radiation by any body. The second is the fact that intensity of radiation drops off as $1/r^2$.

To calculate the non-greenhouse temperature of a planet you need the following pieces of information. You should prompt the user for values to these. To keep things simple use mks units and keep temperatures in Kelvin.

- Radius of the star.
- Surface temperature of the star.
- Orbital semimajor axis of the planet.
- Albedo of the planet.

Use the Stefan-Boltzmann law to determine the energy output per square meter on the stars surface. Make that into a function that takes the needed values. Using the inverse square relationship you can calculate the intensity at the location of the planet (use the ratio of the planet's orbit distance to the stellar radius for this). Make this another function. The star light will cover the planet with an area of πr^2 where r is the planets radius. A fraction of that, determined by the albedo, is reflected. What is not reflected warms the planet. The planet cools through its own thermal radiation from a surface of $4\pi r^2$. Setting the absorbed and emitted values equal allows you to solve the temperature. (Note that the planetary radius should cancel out.) Make a function that takes the incident power and albedo and gives you back a temperature.

10. If you wrote code for project 3.4 you can convert that to use functions. Write one function for a sphere and another for a plane. Each should take information for a ray as well as for the geometry that is being intersected. They should return the value of t at the point of intersection. An appropriate negative value can be used if they do not intersect.

11. In project 3.5, you wrote conditional statements to determine if you had enough ingredients for some simple recipes. The numbers were kept small there in large part because of the length of code and problems with duplication. Redo that problem for the situation where you have five items in your pantry and the recipe involves four ingredients. Use functions to make this manageable.

12. Convert your code from project 3.6 to use functions.

13. You are a waitress at a high end restaurant. You need a script (and appropriate functions) that will record the order for each table and then calculate the taxes and tip (if appropriate) to produce a ticket (bill) for the table. The largest table holds 8 people. You can assume that all the orders per table will be on one ticket (i.e. the same bill). Customers may order just food, just a beverage, or both. You can assume that each customer will, at most, order one food item and one drink item; however, it is possible that they will not order anything. Food and beverage orders are processed separately because the beverage order is handled by the bar (check ID if beverage is alcoholic). Food tax rate is 7%, but there is an extra 2% tax on drinks. If there are more than 6 customers at the table, an 18% tip will be added to the bill. You can charge whatever you like for the food and beverage. You should offer at least 3 different food items, each with their own price. You should also offer at least 2 different beverage items, each with their own price.

14. You are writing a script for a department store that is running a clothing sale. The current sale gives the customer a $30 discount if they order at least $75 of clothing, a $60 discount if they order at least $150 of clothing, and a $75 discount if they order $200 or more. The discount is taken on the clothing purchase — shipping and tax are not included in the total when calculating the discount. If the order totals more than $125 then shipping is free, otherwise there is a $15 shipping charge applied to the order. Write functions to take the order. This function should ask the user for the total clothing cost, then calculate the discount, tax, and shipping. Finally, produce an itemized bill for the order.

15. Write a script that calculates and prints the bill for a cellular telephone company. The company offers two types of plans: residential and commercial. Its rates vary depending on the type of plan. The rates are computed as follows:

   - Residential plan: $40.00 per month plus the first 50 texts are free. Charges for over 50 texts are $0.20 per text.
   - Commercial plan: $70.00 per month plus
     - For texts from 7:00 a.m. to 6:59 p.m., the first 75 texts are free; charges for over 75 texts are $0.10 per text
     - For texts from 7:00 p.m. to 6:59 a.m., the first 100 texts are free; charges for over 100 texts are $0.05 per text.

   Your script should prompt the user to enter an account number, a plan code, the number of texts sent during the day, and number of texts sent during the night (if applicable). A plan code of r or R means residential plan; a plan code of c or C means commercial plan. Treat any other character as an error, displaying a message to the customer.

Your script should contain functions to calculate and return the billing amount for residential plan and commercial plan.

16. Craps is a popular dice game in Las Vegas. Write a program to play a variation of the game, as follows: Ask the player to place a bet on any or all of four sections: the FIELD section; the NUMBER section; OVER or UNDER 7 section; or the 7, 11, or CRAPS section.

    - If the player placed their bet in the FIELD section:
        - The player may be on a 2, 3, 4, 9, 10, 11, or 12.
        - If the total of the 2 dice equals a 2, 3, 4, 9, 10, 11, or 12, the player wins.
        - If the total of the 2 dice equals 5, 6, 7, or 8, the House wins.
    - If the player placed their bet in the NUMBER section:
        - The player may bet on a 4, 5, 6, 8, 9, or 10.
        - If the dice total 4, 5, 6, 8, 9, or 10 the player wins.
        - House wins on all other numbers on that roll.
    - If the player placed their bet in the OVER or UNDER 7 section:
        - Player bet that the total of the 2 dice will be either Under 7 or Over 7.
        - Both over and under lose to the House if a total of 7 is thrown.
    - If the player placed their bet in the 7, 11, or CRAPS section:
        - The player may bet on a 7, or an 11 or any Craps (dice totaling either 2, 3, or 12) coming up on the throw of the dice.
        - House wins if the number selected does not come up.

    Roll two dice. Each die has six faces representing values 1, 2, 3, 4, 5, and 6, respectively. You can generate a random `Int` between 0 and `n` in Scala using `util.Random.nextInt(n)`. Check the sum of the two dice and tell the player if they won or if the house won.

17. Neglecting air resistance, objects that are thrown or fired into the air travel on a parabolic path of the form $x(t) = v_x t$, $y(t) = -\frac{1}{2}gt^2 + v_y t + h$, where v_x and v_y are the components of the velocity, g is the acceleration due to gravity, and h is the initial height. Write a function that is passed the speed, the angle relative to the ground, and the initial height of a projectile and results in the distance the projectile will go before it hits the ground with $y(t) = 0$.

Chapter 5

Recursion for Iteration

Gaining conditionals provided us with a lot of power, we can express more complex logic in our code. Adding functions gave us the ability to break problems into pieces and reuse functionality without retyping code. There is still something very significant that we are missing. Currently, when we write a piece of code, it happens once. We can put that code into a function and then call the function over and over, but it will only happen as many times as we directly call it. We cannot easily vary the number of times that something happens, or make anything happen a really large number of times. This is a problem, because one of the things that computers are really good at is doing the same thing many times without getting bored or distracted. The is a capability that we really need to add to our toolbox. There is more than one way to make something happen multiple times in Scala. One of these ways, RECURSION, we can do with just functions and conditionals, constructs that we have already learned.

5.1   Basics of Recursion

Recursion is a concept that comes from mathematics. A mathematical function is recursive if it is defined in terms of itself. To see how this works, we will begin with factorial. You might recall from math classes that $n!$ is the product of all the integers from 1 up to n. We might write this as $n! = 1 * 2 * \ldots * n$. More formally we could write it like this.

$$n! = \prod_{i=1}^{n} i$$

Both of the formal and informal approaches define factorial in terms of just multiplication, and assume that we know how to make that multiplication happen repeatedly. We can be more explicit about the repetition if we write the definition using recursion like this.

$$n! = \begin{cases} 1 & n < 2 \\ n * (n - 1)! & otherwise \end{cases}$$

In this definition, the factorial function is defined in terms of itself. To describe what factorial is, we use factorial.

To see how this works, let us run through an example using the substitution approach we discussed in section 4.7 and take the factorial of 5. By our definition we get that $5! = 5 * 4!$. This is because 5 is not less than 2. Subsequently, we can see that $4! = 4*3!$ so $5! = 5*4*3!$. This leads to $5! = 5 * 4 * 3 * 2!$ and finally to $5! = 5 * 4 * 3 * 2 * 1$.

This definition and its application illustrate two of the key aspects of recursion. There are two possibilities for the value of this function. Which one we use depends on the value of n. In the case where n is less than 2, the value of the factorial is 1. This is called a BASE CASE. All recursive functions need some kind of base case. The critical thing about a base case is that it is not recursive. When you get to a base case, it should have a value that can be calculated directly without reference back to the function. Without this you get what is called INFINITE RECURSION. There can be multiple different base cases as well. There is no restriction that there be only one, but there must be at least one.

To see why the base case is required, consider what would happen without it. We would still get $5! = 5 * 4 * 3 * 2 * 1!$, but what would happen after that? Without a base case, $1! = 1 * 0!$ and $0! = 0 * (-1)!$. This process continues on forever. That is why it is called infinite recursion.

The second case of the recursive definition demonstrates the recursive case. Not only does this case refer back to the function itself, it does so with a different value and that value should be moving us toward a base case. In this case, we define $n!$ in terms of $(n-1)!$. If the recursive case were to use the same value of n we would have infinite recursion again. Similarly, if it used a value greater than n we would also have infinite recursion because our base case is for small numbers, not large numbers.

What else could we define recursively? We could define multiplication, which is used by factorial, recursively. After all, at least for the positive integers, multiplication is nothing more than repeated addition? As with the factorial, we could write a definition of multiplication between positive integers that uses a math symbol that assumes some type of repetition like this.

$$m * n = \sum_{i=1}^{n} m$$

This says that $m*n$ is m added to itself n times. This can be written as a recursive function in the following way.

$$m * n = \begin{cases} 0 & n = 0 \\ m + (m * (n - 1)) & otherwise \end{cases}$$

This function has two cases again, with a base case for a small value and a recursive case, that is defined in terms of the value we are recursing on using a smaller value of that argument.

We could do the same type of things to define exponentiation in terms of multiplication. We could also use an increment (adding 1) to define addition by higher numbers. It is worth taking a look at what that would look like.

$$m + n = \begin{cases} m & n = 0 \\ 1 + (m + (n - 1)) & otherwise \end{cases}$$

While this seems a bit absurd, it would be less so if we named our functions. Consider the following alternate way of writing this.

$$add(m, n) = \begin{cases} m & n = 0 \\ 1 + add(m, n - 1) & otherwise \end{cases}$$

Now it is clear that as long as we know how to do increment (+1) and decrement (-1) we could write full addition. With full addition we could write multiplication. With multiplication we can write exponentiation or factorial. It turns out that this is not all you can do. It might be hard to believe, but if you have variables, recursion (which simply requires functions and an if construct), increment, and decrement, you have a full model of computation. It can calculate anything that you want. Of course, we do not do it that way because it would be extremely slow. Still, from a theoretical standpoint it is very interesting to know that so much can be done with so little.

5.2 Writing Recursive Functions

We have seen the mathematical side of recursion and have written some basic mathematical functions as recursive functions. Now we need to see how we write these things and more in Scala. The translation from math functions to programming functions is not hard. In fact, little will change from the math notation to the Scala notation.

As before, we will begin with the factorial function. Here is the factorial function written in Scala in the REPL.

```
scala> def fact(n:Int):Int = if (n<2) 1 else n*fact(n-1)
fact: (n: Int)Int
```

We have called the function `fact`, short for factorial. The body of the function is a single if expression. First it checks the value of n to see if it is less than 2. If it is, the expression has a value of 1. Otherwise, it is `n*fact(n-1)`. We can see the results of using this function here:

```
scala> fact(5)
res1: Int = 120
```

We see here that it correctly calculates the factorial of 5.

One significant difference between recursive and non-recursive functions in Scala is that we have to specify the return type of recursive functions. If you do not, Scala will quickly let you know it is needed.

```
scala> def fact(n:Int) = if (n<2) 1 else n*fact(n-1)
<console>:6: error: recursive method fact needs result type
       def fact(n:Int) = if (n<2) 1 else n*fact(n-1)
                                           ^
```

Factorial is an interesting function that is significant in Computer Science when we talk about how much work certain programs have to do. Some programs have to do an amount of work that scales with the factorial of the number of things they are working on. We can use our factorial function to see what that would mean. Let us take the factorial of a few different values.

```
scala> fact(10)
res2: Int = 3628800

scala> fact(15)
res3: Int = 2004310016

scala> fact(20)
res4: Int = -2102132736
```

The first two show you that the factorial function grows very quickly. Indeed, programs that do factorial work are referred to as intractable because you cannot use them for even modest size problems. The third example though shows something else interesting. The value of 20! should be quite large. It certainly should not be negative. What is going on here?

If you remember back to chapter 2, we talked about the way that numbers are represented on computers. Integers on computers are represented by a finite number of bits. As a result, they can only get so large. The built in number representations in Scala use 32 bits for Int. If we changed our function just a bit to use Long we could get 64 bits. Let's see that in action.

```
scala> def fact(n:Long):Long = if (n<2) 1L else n*fact(n-1)
fact: (n: Long)Long

scala> fact(20)
res5: Long = 2432902008176640000

scala> fact(30)
res6: Long = -8764578968847253504
```

The 64 bits in a Long are enough to store 20!, but they still fall short of 30!. If we give up the speed of using the number types hard wired into the computer, we can represent much larger numbers. We can then be limited by the amount of memory in the computer. That is a value not measured in bits, but in billions of bytes. To do this, we use the type BigInt.

The BigInt type provides us with arbitrary precision arithmetic. It does this at the cost of speed and memory. You do not want to use BigInt unless you really need it. However, it can be fun to play with using a function like factorial which has the possibility of getting quite large. Let us redefine our function using this type and see how it works.

```
scala> def fact(n:BigInt):BigInt = if (n<2) 1L else n*fact(n-1)
fact: (n: BigInt)BigInt

scala> fact(30)
res7: BigInt = 265252859812191058636308480000000

scala> fact(150)
res8: BigInt = 57133839564458545904789328865261054003189553578601126418
2548375833179829124845398393126574488675311145377107878746854204162666
2501986845044663559491959220665749425920957357789293253572904449624
7240541679072211844543712226967552000000000000000000000000000000000000
00
```

Not only can this version take the factorial of 30, it can go to much larger values as you see here. There are not that many applications that need these types of numbers, but those that do greatly benefit from this type of functionality.

Now that we have beaten factorial to death, it is probably time to move on to a different example. The last section was all about examples pulled from mathematics. The title of this chapter though is using recursion for iteration. This is a far broader programming concept than the mathematical functions we have talked about. So let us use an example that is very specific to programming.

A simple example to start with is to write a function that will "count" down from a certain number to zero. By count here we it will print out the values, so these functions now have side effects. Like the factorial, we will pass in a single number, the number we want to count down from. We also have to have a base case, the point where we stop counting. Since we are counting down to zero, if the value is ever below zero then we are done and should not print anything.

In the recursive case we will have a value, n, that is greater than or equal to zero. We definitely want to print n. The question is, what do we do after that? Well, if we are counting down, then following n we want to have the count down that begins with n-1. Indeed, this is how you should imagine the recursion working. Counting down from n is done by counting the n, then counting down from n-1. Converting this into code looks like the following:

```
def countDown(n:Int):Unit = {
  if (n>=0) {
    println(n)
    countDown(n-1)
  }
}
```

The way this code is written, the base case does nothing so we have an `if` statement that will cause the function to finish without doing anything if the value of n is less than 0. You can call this function passing in different values for n to verify that it works.

Now let us try something slightly different. What if I want to count from one value up to another? In some ways, this function looks very much like what we already wrote. There are some differences though. In the last function we were always counting down to zero so the only information we needed to know was what we were counting from. In this case, we need to be told both what we are counting from and what we are counting to. That means that our function needs two parameters passed in. A first cut at this might look like the following.

```
def countFromTo(from:Int,to:Int):Unit = {
  println(from)
  if (from!=to) {
    countFromTo(from+1, to)
  }
}
```

This function will work fine under the assumption that we are counting up. However, if you call this with the intention of counting down so that `from` is bigger than `to`, you have a problem. To see why, let us trace through this function. First, let us see what happens if we call it with 2 and 5, so we are trying to count up from 2 to 5. What we are doing is referred to as tracing the code. It is the act of running through code to see what it does. The substitution method effectively does this when we have no side effects, but printing is a side effect, so we need to run through the code in a different way to make certain we get the right behavior. This is an essential ability for any programmer. After all, how can you write code to complete a given task if you are not able to understand what the code you write will do? There are lots of different approaches to tracing. Many involve tables where you write down the values of different variables. For recursive functions you can often just write down each call and what it does, then show the calls it makes. That is what we will do here. We will leave out the method name and just put the values of the arguments as that is what changes.

```
(2,5) => prints 2
    ↓
(3,5) => prints 3
    ↓
(4,5) => prints 4
    ↓
(5,5) => prints 5
```

The last call does not call itself because the condition `from!=to` is `false`.

Now consider what happens if we called this function with the arguments reversed. It seems reasonable to ask the function to count from 5 to 2. It just has to count down. To see what it really does, we can trace it.

```
(5,2) => prints 5
    ↓
(6,2) => prints 6
    ↓
(7,2) => prints 7
    ↓
(8,2) => prints 8
    ↓
```

...

This function will count for a very long time. It is not technically infinite recursion because the `Int` type only has a finite number of values. Once it counts above $2^{31} - 1$ it wraps back around to -2^{31} and counts up from there to 2 where it will stop. You have to be patient to see this behavior. Even if it is not infinite, this is not the behavior we want. We would rather the function count down from 5 to 2. The question is, how can we do this? To answer this we should go back to the trace and figure out why it was not doing that in the first place.

Looking at the code and the trace you should quickly see that the problem is due to the fact that the recursive call is passed a value of `from+1`. So the next call is always using a value one larger than the previous one. What we need is to use +1 when we are counting up and -1 when we are counting down. This behavior can be easily added by replacing the 1 with an `if` expression. Our modified function looks like this.

```scala
def countFromTo(from:Int,to:Int):Unit = {
  println(from)
  if (from!=to) {
    countFromTo(from + (if (from<to) 1 else -1), to)
  }
}
```

Now when the `from` value is less than the `to` value we add 1. Otherwise, we will add -1. Since we do not get to that point if the two are equal, we do not have to worry about that situation. You should enter this function in and test it to make sure that it does what we want.

5.3   User Input

We saw back in chapter 2 that we can call the function `readInt` to read an integer from standard input. Now we want to read multiple values and do something with them.[1] We will start by taking the sum of a specific number of values. We can write a function called `sumInputInts`. We will pass this function an integer that represents how many integers we want the user to input, and it will return the sum of those values. How can we define such a function recursively? If we want to sum up 10 numbers, we could say that sum is the first number, plus the sum of 9 others. The base case here is that if the number of numbers we are supposed to sum gets below 1, then the sum is zero. Let us see what this would look like in code.

```
def sumInputInts(num:Int):Int = {
  if (num>0) {
    readInt()+sumInputInts(num-1)
  } else {
    0
  }
}
```

The `if` is being used as an expression here. It is the only expression in the function so it is the last one, and it will be the result value of the function. If `num`, the argument to the function, is not greater than zero, then the functions value is zero. If it is greater than zero, the function will read in a new value and return the sum of that value and what we get from summing one fewer values.

What if we do not know in advance how many values we are going to sum? What if we want to keep going until some endpoint is reached? We could do this. One problem is determining what represents the end. We need to have the user type in something distinctly different that tells us they have entered all the values they want to sum. An easy way to do this would be to only allow the user to sum positive values and stop as soon as a non-positive value is entered. This gives us a function that does not take any arguments. We do not have to tell it anything. It will return to us an integer for the sum of the numbers entered before the non-positive value. Such a function could be written as follows.

```
def sumInputPositive():Int = {
  val n = readInt()
  if (n>0) {
    n+sumInputPositive()
  } else {
    0
  }
}
```

This time we read a new value before we determine if we will continue or stop. The decision is based on that value, which we store in the variable n. Empty parentheses have been added after the function name for both the declaration and the call. This is a style issue because they are not required. It is considered proper style to use parentheses if the function has side effects and to leave them off if it does not. You will recall from chapter 4 that side effects are the way in which a function changes things that go beyond just returning a value. What does this function do that causes us to say it has side effects? The side effects here are in the

---

[1] Remember that you will need to `import io.StdIn._` before using these input methods.

form of reading input. Reading input is a side effect because it can have an impact beyond that one function. Consider having two different functions that both read input. The order in which you call them is likely to change their behavior. That is because the second one will read input in the state that is left after the first one.

This function does a good job of letting us add together an arbitrary number of user inputs, but it has a significant limitation, it only works with positive values. That is because we reserve negative values as the stop condition. There could certainly be circumstances where this limitation was a problem. How could we get around it? What other methods could we use to tell the function to stop the recursion? We could pick some particular special value like -999 to be the end condition. While -999 might not seem like a particularly common number, this is really no better than what we had before because our function still cannot operate on any valid integer value. We'd like to have the termination input be something like the word "quit". Something special that is not a number.

We can do this if we do not use `readInt`. We could instead use the `readLine` function which will read a full line of input and returns a `String`. You might be tempted to create a method like this:

```scala
def sumUntilQuit():Int = {
  val n = readLine()
  if (n!="quit") {
    n+sumUntilQuit()
  } else {
    0
  }
}
```

If you enter this into a file and then load it into the console, you will get the following error.

```
<console>:8: error: type mismatch;
 found : String
 required: Int
       n+sumUntilQuit()
        ^
```

This is because the function is supposed to return an `Int`, but n is a `String` and when we use + with a `String` what we get is a `String`. Why is n a `String`? Because that is the type returned by `readLine` and Scala's type inference decided on the line `val n = readLine()` that n must be a `String`.

This problem can be easily fixed. We know that if the user is giving us valid input, the only things which can be entered are integer values until the word "quit" is typed in.[2] So we should be able to convert the `String` to an `Int`. That can be done as shown here.

```scala
def sumUntilQuit():Int = {
  val n = readLine()
  if (n!="quit") {
    n.toInt+sumUntilQuit()
  } else {
    0
  }
}
```

---

[2]We generally assume that user input will be valid, as it makes things easier and allows us to focus on the logic of the problem we are solving. Later in this chapter, we will introduce the `try-catch` construct, which will allow us to detect and handle when the user enters something it is not what we were expecting.

Now we have a version of the function which will read one integer at a time until it gets the word "quit".

Summing up a bunch of numbers can be helpful, but it is a bit basic. Let us try to do something more complex. A tiny step up in the complexity would be to take an average. The average is nothing more than the sum divided by the number of elements. In the first version of the function when we enter how many number would be read, this would be trivial to write. We do not even need to write it. The user knows how many numbers there were, just divide by that. This is not so straightforward for the other versions. We do not know how many values were input, and we do not want to force the user to count them. Since we need both a sum and a count of the number of values to calculate an average, we need a function that can give us both.

This is another example of a function that needs to return two values and as before, we will use a tuple to do the job. So we will write a new function called `sumAndCount`, which returns a tuple that has the sum of all the numbers entered as well as the count of how many there were. We will base this off the last version of `sumUntilQuit` so there are no restrictions on the numbers the user can input. Such a function might look like the following:

```
def sumAndCount():(Int,Int) = {
  val n = readLine()
  if (n!="quit") {
    val (s,c) = sumAndCount()
    (s+n.toInt, c+1)
  } else {
    (0, 0)
  }
}
```

If you load this function into the REPL and call it, you can enter a set of numbers and see the return value. If, for example, you enter 3, 4, 5, and 6 on separate lines followed by "quit", you will get this:

```
res0: (Int, Int) = (18,4)
```

This looks a lot like what we had before, only every line related to the return of the function now has a tuple for the sum and the count. We see it on the first line for the result type. We also see it on the last line of both branches of the `if` expression for the actual result values. The last place we see it is in the recursive branch where the result value from the recursive call is stored in a tuple. This syntax of an assignment into a tuple is actually doing pattern matching which will be discussed later in this chapter.

Now we have both the sum and the count. It is a simple matter to use this in a different function that will calculate the average. The function shown below calls `sumAndCount` and uses the two values that are returned to get a final answer.

```
def averageInput():Double = {
  val (sum,count) = sumAndCount()
  sum.toDouble/count
}
```

The one thing that you might at first find odd about this function is that it has two places where `Double` appears. It is the result type and in the last expression the `toDouble` method is called on `sum`. This is done because averages are not generally whole numbers. We have to call `toDouble` on `sum` or `count` because otherwise Scala will perform integer division which truncates the value. We could convert both to `Double`s, but that is not required because numerical operations between a `Double` and an `Int` automatically convert the `Int` to a `Double` and result in a `Double`.

5.4   Abstraction

What if, instead of taking the sum of a bunch of user inputs, we want to take a product? What would we change in `sumAndCount` to make it `productAndCount`? The obvious change is that we change addition to multiplication in the recursive branch of the `if`. A less obvious change is that we also need the base case to return 1 instead of 0. So our modified function might look like this.

```
def productAndCount():(Int,Int) = {
  val n = readLine()
  if (n!="quit") {
    val (s,c) = productAndCount()
    (s*n.toInt, c+1)
  } else {
    (1, 0)
  }
}
```

This is almost exactly the same as what we had before. We just called it a different name and changed two characters in it. This copying of code where we make minor changes is something that is generally frowned upon. You might say that it does not smell right.[3] There are a number of reasons why you would want to avoid doing this type of thing. First, what happens if your first version had a bug? Well, you have now duplicated it and when you figure out something is wrong you have to fix it in multiple places. A second problem is closely related to this, that is the situation where you realize you want a bit more functionality so you need to add something. Again you now have multiple versions to add that into. In addition, it just makes the code base harder to work with. Longer code means more places things can be messed up and more code to go through when there is a problem. For this reason, we strive to reduce code duplication. One way we do this is to include abstraction. We look for ways to make the original code more flexible so it can do everything we want. Abstraction is one of the most important tools in Computer Science and a remarkably powerful concept that you will want to understand. Here we are starting with a fairly simple example.

In order to abstract these functions to make them into one, we focus on the things that were different between them and ask if there is a way to pass that information in as arguments to a version of the function that will do both. For this, the changing of the name is not important. What is important is that we changed the operation we were doing and the base value that was returned. The base value is easy to deal with. We simply pass in an argument to the method that is the value returned at the base. That might look like this.

```
def inputAndCount(base:Int):(Int,Int) = {
  val n = readLine()
  if (n!="quit") {
    val (s,c) = inputAndCount(base)
```

---

[3]Indeed, the term "smell" is the actual terminology used in the field of refactoring for things in code that are not quite right and should probably be fixed.

```
      (s*n.toInt, c+1)
  } else {
    (base, 0)
  }
}
```

The argument base is passed down through the recursion and is also returned in the base case. However, this version is stuck with multiplication so we have not gained all that much.

Dealing with the multiplication is a bit harder. For that we need to think about what multiplication and addition really are and how they are used here. Both multiplication and addition are operators. They take in two operands and give us back a value. When described that way, we can see they are like functions. What we need is a function that takes two Ints and returns an Int. That function could be multiplication or addition and then the inputAndCount function would be flexible enough to handle either a sum or a product. It might look like this.

```
def inputAndCount(base:Int, func:(Int,Int)=>Int):(Int,Int) = {
  val n = readLine()
  if (n!="quit") {
    val (s,c) = inputAndCount(base,func)
    (func(s,n.toInt), c+1)
  } else {
    (base, 0)
  }
}
```

The second argument to inputAndCount, which is called func, has a more complex type. It is a function type. It is a function that takes two Ints as arguments and returns an Int. As with base, we pass func through on the recursive call. We also used func in place of the * or the + in the first element of the return tuple in the recursive case. Now instead of doing s+n.toInt or s*n.toInt, we are doing func(s,n.toInt). What that does depends on the function that is passed in.

To make sure we understand this process we need to see it in action. Let us start with doing a sum and use the longest, easiest to understand syntax. We define a function that does addition and pass that in. For the input we type in the numbers 3, 4, and 5 followed by "quit". Those values are not shown by the REPL.

```
scala> def add(x:Int,y:Int):Int = x+y
add: (x: Int,y: Int)Int

scala> inputAndCount(0,add)
res3: (Int, Int) = (12,3)
```

In the call to inputAndCount we used the function add, which was defined above it, as the second argument. Using a function defined in this way forces us to do a lot of typing. This is exactly the reason Scala includes function literals. You will recall from chapter 4 that a function literal allows us to define a function on the fly in Scala. The normal syntax for this looks a lot like the function type in the definition of inputAndCount. It uses a => between the parameters and the body of the function. Using a function literal we could call inputAndCount without defining the add function. That approach looks like this.

```
scala> inputAndCount(0, (x,y) => x+y)
res4: (Int, Int) = (12,3)
```

One thing to notice about this is that we did not have to specify the types on x and y. That is because Scala knows that the second argument to inputAndCount is a function that takes two Int values. As such, it assumes that x and y must be of type Int.

If you remember back to section 4.5 on function literals, you will recall there is an even shorter syntax for declaring them that only works in certain situations. That was the syntax that uses _ as a placeholder for the arguments. This syntax can only be used if each argument occurs only once and in order. That is true here, so we are allowed to use the shorthand. That simplifies our call all the way down to this.

```scala
scala> inputAndCount(0, _+_)
res5: (Int, Int) = (12,3)
```

Of course, the reason for doing this was so that we could also do products without having to write a second function. The product function differed from the sum function in that the base case was 1 and it used * instead of +. If we make those two changes to what we did above, we will see that we have indeed created a single function that can do either sum or product.

```scala
scala> inputAndCount(1, _*_)
res6: (Int, Int) = (60,3)
```

Not only did we succeed here, we did so in a way that feels satisfying to us. Why? Because our abstraction can be used in a minimal way and only the essential variations have to be expressed. We do not have to do a lot of extra typing to use the abstraction. It is not much longer to call inputAndCount than it was to call sumAndCount or productAndCount. In addition, the only things we changed between the two calls were the changing of the 0 to 1 and the + to *. Those were the exact same things we had to change if we had done the full copy and paste of the functions. This means that the concept we want to express is coming through clearly and is not complicated with a lot of overhead.

The inputAndCount function is what we call a higher order function. That is because it is a function that uses other functions to operate. We provide it with a function to help it do its job. This type of construct has historically been found mostly in functional languages. With the addition of lambda expressions, many mainstream languages also allows this type of funtionality.

You might say we only had to duplicate the code once to have the sum and the product. Is it really worth the effort of our abstraction to prevent that? Is this really all that smelly? The answer, with only a sum and a product, is probably "no". A single code duplication is not the end of the world. However, if I next ask you to complete versions that return the minimum or the maximum, what do you do then? Without the abstraction, you get to copy and paste two more versions of the code and make similarly minor changes to them. With the abstraction, you just call inputAndCount with different arguments. The question of whether it is worth it to abstract really depends on how much expansion you expect in the future. If the abstraction does not take much effort it is probably worth doing to start with because it is often hard to predict if you will need to extend something in the future. You might not feel this in what we do in this book, but it becomes very significant in professional development when you often are not told up front exactly what is wanted of you and even when you are, the customer is prone to change their mind later in the process.

Tail Recursive Functions

Every call to a function typically takes a little memory on what is called the stack to store the arguments, local variables, and information about where the call came from. If a recursive function calls itself many times, this can add up to more than the amount of memory set aside for this purpose. When this happens, you get a stack overflow and you program terminates.

Scala will optimize this away for recursive functions that have nothing left to do after the recursive call. You can convert a recursive function that is not tail recursive to one that is by passing in additional arguments. To see this we can demonstrate how this would be done on the `sumAndCount` function on page 113. The original version is not tail recursive because it does that addition to both the sum and the count after the recursive call. Just doing one of those would prevent the function from being tail recursive. To change this, we can pass in the sum and count as arguments. The produces the following code.

```
def sumAndCountTailRec(sum:Int, count:Int):(Int,Int) = {
  val n = readLine()
  if (n!="quit") {
    sumAndCountTailRec(sum+n.toInt, count+1)
  } else {
    (sum, count)
  }
}
```

Note that when you call this version of the function, you have to provide initial values for `sum` and `count`, both of which should probably be zero. It is possible to hide this by nesting functions, but that is a topic for later.

If a function really needs to be tail recursive, you can tell Scala this with an annotation. Annotations are more advanced features that we do not deal with much in this book. To do this, you first need to `import annotation.tailrec`, then you can put `@tailrec` in from of any function that must be tail recursive. If Scala cannot make the function tail recursive, it will give an error.

5.5 Matching

The `if` expression is not the only conditional construct in Scala. There is a second, far more expressive, conditional construct, `match`. While the `if` construct picks between two different possibilities, based on whether an expression is `true` or `false`, the `match` construct allows you to pick from a large number of options to see if a particular expression matches any of them. The term "matches" here is vague. Indeed, the power of the `match` construct comes from something called PATTERN MATCHING.

The syntax of the `match` expression in its simplest form is as follows.

```
expr match {
  case pattern1 => expr1
  case pattern2 => expr2
  case pattern3 => expr3
```

```
    . . .
}
```

The value of the expression before the `match` keyword is checked against each of the different patterns that follow the `case` keywords in order. The first pattern that matches will have its expression evaluated and that will be the value of the whole match expression.

We can use this to repeat our example from chapter 3 related to the cost of food at a theme park.

```scala
def foodPriceMatch(item:String,size:String):Double = item match {
  case "Drink" => size match {
    case "S" => 0.99
    case "M" => 1.29
    case _ => 1.39
  }
  case "Side" => size match {
    case "S" => 1.29
    case "M" => 1.49
    case _ => 1.59
  }
  case "Main" => size match {
    case "S" => 1.99
    case "M" => 2.59
    case _ => 2.99
  }
  case _ => size match {
    case "S" => 4.09
    case "M" => 4.29
    case _ => 5.69
  }
}
```

When an entire function is a single `match`, it is customary to put the start of the `match` after the equals sign as done here. Inside we have the four different `cases`, each with its own `match` on the `size`. The one thing that might seem odd here is the use of the underscore. An underscore as a pattern matches anything. This is done so that the behavior would agree with what we had in the `if` version where it defaulted to "Combo" as the item and "Large" as the size.

This example shows that the `match` expressions can be nested and the _ can be used to match anything. There is a lot more to the `match` expression though. The following example shows how to use match to give responses to whether you might buy different food items.

```scala
def buy(food:(String, Double)):Boolean = food match {
  case ("Steak", cost) if (cost < 10) => true
  case ("Steak", _) => false
  case (_, cost) => cost < 1
}
```

This is a very limited example, but it demonstrates several aspects of `match` that are worth noting. First, `food` is a tuple, and the cases pull the two items out of that tuple. That is part of the pattern matching aspect. We will use this later on when we have other things that work as patterns. The second thing we see is that if we put a variable name in the pattern, it will match with anything and it will be bound to the value of that thing. In the example, the second element of the tuple in the first and third `cases` is given the name cost. That variable could appear in the expression for the case as in the last case where we will buy anything for under a dollar. It can also be part of an if that "guards" the case. The pattern in the first case will match anything that has the word "Steak" as the food. However, the if means we will only use that case if the cost is less than 10. Otherwise, it falls down and checks later cases.

---

match **versus** switch

If you are familiar with other programming languages you might have heard of a switch statement before. On the surface, match might seem like switch, but match is far more powerful and flexible in ways allowing you to use it more than a switch.

For this chapter, an appropriate example would be to demonstrate recursion using match instead of an if. We can start with a simple example of something like countDown.

```
def countDown(n:Int):Unit = n match {
  case 0 =>
  case i =>
    println(i)
    countDown(i-1)
}
```

This function is not quite the same as what we had with the if because it only stops on the value 0. This makes it a little less robust, but it does a good job of illustrating the syntax and style of recursion with a match. The recursive argument, the one that changes each time the function is called, is the argument to match. There is at least one case that does not involve a recursive call and at least one case that does.

A more significant example would be to rewrite the inputAndCount function using a match.

```
def inputAndCount(base:Int,func:(Int,Int)=>Int):(Int,Int) = readLine() match {
  case "quit" =>
    (base, 0)
  case n =>
    val (s, c) = inputAndCount(base, func)
    (func(s, n.toInt), c+1)
}
```

Here the call to readLine is the argument to match. This is because there is not a standard recursive argument for this function. The decision of whether or not to recurse is based on user input, so the user input is what we match on.

5.6 Bad Input, Exceptions, and the try/catch Expression

At this point, you have probably noticed that if you call readInt and enter something that is not a valid Int it causes the program to crash. If you looked closely at the output when that happens you have noticed that it starts with java.lang.NumberFormatException. This is the error type reported when the program is expecting a number and gets something that does not fit that requirement. The `NumberFormatException` is just one type of `Exception`. There are many others that occur for various reasons, and you can even create your own types of exceptions.

We generally assume that users will input appropriate values to keep examples simple. As such, we do not include code for handling situations when they do not. If you want to be able to make more flexible code or simply deal nicely with users giving invalid inputs you need to use the `try`/`catch` expression. This begins with a `try` block that does what the name implies, it is going to try to execute a piece of code with the knowledge that it might fail. This is followed by a `catch` block with different cases for things that could go wrong. Here is a very basic example where we try to read an integer and give back the value that is read if it works, otherwise, it gives us zero.

```
val num = try {
  readInt()
} catch {
  case _ => 0
}
```

As this shows, the `try`/`catch` is an expression with a value. One point to keep in mind about this is that the values given back by different cases generally need to match the type of the last expression of the `try` block. If you run this code, you will get a warning about it catching all `Throwables`, and it asks if you want to do that. Catching everything with a `case _` is not good form. There are some things that you should not be catching because you cannot do anything to fix them. An example would be an `OutOfMemoryError`. If you run out of memory, just returning zero is not going to save the program. You can fix this by specifying the type on the pattern for the case as shown below.

You could also see the normal printout you get from an exception by giving a name to the exception in the `case` associated with it. All exception objects have a method called `printStackTrace`.

```
val num = try {
  readInt()
} catch {
  case e =>
    e.printStackTrace
    0
}
```

This can be especially useful during debugging as that stack trace includes a significant amount of useful information.

Both of these examples have the pitfall that they give back a value of zero when anything goes wrong. If you really need to have an integer read, and only want to handle `NumberFormatExceptions`, you might consider code like this.

```
def readIntRobust():Int = try {
  readInt()
} catch {
  case e:NumberFormatException =>
    println("That was not an integer. Please try again.")
    readIntRobust()
}
```

This recursive function will call itself repeatedly until the user enters a valid integer. This uses a different type of pattern in the case, one which only matches a certain type. In this case, it only matches an exception of the `NumberFormatException` type. So if some other type of exception were to occur, that would still cause a crash. Such behavior is typically want you want. You should only handle errors that you know how to deal with at that point in the code. This same type of syntax, where you have a variable name followed by a colon and a type, can be used in other places that allow patterns, such as with `match`. Outside of the `try`/`catch` expression, we will not find much use for it in this book as we generally know the types that we are working with, and this pattern is most useful when that is not the case.

5.7 Putting It Together

Back to the theme park. Let's take some of the functions we wrote previously, and put them to use in a recursive function that is part of a script that we could use at the front of the park where they sell admission tickets. The recursion will let us handle multiple groups paying for admission with any number of people per group. It will then add up all the admissions costs so we can get a daily total. In addition, it will keep track of the number of groups and people that come in. The code for this makes the same assumptions we made before about groups in regards to coolers and the water park.

Listing 5.1: EntryGate.scala

```scala
def entryCost(age:Int, cooler:Boolean, waterPark:Boolean):Double = {
  (if (age < 13 || age > 65) 20 else 35) +
  (if (cooler) 5 else 0) +
  (if (waterPark) 10 else 0)
}

def individualAdding(num:Int, cooler:Boolean, waterPark:Boolean):Double = {
  if (num > 0) {
    println("What is the personâĂŹs age?")
    val age = readInt()
    entryCost(age, cooler, waterPark)+individualAdding(num-1, false, waterPark)
  } else 0.0
}

def groupSizeCost():(Int,Double) = {
  println("How many people are in your group?")
  val numPeople = readInt()
  println("Is the group bringing a cooler? (Y/N)")
  val cooler = readLine() == "Y"
  println("Will they be going to the water park? (Y/N)")
  val waterPark = readLine() == "Y"
  (numPeople,individualAdding(numPeople,cooler,waterPark))
}

def doAdmission():(Int,Int,Double) = {
  println("Is there another group for the day? (Y/N)")
  val another = readLine()
  if (another == "Y") {
    val (people, cost) = groupSizeCost()
    val (morePeople, moreGroups, moreCost) = doAdmission()
```

```
      (people+morePeople, 1+moreGroups, cost+moreCost)
  } else (0,0,0.0)
}
```

```
val (totalPeople, totalGroups, totalCost) = doAdmission()
println("There were "+totalPeople+" people in "+totalGroups+
  " groups who paid a total of "+totalCost)
```

This code has two different recursive functions, one for groups and one for people. The doAdmission function recurses over full groups. It asks each time if there is another group and if there is it uses groupSizeCost to get the number of people and total cost for the next group, then adds those values to the return of the recursive call. The groupSizeCost function uses the recursive individualAdding function to run through the proper number of people, ask their age, and add them all up.

One interesting point to notice about individualAdding is that the function takes a Boolean for whether or not there is a cooler, but when it recursively calls itself, it always passes false for that argument. This is a simple way to enforce our rule that each group only brings in one cooler. If the initial call uses true for cooler, that will be used for the first person. All following calls will use a value of false, regardless of the value for the initial call.

5.8   Looking Ahead

This will not be our last look at recursion in this book. Indeed, we have just scratched the surface. We have only used recursion in this chapter to repeat tasks, as a model for repetition. The real power of recursion comes from the fact that it can do a lot more than just repetition. Recursive calls have memory. They do not know what they are doing, they just remember what they have done. This really comes into play when a recursive function calls itself more than once. That is a topic for later, but before we leave this first encounter with recursion here is a little brain teaser for you to think about.

Below is a little bit of code. You will notice that it is nearly identical to the countDown function that we wrote near the beginning of this chapter. Other than changing the name of the method the only difference is that the two lines inside the if have been swapped. Put this function into Scala. What does it do? More importantly, why does it do that?

```
def count(n:Int):Unit = {
  if (n>=0) {
    count(n-1)
    println(n)
  }
}
```

5.9 End of Chapter Material

5.9.1 Problem Solving Approach

While recursion gives us much greater flexibility with the ability to repeat code an arbitrary number of times, it used the same constructs we had learned before and, for that reason, it does not increase the list of options we have for any given line. It only adds new ways to think about and use options we had before. This chapter did include one new type of statement/expression, the `match` conditional expression. That is now included along with `if` as a conditional option. So when you go to write a line of code, the following are your only options.

1. Call a function just for the side effects.

2. Declare something:

   - A variable with `val` or `var`.
   - A function with `def`. Inside of the function will be statements that can pull from any of these rules. The last statement in the function should be an expression that is the result value.
   - A new name for a type with `type`.

3. Assign a value to a variable.

4. Write a conditional statement:

   - An `if` statement.
   - A `match` statement.

5.9.2 Summary of Concepts

- The concept of recursion comes from mathematics where it is used to refer to a function that is defined in terms of itself.

  - All recursive functions must have at least one base case that does not call itself in addition to at least one recursive case that does call itself.
  - Lack of a base case leads to infinite recursion.

- In a programming context, a recursive function is one that calls itself.

  - Recursion is used in this chapter to provide repetition.

    – Typically an argument is passed in to the function, and on each subsequent call the argument is moved closer to the base case.

    – If the argument is not moved toward the base case it can result in an infinite recursion.

- Recursion can also be done on user input when we do not know in advance how many times something should happen.

    – Reading from input is a mutation of the state of the input.

    – Functions that use this might not take an argument if they read from standard input.

    – The base case occurs when a certain value is input.

- It is inefficient to make copies of code that only differ in slight ways. This can often be dealt with by introducing an abstraction on the things that are different in the different copies.

    – With functions, this is done by passing in other arguments that tell the code what to do in the parts that were different in the different copies.

    – Values can easily be passed through with parameters of the proper value type.

    – Variations in functionality can be dealt with using parameters of function types. This makes the abstract versions into higher-order functions.

- There is another type of conditional construct in Scala called `match`.

    – A `match` can include one or more different `cases`.

    – The first `case` that matches the initial argument will be executed. If the `match` is used as an expression, the value of the code in the `case` will be the value of the `match`.

    – The `cases` are actually patterns. This gives them the ability to match structures in the data and pull out values.

        * Tuples can be used as a pattern.

        * Lowercase names are treated as `val` variable declarations and bound to that part of the pattern.

        * You can use _ as a wildcard to match any value that you do not need to give a name to in a pattern.

    – After the pattern in a `case` you can put an `if` guard to further restrict that `case`.

- The `try`/`catch` expression can be used to deal with things going wrong.

    – The block of code after `try` will be executed. If an exception is thrown, control jumps from the point of the exception down to the `catch`.

    – The `catch` will have different `cases` for the exceptions that it can handle.

    – It only makes sense for this to be an exception if all paths produce the same type.

5.9.3 Self-Directed Study

Enter the following statements into the REPL and see what they do. Try some variations to make sure you understand what is going on. Note that some lines read values so the REPL will pause until you enter those values. The outcome of other lines will depend on what you enter.

```scala
scala> def recur(n:Int):String = if (n<1) "" else readLine()+recur(n-1)
scala> recur(3)
scala> def recur2(n:Int,s:String):String = if (n<1) s else recur2(n-1,s+readLine())
scala> recur2(3,"")
scala> def log2ish(n:Int):Int = if (n<2) 0 else 1+log2ish(n/2)
scala> log2ish(8)
scala> log2ish(32)
scala> log2ish(35)
scala> log2ish(1100000)
scala> def tnp1(n:Int):Int = if (n<2) 1 else
  1+(if (n%2==0) tnp1(n/2) else tnp1(3*n+1))
scala> tnp1(4)
scala> tnp1(3)
scala> def alpha(c:Char):String = if (c>'z') "" else c+alpha((c+1).toChar)
scala> alpha('a')
```

5.9.4 Exercises

1. Write functions that will find either the minimum or the maximum value from numbers input by the user until the user types in "quit".

2. Use `inputAndCount` to find the minimum and maximum of numbers that the user enters.

3. Write exponentiation using multiplication. Your function only has to work for positive integer values.

4. If you did 3, most likely you have a function where if you raise a number to the Nth power, it will do N (or maybe $N-1$) multiplications. Consider how you could make this smarter. It turns out that you can make one that does a lot fewer multiplication, $\log_2 N$ to be exact. Think about how you would do this and write code for it.

5. Write a recursive function that will print powers of two up to some power.

6. Write a recursive function that will print powers of two up to some value.

7. Write recursive functions that will print a multiplication table up to 10s. Try to get it running first, then consider how you could make everything line up.

8. Describe the behavior of the last count function in the chapter.

9. Write a recursive function called `isPrime` that returns a `Boolean` and lets you know whether or not a number is prime.

10. Write a recursive function that prints the prime factors of a number.

11. Write a recursive function that lets a user play rock, paper, scissors until either the user or the computer wins 3 times. Hint: use util.random.nextint.

12. Write a recursive function that simulates a coin flip and counts the number of heads and the number of tails that occur in 1000 flips. Hint: use util.random.nextint.

13. Suppose that tuition at a university costs $30,000 per year and increases 5% per year. Write a recursive function that computes the total cost of tuition a student would pay for a 4-year degree if they started 10 years from now.

14. Write a recursive function that takes an integer and returns the number with the digits reversed. Using your recursive algorithm, reverse the following number: 07252015.

15. An efficient method of finding the greatest common divisor, gcd, of two integers is Euclid's algorithm. This is a recursive algorithm that can be expressed in mathematical notation in the following way.

$$gcd(a, b) = \{ \begin{array}{ll} a & b = 0 \\ gcd(b, a \mod b) & otherwise \end{array}$$

Convert this to Scala code.

16. Certain problems can use a bit more information than just the gcd provided by Euclid's algorithm shown in exercise 15. In particular, it is often helpful to have the smallest magnitude values of x and y that satisfy the equation $gcd = xa + by$. This information can be found efficiently using the extended Euclid's algorithm. This is the math notation for that function.

$$eEuclid(a, b) = \{ \begin{array}{ll} (a, 1, 0) & b = 0 \\ (d, x, y) = eEuclid(b, a \mod b), (d, y, x - \lfloor a/b \rfloor * y) & otherwise \end{array}$$

Convert this to Scala. Note that the $\lfloor a/b \rfloor$ operation is naturally achieved by the truncation of integer division for positive a and b.

5.9.5   Projects

1. For this option I want you to write functions that do the basic math operations of addition, multiplication, and exponentiation on non-negative Ints. The catch is that you cannot use +, *, or any functions from math. You only get to call the successor and predecessor functions shown here.

```
def succ(i:Int):Int = i+1
def pred(i:Int):Int = i-1
```

These functions basically do counting for you. So you will define addition in terms of those two, then define multiplication in terms of addition and exponents in terms of multiplication.

Put the functions that you write in a script and have the script prompt for two numbers. Using your functions, print the sum, the product, and the exponent of those two numbers.

2. Write a function that will return a String with the prime factorization of a positive Int. The format for the print should be p^e+p^e+... Here each p is a prime number and e is the how many times it appears in the prime factorization. If you call this function with 120 it would return 2^3+3^1+5^1 because $120 = 2*2*2*3*5$. Remember that a number n is divisible by i if n%i==0. For the prime numbers start counting up

from 2. If you pull out all the factors of lower numbers, you will not find any factors that are not prime. For example, when you pull 2^3=8 out of 120 you get 15 which only has factors of 3 and 5 so you cannot get 4.

3. This project builds on top of project 4.10. You are using the functions from that project and putting them into a format where they can be used for multiple geometric objects. The user inputs a ray first and after that is a series of spheres and planes. You want a function that returns the first object hit by the ray, smallest t value, and the parameter (t) of the hit.

4. An interesting twist in biology over the past few decades is the ability to look at the populations of different species and how they interact with one another. Often, the way in which different populations vary over time can be approximated by simple mathematical expressions. In this project you will use your basic knowledge of conditionals and functions with recursion to examine a simple case where you have two different populations that interact in a predator-prey manner.

The simplest form of this problem is the rabbit and fox scenario. The idea is that each summer you count the population of rabbits and foxes in a certain region. This region is fairly well isolated so you do not have animals coming in or leaving. In addition, the climate is extremely temperate, and there is always enough grass so environmental factors do not seem to impact the populations. All that happens is each year the rabbits try to eat and have babies while not getting eaten, and the foxes try to catch rabbits. We will make up some formulas for what happens to the population from one year to the next, and you will write a program to produce this sequence.

Over the course of each year, the rabbit population will be impacted in the following ways. Some rabbits will be born, some rabbits will die of natural causes, and some rabbits will be eaten. Similarly some foxes will be born and some will die. The number of rabbits eaten depends upon the population of foxes (more foxes eat more rabbits), and the number of foxes who are born and die depends on the number of rabbits because foxes cannot live long or have young without finding rabbits to eat. We can combine these things to come up with some equations that predict the numbers of foxes and rabbits in a given year based on the number in the previous year.

$$R_{n+1} = R_n + A * R_n - B * R_n * F_n$$
$$F_{n+1} = F_n - C * F_n + D * R_n * F_n$$

Here we assume that the natural tendency of rabbit populations is to increase without foxes around and the natural tendency of fox populations is to decrease without rabbits around. The four constants should have positive values. A represents the normal increase in rabbit population without predation. B is the predation rate and is multiplied by both the rabbit population and the fox population because if either one is small, the predation rate is small. C is the rate at which foxes would normally die out without being able to bear young (if they did not have enough food). D is the rate at which fox will bear young when they do have rabbits to feed on. In reality, foxes and rabbits only come in whole numbers, but for numeric reasons, you should use `Doubles` in your program.

The input for your program is the initial rabbit population, R_0, the initial fox population F_0, and the four constants. To start you off, you might try values of 100, 10, 0.01, 0.001, 0.05, and 0.001. The last four numbers are A, B, C, and D, respectively. You can play with these values to try to find some that produce interesting results. Print out the first 1000 iterations. To make it so that you can see your results easily, output only numbers. Never prompt for anything. The advantage of this is that you can create a file that is easy to plot. For plotting, you can input the values into a spreadsheet like Excel. Under Linux you could also use gnuplot. When you run the program you redirect the output to a file then you can run gnuplot and plot it to see what it looks like. If you print 3 numbers per line, "n R F", and put it in a file called "pop.txt" then you can plot that in gnuplot with a command like "plot 'pop.txt' using ($1):($2), 'pop.txt' using ($1):($3)". There are many other options in gnuplot and you can use the help command to see them.

Write this as a script that has the user enter R_0, F_0, A, B, C, and D without any prompts. It should then output only the numbers for plotting so that they can be sent to a file that can be plotted without editing.

5. Suppose that you want to take out a house loan. Your monthly payment for the loan pays both the principal and the interest. The formula for your monthly payment is:

$$monthlyPayment = \frac{loanAmount \times monthlyInterestRate}{1 - (1 + monthlyInterestRate)^{numberOfYears*12}}$$

The formula for your total payment is:

$$totalPayment = monthlyPayment * numberOfYears * 12$$

You can compute the monthly interest by multiplying the monthly interest rate and the remaining principal balance. The principal paid for the month is the monthly payment minus the monthly interest. Write a script that lets a user enter the house loan amount, number of years they want to take to pay off the loan, and annual interest rate. Display the monthly payment, the total payment amount and lastly, an amortization schedule that shows the payment number, interest paid, principle paid, remaining balance for each of the monthly payments to be paid for the duration of the loan.

6. Write two recursive functions that returns the number of vowels in a string. One should use **if** and the other should use **match**. Use the following quote, by Thomas Edison, for your test:

"Our greatest weakness lies in giving up. The most certain way to succeed is always to try just one more tim".

7. What would you do if you have a team of 10 people, but you only need 4 people working on the next project? How many different ways can you create a team of 4 people? You can use the following formula to find the number of ways **p** different people can be chosen from a set of tt team members, where **p** and **t** are non-negative integers and $p <= t$:

$$Combinations(t, p) = t!/p!(t - p!)$$

and where the exclamation point denotes the factorial function. Write a recursive function to implement Combinations(t, p) which determines the number of ways p different people can be chosen from a team of t people. You can assume that Combinations(t, 0) = Combinations(t, t) = 1. You can also assume that Combinations(t, p) = Combinations(t-1, p-1) + Combinations(t-1, p). Using your recursive algorithm, determine Combinations(5, 3) and Combinations(9, 4).

8. Write a recursive function to print a string backwards. Go ahead and try this quote by Thomas Edison for your test: "I have not failed; I've just found ten thousand ways that won't work".

9. Write a recursive function that generates the following pattern of letters. Make sure that the user enters an integer between 1 and 26. If the non-negative integer is 6, then the pattern generated is:

ABCDEF

ABCDE

ABCD

ABC

AB

A

A

AB

ABC

ABCD

ABCDE

ABCDEF

10. Write a script that will calculate a student's GPA with each course properly weighted by the number of hours of credit it counts for. It should prompt for a grade and a number of hours for each class. The grades are letter grades that are converted to a 4-point GPA scale according to the following table. When "quit" is entered as the grade you should stop reading values. The script should then print the cumulative GPA.

Grade	GPA
A	4.000
A-	3.666
B+	3.333
B	3.000
B-	2.666
C+	2.333
C	2.000
C-	1.666
D+	1.333
D	1.000
F	0.000

11. The Mandelbrot set is a fractal in the complex plane. It is described by the simple equation $z_{n+1} = z_n^2 + c$ where $z_0 = 0$, and c is the point in the plane. That means that both c and z are complex values. You can represent them with the type (Double, Double). If the sequence is bounded, the point is in the set. If it heads off to infinity, it is not. Use the solution to exercise 4.3 to write a function that will take a value of c, and a value of z and return how many iterations it takes for the magnitude of z_n to become greater than 4.0 or 1000 if it does not get that big in 1000 iterations.

12. Using a computer, you can estimate the value of π by picking random numbers. In fact, you can find numeric solutions for most integration problems by picking random numbers. This approach to solving such problems is called the Monte-Carlo method. For this option, you will write code to do that using calls to math.random() to generate the random numbers and use that to integrate the area under a curve. You will have a

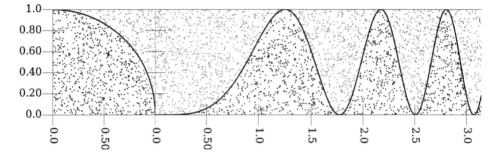

FIGURE 5.1: This figure is associated with project 12. This shows how you can use picking random numbers to estimate the area under curves. This figure shows two curves and random points in square regions that bound those curves. The points that are below the curves are darker than those above. The fraction of points below is proportional to the area under the curve.

script that uses this method to solve two problems, one of which is getting an estimate for the value of π.

The approach is illustrated in figure 5.1. If you can draw a bounding box around a region of a function, you can randomly select points inside of that box and the area under the curve is estimated by

$$A\frac{N_{below}}{N_{total}}$$

where A is the area of the box. The more points you draw randomly, the more accurate the estimate.

Write a recursive function with the following signature.

```
def countUnder(nPnts:Int,xmin:Double, xmax:Double, ymin:Double,
    ymax:Double,f: Double=>Double):Int
```

This function should generate the specified number of random points in the rectangle with the given bounds and return how many of them were below the curve defined by the function f. Note that you can get a random number in the range $[min, max)$ with code like this.

```
val r = min+math.random*(max-min)
```

You want to do this for the x and y axes.

To estimate π use a function of $\sqrt{1-x^2}$ with x and y in the range of $(0, 1)$. The full rectangle has an area of 1 while the area under the curve is $\pi/4$. If you multiply the fraction of points under the curve by 4, you get an estimate of pi.[4]

In addition to printing out your estimate of π, you should print out an estimate of

$$\int_0^\pi (\sin x^2)^2$$

for the same number of random points.

---

[4]Be sure to pay attention to whether you are doing integer or floating-point arithmetic.

13. If you did project 4.1, you can now add another option to it. Write a function that takes a rational number as an (`Int`,`Int`) and returns an (`Int`,`Int`) where all common factors have been removed so that it is in lowest form.

   With that added in, write a script that uses a recursive function to do calculations on rational numbers. The script presents the user with a menu with options for add, subtract, multiply, divide, and quit. If one of the first four options is selected, it should ask for a fraction and perform that operation between the "current" fraction and the one entered. When the program starts, the "current" value will be $\frac{1}{1}$ represented by (`1`,`1`). After each operation you print the new current value, making sure it is always in lowest form.

The web site provides additional exercises and projects with data files that you can process using input redirection.

Chapter 6

Arrays and Lists in Scala

Adding conditionals and functions expanded our capabilities dramatically. Combining it with recursion gives us the ability to do almost anything we want. Technically, we have full computing power. However, there are some ideas that are challenging to express with what we currently know and there are better ways to say them in Scala and programming languages in general.

One of these ideas is the ability to refer to many different objects with one name. In chapter 2 we gained the ability to refer to objects by names with `val` and `var` declarations. Using those, we could make as many names as we want and hard code references to them, but we have to make different names for each declaration if we want to work with the different values. This is a reasonable thing to do if we only need a few variables. For instance, if we wanted to read in five numbers so that we could total them, we would just create five variable declarations. However, what if we wanted to read in 100 numbers, or 1000 numbers? Just the amount of effort necessary to create that many variable declarations is substantial, let alone writing the statement(s) necessary to find the total.

In this chapter we will begin learning about COLLECTIONS in Scala. Collections are types that allow us to store and look up many different values using a single name. Most of this chapter will focus on the most basic collection types, ARRAYS and LISTS.

6.1 Making Arrays

The most basic collections of data in Scala or other languages are **Arrays** and **Lists**. Virtually every language will include one as something of a fundamental aspect of the language. Scala happens to include both. Each are easy access parts of the library.[1] The `Array` and the `List` are what Scala refers to as SEQUENCES. That means that they store a number of different values in a specific order, and you can get to the elements in them by an integer index.

In the last chapter, we used recursion to do things such as calculate the sum or take the product of a bunch of numbers entered by the user. What if we wanted to do both? We could have done both a sum and a product if the user entered the numbers twice, but the user does not want to do that. We had one significant problem with what we were doing. We took the numbers from the user and performed operations on them, but we did not really store them so that we could use them again. It was all a one shot deal. The reason for this was that the types we have so far store a single value or, in the case of a tuple, a fixed number of values. We do not really have a good way to store a variable number of values so that we can work with them and do multiple operations on them. That is exactly what collections will allow us to do. To really understand this, it helps to look at some examples.

We will start by making some `Array`s that have values in them.

```
scala> Array(1,2,3,4)
res0: Array[Int] = Array(1, 2, 3, 4)

scala> Array(1.0,2.0,3.0,4.0)
res1: Array[Double] = Array(1.0, 2.0, 3.0, 4.0)

scala> Array('c','a','t')
res2: Array[Char] = Array(c, a, t)

scala> Array("This","is","a","test")
res3: Array[String] = Array(This, is, a, test)
```

Here we have created four different `Array`s of different types. The syntax for making an `Array` with values in it is to follow the word "`Array`" with a list of parameters for the values we want in the `Array`. As always, after each expression the REPL tells us the name used for this, the types of the expression, and the value of the expression. The type of the expression here shows something that we have never seen before. The type is not just `Array`. The type name `Array` is followed by square brackets that have another type in them. This is called a PARAMETERIZED TYPE. All the collection types in Scala are parameterized. We will talk about this in more detail later in this chapter, but the meaning should be clear. The type `Array[Int]` is an `Array` that can hold integer values in it. As usual, we did not have to tell Scala that the first `Array` was an `Array` of type `Int`, it figured that out. You can override Scala's type inference and specifically tell it the type you want. The example below makes an `Array[Double]` even though all the values in it are integers. This is perfectly valid because all values in `Int` are also valid values of `Double`.

```
scala> Array[Double](1,2,3,4)
res4: Array[Double] = Array(1.0, 2.0, 3.0, 4.0)
```

Such forcing has to match what you pass in. Scala will not allow you to do this for an invalid conversion.

---

[1]All of the types and methods we have been using to date are part of a collection that is generally referred to as the standard library or API. All useful languages come with such libraries that contain code that performs specific tasks and provide standardized solutions to certain problems that are required for a broad range of programs. The usefulness of many languages is often determined by the breadth of their standard libraries. Scala has an extremely complete standard library that you can use. This is at least in part because it can access the Java libraries.

```
scala> Array[Int]("Does","this","work")
<console>:6: error: type mismatch;
 found : java.lang.String("Does")
 required: Int Array[Int]("Does","this","work")
                         ^

<console>:6: error: type mismatch;
 found : java.lang.String("this")
 required: Int Array[Int]("Does","this","work")
                                 ^

<console>:6: error: type mismatch;
 found : java.lang.String("work")
 required: Int Array[Int]("Does","this","work")
                                        ^
```

6.2 Using Arrays

So now you know one way to make arrays, but what can you do with them? The first things we need be able to do are get the values stored in the array and change the values stored in the array. This is done by following the name of the array with parentheses and a number that is the index of the thing we want. Arrays, like with most things in most modern programming languages, are zero indexed. So the first element in the array is at index zero, and, if there are N things in the array, the last one is at index N-1. Here is an example.

```
scala> val arr=Array(7,4,6,3,9,1)
arr: Array[Int] = Array(7, 4, 6, 3, 9, 1)

scala> arr(0)
res5: Int = 7

scala> arr(1)
res6: Int = 4

scala> arr(5)
res7: Int = 1
```

The first statement creates a new variable named **arr** that is an array of integers and gives that array six values to store. The next three commands basically pull out values from that array. The expression **arr(0)** gets the first elements, **arr(1)** gets the second, and **arr(5)** gets the last.

What goes in the parentheses does not have to be a simple integer. It can be any expression of type **Int**. You could do something much more complex to pull out a value.

```
scala> val i = 2
i: Int = 2

scala> arr(2*i-2)
res8: Int = 6
```

The same type of expression can also be used in an assignment expression. So we can alter the values that are stored in an array.

```
scala> arr(4) = 99

scala> arr
res9: Array[Int] = Array(7, 4, 6, 3, 99, 1)
```

Here we see how we can use assignment to change the value of one of the elements of the Array. This might surprise you because we originally declared arr to be a val. Previously we said that you cannot change what is referred to by a variable that is created with val. This does not actually break that rule. You see, the name arr still refers to the same Array. What has changed is the value in the Array. An analogy might be that an Array is a house and the values are the people in it. The variable arr refers to a particular house. People might come and go from the house, but the house itself is the same. We can demonstrate this by trying to change what Array the variable arr references.

```
scala> arr = Array(2,7,5)
<console>:6: error: reassignment to val
       arr=Array(2,7,5)
          ^
```

We call the Array type a MUTABLE type because the values in it can be changed. Types whose internal values cannot be changed are called IMMUTABLE types. This distinction is very significant to us in Scala and in programming in general, as it alters how we deal with data. We will talk more about it because the List type, which we will explore in the next section, happens to be immutable. Indeed, all of the types that we have seen previously are immutable. Once they are created, the values they have never change. The Array is our first example of something that is mutable.

   Not everything about an array can be changed. As we have seen, we can change the values stored in an array. However, we cannot efficiently change how many things are stored in an Array.[2] The number of things an Array holds can be called the length or the size of the Array. In fact, the Array type has methods called length and size which give us this information.

```
scala> arr.length
res10: Int = 6

scala> arr.size
res11: Int = 6
```

When you create an Array you have to specify a length and that length will never change. If you use a var style variable, you can make a new Array with a different length and have

---

[2]There are :+ and +: operators that make new Arrays with one additional element, but you should not use these much.

the name refer to it, but to do that you create a completely new `Array`, you do not alter the size of the old one.[3]

When you are accessing the array, if you try to use an index value that is either negative or too large, Scala will give you an error message saying that you have gone outside the bounds of the array. We can see what that looks like here with attempts to access and change indexes out of bounds.

```
scala> arr(100)
java.lang.ArrayIndexOutOfBoundsException: 100
     at .<init>(<console>:7)
     at .<clinit>(<console>)
     at RequestResult$.<init>(<console>:9)
     at RequestResult$.<clinit>(<console>)
     at RequestResult$scala_repl_result(<console>)
     at sun.reflect.NativeMethodAccessorImpl.invoke0(Native Method)
     at sun...

scala> arr(100) = 0
java.lang.ArrayIndexOutOfBoundsException: 100
     at .<init>(<console>:7)
     at .<clinit>(<console>)
     at RequestResult$.<init>(<console>:9)
     at RequestResult$.<clinit>(<console>)
     at RequestResult$scala_repl_result(<console>)
     at sun.reflect.NativeMethodAccessorImpl.invoke0(Native Method)
     at sun...
```

Now that we know the ground rules for using `Arrays`, we can write some functions that take advantage of them in a useful way. In the last chapter we used recursion to read in a set of numbers that we could then do operations on. One motivation we used for collections was that previously we could not store the values to do two or more operations on them. So two useful functions might be to have one that fills an `Array` from user input and another that does the types of operations we did before, but this time on the `Array`.

We will start with the function to fill the `Array` from user input. This function needs to be passed the `Array` that it will fill. When we are using an `Array`, it will be easiest to use the style where we specify at the beginning how many numbers are to be read. This is because `Arrays` have a fixed size. The `Array` knows its size so that information does not need to be passed. What does need to be passed is the index that the next number will be read into. The termination condition would be that we are trying to read into a location beyond the end of the `Array`. The function to do this might look like this.

```
def fillArray(arr:Array[Int], index:Int):Unit = {
  if (index < arr.length) {
    arr(index) = readInt()
    fillArray(arr, index+1)
  }
}
```

This function is straightforward, but it is worth noting that it does not produce a result. The fact that `Arrays` are mutable means that we can pass in an `Array` and have the function

---

[3]Note that you are not likely to notice this issue in the REPL because you can declare a new `val` that replaces a pre-existing one. However, in a script, you cannot have two things with the same name in the same scope, so it is important to understand this restriction.

mutate the contents. As a result, nothing needs to be returned. The results of this function are held in the modified contents of the `Array`. The following shows this function in action.

```
scala> val numbers = Array(0, 0, 0, 0)
numbers: Array[Int] = Array(0, 0, 0, 0)

scala> fillArray(numbers, 0)

scala> numbers
res12: Array[Int] = Array(2, 3, 4, 5)
```

A new `Array` of integers is created that can hold four values. We call `fillArray` and enter the values 2, 3, 4, and 5. After doing that we can inspect numbers and see that it now holds those values.

Now we need to perform an operation on the contents of the `Array`. We will skip the step of making a special function for just doing addition or multiplication and jump straight to the more abstract and flexible method where we pass in a function to operate on the contents. In addition to the function doing the operation, we also need an `Array` and an integer for the current index. Unlike what we did in the previous chapter, we do not need to pass in a base case value because we know when we are at the beginning or end of the `Array`. The function could look like this.

```
def operateOnArray(arr:Array[Int],index:Int, func:(Int,Int)=>Int):Int = {
  if (index < arr.length-1) {
    func(arr(index), operateOnArray(arr, index+1, func))
  } else {
    arr(arr.length-1)
  }
}
```

If an index at or beyond the last element of the `Array` is passed in, this function results in the last element of the array. Otherwise, it applies the function to the current element and the result of the recursive function on subsequent elements. We can see this in action on the previously defined `Array`, `arr`, in these commands.

```
scala> val arr=Array(7,4,6,3,9,1)
arr: Array[Int] = Array(7, 4, 6, 3, 9, 1)

scala> operateOnArray(arr, 0, _+_)
res13: Int = 30

scala> operateOnArray(arr, 0, _*_)
res14: Int = 4536
```

Already this allows us to see the added power we get from using an `Array`. Having the values stored gives us the ability to operate on them multiple times without having to input them multiple times.

6.3 Lists

Arrays are the built in collection of choice in most non-functional languages. An Array is typically stored as a single block of memory. This makes them fast and efficient for a lot of operations. As we saw though, Arrays are mutable. Functional languages tend to lean away from mutation and other side effects.[4] If you do not allow mutation, Arrays become much less efficient. If you want to change a single value in an immutable Array, you have to make a complete copy of the Array. For this reason, functional languages tend to prefer lists. Technically the List type in Scala is an immutable singly linked lists, a data structure that you should become familiar with in a later course. You do not have to understand the structure in detail to see how to use it.

We can build a List using the same syntax we used to build an Array with initial values.

```
scala> List(7,4,6,3,9,1)
res15: List[Int] = List(7, 4, 6, 3, 9, 1)
```

Like the Array type, the List type is parametric and Scala will figure out the best type if you use this syntax.

Unlike Arrays, Lists do not have to have a fixed length determined at the time it is created. Thus, there is another way to put Lists together when we do not know initially all of the values that will be stored in them. We can efficiently build Lists one element at a time if we add elements to the front of the List. To add elements to a List we use the "cons" operator, ::. Here is an example of adding a single element to an existing List.

```
scala> val lst = List(2,3,4)
lst: List[Int] = List(2, 3, 4)

scala> 1::lst
res16: List[Int] = List(1, 2, 3, 4)
```

We begin by creating a List that contains 2, 3, and 4, then cons a 1 to the beginning of the list. This operation did not alter the original List. Instead it gave us a new list with an extra element at the beginning. Because of the way that Lists work, this is efficient. It did not actually have to copy the whole List to make it work. We can look at lst again to see that it has not changed.

```
scala> lst
res17: List[Int] = List(2, 3, 4)
```

If you build a List with the cons operator, it is common to start with an empty List. There are two ways to represent an empty List in Scala. One is to use what we did before, but not put any arguments in the parentheses. The other is to use Nil. So we can build the same list we had before this way.

```
scala> 1::2::3::4::Nil
res18: List[Int] = List(1, 2, 3, 4)
```

You have to place the Nil at the end because the :: operator needs to have a List on the right hand side. Notice that for this to work, the :: operator is right associative.

---

[4]This was something we first discussed with the difference between val and var. As a general rule, programs are easier to reason about when they do not include mutation. Including mutation also makes a number of things more complex by introducing new classes of errors.

So 1::2::3::Nil is the same as 1::(2::(3::Nil)). This is the opposite of the normal mathematical operators which are left associative. In Scala, any operator that ends with a colon will be right associative.

We can use the cons operator to write a function that builds a List of numbers from user input. This is a recursive method that will read in numbers until it gets to "quit". Going until "quit" works well for Lists because we can easily and efficiently add new elements to a List. That was not the case for Arrays where we needed to have the size of the Array set when we began. The method for doing this is quite simple.

```scala
def inputList():List[Int] = {
  val in = readLine()
  if (in == "quit") Nil else in.toInt::inputList()
}
```

We can see this at work as well if we run it and type in 3, 4, 5, and "quit" on separate lines.

```scala
scala> inputList()
res19: List[Int] = List(3, 4, 5)
```

It is possible to access the elements of a List the same way you do an Array by putting an index inside of parentheses. However, for a List this is generally inefficient. The preferred method, especially in a recursive function, is to use the methods head and tail or with pattern matching. The head method will give you back the first element of the List. The tail method gives you a List of all the elements after the first element. Here are simple examples operating on the lst defined above.

```scala
scala> lst.head
res20: Int = 2
```

```scala
scala> lst.tail
res21: List[Int] = List(3, 4)
```

Using these methods we can write an operateOnList function that mirrors the operateOnArray function like this.

```scala
def operateOnList(lst:List[Int], func:(Int,Int)=>Int):Int = {
  if (lst.tail == Nil) lst.head else
    func(lst.head,operateOnList(lst.tail, func))
}
```

Note that we do not require an index to be passed in. We do not have any +1 or -1 in this function either. That type of behavior comes from the fact that when we recurse, we pass in lst.tail. We can see this function in action here.

```scala
scala> val lst=List(7,4,6,3,9,1)
lst: List[Int] = List(7, 4, 6, 3, 9, 1)
```

```scala
scala> operateOnList(lst, _+_)
res22: Int = 30
```

```scala
scala> operateOnList(lst, _*_)
res23: Int = 4536
```

This function was written using an if statement. When working with lists, it is also common to use pattern matching. The `::` can be used in a pattern to indicate a list with different parts. This particular function can be rewritten as shown here.

```scala
def operateOnList2(lst:List[Int], func:(Int,Int)=>Int):Int = lst match {
  case h::Nil => h
  case h::t => func(h, operateOnList2(t, func))
  case _ => 0
}
```

You might wonder about the last case. This is not required, but if we leave it out we will get a warning telling us that the match is not exhaustive. This is not just the compiler being overly picky either. It turns out that the original method that uses an `if` expression is not completely safe. Try calling it with an empty `List` and you will see why.

6.4   Bigger `Arrays` and `Lists` with **Fill** and **Tabulate**

The simple syntax for creating `Arrays` and `Lists` work very well for short sequences. In the case of `Lists`, you can build a large `List` using `::` and a recursive function, but there are even easier ways to do so with the `fill` and **tabulate** methods. To use the `fill` method you simply specify how many elements you want, and the object value you want to fill each element with. Simple invocations of `fill` might look like the following.

```scala
scala> Array.fill(10)(4)
res24: Array[Int] = Array(4, 4, 4, 4, 4, 4, 4, 4, 4, 4)

scala> List.fill(6)(23)
res25: List[Int] = List(23, 23, 23, 23, 23, 23)

scala> List.fill(6)("hi")
res26: List[String] = List(hi, hi, hi, hi, hi, hi)
```

`Array` and `List` here refer to what are called "companion objects", and like all objects, they can have methods. You can see the various methods using tab completion in the REPL. You should notice something odd about the way `fill` is called here. It takes two arguments, but instead of being passed in one parameter list with a comma between them, they are passed in two separate parameter lists. This is called CURRYING. We will discuss currying in full in section 7.8. The first argument is how many values we want in the resulting sequence, and the second is code that gives the values.

The other thing to note about `fill` is that the second argument is passed in a special way called "pass-by-name". This is a very powerful feature of Scala that will be covered in detail is section 7.9. For now, we will simply show why it matters for `fill`. Consider the following two examples.

```scala
scala> List.fill(5)(math.random)
res27: List[Double] = List(0.3426736320227921, 0.8128658516345523,
    0.6149061001393661, 0.5380617420048777, 0.7075727808692085)

scala> Array.fill(3)(readInt)
res28: Array[Int] = Array(1, 2, 3)
```

The behavior of these two examples might seem obvious. As you can guess, in the second example the user entered 1, 2, and 3 as values. This is what we started off the chapter doing functions to read values into an `Array` or `List`. It turns out that as long as we know how many elements we want, we can do this with a single line of code using `fill`.

To see how pass-by-name is different here, consider the following function that is intended to fill a `List` similar to the first example.

```
def fillList(n:Int,x:Double):List[Double] = if (n<1) Nil else x :: fillList(n-1,x)
```

We call this function passing it how many numbers we want in the `List`, and the value we want to use. Here is what happens if we call it with 5 and `math.random`.

```
scala> fillList(5,math.random)
res29: List[Double] = List(0.5701406699076761, 0.5701406699076761,
    0.5701406699076761, 0.5701406699076761, 0.57014066990076761)
```

Unlike the call to `fill`, all the values in this `List` are the same. If we had called it with `readDouble`, only one value would have been read in, and all the elements would have that one value. Pass-by-name is how `fill` is able to get different values in those examples.

Another way to see the impact of pass-by-name is to have the second argument print something.

```
scala> Array.fill(4){ println("Evaluating argument"); 5 }
Evaluating argument
Evaluating argument
Evaluating argument
Evaluating argument
res30: Array[Int] = Array(5, 5, 5, 5)
```

Note that here the second argument has been put in curly braces only. As a general rule, Scala allows you to do that for an argument list that only has one argument. This is part of the reason `fill` is curried. You can add an extra set of parentheses, but you will still need the curly braces because this is a block of code with two expressions in it. This example is illustrative, but not really all that useful. A more interesting example involves using a `var`.

```
scala> var i=1
i: Int = 1

scala> List.fill(5){ i*=2; i }
res31: List[Int] = List(2, 4, 8, 16, 32)
```

Here the `var` i is initialized to be 1. The value argument multiplies i by 2 and stores that value back in i. Then it gives back i as the value to use. You can see in the output that this gives us powers of 2. While there are situations where this can be helpful, this method of filling an array with powers of two is likely to confuse most people reading your code.

The primary limitation with `fill` is that the code in it generally does not know what element of the sequence is being generated. Sometimes that information can be helpful. That is why the **tabulate** method exists. The **tabulate** method creates a new list or array filling each element with its index that has a function that you supply applied to it. You call `tabulate` much like you do `fill`, only the second argument is not a pass-by-name variable, it is a function that takes an `Int` and results in the type that the sequence should be filled with. A simple example is a `List` that is filled with elements that are the squares of their indices. That can be produced by the following code.

```
scala> List.tabulate(5)(i => i*i)
```

```
res32: List[Int] = List(0, 1, 4, 9, 16)
```

A more complex example would fill an `Array` with the result of evaluating a polynomial on each index.

```
scala> Array.tabulate(6)(x => 3*x*x+5*x-7)
res33: Array[Int] = Array(-7, 1, 15, 35, 61, 93)
```

This fills the array with values of $3x^2 + 5x - 7$ where x is the index in the array: 0, 1, 2, 3, 4, and 5.

Note: You can also make larger `Arrays` by calling `new`. If you have any experience with other languages, you might be familiar with this approach. It is generally not the recommended approach in Scala, as it is prone to cause certain types of exceptions.

6.5 Standard Methods

Now you know how to make `Arrays` and `Lists`. In order to do things with them, you need to know what methods are available on them. One of the strengths of the Scala collections is that they have rich interfaces. An interface is a set of protocols, routines, and tools for building software applications. Scala's interfaces have a lot of different methods in them. We looked at `length` and `size` on the `Array` and `head` and `tail` on the `List`, but this was only scratching the surface. You can actually call either of those on either `Lists` or `Arrays`. However, `length` and `size` are not that efficient for `Lists` while `tail` is inefficient on the `Array`. In this section we will run through a sampling of the other methods that are available to us when working with `Lists` and `Arrays`. We will start with the simple ones.

6.5.1 Basic Methods

We break the methods into a few groups based on what they do. Inside of each group the methods are in alphabetical order. The methods that say they give you a new collection result in a collection of the same type that it is called on. So if you call them on an `Array` you will get back an `Array`. If you call them on a `List` you will get back a `List`. Short examples are shown for each using the `lst` variable defined above at `val lst = List(7,4,6,3,9,1)`. The type `Seq` appears occasionally. You can think of this as an `Array` or a `List`.

- Methods that give you part of a collection

  - `drop(n:Int)` – Takes an `Int` and gives you back a new collection where the given number of elements have been removed from the beginning.

    ```
    lst.drop(2)
    res34: List[Int] = List(6, 3, 9, 1)
    ```

  - `init` – Takes no arguments and produces a new collection with all the elements except the last.

    ```
    scala> lst.init
    res35: List[Int] = List(7, 4, 6, 3, 9)
    ```

– `last` – Takes no arguments and produces the last element in the collection.

```
scala> lst.last
res36: Int = 1
```

– `slice(from:Int, until:Int)` – Takes two arguments which are both integer indexes. It produces a new collection with all the elements beginning with the index of the first argument and ending with the one before the index of the second value.

```
scala> lst.slice(2,4)
res37: List[Int] = List(6, 3)
```

– `splitAt(n:Int)` – Takes an `Int` for the index of a location to split the collection at. It produces a tuple of two new collections where the first has the first `n` elements and the second has the rest.

```
scala> lst.splitAt(3)
res38: (List[Int], List[Int]) = (List(7, 4, 6),List(3, 9, 1))
```

– `take(n:Int)` – Takes an `Int` and gives back a new collection with that many elements from the beginning of this collection.

```
scala> lst.take(3)
res39: List[Int] = List(7, 4, 6)
```

– `takeRight(n:Int)` – Like `take`, but pulls the last `n` elements.

```
scala> lst.takeRight(3)
res40: List[Int] = List(3, 9, 1)
```

- Boolean tests

  – `contains(elem:Any)` – Takes an element and gives the result of whether or not the collection contains an element equal to it.

  ```
  scala> lst.contains(8)
  res41: Boolean = false

  scala> lst.contains(3)
  res42: Boolean = true
  ```

  – `endsWith(that:Seq[B])` – Takes a collection of elements and tells whether the current collection ends with elements equal to those in the collection passed in.

  ```
  scala> lst.endsWith(List(3,9,1))
  res43: Boolean = true

  scala> lst.endsWith(List(3,8,1))
  res44: Boolean = false
  ```

  – `isEmpty` – Tells whether or not the collection is empty.

  ```
  scala> lst.isEmpty
  res45: Boolean = false
  ```

– nonEmpty – The opposite of isEmpty.

```scala
scala> lst.nonEmpty
res46: Boolean = true
```

– startsWith(that:Seq[B]) – Takes a collection of elements and tells whether the current collection starts with elements equal to those in the collection passed in.

```scala
scala> lst.startsWith(List(7,5,6))
res47: Boolean = false

scala> lst.startsWith(List(7,4,6))
res48: Boolean = true
```

- Search for something

– indexOf(elem:A) – Takes an element and returns the index of the first element in the collection equal to the value passed in. Gives back -1 if no matching element is found.

```scala
scala> lst.indexOf(3)
res49: Int = 3

scala> lst.indexOf(8)
res50: Int = -1
```

– lastIndexOf(elem:A) – Takes an element and returns the index of the last element in the collection equal to the value passed in. Gives back -1 if no matching element is found.

```scala
scala> lst.lastIndexOf(4)
res51: Int = 1
```

- Other simple methods of note

– diff(that:Seq[A]) – Takes an argument that is a sequence of the same type as what this is called on and produces the multiset difference between the two. This means that it will give you back all the elements that were in the original collection that do not have a match in the argument collection.

```scala
scala> lst.diff(List(1,2,3,4))
res52: List[Int] = List(7, 6, 9)

scala> lst.diff(Array(4,5,6))
res53: List[Int] = List(7, 3, 9, 1)
```

– distinct – Takes no arguments and produces a new collection that only contains the unique members of this collection, so all duplicates are removed from the new collection.

```scala
scala> List(1,2,3,8,5,1,2,8,2).distinct
res54: List[Int] = List(1, 2, 3, 8, 5)
```

– `mkString` – Can be called with zero, one, or three arguments. It builds a single long string from the string representations of the elements. If no argument is provided then nothing is put between the strings for the elements. If one argument is specified, it should be a string that is used to separate the element strings. If three arguments are specified the middle is a separator and the first and last are strings to put before and after the elements.

```scala
scala> lst.mkString
res55: String = 746391

scala> lst.mkString("; ")
res56: String = 7; 4; 6; 3; 9; 1

scala> lst.mkString("[", ", ", "]")
res57: String = [7, 4, 6, 3, 9, 1]
```

– `patch` – This powerful method allows you to produce new sequences where elements have been removed from and inserted into the original sequence. It takes three arguments, an index to start patching at, the elements to put in the patch, and the number of elements to replace with the patch. If the replacement is an empty collection, it simply removes the specified number of elements. If the elements to remove is given as zero, it will only insert the new elements at the given location.

```scala
scala> lst.patch(3,Nil,2)
res58: List[Int] = List(7, 4, 6, 1)

scala> lst.patch(3,List(10,11,12),0)
res59: List[Int] = List(7, 4, 6, 10, 11, 12, 3, 9, 1)

scala> lst.patch(3,List(10,11,12),3)
res60: List[Int] = List(7, 4, 6, 10, 11, 12)
```

– `reverse` – Takes no arguments and produces a new collection with the elements in the reverse order.

```scala
scala> lst.reverse
res61: List[Int] = List(1, 9, 3, 6, 4, 7)
```

– `toArray, toList` – Take no arguments and makes a new collection of the type specified with the elements in the current collection.

```scala
scala> lst.toArray
res62: Array[Int] = Array(7, 4, 6, 3, 9, 1)
```

– `zip(that:Iterable[B])` – Takes another collection as an argument and produces a collection of tuples where the first element comes from the collection this is called on and the second comes from the collection passed in. The length of the result is the shorter of the two.

```scala
scala> lst.zip(lst.reverse)
res63: List[(Int, Int)] = List((7,1), (4,9), (6,3), (3,6), (9,4), (1,7))

scala> lst.zip(Array(1, 2, 3, 4))
res64: List[(Int, Int)] = List((7,1), (4,2), (6,3), (3,4))
```

– `zipWithIndex` – Takes not arguments and produces a new collection of tuples where the first is an element from the collection and the second is its index.

```scala
scala> lst.zipWithIndex
res65: List[(Int, Int)] = List((7,0), (4,1), (6,2), (3,3), (9,4), (1,5))
```

The methods listed above will work on any type of sequence. So they will work on a `List[Int]`, a `List[String]`, an `Array[Double]`, or a `List[Array[Double]]`. There are a few methods provided that have some special requirements for the type of things in the list. They require that certain operations be defined. These methods, which are self-explanatory, are `min`, `max`, `sum`, and `product`. The `min` and `max` methods will work for types that can be ordered. That includes not just things like `Int` and `Double`, but also `Strings` and many other types where an ordering makes sense. The `sum` and `product` methods require that the type of the collection be numeric.

While we wrote `operateOnList` and `operateOnArray` to do sums and products of those collections, in Scala we can simply call the sum or product methods as is seen here.

```scala
scala> lst.sum
res66: Int = 30

scala> lst.product
res67: Int = 4536
```

The requirement that the values be numeric means that while you can concatenate `Strings` with +, you cannot put them together with `sum`. For a `List[String]` or `Array[String]`, you should use `mkString` to concatenate the values.

6.5.2 Higher-Order Methods

While you might feel like the list of methods shown here is rather long and gives us many capabilities, we have not yet hit on the real power of the Scala collections. All of these methods have taken normal values for arguments. Just like our first recursive methods, they can be made more powerful by adding some abstraction and making them higher order methods. Below is a list of many of the higher order methods that are part of the sequences in Scala. The type `A` is the type that is contained in the `List` or `Array`. The type `B` could be any other type. Once again, these examples made use of the declaration `val lst = List(7,4,6,3,9,1)`.

- `count(p:(A)=>Boolean)` – Takes a function that will operate on an element and result in a `Boolean`. Returns the number of elements in the collection for which this returns `true`.

```scala
scala> lst.count(_ > 5)
res68: Int = 3
```

- `dropWhile(p:(A)=>Boolean)` – Takes a function that will operate on an element and resuts in a `Boolean`. Produces a new collection that contains all elements starting with the first one for which that function is false.

```scala
scala> lst.dropWhile(_ > 3)
res69: List[Int] = List(3, 9, 1)
```

- `exists(p:(A)=>Boolean` – Takes a function that will operate on an element and results in a `Boolean`. It has a value of `true` if there is some element in the collection for which the function is `true`.

```
scala> lst.exists(x => x>4 && x<7)
res70: Boolean = true

scala> lst.exists(_ % 5 ==0) // Check if any elements are divisible by 5.
res71: Boolean = false
```

- `filter(p:(A)=>Boolean)` – Takes a function that will operate on an element and result in a `Boolean`. Produces a new collection that contains only the elements for which the function is `true`.

```
scala> lst.filter(_ < 5)
res72: List[Int] = List(4, 3, 1)

scala> lst.filter(x => x>=4 && x<=7)
res73: List[Int] = List(7, 4, 6)

scala> lst.filter(_ % 2 == 0)
res74: List[Int] = List(4, 6)
```

- `filterNot(p:(A)=>Boolean)` – Takes a function that will operate on an element and result in a `Boolean`. Produces a new collection that contains only the elements for which the function is `false`.

```
scala> lst.filterNot(_ < 5)
res75: List[Int] = List(7, 6, 9)

scala> lst.filterNot(x => x>=4 && x<=7)
res76: List[Int] = List(3, 9, 1)

scala> lst.filterNot(_ % 2 == 0)
res77: List[Int] = List(7, 3, 9, 1)
```

- `flatMap(f:(A)=>Seq[B])`[5] – Takes a function that will operate on an element and return a collection. Returns a new collection built from all the result collections appended together.

```
scala> lst.flatMap(n => if (n<6) lst.take(n) else Nil)
res78: List[Int] = List(7, 4, 6, 3, 7, 4, 6, 7)

scala> val str = "A test String."
scala> lst.flatMap(str.take)
res79: List[Char] = List(A, , t, e, s, t, , A, , t, e, A, , t, e, s, t, A, ,
    t, A, , t, e, s, t, , S, t, A)
```

- `forall(p:(A)=>Boolean)` – Takes a function that will operate on an element and result in a `Boolean`. Its value is `true` if the function is `true` for all elements of the collection.

---

[5]The result type of the function `f` that is passed into `flatMap` is technically a `GenTraversableOnce[B]`. This is more general than a `Seq[B]`, but for now the difference is not important.

```
scala> lst.forall(_ > 2)
res80: Boolean = false

scala> lst.forall(_ < 10)
res81: Boolean = true
```

- `foreach(f:(A)=>Unit)` – Takes a function that operates on an element and applies it to all elements in the collection. The result type is `Unit`. This method is called only for the side effects.

```
scala> lst.foreach(n=>println(2*n))
14
8
12
6
18
2
```

- `indexWhere(p:(A)=>Boolean)` – Takes a function that will operate on an element and result in a `Boolean`. It has the value of the index of the first element for which the function is `true`.

```
scala> lst.indexWhere(_ % 2 == 0)
res82: Int = 1

scala> lst.indexWhere(_ > 7)
res83: Int = 4
```

- `lastIndexWhere(p:(A)=>Boolean)` – Takes a function that will operate on an element and result in a `Boolean`. It gives the value of the index of the last element for which the function is `true`.

```
scala> lst.lastIndexWhere(_ % 2 == 0)
res84: Int = 2
```

- `map(f:(A)=>B)` – Takes a function that operates on an element and results in something. It produces a new collection that contains the results of applying that function to all the contents of the original collection.

```
scala> lst.map(_ + 1)
res85: List[Int] = List(8, 5, 7, 4, 10, 2)

scala> lst.map(_ * 2)
res86: List[Int] = List(14, 8, 12, 6, 18, 2)

scala> lst.map(i => i * i)
res87: List[Int] = List(49, 16, 36, 9, 81, 1)
```

- `partition(p:(A)=>Boolean)` – Takes a function that will operate on an element and result is a `Boolean`. Produces a tuple with two new collections. The first contains only the elements for which the function is true and the second is the rest. This is like doing `filter` and `filterNot` at once.

```
scala> lst.partition(_ < 5)
res88: (List[Int], List[Int]) = (List(4, 3, 1),List(7, 6, 9))

scala> lst.partition(_ % 2 == 0)
res89: (List[Int], List[Int]) = (List(4, 6),List(7, 3, 9, 1))
```

- takeWhile(p:(A)=>Boolean) – Takes a function that will operate on an element and result in a Boolean. Produces a new collection that contains all elements all the elements at the beginning for which that function is true.

```
scala> lst.takeWhile(_ > 3)
res90: List[Int] = List(7, 4, 6)
```

While the first set of methods was straightforward, this group could use a bit more explanation. We will focus on map and filter because they are very standard in the functional programming world. Imagine we have a list or an array of people's names. This list might be very long, but for our sample code we will use just a few so you can see the illustration. The names are in the normal format that people write them: "first middle last" with the middle name being optional or potentially having multiple middle names. For programming purposes we would like them to be in the format "last, first middle" because we often care more about the last name. What we have described here is the application of a function to every element of a sequence to get back a new sequence with the modified elements. That is exactly what the map method does for us. All we have to do is figure out how we would write a function to do that transformation, and then we can use it with map. The transformation we want to do is to take the last word in the name and move it to the front with a comma and a space between it and the first name. Basically, we want to split up the string on the last space. The methods listed above could do this nicely, if only they worked on a String. As it turns out, they do. We can treat a String as a sequence of characters. So we can use lastIndexOf to find the last space and splitAt to make two Strings. We can see this on a particular string here.

```
scala> val name="Mark C. Lewis"
name: String = Mark C. Lewis

scala> name.splitAt(name.lastIndexOf(' ')+1)
res91: (String, String) = (Mark C. ,Lewis)
```

Now all we need to do is put that back together in the opposite order with a comma and a space.

If we were going to be reorganizing names like this frequently, we could put this code in a stand alone function. If not, we can use a function literal to define it on the fly. For this example, we will use the latter approach.

```
scala> val names=List("Mark C. Lewis", "Lisa L. Lacher", "Jason C. Hughes",
     |     "Glen R. Stewart","Jen Hogan")
names: List[String] = List(Mark C. Lewis, Lisa L. Lacher, Jason C. Hughes, Glen R.
    Stewart, Jen Hogan)

scala> val lnames=names.map(n => {
     |     val (rest,last)=n.splitAt(n.lastIndexOf(' ')+1); last+", "+rest.trim } )
lnames: List[String] = List(Lewis, Mark C., Lacher, Lisa L., Hughes, Jason C.,
    Stewart, Glen R., Hogan, Jen)
```

So the use of `map` allowed us to complete a task in a quick and short way that could have taken us a fairly large amount of code using other approaches.

If you had a long list of these names, you might want to do something like find all the people who have last names beginning with a particular letter. For example, we might want to find everyone whose last names starts with an 'H'. For this we would use the `filter` function. The `filter` function will select out values that satisfy some condition and give us a sequence with only those values.

```scala
scala> lnames.filter(_.startsWith("H"))
res92: List[String] = List(Hughes, Jason C., Hogan, Jen)
```

Thanks to `filter` and the `startsWith` method, this is a very simple one liner.

6.5.3   reduce and fold

There are some other methods that were not listed above because they take a bit more explanation. They should not be too hard to understand if you have followed up to this point because they do exactly what we did ourselves earlier in this chapter with the `operateOnArray` and `operateOnList` functions. We will start with the `reduce` methods, `reduceLeft` and `reduceRight`. These methods take a function that operates on two elements of the sequence and results in a value of that type. The `reduce` methods repeatedly apply the function to successive elements the same way the `operateOnArray` and `operateOnList` methods did. The difference between the two is whether they apply the operations moving from left to right of from right to left. If the operation is associative, like addition or multiplication, then the order does not impact the result, only potentially the efficiency.[6] For non-associative operations it can matter. Here are a few examples using the lst variable that we defined earlier which was `val lst = List(7,4,6,3,9,1)`.[7]

```scala
scala> lst.reduceLeft(_ + _) // Means 7+4+6+3+9+1
res93: Int = 30

scala> lst.reduceLeft(_ * _) // Means 7*4*6*3*9*1
res94: Int = 4536

scala> lst.reduceLeft(_ - _) // Means 7-4-6-3-9-1
res95: Int = -16

scala> lst.reduceRight(_ - _) // Means 7-(4-(6-(3-(9-1))))
res96: Int = 14
```

The first two calls do what we have seen before taking the sum and product of the list. The last two use a difference, which is not associative, and show how the results of `reduceLeft` and `reduceRight` are different. The `reduceLeft` method gives us $((((7-4)-6)-3)-9)-1$. The `reduceRight` method gives us $7-(4-(6-(3-(9-1))))$.

In the last chapter, we made a recursive method that could do these types of operations with user input. That differed from what we did in this chapter in that we had to pass in an extra value to use when the user terminated the recursion. We did not do anything special with that, but it could have opened other possibilities. The `reduce` operations have to operate on the element type of the collection for both arguments because the first invocation of the function uses two elements from the collection. If we specify the base value, the

---

[6]The `reduceLeft` method is much more efficient on a `List` than `reduceRight` because of the way that `List`s are stored in the memory of the computer.

[7]There is also a general `reduce` method that should only be called using associative functions.

function can take one argument of a different type than the elements in the sequence as long as it has a matching result type. If we say that the sequence has type A, and we want a function that will produce type B, then the function has the form (B,A) => B, and we have to provide a first value of type B. We can run through the sequence applying this function to each of the A values until we get a final B at the end.

This functionality is provided by the `foldLeft` and `foldRight` methods.[8] These methods use currying, which was introduced back in section 6.4 and will be described in detail in section 7.8. The first argument list takes the base value to use on the first application. The second argument list is the function to apply. The types A and B do not have to be different so we can use `foldLeft` to sum up a list of integers like this.

```scala
scala> lst.foldLeft(0)(_ + _)
res97: Int = 30
```

However, unlike the `reduce` methods, we can do other things like count up the total number of characters in a sequence of `Strings`. A `reduce` method on a sequence of `Strings` could only give us back a `String`, but the `fold` methods can give us back another type. Here we do exactly that. We show both the longer function literal syntax as well as the shorter version.

```scala
scala> val wordList=List("How","many","characters","do","we","have")
wordList: List[String] = List(How, many, characters, do, we, have)

scala> wordList.foldLeft(0)((count,word) => count + word.length)
res98: Int = 25

scala> wordList.foldLeft(0)(_ + _.length)
res99: Int = 25
```

6.5.4 Combinatorial/Iterator Methods

There are some other methods on sequences that, for reasons of efficiency and memory limitations, work a bit differently. These methods give us an object that provides access to multiple different collections that are created from the original collection. The object that they give us is called an `Iterator`, and specifically they give us an `Iterator[List[A]]` or an `Iterator[Array[A]]`. The `Iterator` type is more primitive than the `List` or `Array`. You can only go through it once because it is consumed as you go through it. The reason for using an `Iterator` on these methods is generally for performance and memory benefits. The fact that the `Iterator` consumes things as it goes means that it does not have to store all the contents at once. In fact, only one needs to exist at any given time. When it moves from one to the next, it can forget the last and make the next one.

The methods in this category are listed here. To show the values of the outputs a call to `foreach(println)` is used. Without this, all that would be shown is `Iterator[List[Int]]` = non-empty iterator.

- `combinations(n:int)` – Generates all combinations of the elements of this sequence of length n.

  ```scala
  scala> lst.combinations(3).foreach(println)
  List(7, 4, 6)
  ```

---

[8]The `foldRight` method needs a function of type (A,B) => B. There is also a basic `fold` method, but because the order it runs through the collection is not specified, it cannot work with any type B.

```
List(7, 4, 3)
List(7, 4, 9)
List(7, 4, 1)
List(7, 6, 3)
List(7, 6, 9)
List(7, 6, 1)
List(7, 3, 9)
List(7, 3, 1)
List(7, 9, 1)
List(4, 6, 3)
List(4, 6, 9)
List(4, 6, 1)
List(4, 3, 9)
List(4, 3, 1)
List(4, 9, 1)
List(6, 3, 9)
List(6, 3, 1)
List(6, 9, 1)
List(3, 9, 1)
```

- `grouped(size:Int)` – Runs through the sequence, grouping items into groups of the specified `size`.

```
scala> lst.grouped(2).foreach(println)
List(7, 4)
List(6, 3)
List(9, 1)

scala> lst.grouped(3).foreach(println)
List(7, 4, 6)
List(3, 9, 1)
```

- `inits` – Provides an iterator going from the full sequence to an empty one, removing elements from the end. As the name implies, it is related to the `init` method, which returns a single sequence without the last element.

```
scala> lst.inits.foreach(println)
List(7, 4, 6, 3, 9, 1)
List(7, 4, 6, 3, 9)
List(7, 4, 6, 3)
List(7, 4, 6)
List(7, 4)
List(7)
List()
```

- `permutations` – Lets you run through all the permutations of the sequence. Note that the example here only does permutations of the first three elements as there are $6! = 720$ different ones for the full list. This also demonstrates why this method return and `Iterator`. As was shown back in chapter 4, the factorial function grows very quickly, so calling `permutations` on even a modest sized sequence would consume more memory than any computer has if all the permutations were actually created at once.

```
scala> lst.take(3).permutations.foreach(println)
```

```
List(7, 4, 6)
List(7, 6, 4)
List(4, 7, 6)
List(4, 6, 7)
List(6, 7, 4)
List(6, 4, 7)
```

- `sliding(size:Int)` – Provides an iterator that gives the effect of sliding a window of a certain `size` across the sequence.

```
scala> lst.sliding(2).foreach(println)
List(7, 4)
List(4, 6)
List(6, 3)
List(3, 9)
List(9, 1)

scala> lst.sliding(3).foreach(println)
List(7, 4, 6)
List(4, 6, 3)
List(6, 3, 9)
List(3, 9, 1)
```

- `tails` – Gives an iterator that runs through subsequences starting with the full sequence and ending with an empty one removing one element from the left end each step. This is basically the reverse of `inits` and is closely related to the `tail` method.

```
scala> lst.tails.foreach(println)
List(7, 4, 6, 3, 9, 1)
List(4, 6, 3, 9, 1)
List(6, 3, 9, 1)
List(3, 9, 1)
List(9, 1)
List(1)
List()
```

If you want to do something more involved with the values given by the `Iterator`, you can convert it to a `List` or an `Array` with `toList` or `toArray`. However, if there are many elements, as is the case for `lst.permutations` you might want to filter it or do something else to narrow things down before doing that conversion. For example, we could filter down the permutations of `lst` so that we only get those which start with the three largest value. That cuts out the majority of the results, so it can be safely converted from an `Iterator` to a `List`.

```
scala> lst.permutations.filter(_.take(3).sum > 20).toList
res100: List[List[Int]] = List(List(7, 6, 9, 4, 3, 1), List(7, 6, 9, 4, 1, 3),
    List(7, 6, 9, 3, 4, 1), List(7, 6, 9, 3, 1, 4), List(7, 6, 9, 1, 4, 3),
    List(7, 6, 9, 1, 3, 4), List(7, 9, 6, 4, 3, 1), List(7, 9, 6, 4, 1, 3),
    List(7, 9, 6, 3, 4, 1), List(7, 9, 6, 3, 1, 4), List(7, 9, 6, 1, 4, 3),
    List(7, 9, 6, 1, 3, 4), List(6, 7, 9, 4, 3, 1), List(6, 7, 9, 4, 1, 3),
    List(6, 7, 9, 3, 4, 1), List(6, 7, 9, 3, 1, 4), List(6, 7, 9, 1, 4, 3),
    List(6, 7, 9, 1, 3, 4), List(6, 9, 7, 4, 3, 1), List(6, 9, 7, 4, 1, 3),
    List(6, 9, 7, 3, 4, 1), List(6, 9, 7, 3, 1, 4), List(6, 9, 7, 1, 4, 3),
    List(6, 9, 7, 1, 3, 4), List(9, 7, 6, 4, 3, 1), List(9, 7, 6, 4, 1, 3),
```

```
List(9, 7, 6, 3, 4, 1), List(9, 7, 6, 3, 1, 4), List(9, 7, 6, 1, 4, 3),
List(9, 7, 6, 1, 3, 4), List(9, 6, 7, 4, 3, 1), List(9, 6, 7, 4, 1, 3),...
```

These methods provide an easy way to do some more interesting work with data. For example, using `sliding`, you can quickly smooth noisy data by averaging a data series over a window as the following does with random numbers. Note how the output never goes below 0.4 or above 0.82, while the original random data inevitably had values much closer to 0.0 and 1.0.

```
scala> Array.fill(20)(math.random).sliding(5).map(_.sum/5).toArray
res101: Array[Double] = Array(0.40686956898234017, 0.42010662219123807,
    0.460068215933166, 0.5978974398508441, 0.6977117936915517, 0.711388784520442,
    0.815578490945466, 0.7546589297206381, 0.6257741024399198, 0.6461792586151993,
    0.6723958342393901, 0.6599362542338169, 0.7182061307254338,
    0.7920396281126222, 0.7454312388627433, 0.6635182202052169)
```

Using `combinations` or `permutations`, you could run through all possible options of some type and find the one that was the best in some way. For example, if I wanted to find subgroups of numbers that have a small sum, but a large product, I could do something like the following, which finds any groups of 3 numbers from `lst` that have a sum less then 15, but a product greater than 30.

```
scala> lst.combinations(3).filter(s => s.sum<15 && s.product>30).foreach(println)
List(7, 4, 3)
List(7, 6, 1)
List(4, 6, 3)
List(4, 9, 1)
```

6.6   Complete Grades Script/Software Development

In chapter 4 we spent some time writing functions to calculate averages in a class. We now have the ability to put together everything that we have learned and write a script that will keep track of the grades for a single student and be able to calculate that student's average. This could be extended to multiple students, but we will not do that here.

When we solve a large problem, there are certain steps that must be taken. The first is to get a clear description of what problem you are solving. This step is called ANALYSIS. The analysis step does not involve any coding or even thinking about coding. We just want a description that has enough detail that we know what it is we are supposed to create. Once we have that, we can move on to the next step, which is called DESIGN. In the design step we figure out how we are going to go about solving the problem described in the analysis. The design step does not involve coding either, but it specifies what we will do when we get to the coding. The design phase is where we do our problem decomposition. We think about how to break the problem down and what pieces we will need to put together to get a single, whole solution to the problem. After we have a design, we get to the IMPLEMENTATION phase.[9] This is where we will actually write the program. Ideally, by the time that we get to implementation we have worked out most of the problems we will run into. The ability to see the problems we will encounter requires a lot of experience. As novice programmers,

---

[9]These phases are a part of the Software Development LifeCycle (SDLC).

you should try to spend time on design, thinking about how to break the problem down, but do not become so intent on having a complete design that you stall out and never get to implementation.[10]

After you have an implementation you have to make sure it works and fix any problems. This is called the TESTING and DEBUGGING phase because it is where you test the code to find all the bugs and fix them. For real software, after all the bugs are fixed and the software has been rolled out, or deployed, you worry about MAINTENANCE. This is where you continue to make alterations to the software to improve it or address customer requests.

At the level you are at now, you typically do not do much analysis or maintenance. You do not do much analysis because your instructor or this book will give you a fairly complete description of what you are doing. You do not do maintenance because you do not have a customer base that is using the software. A little of each of these can still creep into your work when you make sure you fully understand the description of a problem and then perhaps if you are given a later problem that incorporates code you had done previously.

You should spend a significant amount of time working on design and debugging. In fact, the implementation phase is often the shortest phase of the software creation process.[11] If done properly, most of the thinking goes into the design phase so implementation is simply typing in what you had designed. Testing and debugging is extremely variable in time. It is possible to write your program and have everything work perfectly the first time. That gets less likely as programs get longer and is especially rare for novice programmers, though the static type checking in Scala will help to make it so that if everything in your program checks out it has a better chance of working. When things do not work, debugging can take arbitrarily long periods of time and is often the longest phase of software development.

In real software development, these different phases are not always so distinct, and they are rarely done in sequential order. It is common that in one phase of development you realize something that was missed in an earlier phase. Some approaches to software development mix the phases together so that the developer is involved in more than one at any given time.

At this point we have a very vague description of the problem that we are going to solve. For that reason, we need to do some analysis up front so that we have a clear description of exactly what we are going to solve. This script is going to use a text menu to give the user options. The options will include the ability to add grades of different types as well as an option to print the current grades and average. Types of grades are tests, quizzes, and assignments. For the final average, the tests and assignments count 40% each and the quizzes make up the remaining 20%. The lowest quiz grade should be dropped. The menu will have numbers by each option and users enter a number to select that option. That is probably a sufficient analysis for our needs here. We can start working on the design and if we hit anything we have questions on, we can go back and add some detail to the analysis before we proceed further with the design.

So how are we going to break this problem down? At the top level we have code that repeats over and over and each time a menu is printed, the user selects an option, and we execute code to perform the selected option. The only way we currently know how to do something repeatedly an unspecified number of times is with a recursive function. So we will need a recursive function that contains all this other functionality. That function will call itself until we hit the condition where it exits. There is a natural decomposition at this

---

[10]This type of behavior has been referred to as "paralysis by analysis", though it is more likely to happen during the design phase than the analysis phase.

[11]At least if you do things properly the implementation will be short. I like to tell my students that hours of coding can save minutes of thinking. The point of this is that if you do not sit and think about design before you go to the keyboard and write code you are likely to spend hours hashing out in code the design that you could have nailed down in a few minutes by thinking about it to begin with.

point. Printing the menu is an obvious function. We have already seen that we can write functions for taking straight averages or averages dropping the minimum grade as well as one to calculate the final average.

That breaks down the functionality. If you were to run to the keyboard now and start typing you would quickly realize that something significant was left out of the design: how we are storing the grades. There are many ways that we could do this, but given that our only functionality is to add grades and calculate averages, the solution is fairly clear. Our only real options at this point are `Arrays` and `Lists`. `Arrays` have a fixed size and are inefficient to add to. That is less than ideal for this application because we will be adding grades and have not specified how many there can be. So `Lists` make more sense. We can use three `Lists` that store the three different types of grades. If we do this in an imperative style, the `Lists` will be declared with `var` before the main recursive function. If we use a functional style, they will be passed into the next call for the iteration. We will look at both to illustrate the differences. Most of the functions will not be altered by this distinction.

That gives us a decent design for this little application. Now we turn to the implementation. Here is a listing of the complete script using an imperative approach with `var` declarations for the `Lists`.

Listing 6.1: GradesImperative.scala

```scala
import io.StdIn._

def average(nums:List[Double]):Double = nums.sum/nums.length

def averageDropMin(nums:List[Double]):Double = (nums.sum-nums.min)/(nums.length-1)

def fullAve(tests:Double, assignments:Double, quizzes:Double):Double = 0.4*tests +
    0.4*assignments + 0.2*quizzes

def courseAverage(tests:List[Double],assns:List[Double],
    quizzes:List[Double]):Double = {
  val aveTest=average(tests)
  val aveAssn=average(assns)
  val aveQuiz=averageDropMin(quizzes)
  fullAve(aveTest,aveAssn,aveQuiz)
}

def printMenu:Unit = {
    println("""Select one of the following options:
1. Add a test grade.
2. Add a quiz grade.
3. Add an assignment grade.
4. Calculate average.
5. Quit.""")
}

var tests = List[Double]()
var quizzes = List[Double]()
var assignments = List[Double]()

def mainGrades:Unit = {
  printMenu
  readInt() match {
    case 1 =>
      println("Enter a test grade.")
```

```
      tests = readDouble() :: tests
    case 2 =>
      println("Enter a quiz grade.")
      quizzes = readDouble() :: quizzes
    case 3 =>
      println("Enter an assignment grade.")
      assignments = readDouble() :: assignments
    case 4 =>
      println("The average is "+courseAverage(tests,assignments,quizzes))
    case 5 =>
      return
    case _ =>
      println("Unknown option. Try again.")
  }
  mainGrades
}
```

```
mainGrades
```

The calculation of the average is accomplished by the first four functions, beginning with general functions for finding the average of a `List` of numbers as well as the average with the minimum dropped. This is followed by the function that prints the menu. The only thing of interest here is the use of the triple quote string so that the string can be made to span multiple lines. If you do not do this, you will have to use `\n` to denote line breaks.

Those functions are the same in both imperative and functional solutions. In this imperative version, the `printMenu` function is followed by the declaration of three variables. They are all `vars`, and they are of the type `List[Double]`. They have to be `vars` because the `List` itself is immutable. So, whenever something is added we have to change the variable reference and point to a new `List`. Remember, this can be done efficiently for the `List` type so that is fine.

If you type this code in and play with it a bit, you will probably find that you can make it behave poorly. If you print the grades without entering a quiz grade, the code will crash because the `min` method on a `List` requires that there be at least one element to find the minimum of. If you print the grades when there are no tests or assignments, or only one quiz grade, you can get it so that when you ask for the average you get an output like this.

```
The average is NaN
```

This should lead you to two questions. First, what is this thing called NaN? Second, what is causing this? Technically, the second matters more because we need to know that to fix it. However, knowing the answer to the first can help you figure it out if you were not able to do so from your testing. The `Double` type, as we have discussed previously, is used to represent numeric values that have fractional parts. It has a rather large range and good precision, but it is technically a fixed precision type. You cannot represent any real number with it. In addition, some mathematical operations produce results that are not well defined. Consider the following.

```
scala> 1.0/0.0
res102: Double = Infinity

scala> -1.0/0.0
res103: Double = -Infinity
```

You should have learned in a math class that you cannot divide by zero. If you do this with

`Ints` you get an error. However, if you do it with `Doubles` you get what you see here. Any operation that goes outside the range of the `Double` will also get you either `Infinity` or `-Infinity`, depending on which side of the range it goes out on as is shown here.

```scala
scala> 1e150*1e200
res104: Double = Infinity
```

What does all of this have to do with `NaN`? Well, `NaN` stands for "Not-a-Number", and it is the result of an operation when it is not clear what the result should be. One way to get this is to add `Infinity` and `-Infinity` or do other operations with `Infinity` or `-Infinity` that the computer cannot know the result of. Consider this example.

```scala
scala> 1e150*1e200-1.0/0.0
res105: Double = NaN
```

In general the computer cannot know if this would be out of range on the positive or negative so it goes to `NaN` because it gives up. There is one other way to get a `NaN`: divide zero by zero.

```scala
scala> 0.0/0.0
res106: Double = NaN
```

This is what is happening to cause our error, and it happens when zero test or assignment grades or only one quiz grade have been entered.

If we have no test or assignment grade, the normal average will return $0.0/0$, as the sum is the `Double` 0.0 and the length is the `Int` 0. If the list of quiz grades only has one element, then that one is the minimum. So the function returns $(x-x)/(1-1)$ where x is the one grade as a `Double`. Clearly that is going to be a `NaN`. So how do we fix this? We need to make the `average` and the `averageDropMin` functions a bit smarter so that they have special cases in the instances when they should not do division. We can do this by putting `if` expressions in both.

```scala
def average(nums:List[Double]):Double =
  if (nums.isEmpty) 0 else nums.sum/nums.length

def averageDropMin(nums:List[Double]):Double =
  if (nums.isEmpty || nums.tail.isEmpty) 0 else
  (nums.sum-nums.min)/(nums.length-1)
```

If you use these versions of the functions things should work well.

Now to the functional alternative. The imperative version uses three `var` declarations outside of the function. These declarations can be seen by anything in the script that follows them. We typically refer to this as the GLOBAL SCOPE. For small scripts this might not be a problem, but you do not want to get into the habit of making things global, especially if it is a `var` or mutable. Anything that is global in a large program can be messed up by any line of code, making it very hard to track down certain types of errors. It is better to have our variables declared in a LOCAL SCOPE, inside of a function.

If we simply declare the variables inside of `mainGrades`, it will not work because when you make a recursive call, you get new versions of all of the local variables. Instead, we want to pass the three lists into the function as arguments. The arguments to a function act just like local `val` declarations except that the value is provided by the calling code. This new version will not include any `var` declarations, nor will it have any mutable values as everything is stored in immutable `Lists`. So other than the printing, it is completely functional. Here is what such a function might look like and how we would call it makes the script work.

```scala
def mainGrades(tests:List[Double],assignments:List[Double],
    quizzes:List[Double]):Unit = {
  printMenu
  readInt() match {
    case 1 =>
      println("Enter a test grade.")
      mainGrades(readDouble() :: tests,assignments,quizzes)
    case 2 =>
      println("Enter a quiz grade.")
      mainGrades(tests,assignments,readDouble() :: quizzes)
    case 3 =>
      println("Enter an assignment grade.")
      mainGrades(tests,readDouble() :: assignments,quizzes)
    case 4 =>
      println("The average is "+courseAverage(tests,assignments,quizzes))
      mainGrades(tests,assignments,quizzes)
    case 5 =>
    case _ =>
      println("Unknown option. Try again.")
      mainGrades(tests,assignments,quizzes)
  }
}


mainGrades(Nil, Nil, Nil)
```

Instead of mutating values, this version passes the newly built List into the recursive call where it is a completely new argument value for the next level down in the recursion. The options that do not change the Lists simply pass through the same values. The initial call uses empty lists for all three arguments.

6.7   Playing with Data

Lists of grades make a fairly simple, and easy to understand example, but they do not really highlight a lot of the power of computers and programming. The lists of grades are generally short enough that you could do the calculations by hand without too much difficulty. You are not going to have thousands of grades in a class. Where the power of the computer really stands out is when the amount of data, or the number of calculations is so large that you would never want to do things by hand. In this section, we will look at processing such a data set.

The data set that we are going to work with is a list of baby names maintained by the United States Social Security Administration. You can get these files for yourself at http://www.ssa.gov/OACT/babynames/limits.html. In particular, we will work with the state data. You should download the ZIP file, and unzip it in a directory where you want to work. Note that the ZIP file contains separate files for each state, so you really do want them to be in some separate directory.

The format of the files is fairly simple. Here are the first ten lines of the file for North Dakota.

```
ND,F,1910,Mary,85
ND,F,1910,Alice,61
ND,F,1910,Margaret,61
ND,F,1910,Helen,59
ND,F,1910,Anna,58
ND,F,1910,Florence,58
ND,F,1910,Gladys,50
ND,F,1910,Mildred,50
ND,F,1910,Myrtle,48
ND,F,1910,Ruth,47
```

You can see that each line has five values separated by commas. Those values are the state code (which should be the same for every line in a given file), a letter for gender, the year, the name, and the number of children of that gender given that name in that year. What we want to be able to do is read in the entire file and then answer some questions about the data in it. To make things more concrete, say we want to find the most common name for males and females in each year.

We know what we want to do, and we know what the data file looks like. Now we need to take the problem and break it into pieces that we can solve. One immediate way to break the problem down is to think of reading the data and finding the common names in the data as separate problems.

6.7.1 Reading the Data

The problem of reading a large data file can be viewed as reading one piece and doing it many times. The only way that we currently know how to read data like this is with `readLine`. That will read the full line as a `String`. Once we have that, we need to break it up on the commas and get the pieces. There is a handy method on `String` called `split` that can do exactly that. It takes the delimiter to split on, and returns an `Array[String]`. The following function takes a line of text and returns a tuple of (`String`, `String`, `Int`, `String`, `Int`) with the data from that line. That tuple type is going to be used a lot in this code, so we will use a `type` declaration to give it a shorter name.

```
type NameData = (String, String, Int, String, Int)

def parseLine(line:String):NameData = {
  val parts = line.split(",")
  (parts(0), parts(1), parts(2).toInt, parts(3), parts(4).toInt)
}
```

To use this function, we need to read all the lines from the file and then call that function on each line. The first part of that is nicely done with `fill`, and the second part is a perfect application of the `map` method, giving us code like the following.

```
val lines = Array.fill(43858)(readLine)
val data = lines.map(parseLine)
```

The "magic number" 43858 was the number of lines in the file for North Dakota at the time this was written. You should replace it with the proper number for whatever file you are reading in. We refer to this as a magic number because the way it appears in the code does not really tell us much about what it is for. Such things often make code hard to maintain, for that reason, it is generally advised that you create named constants instead of having such magic numbers in your code as shown here.

```
val linesInFile = 43858
```

```
val lines = Array.fill(linesInFile)(readLine)
val data = lines.map(parseLine)
```

One of the keys to effective time usage when you are writing software is to frequently run the program to make sure that what you have written is correct and the program does what you expect. You could certainly go to the command line and run your code, but you would find there are two problems. First, you have to manually enter over 40 thousand lines of text. Second, even after you do so, the program does not print anything, so it is very hard to tell if what you have written is correct. To fix the second problem, you can temporarily add the following line to the bottom of your script.

```
data.foreach(println)
```

This will print out all the data that was actually read with one record per line. It will be 43,858 lines, so you are not going to check them individually to make sure they are all correct, but you want to see that they have appropriate values in them.

To deal with the problem of needing to enter many lines of input, we are going to use input redirection. You could change the value of **linesInFile** briefly to be a smaller number of lines that you can enter by hand, but eventually we will want to read the whole file. Input redirection is discussed in Appendix A. The idea is that we want the **readLine** command to read from a file instead of the keyboard. Another way to say that is that we want the file to be the source for standard input. This is done with the less than symbol on the command line. If your program was named "CommonNames.scala", and you wanted to run this on the North Dakota names file, you could enter the following on the command line.

```
scala CommonNames.scala < ND.TXT
```

This assumes that the ND.TXT file is in the same directory as your Scala code. If it is, and you have written the code correctly, running this should flood your screen with a bunch of output ending with the records from the end of the file.

6.7.2   Finding Maximum Values

Now that we have all the data read in, we can turn to the issue of finding the most common name in each year. This also breaks down nicely into sub-problems. We need to find all the years in the date. For each year, we need to get the desired date for just that year. Given a group of data, we need to find the most common names.

We will start with the last sub-problem. There are many ways that we could break things down, and how we choose to do so impacts the code that will be written for each sub-problem. The task of finding the most common names makes a good function, we just need to decide what exactly that function should do.

Recall that our original problem statement said to find the most common names for males and females. Given this description, it is tempting to write a function that would take all the data from one year, and return both the most common male names and the most common female names.[12] The problem with that approach is that it is not very reusable. What if later we were given the task of figuring out the most common name in each year regardless of gender, or we were processing a file that only contained a single gender? Coding in the gender distinction at the lowest level in our solution gives us code that is not very flexible. It might be great for the problem at hand, but we would rather make something that is

---

[12]Note that those are both intentionally left plural because there could be ties.

more generally useful. As you might recall from chapter 4, the best functions are the ones we can use in many situations. At this point, you might not actually reuse your functions, but it is still good to keep this objective in mind so that it becomes second nature later on.

Here is the code for such a function.

```scala
def mostCommonNames(nameData:Array[NameData]):Array[NameData] = {
  val maxTimes = nameData.map(_._5).max
  nameData.filter(_._5 == maxTimes)
}
```

This function uses a call to `map` to get only the name frequencies. That is what the `_._5` is doing as the frequency column was the 5th column in the data file and the 5th element of the tuple. Then it calls `max` to get that largest frequency. The function ends by calling `filter` to get all the values that have that frequency. If we did not care about ties, and only wanted one name with the maximum frequency, this could have been done more efficiently using `reduceLeft` as shown here.

```scala
def mostCommonName(nameData:Array[NameData]):NameData = {
  nameData.reduceLeft((nd1,nd2) => if (nd1._5 > nd2._5) nd1 else nd2)
}
```

The way this is written, it will result in the last element in the `Array` that had the highest frequency. Changing the `>` to a `>=` or swapping the `nd1` and `nd2` in the comparison will result in the first of the ties instead of the last. We want to find all ties, so we will stick with the first version.

The task of finding all the years in the data file is fairly simple. You could look in the file and see what years are present, but we can easily get that information from the data itself using a single line.

```scala
val years = data.map(_._3).distinct
```

This line uses `map` to pull out all the years, then makes a call to `distinct` to remove all the duplicate entries.

Once we have the years, another line can give us the most common female or male names. Here is a line giving the most common female names.

```scala
val femaleNames = years.flatMap(year =>
  mostCommonNames(data.filter(nd => nd._2=="F" && nd._3==year)))
```

This code does a `flatMap` across all the years and for each year it calls our `mostCommonNames` function using that data that has been filtered down to only females in that year.[13] To understand why `flatMap` is used here, we need to think about the types of the various expressions in this code. The `mostCommonNames` function results in an `Array[NameData]`. Also, the value `year` is an `Array[Int]`. If we used a regular `map` instead of a `flatMap`, the results would be an `Array[Array[NameData]]`, where each subarray was for a different year. There could be times when you want this, but in this case, we do not need it, and it is a bit harder to work with. So instead, we use `flatMap` to flatten out that `Array[Array[NameData]]` to just an `Array[NameData]`.

We can add a similar line for males along with some lines to print those final results, and we get a complete script that looks like the following.

---

[13]There is a more efficient way to do this using a method called `groupBy`. We have not covered that method because it uses a type called a `Map` that will not be introduced in this volume.

Listing 6.2: CommonNames.scala

```scala
import io.StdIn._

val linesInFile = 324133 //43858

// Define shortened name for the tuple that represents each line
type NameData = (String, String, Int, String, Int)

// This function parses a line of text from the file to a NameData
def parseLine(line:String):NameData = {
  val parts = line.split(",")
  (parts(0), parts(1), parts(2).toInt, parts(3), parts(4).toInt)
}

// Read the contents of the file and parse all the lines
val lines = Array.fill(linesInFile)(readLine)
val data = lines.map(parseLine)

// Function to find the most common names in aa Array
def mostCommonNames(nameData:Array[NameData]):Array[NameData] = {
  val maxTimes = nameData.map(_._5).max
  nameData.filter(_._5 == maxTimes)
}

// Find the years the the data that has been read
val years = data.map(_._3).distinct

// Find most common female and male names
val femaleNames = years.flatMap(year =>
  mostCommonNames(data.filter(nd => nd._2=="F" && nd._3==year)))
val maleNames = years.flatMap(year =>
  mostCommonNames(data.filter(nd => nd._2=="M" && nd._3==year)))

// Output the results
femaleNames.foreach(println)
maleNames.foreach(println)
```

You can pick a particular state file and adjust the value of `linesInFile` and run it to see it work.

6.8   End of Chapter Material

6.8.1   Problem Solving Approach

Despite the length of this chapter, no new valid statement types were introduced. Instead, a significant amount of flexibility was added in what you might do with the existing statement types. In particular, you can now use expressions with collection types and call methods on those collection types. You also learned that types themselves can be more complex with type parameters.

6.8.2 Summary of Concepts

- `Array`s and `List`s are collections that allow us to store multiple values under a single name. They are sequences, meaning that the values have an order and each one is associated with an integer index. Indexes start at zero.

- Both can be created in a number of ways.

  - `Array(e1,e2,e3,...)` or `List(e1,e2,e3,...)` to make short collections where you provide all the values.

  - Use the `::` operator with `List`s to append elements to the front and make a new `List`. Use `Nil` to represent an empty `List`.

  - `Array.fill(num)(byNameExpression)` will make an `Array` of the specified size with values coming from repeated execution of the specified expression. This will also work for `List`s.

  - `Array.tabulate(num)(functionOnInt)` will make an `Array` with `num` elements with values that come from evaluating the function and passing it the index. Often function literals are used here. This will also work with `List`.

- `Array`s have a fixed size and are mutable. Access and mutate using an index in parentheses: `arr(i)`.

- `List`s are completely immutable, but you can efficiently add to the front to make new `List`s. They can be accessed with an index, but using `head` and `tail` in recursive functions is typically more efficient.

- The Scala collections, including `Array` and `List`, have lots of methods you can call on to do things.

  - There are quite a few basic methods that take normal values for arguments and give you back values, parts of collections, of locations where things occur.

  - The real power of collections is found in the higher-order methods. The primary ones you will use are `map`, `filter`, and `foreach`.

  - There are a handful of methods that give you back various pieces of collections, including combinatorial possibilities such as `permutations` and `combinations`.

6.8.3 Self-Directed Study

Enter the following statements into the REPL and see what they do. Try some variations to make sure you understand what is going on. Some of these are intended to fail. You should understand why.

```
scala> val arr = Array(1, 2, 3)
scala> arr(0)
scala> arr(1)
scala> arr(0) = 5
scala> arr.mkString(" ")
scala> val lst = List(2, 3, 5, 7, 11, 13, 17)
scala> lst.head
scala> lst.tail
scala> val lst2 = 1 :: 2 :: 3 :: 4 :: 5 :: Nil
scala> lst.zip(lst2)
```

```
scala> lst.zipWithIndex
scala> val arr3 = Array.fill(5)(readLine())
scala> val lst3 = List.tabulate(20)(i => i*i)
scala> def printList(lst:List[Int]):Unit = lst match {
  case h::t =>
    println(h)
    printList(t)
  case Nil =>
}
scala> printList(lst3)
scala> lst3.mkString(" ")
scala> lst3.map(i => math.sqrt(i))
scala> arr3.map(s => s.length)
scala> arr.permutations.foreach(println)
scala> lst.sliding(4)
scala> val numWords = lst.map(_.toString)
```

6.8.4 Exercises

1. Think of as many ways as you can to make an `Array[Int]` that has the values 1-10 in it. Write them all to test that they work. The first two should be easy. If you work you can get at least three others knowing what has been covered so far.

2. Think of as many ways as you can to make a `List[Int]` that has the values 1-10 in it. Write them all to test that they work. The first two should be easy. If you work you can get at least three others knowing what has been covered so far.

3. Make a `List[Char]` that contains 'a'-'z' without typing in all the characters. (Use `toChar` to make this work.)

4. Given two `Array[Double]` values of the same length, write a function that returns their by element sum. This is a new `Array` where each element is the sum of the values from the two input arrays at that location. So if you have `Array(1,2,3)` and `Array(4,5,6)` you will get back `Array(5,7,9)`.

5. Write a function that produces a sequence of prime numbers. You can use the `isPrime` function that you wrote for exercise 5.9. The function should take an argument for how many prime numbers need to be in the list.

6. Write a function that takes a number of values and returns the average excluding the largest and smallest values.

7. Write a program that asks the user for the total sales receipts for each day of the week and store the amounts in an array. Then calculate the total sales for the week and display the result.

8. Write a program that asks the user for the total sales receipts for each day of the week and store the amounts in a list. Then calculate the total sales for the week and display the result.

9. Write a function that takes a sequence (List or Array) of grades and returns the average with the lowest two grades dropped. Make sure this function behaves reasonably even for smaller input sets.

10. A list of baby names is maintained by the United States Social Security Administration. You can get these files for yourself at `http://www.ssa.gov/OACT/babynames/limits.html`. Download this file and select the state in which you were born. Write a script to find how many people in your state were given your name. Then find out how many males and how many females were born since the year 2000.

11. Write a script that reads in a number of integers between 1 and 100 and stores them in a list and then counts the number of times each number occurs.

12. Write several versions of code that will take an `Array[Int]` and return the number of even values in the `Array`. Each one will use a different technique. To test this on a larger array you can make one using `Array.fill(100)(util.Random.nextInt(100))`

    (a) Use a recursive function.

    (b) Use the count higher-order method.

    (c) Use the filter higher-order method in conjunction with a regular method.

    (d) Use the map higher-order method in conjunction with a regular method.

13. Write a version of `getAndClear` that uses a `List` and completely removes the value at the specified index.

14. Write a script that reads in a series of numbers from the user and stores them in a list. Then pass this list to a function that will remove all of the duplicates in the list. Print out the list to make sure that all duplicates were removed.

15. You can treat a `String` as a sequence of `Char` and use all the methods that we introduced here for `List` and `Array` on it. Using that, what code would you use to do the following? For all of these assume that `str` is a `String` provided by the user through `readLine()`.

    (a) Get a version of the `String` with all the spaces removed.

    (b) Get a version of the `String` with all vowels removed.

    (c) Split off the first word of the `String`. Write code that gives you a (`String`, `String`) where the first element is everything up to the first space in `str` and the second one is everything else.

    (d) Convert to a sequence that has the integer values of each of the characters in `str`.

    (e) Write code that will take a sequence of `Int`, like the one you just made, and give you back a String. (Note that if you get a sequence of `Char` you can use `mkString` to get a simple `String` from it.)

6.8.5 Projects

1. We can express a general polynomial as

$$A_n x^n + A_{n-1} x^{n-1} + A_{n-2} x^{n-2} + \ldots + A_2 x^2 + A_1 x + A_0$$

where the A values are the coefficients on different terms. You can view a sequence of `Doubles` as a polynomial by saying that the value at index n is the value of A_n. Using this representation, `Array(1.0,0.0,2.0)` would represent $2x^2+1$. Note that the order when you look at the `Array` is the opposite of the order seen in the polynomial.

Using this representation of polynomial, write functions that do addition and subtraction of two polynomials. You can assume that the `Arrays` passed in are the same length. Also write functions that gives you the derivative of a polynomial and the anti-derivative.[14] Note that the anti-derivative function will need and extra argument for the value of the constant term.

To help you learn, try to write these functions using both recursion with an index and using standard methods on sequences. For the former option try doing it with `Lists` as well using `::` to build the `Lists` in the recursion. For the latter a hint when you write `add` and `subtract` is to consider using the `zip` method.

If you want an extra challenge, try writing multiplication as well. This is a significantly harder problem, but if you can do it with both recursion and the built in methods you should feel good about your understanding.

2. This project has you implement a simple form of encryption and decryption. You should write at least two functions called `encrypt` and `decrypt`.

   The form of encryption we are using is offset encryption where you simply offset letters in the alphabet. If you pick an offset of 1, then if there is an `'a'` in the original message, the encrypted message will have a `'b'`. For a first cut, write these functions so they just add or subtract a specified offset to all characters. (Note that when you do arithmetic on a `Char` you get an `Int`. Use the `toChar` method to get back that type. So `('a'+1).toChar` will give you `'b'`.) Your methods will both take a `String` and an `Int` and return a `String`. The `Int` passed in is the offset to apply to the characters.

   This approach is easy, but it produces non-letters in certain situations like when you do an offset on `'z'`. Refine your functions so that they skip any characters that are not letters and wraps around so that if the offset goes above `'z'` it comes back to `'a'`. You have to handle capital letters as well. (Note that the `Char` type has methods called `isLetter`, `isUpper`, and `isLower` that can help you with this. You can also do comparisons and math with the `Char` type.)

   If you want an extra challenge and you want to make it harder for someone to try to break your code, revise your methods so that they take an extra `Int` and insert random characters at intervals of that spacing. So if the value is 5, you will insert an extra random character between every fifth character from the message. You can generate a random letter character with the expression `('a'+util.Random.nextInt(26)).toChar`.

   For this whole problem remember that you can treat `Strings` as sequences. So, you can use them like an `Array[Char]` except that they are not mutable.

3. For this project you will be doing some simple data analysis. If you go to the book website and look at the page for this chapter you will find a text file for this exercise that has values in it. It is weather data for 2/17/2010 in San Antonio, TX from the Weather Underground (`http://www.wunderground.com`). It has one line for each hour and each line has a single number that is the temperature for that hour. Write a function that will read this into an `Array[Double]` using redirection and `readDouble`. Then write functions to calculate the following: standard deviation and change per hour. The standard deviation will return a `Double`. The change will return an `Array[Double]` that is one shorter than the original data set.

4. Write an anagram solver. Anagrams are a type of word play in which the letters of a word or phrase are re-arranged to create a new word or phrase. Each letter may only

---

[14]You might recognize this better as an indefinite integral.

be used once. Anagrams have a long history. In ancient times, people used anagrams to find the hidden and mystical meaning in names. Your script should read in two strings and determine if they are anagrams. Two words are anagrams if they contain the same letters. For example, Mary and Army are anagrams. An example of phrases which are anagrams is "Tom Marvolo Riddle" and "I am Lord Voldemort". Notice that blank spaces can be variable and capitalization is not important.

5. Blackjack is one of the most popular card games in a casino. In this project, you will write a two person blackjack game in which the player will try to beat the dealer (the computer) in one of three ways: 1) The player gets 21 points on their first two cards, without the dealer getting blackjack (note that the dealer wins on a tie); 2) The player gets a final score higher than the dealer without exceeding 21; or 3) the dealer's hand exceeds 21. First, the computer deals out two cards to both the player and the dealer (itself). The player's cards are both face-up, but the dealer only has one card face-up and the other is face-down. Next, the dealer checks if either the player or the dealer got a blackjack. If no one got a blackjack, the player gets a chance to add some cards to their hand. The player must decide whether they want a "Hit", which means the dealer deals them another card which is added to their hand, or to "Stand", which means to stop dealing cards to the player. If the player "Hits" and they go over 21, then they lose, and the game is over. If not, then the player gets to decide again whether to "Hit" or "Stand". This process continues until the player either loses or "Stands". If the player has not gone over 21 and "Stands", then the dealer gets a chance to draw cards. At this point, the dealer shows all their cards to the player. The general rule of the game that a dealer must follow is that if the dealer's hand is less than or equal to 16, the dealer must "Hit" and add another card to their hand. The dealer must continue to "Hit" until either the dealer's total is greater than or equal to the player's total (in which case the dealer wins), or the dealer has gone over 21 (in which case the player wins).

6. This project is to be built on top of project 5.3. For this one you make functions that will take a `List` or `Array` of the values used to represent spheres and planes and return the impact parameter of the first one the ray hits and the information on that geometry.

7. For this project, you will write a function that could help people playing Scrabble®. You will want to break this down so you have multiple functions in the end, but the last one is what really matters. If you go to the book's web site, the page for this chapter will have a data file for this exercise called 2of12.txt. This is a file with a list of 41238 English words. There is one word per line. Using input redirection and `readLine` you will read the contents of this file into an `Array[String]`.

Once you have this file read in, your goal is to give back words that the person could likely make using letters he/she has to play. Your end goal is a function that takes the list of the dictionary words and a `String` of the letters the person has to play, which returns all the words that can be made from their letters plus one other letter. The one other letter is something you do not know that should have been played somewhere on the board. For our purposes here we do not care what it is. You should give back any word that can be formed with some subset of the player's letters plus one extra.

To do this you really want to use functions to break the problem down. Consider what higher order methods could be helpful and what types of functions you need to use with them. To make this all work with input redirection download and edit the 2of12.txt file and insert a top line where you put the letters the player has in Scrabble.

The script will read that line to get the letters and then read the proper number of lines to get all the words.

Lastly, to help you solve this problem consider using the `diff` method. This method tells you the difference between two sequences. Play around with calling `diff` using different strings to see what it does. Start with `"aabbcc".diff("abcde")` and `"abcde".diff("aabbcc")`.

8. Now that you can keep store multiple values in `Lists` and `Arrays`, you can push the recipe program to use an arbitrary number of items. Make a script that asks the user for items in their pantry. It should ask for item names and how much they have until the user enters a name of "quit". After that, the program will let the user run through recipes, entering one recipe name, followed by a number of ingredients, followed by the name and amount of each ingredient. It should then tell the user if they have enough to make that recipe and ask if they want to check another recipe.

9. Every semester you get to build a schedule of the courses that you will take in the following semester. This is the start of some projects at the end of which you write a program that will help you do so. For this first part, you want to have the user input courses he/she is interested in taking as well as how much they want to take those courses. The user should then tell the program how many courses should be taken the next semester and the minimum score for the "want to take" sum that will be accepted. The program should print out all of the schedule combinations with that many courses that make the cut. (Hint: Use `combinations` to run through all the possibilities. For each possibility, you can use `map` and `sum` to get the total "score" for that option.)

There are additional exercises and projects with data files to work on posted on the book's web site.

Chapter 7

Type Basics and Argument Passing

We have only begun to scratch the surface when it comes to types in Scala. By the end of this book we will see that not only are there many types provided by the Scala libraries, we can define our own types and give them names. Still, it is worth being more explicit about what a type is and some of the details of types in Scala.

Let us begin with the question of what is a type. A type is a set of values and the associated operations that can be performed on those values. Consider the type `Int`. There are a set of different values that we can represent with an instance of `Int`. They are all the integer values between -2147483648 and 2147483647. There is also a set of operations that you can perform on an `Int` or on pairs of them such as addition and subtraction.

In chapter 6 we began to see that there is more to types in Scala than the basics like `Int`. The types `List` and `Array` are not complete on their own. They are parametric types. To be complete they have to have type parameters. So `List[Int]` is a complete type as is `Array[String]`. `List` and `Array` are far from the only parametric types. Many of the types in the Scala library are parametric as it provides them with much greater flexibility.

7.1 Scala API

Chapter 6 presented a number of different methods that you can call on the `Array` and `List` types. It also hinted at the fact that there are others that had not been mentioned. There are also many other types in the Scala libraries, far more than can be covered well in a single book, and each of those has its own set of methods. You might have wondered how you would find out about all of these methods and types. The complete list of all the

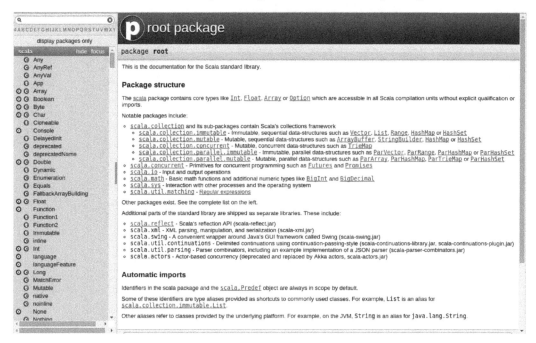

FIGURE 7.1: This figure shows the primary Scala API as it will appear when you first go to it. The left panel shows the types organized by package. The right panel shows the full information for whatever you have selected.

types available in the Scala standard library and methods for them can be found in the Applications Programmer Interface (API). You can find the Scala API at the main Scala web site, `http://www.scala-lang-org`. While the API has a lot of elements that go beyond your current knowledge, it is worth giving you a quick overview of things to help you read it and possibly find things in it.

The link to the API on the main page takes you to the core API. Under "Documentation" you can find APIs for separate modules. If you go to the API, you will see something like what is shown in figure 7.1.[1] The API is divided into two main segments. On the left hand side is a list of entities that are declared in the main library. On the right is a panel showing information about whatever you have selected. This starts off on the root package.

The list on the left has names with little circles to the left of them. Each circle has either an "O", a "C", or a "T" in it. These stand for "object", "class", and "trait". The distinction between a class and a trait is not significant to us in this book. We have been, and will generally continue to refer to these things at "types". While we do not focus on the details of the "object" declaration in Scala in this book, these create what are called singleton objects. As the name implies, this creates an individual object, whereas the type declarations, both classes and traits, produces blueprints from which we can INSTANTIATE many objects, often called INSTANCES.

The objects, classes, and traits are organized into groups called packages by the type of functionality that they provide. At the top is a package called **scala**. The most fundamental types, including Int, Double, and Char, are stored in this package. This package is **imported**

---

[1]It is interesting to note that the API is created by a tool called **scaladoc** that can be run on general Scala code to produce this style of web-based documentation. The **scaladoc** program does not work on scripts, so only the code written at the very end of this book using the application style of Scala programming would be able to benefit from it.

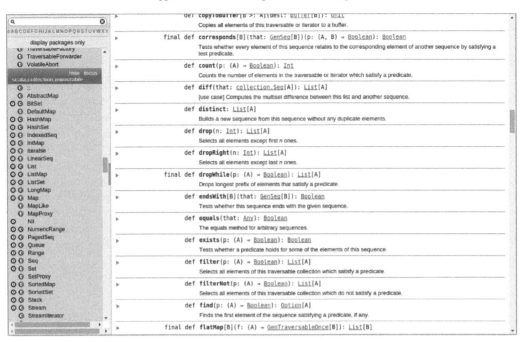

FIGURE 7.2: This figures shows the API with the List class selected.

by default in all Scala programs so that you can use things in it without having to use an `import` statement. Though we did not go into details, you have already seen the `scala.io` package. We are using part of it in the `import io.StdIn._` statements that has gone at the top of most of our scripts. The `imports` in Scala are nested, so we can say just `io.StdIn._` instead of `scala.io.StdIn._` due to the fact that the `scala` package has already been imported. You can use the longer version if you wish. Both work equally well. You could scroll down to the `scala.io` package, or type "StdIn" into the search bar at the top left of the API to see the `StdIn` object. If you click on that, you will see the various read methods that we have been using.

figure 7.2 shows the API after selecting `List` from the `scala.collection.immutable` package and scrolling down a ways into the methods. In the visible section, you can see a number of the methods that were introduced in chapter 6, including `distinct`, `drop`, `exists`, and `filter`. Clicking on any of those methods will reveal a longer description with more information.

The API contains a number of things that go beyond our current knowledge of Scala, so you should not worry if there are things that do not make sense. It is still a useful resource that you should refer to when you are coding. Note that your goal as a programmer is not to memorize the entire API. That would actually be a very poor use of your time. You want to have a feel for what is in the API so that you have a general idea of what is there and are able to look things up quickly when you need to use something you are not familiar with. You will come to know the parts that you use regularly without memorization. Indeed, one of the primary advantages of practice in programming is that it will help you to learn the details without memorization.

7.2   The Option Type

If you start digging through the API much you will find that the type `Option`, which is in the top level `scala` package, comes up a fair bit. The idea of an `Option` type is that they should be used when you are not certain if you will have a value or not. An `Option[A]` can either be `Some[A]` or `None`. If it is `Some`, it will have a value in it that can be retrieved with `get`. If there was no value, then you get `None` and know that it does not have a value. An example of a method that does this is the `find` method on a sequence. You can call `find` on either a `List` or an `Array` and pass it a function that takes an element and returns a `Boolean`. The `find` method is supposed to return the first element for which the function returns `true`. That description sounds simple enough, but what if there is not one? What happens if `find` runs through the whole sequence and nothing makes the function return `true`? For that reason, the `find` method returns `Option[A]` where A is the type contained in the sequence. So if nothing works, it can return `None`.

Let us run through two examples to see how this works using the declaration `val lst = List(7, 4, 6, 9, 3, 1)`.

```scala
scala> lst.find(_ < 4)
res0: Option[Int] = Some(3)

scala> lst.find(_ < 1)
res1: Option[Int] = None
```

The list does contain elements that are less than 4. The first of these is 3 so we get back `Some(3)`. However, it does not contain anything less than 1 so that call returns `None`. You might wonder how you can use this type of result. There are actually a few different approaches that can be employed. Which you use depends a lot on what you wanted to do with the result or based on the result.

The simplest thing to do is to use the `getOrElse` method of the `Option` type. As the name of this method implies, it will get the value that is stored if it is called on a `Some` or give back some default value if it is called on a `None`. Here we can see how this might be used on our two previous examples with `find`.

```scala
scala> lst.find(_ < 4).getOrElse(0)
res2: Int = 3

scala> lst.find(_ < 1).getOrElse(0)
res3: Int = 0
```

This approach works well if you have a good default value that you want to use in the case where nothing was found. It is also good idiomatic Scala. However, this approach does not work well if you actually want to perform additional logic in the situation where there is a value or when you want to do different things based on whether or not there is a value.

In the situation where you had more processing that was only supposed to happen in the case where you have a `Some`, you can use the `map` or `flatMap` methods of `Option`. Calling `map` on a `Some` applies the function to the stored value and results in a `Some` with the result of that function. Calling `map` on a `None` does nothing and results in a `None`. The behavior is exactly what you would expect if `Some` were a `List` with one element and `None` were a `Nil`.

Imagine that we wanted to square the value that we found, but only if we found it. Then map would be a good fit.[2]

```
scala> lst.find(_ < 4).map(i => i*i)
res4: Option[Int] = Some(9)

scala> lst.find(_ < 1).map(i => i*i)
res5: Option[Int] = None
```

As you can see, the one challenge with map, or flatMap, is that you get back another object of type Option. You can go through many maps, flatMaps, and even filters with an Option, but at some point you will likely need to use one of the other approaches to unwrap the contents.

If you actually want to do something in the case where you have a None then you need to turn to basic conditionals. The more idiomatic way to do this in Scala is using match. This little example takes the result of a find and tells us what it found or else tells us that nothing was found.

```
scala> lst.find(_ < 4) match {
     | case Some(i) => "Found "+i
     | case None => "Nothing found"
     | }
res6: String = Found 3
```

Note that this is using the pattern matching capabilities of the match expression to get the value out of the Some.

The Option type also has an isEmpty method that is true for None and false for Some as well as a get method that will return the value in a Some or crash if called on None. This allows you to use an if as well. This is the least idiomatic method of dealing with an Option and it is generally frowned upon. Code using if that does the same thing as the match above would look like this.

```
scala> val result = lst.find(_ < 4)
result: Option[Int] = Some(3)

scala> if (result.isEmpty) "Nothing found" else "Found "+result.get
res7: String = Found 3
```

Note that the result of the call to find needs to be given a name. This is because without pattern matching, it is used twice in the code, and calling it twice would be inefficient.

7.3 Parametric Functions

Types like List, Array, and Option are not the only things that can have parameters. Functions can have parameters as well. This is something that is not used much in this book as it is more important in later courses, but it can help you to read the API if you are introduced to the syntax. We would make a function parametric if we want it to work with

---

[2]You would use flatMap instead of map if the function you were applying was going to result in an Option. You likely do not want an Option[Option[A]] for whatever A type you are using, so flatMap flattens it out to only a single Option.

multiple different types. The simplest example of a PARAMETRIC FUNCTION is the identity method shown here.

```scala
scala> def ident[A](a:A)=a
ident: [A](a: A)A
```

This method takes an argument of any type and returns the value passed in. While there are not many situations where you would want to do this, it demonstrates the syntax of parametric functions. You simply place a type parameter name in square brackets between the function name and the normal argument list. It also demonstrates the power of parametric functions, especially if we put it to use as we do here.

```scala
scala> ident(3)
res1: Int = 3

scala> ident(3.3)
res2: Double = 3.3

scala> ident("3.3")
res3: java.lang.String = 3.3
```

First, this one function works just fine with an `Int`, a `Double`, and a `String`. That is pretty good. Even better, it worked with all those types without us telling it the types. Parametric functions can almost always infer the types of the parameters.

Here is a slightly more complex example though it really does not do much more, a function that takes two arguments and returns a tuple with those two values in it.

```scala
scala> def makeTuple[A,B](a:A,b:B) = (a,b)
makeTuple: [A,B](a: A,b: B)(A, B)
```

This demonstrates how a function can have multiple parameters. They appear as a comma separated list in the square brackets. A last simple example would be a function that makes `Lists` with three elements.

```scala
scala> def threeList[A](a1:A,a2:A,a3:A) = List(a1,a2,a3)
threeList: [A](a1: A,a2: A,a3: A)List[A]
```

The main reason for introducing these is that they help us understand something we saw in the last chapter, the `fold` methods. We said that `fold` was very much like one of the recursive functions we wrote in the last chapter where we pass in both a base value and a function. However, the function that we wrote only works with `Ints`. The `fold` methods work on sequences of any type and what is more, they can return a different type. With the use of parameters we can write a function with this same capability.

```scala
def ourFold[A,B](lst:List[A],base:B)(f:(A,B)=>B):B = {
  if (lst.isEmpty) base
  else f(lst.head,ourFold(lst.tail,base)(f))
}
```

Like the one in the API, this method is curried, a topic we will discuss in detail in section 7.8. It turns out that doing so helps with type inference and allows us to not have to specify types on the function. We can see this working with `lst` using two different processing functions.

```scala
scala> ourFold(lst,0)(_+_)
res0: Int = 30
```

```
scala> ourFold(lst,"")(_+" "+_)
res2: java.lang.String = 7 4 6 3 9 1
```

The first one takes the sum as we have seen several times already. This does not really exercise the ability to have different types because everything involved is an `Int`. The second example though puts those `Int`s together in a `String`, effectively making use of that second type.

There was another example we saw previously that could benefit from parametric types. Back in chapter 4 we looked at doing function composition. At the time we only worked with mathematical functions and limited ourselves to the `Double` type. By this point you should see there is a lot more to functions than numbers. We really should be able to compose two functions, `f` and `g`, as long as the output of function `g` is a type that we can pass into function `f`. So there are three types involved here, the type passed into `g`, the type returned by `g` then passed into `f`, and the type returned by `f`. Thanks to parametric types, we can write such a function in one line of Scala.

```
def compose[A,B,C](f:(B)=>A,g:(C)=>B):(C)=>A = (x)=>f(g(x))
```

7.4 Subtyping

So far we have generally talked about types as if they are completely independent and unrelated. We have written functions that might work with an `Int` or a `String` or a `Tuple`. The one exception to this was in section 7.2, where it was clear that `Option`, `Some`, and `None` are somehow related. By adding parameters we were able to make functions that could work with any type, but this still did not imply any relationship between the types. In reality, Scala, like most object-oriented languages, supports SUBTYPING. A type `B` is a SUBTYPE of type `A` if any place where we would want to use an object of type `A`, we can use an object of type `B` instead.

The concept of subtyping is dealt with in detail in *Object-Orientation, Abstraction, and Data Structures Using Scala* [1] in the chapter on inheritance, but for now you should be aware of the term because there will be times when it will come up. The `Option` type was our first example. If you wrote a function that took an `Option[Int]` as its argument, you could have passed in a `Some[Int]` or a `None` and it would have worked. That is because `Some` and `None` are both subtypes of `Option`.

Some of the general subtype relationships in Scala are shown in figure 7.3. This is a UML[3] Class diagram. The boxes represent types. The arrows point from a subtype to a supertype.

At the top of the figure there is an ultimate supertype called `Any`. Every object in Scala is an instance of `Any`. Since all values in Scala are objects, everything is an instance of `Any`, either directly or indirectly. The types like `Int` and `Double` that we learned about back in chapter 2 are on the left hand side of figure 7.3 under a type called `AnyVal` which is a subtype of `Any`. On the right hand side there is another type called `AnyRef` that has a bunch of unspecified types below it. All the other types we have or will talk about in this book fall somewhere under `AnyRef`.

---

[3]The Unified Modeling Language (UML) is a general purpose modeling language used in object-oriented software development.

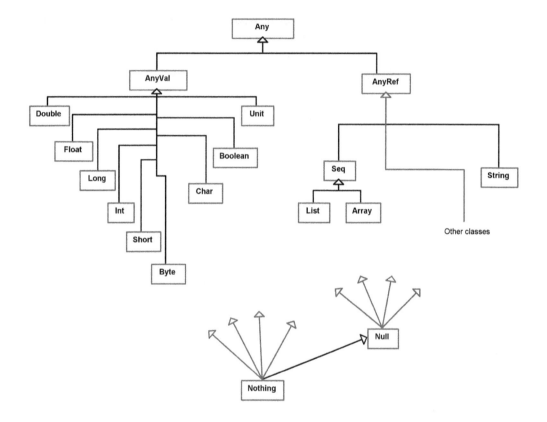

FIGURE 7.3: Diagram of general subtype relationships in Scala. This figure has been adapted from a similar figure in *Programming in Scala* by Odersky, Spoon, and Venners [3].

Near the bottom on the right side of the figure is a type called `Null`. This is a subtype of all the `AnyRef` types. One implication of this is that any variable of an `AnyRef` type, or any function/method with a result that is a subtype of `AnyRef` can have a value of `null`. This exists in Scala largely to maintain backward compatibility with Java. However, as a general rule you should try to avoid using `null` in your programs as they are very prone to cause it to crash.

At the very bottom of the diagram is a type called `Nothing` which is a subtype of all other types in Scala. There is no value of type `Nothing`. This type exists to make the type system complete and to handle situations when functions do not result in anything. We are not going to do anything in this book that requires you to use the `Nothing` type explicitly, but there are times when it might appear anyway, and understanding what it is will help you figure out what is going on. Consider the following examples that make an empty `List` and an empty `Array`.

```
scala> List()
res8: List[Nothing] = List()

scala> Array()
res9: Array[Nothing] = Array()
```

Note the type that was inferred by Scala. You might also run into errors that mention `Nothing`. Do not freak out when you see them. Instead realize that something in your code is probably not giving Scala enough information about types for it to figure out something else.

7.5 Variable Length Argument Lists

Have you wondered how you were able to construct Lists and Arrays with calls like this?

```
val lst = List(7,4,6,3,9,1)
val lst2 = List(1,2,3)
```

What exactly is going on here? It turns out that `List` in this usage is one of those singleton objects mentioned in section 7.1, and they have put code in it that allows it to be called like a function. More importantly for us now, is that we are able to pass a variable number of arguments in this call. In the first case we have passed six arguments. In the second we have passed three arguments. As you have seen, it would work just as well with any other number. The functions we have written cannot do this. Every function we have written so far has had an argument list with a specific number of values in it. So how can we make it so our function will let us pass in a variable number of arguments? In Scala, this is not hard to do. Simply add an asterisk after the last type of the last argument in the parameter list. This will allow the caller to pass zero or more of that type to create a VARIABLE LENGTH ARGUMENT LIST.[4]

A convenient place to do this would be an average function like we might use for calculating a grade. Such a method might look something like this.

```
def average(nums:Double*):Double = ???
```

---

[4]These are often referred to by the shorter name, varargs.

This would allow us to call the function in any of the following ways or with far more or fewer arguments.

```
average(1,2)
average(2,3,5,7,11,13)
average(100,95,63,78)
```

The question is, how do we use nums inside of the function? The answer to this is that we treat nums just like a List or an Array. Technically, it is a Seq, and all of the methods discussed in chapter 6 are available on it. These methods will make it very easy to write the average function.

```
def average(nums:Double*):Double = nums.sum/nums.length
```

That is it. Nothing more is needed because we can use sum on a sequence of Double values, and we can get the length of any sequence.

What if we want the average with the minimum value dropped? That does not add much complexity because there is a min method that we can call as well.

```
def averageDropMin(nums:Double*):Double = (nums.sum-nums.min)/(nums.length-1)
```

We could use these along with the fullAve function from section 6.6 to make a revised version of courseAverage.

```
def courseAverage(test1:Double,test2:Double,assn1:Double,
    assn2:Double,assn3:Double,quiz1:Double,quiz2:Double,
    quiz3:Double,quiz4:Double):Double = {
  val aveTest=average(test1,test2)
  val aveAssn=average(assn1,assn2,assn3)
  val aveQuiz=averageDropMin(quiz1,quiz2,quiz3,quiz4)
  fullAve(aveTest,aveAssn,aveQuiz)
}
```

This code makes nice use of the fact that average can be called with different numbers of arguments. However, it is fixed with a specific number of grades of each type. Only the last argument to a function can be a variable length. That should not be surprising as otherwise Scala would not know where the arguments should stop counting as one variable length list and start counting as the next. We can also make this function more flexible by passing in Lists of these different values instead as was done in section 6.6. Then the function signature would look more like this.

```
def courseAverage(tests:List[Double],assns:List[Double],
    quizzes:List[Double]):Double = {
  ...
}
```

Unfortunately, if you try to write this in the first way that comes to mind you will find that Scala does not like it.

```
def courseAverage(tests:List[Double],assns:List[Double],
    quizzes:List[Double]):Double = {
  val aveTest = average(tests)
  val aveAssn = average(assns)
  val aveQuiz = averageDropMin(quizzes)
  fullAve(aveTest,aveAssn,aveQuiz)
}
```

You will get an error that looks like this.

```
<console>:11: error: type mismatch;
 found   : List[Double]
 required: Double
        val aveTest=average(tests)
                            ^
<console>:12: error: type mismatch;
 found   : List[Double]
 required: Double
        val aveAssn=average(assns)
                            ^
<console>:13: error: type mismatch;
 found   : List[Double]
 required: Double
        val aveQuiz=averageDropMin(quizzes)
                                   ^
```

This is because `List[Double]` is not the same type as `Double*`. However, the two are very similar in practice, and it seems that you should be able to quickly and easily use the `List`s in a place that calls for a variable length argument. Indeed, you can. You just have to tell Scala that is what you are doing. You do this with a syntax much like specifying the type of something with a colon after the name followed by the type. You do not use `Double*` as the type though, instead you use `_*` because Scala really does not care about the specific type.

```
def courseAverage(tests:List[Double],assns:List[Double],
    quizzes:List[Double]):Double = {
  val aveTest = average(tests:_*)
  val aveAssn = average(assns:_*)
  val aveQuiz = averageDropMin(quizzes:_*)
  fullAve(aveTest,aveAssn,aveQuiz)
}
```

Now we have a function that computes an average with `List`s; so, it is flexible in the number of grades passed in and rather simple to call and use in a larger program. It still has the bugs that we fixed back in section 6.6 related to lists with zero or one element, but from the discussion there, you should be able to easily fix them.

7.6 Mutability and Aliasing

In chapter 6 we saw that `List` is immutable while `Array` is mutable. It has been stated that the functional style will tend to use immutable data and that while mutable data has significant benefits for some operations, it is also potentially less safe. You might wonder why it is less safe though. After all, it should not be too hard to make sure that you do not make changes to an `Array` when you should not. That is true in small programs, but it gets harder and harder as the programs get larger and for reasons that you probably do not fully grasp at this point. The goal of this section is to show you one of the subtle challenges with mutable data.

A big part of the problem comes from something called ALIASING. This is when two dif-

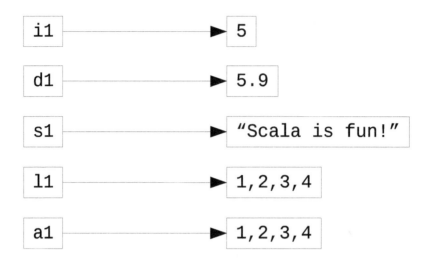

FIGURE 7.4: Simple image of memory layout for a few variables. The objects on the right side are overly simplified, especially the list, but this portrays sufficient detail for understanding what is happening at this point.

ferent names refer to the same object. To understand this, we need to refine our view of what is happening in the memory of the computer. Consider the following variable declarations.

```
var i1 = 5
var d1 = 5.9
var s1 = "Scala is fun!"
var l1 = List(1,2,3,4)
var a1 = Array(1,2,3,4)
```

All of these have been made as **vars** just to give us the option of altering them below. Normally we would want to use a **val** unless we explicitly found a reason why we had to change them. In this case the reason is just for illustration.

The memory layout looks something like what is shown in figure 7.4. This is an idealized representation. In reality memory is all linear and has structure to it that will be discussed in chapter 13. For now, this view is sufficient. On the left are boxes that represent the variables. In Scala the best mental picture is to see the variables as boxes that store references to values.[5]

What happens to this picture if we change the value of two of the variables? For example, say that now we execute these two lines of code.

```
i1=8
s1="New string."
```

The first one changes i1 so that it references the value 8 instead of 5. The second one changes the s1 so that it references this alternate **String**. What would this look like in

---

[5]In some situations Scala will put the actual values in the boxes for the variables, but that is an optimization that you do not really have control over, and it will not alter the behavior; so, it is good to picture the memory like this.

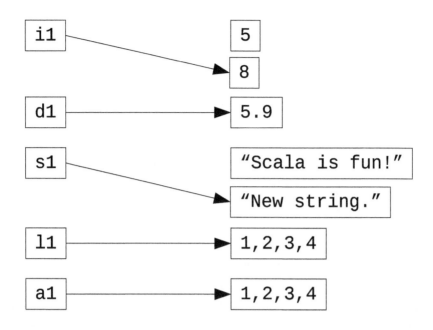

FIGURE 7.5: Modified view of the memory after two variables have been assigned different values.

memory? The result is shown in figure 7.5. The 5 and the `String` "Scala is fun!" have not been changed. They remain in memory just like before. Because nothing references them, these objects will be collected and disposed of automatically by the GARBAGE COLLECTOR, a part of the system that is responsible for freeing up memory that is no longer in use. What has changed is the references in `i1` and `s1`. They now point to new objects in new chunks of memory that hold these new values. Both the `Int` and `String` types are immutable. As a result, once the objects are created, nothing can change them. They do not allow any operations that would alter their values. If we had made these variables using `val` we would not be able to change where the arrows point either.

Indeed, all of the types used here are immutable with the exception of the `Array`. So `Array` is the only one where we can change the value in the box on the right. To see how this can matter we will consider only the `List` and the `Array` variables and add two more lines of code to the picture.

```scala
val l2 = l1
val a2 = a1
```

This gives us a slightly different picture that is shown in figure 7.6. Now we have two different variables that point to each of the `List` and `Array` objects. These second references to the same object are often called aliases. Aliasing is where mutability often starts to cause problems. If you have not figured out why yet, perhaps this code will make it clear.

```scala
scala> a2(3) = 99

scala> a1
res1: Array[Int] = Array(1, 2, 3, 99)
```

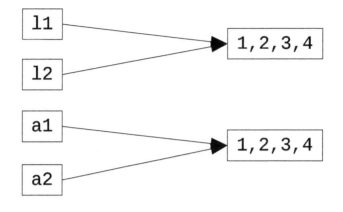

FIGURE 7.6: What the memory looks like after the introduction of l2 and a2.

In the first line we change the value in the last slot in a2. On the next line we look at the value of a1. Note that it is changed. We did not explicitly do anything with a1, but the value is changed because a2 is an alias of a1. You cannot do this with l2 because the List type is immutable, the assignment statement will not compile for the List.

There are times when the ability to make an alias and change it is great. There are some tasks for which this can make things much more efficient. However, if you lose track of what is going on and what names are aliases of what other names you can run into serious problems.

At this point it might be easy to think that all you have to do is avoid making aliases. If you just do not write lines like val a2 = a1 then you would be fine, right? While that type of aliasing might be avoidable, the next section shows why it is pretty much impossible to avoid aliasing all together.

7.7 Basic Argument Passing

Now that we have looked at the way memory is organized for variable declarations and the aliasing issue, we should take another look at what happens when you call a function. The two are actually very similar. When you pass values into a function, the function has local vals with the local argument names, but they reference the same objects that the outside variables referenced. Consider the following code from a script.

```scala
def getAndClear(arr:Array[Int],index:Int):Int = {
  val ret = arr(index)
  arr(index) = 0
  ret
}

val numbers = Array(7,4,9,8,5,3,2,6,1)
val place = 5
val value = getAndClear(numbers,place)
```

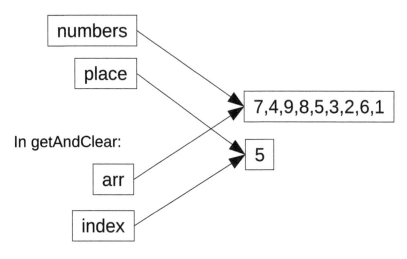

FIGURE 7.7: The memory layout from the `getAndClear` script. The arguments passed into the function become aliases for the objects created at the top level of the script.

The function is passed an `Array` of `Ints` and an `index` for a location in that array. It is supposed to return the value at that location and also "clear" that location. The meaning of "clear" here is to store a zero at that location. To see how this works look at figure 7.7 which shows the arrangement of memory. The function is defined and the variables `numbers` and `place` are both declared and initialized. We get new objects for them to reference.

When the `getAndClear` function is called, `numbers` and `place` are passed in to be the values of the arguments `arr` and `index`. While the code is in `getAndClear`, `arr` is an alias for the object referred to by `numbers` and `index` is an alias for the object referred to by `place`. This is significant because the `getAndClear` method has a side effect. It modifies one of the values stored in the `Array` that is passed to it. When the code gets down to "Other stuff", the memory looks a bit different as shown in figure 7.8. At that point, the function has finished so its local variables are no longer needed. In addition, the 6th element in the array `numbers` has been changed. The function managed to change the value for the `Array` as seen outside that function. With a name like `getAndClear` you could probably figure out that this type of side effect might happen. However, when you pass any mutable value into a function, you need to be aware that the function could change your value, even if you do not really want it to.

How would this be different with a `List`? Well, the `List` type is immutable; so, it would be impossible for the function to change the list. We can't write a line like `arr(index) = 0` and still have the code compile. If you wanted it changed, the function would need to return a tuple including both the value gotten and a new `List` with that value changed. We will quickly go through a few ways to do this.

In top level of script after getAndClear:

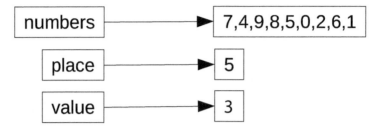

FIGURE 7.8: The configuration of memory after `getAndClear` has executed.

```
def getAndClear(lst:List[Int],index:Int):(Int,List[Int]) = {
  (lst(index), lst.zipWithIndex.map( tup => {
    val (n,i) = tup
    if (i == index) 0 else n} ) )
}
```

There is only one expression in this function, the tuple to return. The first element in the tuple is `lst(index)`. It looks up the proper value and uses the value there. Remember that looking things up in this way on a `List` is not really efficient, but we have to do so if given a `List` and an index. The second part of the tuple is a bit more complex. It calls the method `zipWithIndex` on `lst`. This returns a new `List` where each element is a tuple that contains an element from the `List` along with the index of that element. We have to do this because we need the index in order to decide if it is the one to be cleared or not. If we were clearing based on value instead of position this would not be needed.

The method `map` is immediately called on the `List` of tuples, and it is given a function literal that takes one argument called `tup`, short for tuple. This function literal uses curly braces and has two lines. The first line is a `val` that pulls the two parts out of the tuple. The number from the `List` is called `n` and the index is called `i`. The function literal finishes with an `if` expression that checks the index and gives the value of 0 or `n` based on if `i` is the same as `index`.

Substituting for a value in a `List` is a common enough operation that the API contains a method to do it for us. So we can write a simpler version of `getAndClear` for a `List` that uses the `updated`. This method takes an index and a new value and returns a modified list.

```
def getAndClear2(lst:List[Int],index:Int):(Int,List[Int]) = {
  (lst(index), lst.updated(index,0))
}
```

Another approach would be to pull off the elements before and after the index and then stick them back together with a 0 between them.

```
def getAndClear3(lst:List[Int],index:Int):(Int,List[Int]) = {
  (lst(index), lst.take(index):::(0::lst.drop(index+1)))
}
```

This version uses the cons operator, `::`, which we saw before. It adds one new element to the front of a `List` and gives that back. It also uses an operator that we have not seen previously, the : : : operator. This does roughly the same thing as : : except that the argument on the left is a full `List` that should be appended to the front of the second `List`.

Lastly, we could write this using recursion. Here is one way to do that.

```
def getAndClear4(lst:List[Int],index:Int):(Int,List[Int]) = {
  if (index == 0) (lst.head, 0::lst.tail)
  else {
    val (n, rest) = getAndClear4(lst.tail, index-1)
    (n, lst.head::rest)
  }
}
```

Of course, because the `List` version is having to rebuild the `List`, it could go a step further and truly clear the value. It could remove it from the `List` instead of overwriting it with zero. This would be done with `filter` plus a `map` on the first version. Using `updated` would not make sense for that, so the second version would not be viable. The third and fourth versions are very easy to change. We leave them as an exercise for the reader to create and try.

An interesting question now becomes which of these is better, the `Array` or the `List`? Is it better to use an `Array` which can be mutated or a `List` which cannot? The answer is that it depends on whether you want a function to be able to mutate the data. If the function is supposed to mutate the data then there can be a speed benefit to passing something mutable. However, if you do not want the data mutated then there is a problem with passing mutable data. When you pass mutable data into a function, you have no control over whether it is mutated. In situations where you do not want your data mutated, this requires that you make a DEFENSIVE COPY. To protect your version of the data, you make a copy of it and pass the copy into the function.

For this reason, the question of whether you want to use a mutable or immutable structure depends a lot on the way in which you use that data. If it is going to be mutated frequently then it might make sense for it to be mutable. However, if it is not going to be mutated frequently then immutable likely makes sense so you do not have to make defensive copies. Even in situations where there might be a fair bit of mutation, you need to consider how much defensive copying will be needed if you choose mutable data.

There is another aspect of normal argument passing that is very significant to understand. Arguments to functions are normally evaluated before they are passed to the function. Imagine that you have the following function.

```
def successorWithPrint(i:Int):Int = {
  println("In the function.")
  i+1
}
```

Now we call this function with an argument that needs to be evaluated.

```
scala> successorWithPrint(2+3)
In the function.
res0: Int = 6
```

This call evaluates 2+3, then passes that value 5 into the function. We can make this clear if the argument has a side effect.

```
scala> successorWithPrint({println("argument"); 2+3})
argument
In the function.
res1: Int = 6
```

Here you can see that the print in the argument happens before the print in the function, demonstrating that the argument is evaluated before it is passed into the function for normal argument passing.

Code Blocks as Arguments

Putting multiple expressions in an argument, as with `print("argument"); 2+3` above, requires the use of curly braces to make a code block. When there is only one argument in a list, having both parentheses and curly braces, as was used above, is somewhat redundant. For the reason, one of the features of the Scala syntax is that in situations like this you can leave off the parentheses. So that call could also be written in the following way.

```
successorWithPrint {println("argument"); 2+3}
```

7.8   Currying

In chapter 6 we came across a number of methods that used something called currying, where arguments could be separated into different argument lists. This is a common feature of functional programming languages. In case you are wondering, the name has nothing to do with spices. It comes from the name of Haskell Curry, a mathematician and logician. In this section we want to show you not just how to call functions/methods that use currying, but how to write your own such methods and why you would want to.

Let's start off by writing our own curried functions. This first example shows how we can make a curried function using the normal Scala function type and function literal syntax.

```
scala> def add(x:Int):(Int)=>Int = (y:Int)=>x+y
add: (x: Int)(Int) => Int

scala> add(5)(6)
res0: Int = 11

scala> val plus5 = add(5)
plus5: (Int) => Int = <function1>

scala> plus5(6)
res1: Int = 11
```

The first input defines a function `add` that takes an integer and returns a function which takes an integer and returns an integer. The `add` function is a higher-order function because it returns a function. The second input to the REPL shows that we can use `add` as a curried function passing it two arguments separately. In the third line, we call **add** and pass it the argument 5. We store the result of this in a variable named **plus5**. Note that **plus5** is actually a function that takes an integer and returns an integer. At the end we call **plus5** and pass it the value 6 which gives us back 11.

This example shows what it means to curry a function. There can be times when this approach is quite useful. For that reason, Scala provides a shortcut for doing it. If you define a function and follow it with more than one argument list, the function is curried. So, the above could be entered like this instead.

```
scala> def add(x:Int)(y:Int) = x+y
add: (x: Int)(y: Int)Int

scala> add(5)(6)
res2: Int = 11

scala> val plus5 = add(5)_
plus5: (Int) => Int = <function1>

scala> plus5(6)
res3: Int = 11
```

The only difference here is the underscore where the second argument list would go when we make **plus5**. This tells Scala that you really meant to only specify the first argument.

Hopefully these examples made it clear that currying is not hard to do in Scala. Still, you might be wondering why you would bother. We have actually seen two things in this chapter that can provide the motivation to curry a function.[6] The first was in the section on variable length argument lists. We said that we could not use variable length arguments for tests, assignments, and quizzes because a variable length argument can only be used with the last argument in an argument list. One way around this is to curry the function making each type of grade be it's own argument list. That would produce the following function.

```
def courseAverage(tests:Double*)(assns:Double*)(quizzes:Double*):Double = {
  val aveTest = average(tests:_*)
  val aveAssn = average(assns:_*)
  val aveQuiz = averageDropMin(quizzes:_*)
  fullAve(aveTest,aveAssn,aveQuiz)
}
```

Notice that you still need the :_* when passing the arguments through to other functions that have variable length argument lists. This version allows us to invoke the function like this.

```
courseAverage(80,95,100)(100,80)(56,78,92,76)
```

So currying gives us the ability to have more than one variable length argument list in a single function.

The other place in this chapter where we have introduced something that might benefit from currying was right before the section on currying in the aside about code blocks as arguments. The syntax where you leave off parentheses and only use curly braces only works when an argument list has a single argument. Without currying, this means that we could never take advantage of this feature for any function that needs more than one argument.

---

[6]There are actually several other reasons why one might choose to curry a function that go beyond the scope of this book.

With currying, we can simply break up the arguments so that any argument that might need a block of code occurs on its own. Indeed, this is one of the reasons why `Array.fill` and `List.fill` are curried. It is common to want to have more than a single expression in the second argument. Having that second argument stand alone as its own argument list allows the parentheses to be left off as we see here in this code that produces an `Array` with the squares of powers of two.

```scala
scala> var i = 1
i: Int = 1

scala> Array.fill(10) {
     | i *= 2
     | i*i
     | }
res4: Array[Int] = Array(4, 16, 64, 256, 1024, 4096, 16384, 65536, 262144, 1048576)
```

This is most useful in situations like `fill` where an argument is passed by name. We will explore this more in the next section.

It is interesting to note that both of the uses of currying presented here is actually related to the original purpose of currying, which was to partially evaluate functions. These usages allow us to better take advantage of features of the Scala language. If you do enough work in Scala, you will likely find this is the case for the vast majority of the uses of currying that you use and create.

7.9   Pass-By-Name

Section 7.7 looked at the way in which arguments are passed in Scala. This is a style called PASS-BY-VALUE. The function gets a copy of what the variable was in the calling code. The function cannot change that variable, it can only change what it refers to if what it refers to is mutable. Some languages provide an alternate way of passing things called PASS-BY-REFERENCE. When something is passed by reference, the function has the ability to change the variable in the calling function itself. The fact that all variables in Scala basically are references blurs the line between the two, but fundamentally Scala only allows pass-by-value, and the function can only modify things seen to the outside if the variable refers to a mutable object.

While Scala does not provide a true pass-by-reference, it does provide a passing style that few other languages provide, PASS-BY-NAME, which was first mentioned in section 6.4. The idea of pass-by-name is that the argument is passed not as a value, but as a THUNK that is basically a set of code that will be executed and give a value when the parameter is used in the function. You can imagine pass-by-name as automatically creating a function that takes no argument and returns a value that will be executed each time the argument is used. To help you understand this and see the syntax, we will give a simple example. We will start making a basic increment function in the way we are used to doing.

```scala
scala> def incr(n:Int):Int = {
     | println("About to increment.")
     | n+1
     | }
incr: (n: Int)Int
```

The print statement is just to help us keep track of what is happening when. Now we will write the same function again, but this time pass the argument by name instead.

```scala
scala> def incrByName(n : =>Int):Int = {
     | println("About to increment.")
     | n+1
     | }
incrByName: (n: => Int)Int
```

The syntax for passing an argument by name is to put a rocket before the type of the argument. This is the same arrow used in function literals. If we were to put empty parentheses in front of it to get `() => Int` we would have the type of a function that takes no arguments and returns an `Int`. The by-name argument is much the same only when you call the function you do not have to explicitly make a function for the argument, any collection of code will work. To start off, we will call this function in the simplest way we possibly could.

```scala
scala> incr(5)
About to increment.
res0: Int = 6

scala> incrByName(5)
About to increment.
res1: Int = 6
```

No surprises here. They appear to both do the same thing. Both print the statement and give us back 6. However, the two are not really doing the same thing. To make that clear, we will call them again and have the argument be a block that includes a print.[7]

```scala
scala> incr {println("Eval"); 5}
Eval
About to increment.
res2: Int = 6

scala> incrByName {println("Eval"); 5}
About to increment.
Eval
res3: Int = 6
```

Now it is clear that they are not doing the same thing. When you pass an argument by value, it is evaluated before the function starts so that you can have the value to pass through. When you pass by name, the evaluation happens when the parameter is used. That is why the line "Eval" printed out after "About to increment." in the second case. The `println` in the function happens before the value of n is ever accessed.

Using pass-by-name gives you the ability to do some rather interesting things when mutations of values comes into play. To see that, we can write a slightly different function.

```scala
scala> def thriceMultiplied(n : => Int):Int = n*n*n
thriceMultiplied: (n: => Int)Int
```

It might seem that this method should be called cubed, but as we will see, that name might

---

[7]Note that we are making use of the fact that an argument list with a single argument can skip the parentheses when a block of code is passed.

not always apply for it because it uses pass-by-name. To start with, let us call it with an expression that gives a simple value, but prints something first.

```scala
scala> thriceMultiplied {println("Get value."); 5}
Get value.
Get value.
Get value.
res4: Int = 125
```

Note that the `println` statement happened three times. This is because the value n was used three times in the function. Had the parameter been passed by-value, that would not have happened. It would have printed once, when the function was called.

This particular call still gave us a valid cube though. So why was the function not simply called cube? The reason is that if the code in the argument does the right thing, the result will not be a cube. Here is an example.

```scala
scala> var i=5
i: Int = 5

scala> thriceMultiplied {i+=1; i}
res5: Int = 336
```

336 is not a perfect cube. So how did we get this value? In this example we introduced a var. The code in the pass-by-name argument alters this var. As a result, every time that n is used, the value given by n is different. The first time we use n the value is 6 because the original 5 was incremented and then returned. The next time it is 7 because the 6 that it was set to the previous time is incremented again. The last time it is 8. So the answer is $6 * 7 * 8 = 336$.

7.10   Multidimensional Arrays

When we first introduced `Arrays` and `Lists` back in chapter 6, we saw that these types are parametric. That means that the type requires a type argument to be fully defined. So you cannot have just an `Array` or just a `List`. Instead you have `Array[Int]` or `List[String]`. Each of these is a type in Scala. The parameter for these parametric types can be any type.[8] If you put these together you can build things like `Array[Array[Double]]`, `List[List[String]]`, or `List[Array[Int]]`. You do not have to stop there though. Scala will be perfectly happy with a `List[Array[List[List[Array[Int]]]]]`. It is not clear what you would want such a type for, but if you find a use, Scala will support it.

In the case of `Arrays` of `Array` types, we have a special term for them. They are called multidimensional arrays. This is because of how you might picture them in your head. You can picture a normal `Array` as a row with multiple bins that each store a value. An `Array[Array[Int]]` could be pictured as a table of integers that has rows and columns. Such a table could be said to be two dimensional. If you had an `Array[Array[Array[Int]]]` you could picture it as a cube of values in three dimensions. In general all these things can be applied to `Lists` just as well as `Arrays`, but the term multidimensional list is not nearly as common, and the fact that it is inefficient to do direct indexing on `Lists` means that most applications that have multidimensional data structures will not create them with `Lists`.

---

[8]There can be restrictions on parametric types, but the collections generally allow any type to be used.

So, how can you create and use multidimensional arrays? The most basic syntax mirrors what we used to originally create normal `Arrays`.

```scala
scala> val tda1 = Array(Array(1,2,3),Array(4,5,6))
tda1: Array[Array[Int]] = Array(Array(1, 2, 3), Array(4, 5, 6))
```

In this usage the number of elements in each sub-array does not have to be the same. When the lengths are different, they are called ragged arrays. When they are all the same, they are called rectangular arrays.

If you are interested in making large rectangular arrays, you should use either the `fill` method or the `tabulate` method. For `fill`, all you have to do is pass in multiple arguments in the first parameter list.

```scala
scala> val tda2 = Array.fill(10,10)(0)
tda2: Array[Array[Int]] = Array(Array(0, 0, 0, 0, 0, 0, 0, 0, 0, 0),
Array(0, 0, 0, 0, 0, 0, 0, 0, 0, 0), Array(0, 0, 0, 0, 0, 0, 0, 0, 0, 0),
Array(0, 0, 0, 0, 0, 0, 0, 0, 0, 0), Array(0, 0, 0, 0, 0, 0, 0, 0, 0, 0),
Array(0, 0, 0, 0, 0, 0, 0, 0, 0, 0), Array(0, 0, 0, 0, 0, 0, 0, 0, 0, 0),
Array(0, 0, 0, 0, 0, 0, 0, 0, 0, 0), Array(0, 0, 0, 0, 0, 0, 0, 0, 0, 0),
Array(0, 0, 0, 0, 0, 0, 0, 0, 0, 0))
```

In the case of `tabulate` you pass multiple arguments to the first parameter list and also use a function that takes the same number of arguments for the second parameter list.

```scala
scala> val tda3 = Array.tabulate(10,10)((i,j) => i*j)
tda3: Array[Array[Int]] = Array(Array(0, 0, 0, 0, 0, 0, 0, 0, 0, 0),
Array(0, 1, 2, 3, 4, 5, 6, 7, 8, 9), Array(0, 2, 4, 6, 8, 10, 12, 14, 16, 18),
Array(0, 3, 6, 9, 12, 15, 18, 21, 24, 27), Array(0, 4, 8, 12, 16, 20, 24, 28,
32, 36), Array(0, 5, 10, 15, 20, 25, 30, 35, 40, 45), Array(0, 6, 12, 18, 24,
30, 36, 42, 48, 54), Array(0, 7, 14, 21, 28, 35, 42, 49, 56, 63), Array(0, 8,
16, 24, 32, 40, 48, 56, 64, 72), Array(0, 9, 18, 27, 36, 45, 54, 63, 72, 81))
```

Note that the number of arguments you pass into the first argument list of `tabulate` or `fill` determines the dimensionality of the resulting structure. These examples made two dimensional arrays, but you can use up to five arguments with this usage.

You can also use `fil` or `tabulate` to make non-rectangular arrays by having the code block/function passed as the second argument build `Arrays` of different lengths. That technique is used here with `fill` to make a truly ragged `array` with rows of random lengths[9] and with `tabulate` to create a triangular 2-D array.

```scala
scala> val tda4 = Array.fill(10)(Array.fill(util.Random.nextInt(10)+1)(0))
tda4: Array[Array[Int]] = Array(Array(0, 0, 0, 0), Array(0, 0, 0, 0, 0, 0, 0),
Array(0, 0, 0, 0, 0, 0, 0, 0, 0, 0), Array(0), Array(0, 0, 0, 0), Array(0, 0,
0, 0), Array(0, 0, 0, 0, 0, 0, 0, 0, 0, 0), Array(0, 0, 0, 0), Array(0, 0, 0,
0), Array(0, 0, 0, 0, 0, 0))
```

```scala
scala> val tda5=Array.tabulate(10)(i => Array.fill(i+1)(0))
tda5: Array[Array[Int]] = Array(Array(0), Array(0, 0), Array(0, 0, 0),
Array(0, 0, 0, 0), Array(0, 0, 0, 0, 0), Array(0, 0, 0, 0, 0, 0),
Array(0, 0, 0, 0, 0, 0, 0), Array(0, 0, 0, 0, 0, 0, 0, 0),
Array(0, 0, 0, 0, 0, 0, 0, 0, 0), Array(0, 0, 0, 0, 0, 0, 0, 0, 0, 0))
```

Note that the output here has a different number of elements in each of the sub-arrays.

To access the elements of a 2-D `Array`, simply put two sets of parentheses after the

---

[9]To get random integers, we use the `nextInt` method in the `Random` object in the `util` package.

`Array` name with two different indices in them. For example, we can pull values out of `tda2` in this way.

```
scala> tda3(3)(4)
res0: Int = 12
```

The 2-D `Array` tda3 was created to be something like a multiplication table. This particular expression pulled off the element at position 3,4 which is 12. To understand why you use this syntax, think of the types shown here.

```
scala> tda3
res32: Array[Array[Int]] = Array(Array(0, 0, 0, 0, 0, 0, 0, 0, 0, 0), Array(0, 1,
    2, 3, 4, 5, 6, 7, 8, 9), Array(0, 2, 4, 6, 8, 10, 12, 14, 16, 18), Array(0, 3,
    6, 9, 12, 15, 18, 21, 24, 27), Array(0, 4, 8, 12, 16, 20, 24, 28, 32, 36),
    Array(0, 5, 10, 15, 20, 25, 30, 35, 40, 45), Array(0, 6, 12, 18, 24, 30, 36,
    42, 48, 54), Array(0, 7, 14, 21, 28, 35, 42, 49, 56, 63), Array(0, 8, 16, 24,
    32, 40, 48, 56, 64, 72), Array(0, 9, 18, 27, 36, 45, 54, 63, 72, 81))

scala> tda3(3)
res33: Array[Int] = Array(0, 3, 6, 9, 12, 15, 18, 21, 24, 27)

scala> tda3(3)(4)
res34: Int = 12
```

The original value is an `Array[Array[Int]]`. When you index into an `Array` you get back an instance of the type it stores. In that case, that means that `tda(3)` has the type `Array[Int]`. Indexing into that gets down to a single `Int`.

The advantages and restrictions of the `List` and `Array` types that were discussed in chapter 6 apply to higher dimensional cases as well. Once you have made an `Array`, the values can be changed, but the length cannot. Similarly, you can make new `List`s by efficiently adding to the head of old ones, but you cannot mutate values in a `List` even if it has a higher dimension.

7.11 Classifying Bugs

By this point you have certainly learned that simply writing your code does not mean that it is correct and that it will work. Errors in programs are commonly referred to as BUGS. The term bug has a long history in the engineering field. It came into the field of computer science early because one of the first ones was a moth. The earliest computers were huge. They took up rooms the size of gymnasiums. An early malfunction in one turned out to be a moth that had flown in and was causing a short circuit.

While the term "bug" has stuck around, the implications of this term no longer fit. The term and its history imply that it is something beyond the control of the programmer that just put itself into the code and now the programmer has to search for it. In reality, virtually all modern bugs are really mistakes on the part of the programmer. They are things that the programmer put into the code and now the programmer needs to correct.

Not all bugs are the same. There are three fundamentally different types of errors.

1. SYNTAX ERROR – Error in the structure of the code that is found by the compiler. If

you have a syntax error it means that your program is not valid for the language you are writing in, for our purposes that is Scala.

2. RUNTIME ERROR – Error that causes the program to crash while running.

3. LOGIC ERROR – Error that does not crash the code, but causes it to produce the wrong answer.

Each of these deserves some discussion. We will also talk about how they compare to one another and the way in which they impact programmers.

When you are first learning to program, the errors that you likely run into the most are syntax errors. These can be as simple as typos or misspellings. They can also be more complex like type mismatch errors or calling functions or methods with the wrong number of arguments. The common element that makes something a syntax error is that it is discovered by the compiler when it is trying to translate the human written code into a format that is more computer friendly. Different programming languages do different amounts of checking for errors at compile time. This is often called STATIC CHECKING because it can be done without actually running the program.

Scala does a significant amount of static checking for errors.[10] The way we have been running our programs in the REPL or with scripts, the compile stage is not clearly separated from the running of the program. The Scala system is running a compiler in the background and then executing the results of the compile. The syntax errors display a message like the following:

```
timing.scala:5: error: not found: value This
This line will not compile
^

one error found
```

They tell you the name of the file along with a line number. Then they describe the error and show the line and where on the line the error occurred.

The second type of bug is a runtime error. This type of error occurs when everything is syntactically correct and the program compiles and runs. However, during the run this error causes the program to crash. In Scala, a runtime error will produce a message that looks similar to the following:

```
java.lang.ArrayIndexOutOfBoundsException: -1
        at Main$$anon$1.<init>(timing.scala:5)
        at Main$.main(timing.scala:1)
        at Main$.main(timing.scala)
   ...
```

This message tells you what went wrong and then prints something called a stack trace that includes line numbers. In this simple example, the code failed on line 5. The stack trace shows the line of code that was currently executing at the top, followed by the line that called that function, followed by the line that called that function, etc. Sometimes the top line is not in your code, but in a library. That generally means it is caused by something you passed into a library call. You can look down the stack trace to find the first line that is in your code for guidance on what is causing the problem.

There are many different reasons why runtime errors happen, and it might be dependent on user input. So a runtime error might not be found from running the program once or

---

[10]The significant static checking of types was a major influence in the selection of Scala for this textbook. Other introductory languages, like Python, do significantly less static checking of programs.

twice. To reduce the number of runtime errors in a program you have to run it with multiple different test inputs. This will be discussed in more detail in section 8.4. In a general sense, it is impossible to prove that code has no runtime errors.

The last type of error is a logic error. An error is a logic error if the code compiles and runs to normal termination, but provides an incorrect output or behaves incorrectly. These errors can come in all types of forms, and there is no message that tells you there is a problem. You know there is a problem by checking the output or behavior of the program to see if it matches expectations. Like runtime errors, logic errors might not occur for all inputs. There might be only specific inputs that trigger the error.

If you have a choice, you want errors of types earlier on this list. As a novice programmer you probably get tired of dealing with syntax errors and find them frustrating. However, the reality is that syntax errors are by far the best type of error. This is because syntax errors give you the most information on how to fix them and are detected by the compiler in a way that does not depend on inputs. Your second choice would be a runtime error for some of the same reasons, it provides you with some information related to what is wrong and, as a result, helps you fix the problem. By contrast, logic errors provide no information on how to fix them or even where they are. This makes logic errors very challenging to fix, especially in larger programs.

Different languages do more or less to help you with error detection. One of the significant advantages of Scala is that it is designed to maximize the number of errors that are syntax errors and reduce the number of runtime and logic errors. Part of this is due to the type checking system of Scala. The language does significant static type checking to make sure that all the values you are using are of a type that is appropriate for the usage.

Many of the higher order functions and methods in Scala also help to prevent common errors that programmers face in other languages. This is also true of rules such as the requirement to initialize variables at declaration and the general preference of `val` declarations over `var` declarations. To understand this, simply consider the following line of code.

```
val dbl = nums.filter(_>0.5).map(_*2)
```

This line of code will give us a new collection of numbers that are twice the magnitude of the elements but only for the elements that were originally bigger than 0.5. There are not too many ways to mess up this line without having it be a syntax error. It is possible to get a logic error if you mistype the greater than or the multiplication, but that is something no programming language can really fix. The programmer has to correctly communicate the key logic, but this line does not have all that much other than that key logic. Despite this brevity, it is fairly easy to read for anyone who has experience with `map` and `filter`.

To understand the real value of this line of code, you have to consider the alternative, which is what you would have to write in most other programming languages that do not provide `map` and `filter` or if you simply choose not to use them. We will write equivalent code that is specific for an `Array`. The Scala code works equally well for any sequence, but we will ignore that advantage for now.

```
var cnt = 0
for (i <- 0 until nums.length) {
  if (nums(i) > 0.5) cnt += 1
}
val dbl = Array.fill(cnt)(0.0)
cnt = 0
for (i <- 0 until nums.length) {
  if (nums(i) > 0.5) {
    dbl(cnt) = nums(i)*2
```

```
      cnt += 1
    }
}
```

The first loop counts how many elements are greater than 0.5. This is required because `Arrays` have to be given a size when they are created. Once we know how many there will be, we can make the `Array`. The second loop fills in the `Array` with values that are twice those in the original `Array`.

Clearly this second version is longer. More importantly, there are a lot more places where typos become runtime or logic errors. The reality is that the one line version is doing basically this same thing. However, most of the code is in the libraries and is not rewritten by each programmer every time. This works, in large part, because of the first-class functions and the ease with which function literals can be written in Scala.

7.12 End of Chapter Material

7.12.1 Problem Solving Approach

You can make functions that take variable numbers of arguments and you can pass arguments into functions using pass-by-name.

7.12.2 Summary of Concepts

- You can make the last parameter of a function accept a variable number of arguments, var args, by putting an asterisk after the type. The name given to that parameter functions as a sequence. To pass a real sequence into a function in place of a var args parameter, follow it with :_* to tell Scala you want that seen as a * argument of some type.

- When two variables refer to the same object in memory we call them aliases.

- Mutable objects can cause problems with aliases because changes to the object can be made using any of the aliases, and they will be reflected in all of them.

- By default, arguments in Scala are passed by-value. The value of the reference is passed into the function. This automatically makes an alias of the object. If you want to protect a mutable object from unintended changes when it is passed to a function you need to make a defensive copy.

- Scala also allows pass-by-name parameters. Instead of passing a value, these pass a chunk of code called a thunk that is evaluated every time the parameter is used.

- The type parameters on collections can themselves be other collections. This allows for the creation of multidimensional `Arrays` and `Lists` in Scala. The `fill` and `tabulate` methods can produce these by passing the proper number of arguments into the first argument list.

7.12.3 Self-Directed Study

Enter the following statements into the REPL and see what they do. Try some variations to make sure you understand what is going on. Some of these are intended to fail. You should understand why.

```scala
scala> val arr = Array(1, 2, 3)
scala> arr(0)
scala> arr(1)
scala> arr(0) = 5
scala> arr.mkString(" ")
scala> val lst = List(2, 3, 5, 7, 11, 13, 17)
scala> lst.head
scala> lst.tail
scala> val bigArr = new Array[String](100)
scala> bigArr(0) = "A string"
scala> bigArr(1) = "Another string"
scala> bigArr(0).length
scala> bigArr(2).length
scala> val lst2 = 1 :: 2 :: 3 :: 4 :: 5 :: Nil
scala> lst.zip(lst2)
scala> lst.zipWithIndex
scala> val arr3 = Array.fill(5)(readLine())
scala> val lst3 = List.tabulate(20)(i => i*i)
scala> def printList(lst:List[Int]) = lst match {
  case h::t =>
    println(h)
    printList(t)
  case Nil =>
}
scala> printList(lst3)
scala> lst3.mkString(" ")
scala> lst3.map(i => math.sqrt(i))
scala> arr3.map(s => s.length)
scala> bigArr.filter(s => s!=null)
scala> arr.permutations.foreach(println)
scala> lst.sliding(4)
scala> def printWords(copies:Int,words:String*) {
  words.foreach(w => println(copies*w))
}
scala> printWords(3,"Hi")
scala> printWords(2,"This","is","a","test")
scala> val numWords = lst.map(_.toString)
scala> printWords(4,numWords)
scala> printWords(4,numWords:_*)
scala> val arrAlias = arr
scala> arrAlias.mkString(" ")
scala> arr.mkString(" ")
scala> arrAlias(0) = 42
scala> arr.mkString(" ")
scala> def alterArray(a:Array[Int]) {
  a(0) = 99
  a(1) = 98
  a(2) = 97
}
scala> alterArray(arr)
```

```
scala> arr.mkString(" ")
scala> def getIndex[A](index:Int,a:Array[A]):Option[A] = {
  if (index<a.length) Some(a(index)) else None
}
scala> getIndex(1,arr)
scala> getIndex(4,arr)
scala> lst.find(_ == 5)
scala> lst.find(_ == 9)
scala> def recursiveWhile(cond: =>Boolean)(body: =>Unit) {
  if (cond) {
    body
    recursiveWhile(cond)(body)
  }
}
scala> var i=0
scala> recursiveWhile(i<10) {
  println(i)
  i += 1
}
```

7.12.4 Exercises

1. Think of as many ways as you can to make an `Array[Array[Int]]` that has the values 1-10 in it. Write them all to test that they work. The first two should be easy. If you work you can get at least three others knowing what has been covered so far.

2. Create a 2D array of 4 rows and 5 columns and store the row_index+column_index+5 in each element. Next sum the array by rows and then by columns and print the rows total and the columns total.

3. Write a function called smallestNumber that finds the smallest number in a 2D array and returns the row and column index. Ignore the possibility of a value occurring multiple times.

4. Reverse the diagonal of an array. Create a 2D array that has the same number of rows and columns, for example, the array "arr" has 3 rows and 3 columns. Write a script that will reverse the contents of the diagonal elements. That means you need to swap `arr(0)(0)` with `arr(3)(3)` and swap `arr(1)(1)` with `arr(2)(2)`. To reverse the opposite diagonal you would swap `arr(0)(3)` with `arr(3)(0)` and swap `arr(1)(2)` with `arr(2)(1)`.

5. Write a script that creates a 12X12 Multiplication table. Be sure to show the numbers 1 through 12 as both row and column labels.

6. Given two 2D `Array[Array[Double]]` values of the same length, write a function that returns their by element sum. This is a new `Array` where each element is the sum of the values from the two input arrays at that location. So if you have `Array(Array(1,2,3))` and `Array(Array(4,5,6))` your new array will have `Array(Array(5,7,9))`.

7. Write a function that takes a variable number of grades and returns the median.

8. Write a script that prompts the user for the numbers for a one-dimensional array of size 25. Next, write a function that takes that one dimensional array and places the contents into a two-dimensional array of 5 rows and 5 columns.

7.12.5 Projects

1. This project calculates a very simplified golf handicap. The idea of a golf handicap came about in 1911. The handicap would allow golfers who had different abilities to play together in a more evenly-matched way. The formula used to calculate a handicap is as follows:

$$(Score - CourseRating) \times 113/SlopeRating$$

To calculate your golf handicap you need at least 5 18-hole scorecards or 10 9-hole scorecards. A score card will look something like the following:

Hole	1	2	3	4	5	6	7	8	9	OUT
Par	5	4	3	4	5	4	3	4	4	36
Score	7	5	3	5	8	6	4	5	4	47

Hole	10	11	12	13	14	15	16	17	18	IN	TOTAL
Par	4	4	4	3	5	4	4	3	5	36	72
Score	5	6	4	5	7	7	5	3	7	49	96

Par indicates the predetermined number of strokes that a 0-handicap golfer should take to complete the hole. Although a course's actual course rating may be different than the total par listed on the score card, we will assume that the Course Rating = Total par for the course. The Slope Rating shows the difficulty of a course for an average golfer. We will assume that our course's slope rating is 130. Score indicates how many strokes you took to complete the hole.

First, you will need to record the scorecards. Next, you will need to determine your Adjusted Gross Score (AGS). We will assume that the maximum number of strokes you can take on any hole is 8; so, you will need to adjust the scorecards accordingly (both the per hole score and the total for each round). Now that you have the AGS, you will need to determine your handicap differential. The equation for a Handicap Differential is the Course Rating minus your AGS, multiplied by 113, and divided by the Slope Rating, or $(AGS - CourseRating)x113/SlopeRating$.[11] For example, say your AGS is 96, the Course Rating is 72, and the Slope Rating is 130. You would have the equation $(AGS - CourseRating)x113/SlopeRating$, or $(96 - 72)x113/130$, which results in a Handicap Differential of 12.1. Next, we need to calculate our handicap index. The formula for your Handicap Index is the sum of your differentials divided by the number of differentials, multiplied by 0.96, or $(SumofDifferentials/NumberofDifferentials)x0.96$. Now that you have your Handicap Index, you can total up your Course Handicap using the formula: $(HandicapIndex)x(SlopeRating)/113$. Round to the nearest whole number.

2. This project can be used to assign seats at a dine-in movie theater. The theater has 20 rows, with 9 seats in each row. Row A is for handicapped individuals only. Show the theater patron the seating plan and then ask the theater patron if they need a handicapped accessible seat. Then ask them for their desired seat. The seating display should look something like this:

---

[11]http://www.usga.org/Rule-Books/Handicap-System-Manual/Rule-10/

	1	2	3	4	5	6	7	8	9
Row A	*	*	*	X	X	*	*	*	*
Row B	X	X	*	*	*	*	*	*	*
Row C	*	*	X	*	*	X	X	*	X
Row D	*	*	*	*	*	*	*	*	*
Row E	*	*	X	X	*	*	*	X	X
Row F	X	*	*	*	*	*	*	*	*
.	.	.	.	.	.	.	.	.	.
.	.	.	.	.	.	.	.	.	.
.	.	.	.	.	.	.	.	.	.
Row T	*	*	X	X	X	X	*	*	X

Seats that are already reserved are marked with an X. Seats that are available are marked with a *.

3. Both you and your friend decide to go bowling. Bowling is a game where a ball is rolled down an alley in an attempt to knock down pins. Write a script that calculates the total score of both of your games. One game of bowling consists of 10 frames, with a minimum score of zero and a maximum of 300. Each frame consists of two chances to knock down ten pins. If all ten pins are knocked down on your first ball you get a "strike". A strike is denoted by an X on the score sheet. If it takes you two rolls to knock down all ten pins, it is called a "spare". A spare is denoted by a \on the score sheet. If there is at least one pin still standing after two rolls, it is called an "open frame". Here are the scoring rules:

   - Open frames are scored at face value.

   - A strike is worth 10, plus the value of your next two rolls. For example, if you knock down 4 pins on your next first roll and zero pins on your next second roll, the score for that frame will be 10+4+0=14. However, if you get another strike on your next first roll and another strike on your next second roll, the score for that frame will be 10+10+10=30. The minimum score for a frame in which you get a strike is 10 and the maximum score is 30.

   - A spare is worth 10, plus the value of your next roll. The minimum score for a frame in which you get a spare is 10 and the maximum score is 20 (a spare and a strike on your next roll).

   Each frame consists of up to 2 rolls, except for the tenth frame which could consist of 3 rolls if the player gets a strike on their first roll of the frame. Read in the bowling marks for each frame that you and your friend recorded, then display the score sheet with the game totals displaying the frame, each roll result (X, \, or a number), frame score, and a running total for each player.

4. The Galton box, also known as the bean machine, is a device created by Sir Francis Galton. It consists of a vertical board with evenly spaced pegs in a triangular pattern. Each time a ball hits a peg, it will fall to either the left or right of the peg. Piles of balls are accumulated in the slots at the bottom of the board. Write a script that simulates the bean machine that has 8 slots in which you drop 20 balls. You should simulate the falling of each ball by printing the ball's path. For example, LLRRLLR indicates each direction the ball fell before it landed in slot 4 on a bean machine with 8 slots. Print the final buildup of the balls in the slots in a histogram.

5. Write a function that takes a `Double*` and returns the median value (not the mean)

as a `Double`. Your solution is not expected to be efficient. This is more of a test of your ability to do logic using the methods that were presented in this chapter.[12]

6. Write a function that takes an `Int*` and returns the mode of the values. As with the previous project, the goal here is more for you to think about how to make a solution than to make an efficient solution.

There are additional exercises and projects with data files to work on posted on the book's web site.

---

[12]The API includes a number of methods for sorting. Those could be used to make a more efficient and shorter solution to this problem, but the goal here is for you to do this without sorting.

Chapter 8

Loops

We saw in chapter 5 how we could use recursion to produce an iterative behavior where something was done multiple times. We also saw in chapter 6 that we can use collections to make certain operations happen multiple times. These approaches are the primary methods used in functional languages, and they often work well to provide the functionality we need in Scala. Most languages, including Scala, provide other constructs called loops that are designed specifically for creating this iterative behavior. In this chapter we will explore the loop structures present in Scala and see how we can use these in our programs. We will start by repeating some of the things that we did previously using recursion.

8.1 `while` Loop

The most basic looping construct is the `while` loop. The name tells you roughly what it does. You should keep repeating something while some condition is true. The syntax of the `while` loop is as follows.

```
while (condition) statement
```

The `condition` can be any expression that evaluates to a `Boolean` type. The `statement` is very commonly a block of code, so you will typically see the `while` followed by curly braces.

To see an example of this, let us use a `while` loop in a function that builds a `List` of numbers input by the user. The `List` will end when the user enters "quit". This is exactly like the example that we did with recursion. We could not do it with the collections because we did not know in advance how many numbers there would be.

```scala
def readInts:List[Int] = {
  var lst = List[Int]()
  var input = readLine()
  while (input != "quit") {
    lst = input.toInt :: lst
    input = readLine()
  }
  lst
}
```

This code is distinctly imperative. We have to declare two variables with `var` to make it work. In fact, the `while` loop and it's partner, the `do-while` loop that we will discuss in the next section, are only usable as statements. They are not expressions and cannot be used in places that need to have a value as all they produce is `Unit`. The fact that the `while` loop has to be used in an imperative way is somewhat implicit in the way it works. For the code inside of the `while` loop to execute, the condition must be `true` originally. In order for the loop to finish, the value of the condition has to change to `false`. This change requires the mutation of data at some point; so, it is impossible to use the `while` loop in a completely functional way.

Another thing to note about the `while` loop is that it is a PRE-CHECK loop. This means that the condition is checked before the body of the loop is executed. As a result, it is possible that the contents of the loop will never execute. If the condition is `false` when the loop is first reached, the body of the loop will never execute.

Let us look at another example of the `while` loop. One of our first examples of using recursion to get iteration was the factorial. We can rewrite factorial using a `while` loop in the following way.

```scala
def factorial(n:Int):Int = {
  var product = 1
  var i = 1
  while (i <= n) {
    product *= i
    i += 1
  }
  product
}
```

We declare two variables named `product` and `i` at the beginning and initialize both to 1. The condition on the `while` loop causes it to iterate as long as `i` is less than or equal to `n`. Inside the loop, the value product is multiplied by `i` and `i` is incremented by one. This is an example of a counter controlled loop.

The `*=` and `+=` operators are examples of assignment operators. They provide a handy shorthand for when you want to apply a mathematical operation to a value and store the result back in the original variable. You can follow any operator by an equal sign and Scala will see it as a compound operation that performs the specified operator and stores the value back. The storing of the value is a mutation operation. As such, these operators have to be used with mutable data. That either requires `var` declarations, or mutable constructs such as `Arrays`.

This function also shows another element that is common to most `while` loops and which can lead to a common form of bug. The line `i += 1` is the incrementor (iterator) of the loop. It is what moves us from one case to the next in the loop so that the loop eventually terminates. The common bug is to accidentally leave out the iterator. Consider what happens in this code if you do that. Without the line `i += 1`, the value of i will never change. As a result, the condition will never change either and if it is `true` to begin with, it will be `true` forever. This leads to what we call an infinite loop, a loop that never exits. This type of error is easy to put into the code with a `while` loop, because the loop does not include anything in its structure to remind you to include the incrementor. The fact that the `while` loop requires mutation and is a bit prone to bugs means that it is something we will only use when we find that we have a good reason to. The reason will generally be that we want to repeat some action multiple times, but we have no way of knowing how many before the loop starts, and we do not want to write a recursive function.[1]

8.2 `do-while` Loop

Scala provides a second construct that is very closely related to the `while` loop. It is the `do-while` loop. The syntax for the `do-while` loop is as follows.

```
do {
  statements
} while (condition)
```

The curly braces are not technically required here either, but it is rare to see a `do-while` loop without them. Again, the statement does very much what it says that it does. It will do the statements `while` the condition is `true`.

Given how very similar this sounds to the normal `while` loop, you might wonder what the difference is. The difference is implied by the layout. The normal `while` loop checks the condition before it executes the statements in the body of the loop. The `do-while` loop checks the condition after it executes the body of the loop. As a result, the body of a `do-while` loop will always execute at least once.

The `do-while` loop is not used that often in programming. The only times it is used are in situations where the POST-CHECK nature of the loop is helpful and you want the contents of the loop to always happen once. A common example of this is in menu based applications where you need to read what the user wants to do and then act on it. The decision of whether or not the loop should be executed again is based on what option the user picks.

The `mainGrades` function from the end of chapter 6 was an example of this. In that chapter we wrote the program using recursion because that was the only method we knew for making the program execute the same code repeatedly for an unknown number of times. This function can be converted over to use a `do-while` loop, and the result might look like the following.

```
def printMenu:Unit = {
  println("""Select one of the following options:
1. Add a test grade.
```

---

[1]If a recursive function is not "tail recursive", then the `while` loop will be superior when the action needs to happen more than several thousand times.

```
  2. Add a quiz grade.
  3. Add an assignment grade.
  4. Calculate average.
  5. Quit.""")
}

def mainGrades {
  var tests = List[Double]()
  var assignments = List[Double]()
  var quizzes = List[Double]()
  var option = 0

  do {
    printMenu
    option=readInt()
    option match {
      case 1 =>
        println("Enter a test grade.")
        tests ::= readDouble()
      case 2 =>
        println("Enter a quiz grade.")
        quizzes ::= readDouble()
      case 3 =>
        println("Enter an assignment grade.")
        assignments ::= readDouble()
      case 4 =>
        println("The average is "+
          courseAverage(tests,assignments,quizzes))
      case 5 =>
      case _ =>
        println("Unknown option. Try again.")
    }
  } while (option != 5)
}
```

Whether you use this or the code in section 6.6 is primarily a question of style. Most developers would probably write this version by default, but that is mainly because most developers have a background in imperative programming and will tend to favor this approach for reasons of familiarity. The form using recursion would be preferred by most programmers used to functional programming. Like the `while` loop, the `do-while` loop requires mutation and is prone to the problem of forgetting to increment which leads to an infinite loop.

8.3   `for` **Loop**

The `while` loop is logically the simplest loop, but it is not the most commonly used loop in most languages. In languages that provide a `for` loop, it is typically the most commonly used loop. The `for` loop in Scala is a bit different from that provided in many other languages, but you will probably find that it is the one that you turn to the most when you are putting iteration into your code.

In most languages, the natural usage of the `for` loop is to count so we will start with that. A `for` loop that counts from 1 to 10 and prints out the values would be written as follows in Scala.

```scala
for (i <- 1 to 10) {
  println(i)
}
```

The name `i` is a variable name just like you would get with a `val` declaration for Scala. As such, you can call it whatever you want. For counting loops it is very common to use names such as `i`, `j`, and `k`. However, anything will work and as with other variable names, it is better if your choice communicates something to those reading the code to make it easier for them to understand. After the variable name is an arrow pointing to the left made from a less than and a hyphen or minus sign. You will see this in all of your `for` loops in Scala. You can read the `<-` as "in". After that is a nice expression that you should be able to read. We will talk more about exactly what that means shortly. You can read this `for` loop as something like "for i in 1 to 10".

As we saw in chapter 6, the indexes in collections in Scala do not start counting at one. Instead, they start counting at zero. As such, you often would not count from 1 to 10. Instead, we would normally count from 0 up to 9. This could be expressed in Scala by replacing 1 to 10 with 0 to 9. However, it is very common that you want to start at zero and express the number of elements you want to go through. For this reason you may want to use `until` instead of `to`, thus the following also works in Scala.

```scala
for (i <- 0 until 10) {
  println(i)
}
```

Using `until` causes the counting to finish one before the last value listed.

The `for` loop in Scala is not just about counting though. Indeed, this usage is something of a special case in Scala. In general, the expression to the right of the `<-` in a `for` loop in Scala can evaluate to any type of collection. In other languages, this type of loop that runs through the elements of a collection is often called a FOR-EACH loop because it does something for each element of the collection. That might remind you of the `foreach` method from section 6.5. This is more than a passing resemblance.

To illustrate this usage of the `for` loop, consider the following code in the REPL.

```scala
scala> List.tabulate(10)(i => i*i)
res0: List[Int] = List(0, 1, 4, 9, 16, 25, 36, 49, 64, 81)

scala> for (elem <- res0) {
     | println(elem)
     | }
0
1
4
9
16
25
36
49
64
81
```

In this case, the `for` loop actually does exactly the same thing that `foreach` does and runs the code inside the loop on each of the different elements in the list. What code you put in the `for` loop can be arbitrarily complex. The `println` statements shown here just make simple examples.

A more general beginning description of the syntax of the `for` loop would be the following.

```
for (name <- collection) statement
```

The `name` can be any valid Scala name. The collection can be any expression that results in a collection. The statement can be anything that you want to have happen. Frequently a code block is used, and you will see multiple statements in curly braces.

Let us use a `for` loop to evaluate a polynomial. We will treat an `Array[Double]` as the polynomial where each element in the array is a coefficient of a term in the polynomial. For example, the polynomial $3x^3 + 6x^2 - 4x + 7$ is represented as `Array(3.0,6.0,-4.0,7.0)`. We want the function to evaluate the polynomial for a particular value of x that will also be passed into the function. We will write this in several ways. The first one will use a counting loop.

```scala
def evalPolyCount(coefs:Array[Double],x:Double):Double = {
  var ret = 0.0
  for (i <- 0 until coefs.length) {
    ret += coefs(i)*math.pow(x, coefs.length-1-i)
  }
  ret
}
```

This will work, but it is particularly inefficient. The use of math.pow for small integer exponents is very inefficient. Walking through the `Array` with the index is not bad, but if we decided to use a `List` for the coefficients that would change.

Recall that the `for` loop is intended to go through the elements of a collection. As such, we could just run through the elements of `coefs` and perform the math. The only challenge in this is that we were using the index, `i`, to calculate the power of `x` as well. We could get rid of that and remove the use of `pow` if we simply went through the `Array` in the reverse order. Putting that logic into code produces this.

```scala
def evalPolyReverse(coefs:Array[Double],x:Double):Double = {
  var ret = 0.0
  var power = 1.0
  for (c <- coefs.reverse) {
    ret += c*power
    power *= x
  }
  ret
}
```

This version does not count with an index. Instead it runs through the array elements. Each value in the `Array` is put into `c` and then the return value is incremented. A separate variable called `power` is created with a value of 1.0 and each time through it is multiplied by x. This provides us with a running power of x and removes the need to call `math.pow`.

This function is also perfectly correct. It's main drawback is that in order to do the powers of x properly, the `Array` had to be reversed. Given this usage, that will create a completely new `Array` and copy the elements across in the reverse order. While that is also inefficient, this does allow us to nicely illustrate the usage of the `for` loop to run through any collection, even one that is created through operations on other collections. This loop is also very imperative as we have two `var` declarations that are being mutated. So far we have only used the `for` loop as a statement, so this type of mutation is almost inevitable. In section 8.3.2 we will see how we can use the `for` loop as an expression, something we could not do with `while` or `do-while`.

8.3.1   Range Type

Now that you have seen that the `for` loop really just runs through a collection, you might wonder about the counting usage with something like this.

```
for (i <- 0 until 10) { ...
```

To understand this it might help to type some expressions into the REPL and see what is really going on. Here we have done that.

```
scala> 1 to 10
res1: scala.collection.immutable.Range.Inclusive = Range(1, 2, 3, 4, 5, 6, 7, 8,
    9, 10)

scala> 0 until 10
res2: scala.collection.immutable.Range = Range(0, 1, 2, 3, 4, 5, 6, 7, 8, 9)
```

The expressions `1 to 10` and `0 until 10` give us back values that actually are collections. Specifically, they produce instances of the **Range** type, which is a subtype of sequence. You already know that Scala allows infix notation for methods that take only one argument. In this case, `1 to 10` is really `1.to(10)`. You could type it using that format, but there is no reason to.

The **Range** type in Scala gives us a simple way to use the for-each style loop that Scala provides to do the counting operations that most programmers are used to using the `for` loop to accomplish. You can use `to` and `until` with other integer types including the **Char** type. So the expression `'a' to 'z'` will give you a collection of all of the lower case letters.

What if you want to count down? The i **Range** type has a method called by defined on it that will allow you to specify the step between elements in the **Range**. So, if you want to go from 10 down to 1, you can use `10 to 1 by -1`. We can use this to get a version of the polynomial evaluation function that uses the index counting, but does not require `math.pow` and instead keeps track of the exponent of x.

```
def evalPolyCountDown(coefs:Array[Double],x:Double):Double = {
  var ret = 0.0
  var power = 1.0
  for (i <- coefs.length-1 to 0 by -1) {
    ret += coefs(i)*power
    power *= x
  }
  ret
}
```

This version has most of the advantages of both of the previous versions. It does not have to reverse the **Array**, nor does it require the use of `math.pow`. The downside is that it still would not translate well to a **List**, and it is still very imperative.

If you use `by` to specify the step, you can also use **Range**s of the **Double** type. Other, less standard, numeric types like **BigInt** will also work nicely with the **Range** type. The fact that the **Range** type is really a collection means that all of the methods that were discussed in chapter 6 are available for them. This leads to a concise way of expressing factorial.

```
scala> (1 to 5).product
res21: Int = 120
```

In addition to `product` and `sum`, you can also apply `map`, `filter`, or other operations to instances of the `Range` type.

There are some times when you want to count through the indices of a collection like an `Array`. You could do this with code like the following assuming that you have an `Array` called `a`.

```
for (i <- 0 until a.length) ...
```

You can also use the `indices` method on the collection. Calling `a.indices` will give you a `Range` that goes from the first index to the last one. So this loop could also be expressed in this way.

```
for (i <- a.indices) ...
```

Not only is this shorter, it is slightly less error prone in that you cannot accidentally start at 1 instead of 0, nor can you accidentally use `to` instead of `until`.

8.3.2 yield

The `while` loop is a statement only and cannot be used as an expression. This is not true of the `for` loop. You can cause the `for` loop to produce a value so that it can be used as an expression. This is done by putting the `yield` keyword right after the close parentheses of the `for` loop. When you use `yield`, instead of a statement you need to have an expression. The result of the `for` expression is a new collection with all of the yielded values in it. The following shows a simple example of this.

```
scala> for (i <- 1 to 10) yield i*i
res5: scala.collection.immutable.IndexedSeq[Int] =
Vector(1, 4, 9, 16, 25, 36, 49, 64, 81, 100)
```

Another slightly different example shows how you could use a `for` loop to make a collection that is filled with values read in from input.

```
val nums = for (i <- 1 to 10) yield readInt()
```

This could work as an alternative to `fill` if you find it more readable.

You should note that the example gives a result with a type we have not seen before. The general type is listed as an `IndexedSeq[Int]` and the specific object is a `Vector[Int]`. Do not let these different types throw you off. For our purposes, we will use them just like we would the `Array` type. The difference between the `Array` and the `Vector` is that the `Vector` is immutable. You can index into it efficiently like an `Array`, but like a `List`, you are not allowed to change the values of the elements. All the standard functions that we saw earlier for `Array` and `List` will work on these types as well. If you really need an `Array` or a `List`, you can call `toArray` or `toList` on either a `Vector` or a `Range`.

```
scala> (1 to 10).toList
res6: List[Int] = List(1, 2, 3, 4, 5, 6, 7, 8, 9, 10)
```

8.3.3 `if` Guards

The `for` loop in Scala also allows conditionals. After the first generator, you can put an `if` that is followed by a condition. The `for` loop will only happen for those instances where the condition is `true`. This can lead to a more compact syntax than putting an `if` inside of the `for` loop. It can also be more efficient. Most importantly, it can be useful when you have a `yield` so that you do not add unwanted elements to the value of the expression.

As an example of this, and other aspects of the `for` loop, let us consider having a sequence of points in 2-D that are stored as (`Double,Double`). We want an expression that will give us back a sequence that has the distances to those points from the origin. The catch is that we only want the distances that are less than one. Without the `if` guard, this would require two steps. One would calculate the distances and a second would filter out the large values. The `if` guard lets us do this in a single loop.

```
for ((x,y) <- points; if magnitude(x,y)<1.0) yield magnitude(x,y)
```

This example was written assuming a function called `magnitude` that might look like the following.

```
def magnitude(x:Double,y:Double):Double = math.sqrt(x*x+y*y)
```

The beginning of this loop illustrates how you can use a pattern on a tuple to pull out the two elements in the point. This is actually one of the great strengths of the `for` loop in Scala that helps simplify your code.

Syntax Note

Note that you do not have to include parentheses after the `if` in an `if` guard. You can, but unlike a normal `if` it is not required.

The one significant drawback of this approach is that the magnitude function is called twice. The `sqrt` function can be expensive so this is less than ideal. We will see how to get around that shortly.

8.3.4 Multiple Generators

The `for` loop in Scala also supports the ability to iterate across multiple collections in a single loop. This can be done by putting more than one `variableName <- collection` in the parentheses. We call these generators. Each of these that you put into the `for` loop will generate values from the collection to run through the logic. The first generator will pull the first value from its collection. A second generator will then run through all of its values before the first one goes on to the second option. So the number of times the body of the loop happens goes as the product of the number of elements in the collections for the generators, not the sum. To help you understand this, consider the following example.

```
scala> for (i <- 1 to 5; c <- 'a' to 'c') println(i+" "+c)
1 a
1 b
1 c
2 a
2 b
2 c
```

```
3 a
3 b
3 c
4 a
4 b
4 c
5 a
5 b
5 c
```

You can see that the character variable, c, goes through all of its values for each value that i takes on. As a result, there are 15 lines printed.

You might wonder why you would want to do this. Consider again the example of using a 2-tuple to represent points. You might want to make a collection of all the points on a grid in some range of x and y values with a particular spacing in the grid. You could do that with the following code.

```
val xmin = -1.5
val xmax = 0.5
val xstep = 0.01
val ymin = -1.0
val ymax = 1.0
val ystep = 0.01
val pnts = for (x <- xmin to xmax by xstep;
  y <- ymin to ymax by ystep) yield (x,y)
```

The output from the last line will appear in the REPL as something like the following.

```
pnts: scala.collection.immutable.IndexedSeq[(Double, Double)] =
Vector((-1.5,-1.0), (-1.5,-0.99), (-1.5,-0.98), (-1.5,-0.97),
(-1.5,-0.96), (-1.5,-0.95), (-1.5,-0.94), (-1.5,-0.93), (-1.5,-0.92),
(-1.5,-0.91), (-1.5,-0.9), (-1.5,-0.89), (-1.5,-0.88), (-1.5,-0.87),
(-1.5,-0.86), (-1.5,-0.85), (-1.5,-0.84), (-1.5,-0.83), (-1.5,-0.82),
(-1.5,-0.81), (-1.5,-0.8), (-1.5,-0.79), (-1.5,-0.78), (-1.5,-0.77),
(-1.5,-0.76), (-1.5,-0.75), (-1.5,-0.74), (-1.5,-0.73), (-1.5,-0.72),
(-1.5,-0.71), (-1.5,-0.7), (-1.5,-0.69), (-1.5,-0.68), (-1.5,-0.67),
(-1.5,-0.66), (-1.5,-0.65), (-1.5,-0.64), (-1.5,-0.63), (-1.5,-0.62),
(-1.5,-0.61), (-1.5,-0.6), (-1.5,-0.59), (-1.5,-0.58), (-1.5,-0.57),
(-1.5,-0.56), (-1.5,-0.55), (-1.5,-0.54), (-1.5,-0.53), (-1.5,-0.52),
(-1.5,-0.51), (-1.5,-0.5), (-1.5,-0.49), (-1....
```

In this case, the output is truncated before it even gets to the second value of x.

8.3.5 Patterns in for Loops

One example above used the pattern (x,y) to the left of the <- in a for loop. You can use any pattern that you want in that position of a for loop. What makes this truly useful in for loops is that any value that does not match the pattern is skipped.

Our main usage for this will be to quickly and easily pull values out of tuples. However, we can present one other interesting usage here that uses the fact that collections can be used as patterns. Consider the following code that makes an Array[Array[Double]] where each of the contained Arrays has a variable length between 3 and 9.

```
scala> val twoD = Array.fill(100){
    | Array.fill(util.Random.nextInt(7)+3)(math.random)
```

```
  | }
twoD: Array[Array[Double]] = Array(Array(0.9714402463829903, 0.14015197447391436,
    0.8524582916143384, 0.6162004743306447, 0.620366190244299, 0.36698269639501,
    0.46318519546397396), Array(0.6436214632596926, 0.48145976017298175,
    0.5205354884596076, 0.20188086494076174, 0.9186534118857578,
    0.206412655336915), Array(0.41326520865491023, 0.5388572013936772,
    0.3835287127371739, 0.840667735649998, 0.5776048750341035, 0.8564378792435797,
    0.33358311231736193), Array(0.8386133676386185, 0.19634635871412187,
    0.85047321636848, 0.8920110191832437, 0.22432093122102714, 0.9053781210756321,
    0.7642421256500077), Array(0.7958975255688977, 0.30398364976466374,
    0.8810424486159291, 0.1328719423800543, 0.7129174104031204),
    Array(0.6067234631262645, 0.5276942206810142, 0.06504059155788122,
    0.4145379572950526...
```

You can imagine a situation where this is data that you got from some data set and you only care about the entries with only three values in them and for those you only want the average value. The following `for` loop would provide exactly that data.

```
scala> for (Array(d1,d2,d3) <- twoD) yield (d1+d2+d3)/3
res6: Array[Double] = Array(0.7844266684446944, 0.4057923197637461,
    0.44310232980470454, 0.5634809009372609, 0.576642991638965,
    0.3789396949661376, 0.5706536514773105, 0.5844720273258665,
    0.3445436835569556, 0.3547819380526076, 0.5996534540605474,
    0.38416980809208406, 0.8018516553113365, 0.2244482193993954,
    0.5098449834833878, 0.6578966352311121)
```

Note that the array `twoD` has a length of 100, but `res1` has only 16 elements. That is because the other 84 had more than three elements in them. The pattern `Array(d1,d2,d3)` matches only `Arrays` that have exactly three elements in them. Those three elements are bound to the names `d1`, `d2`, and `d2`.

8.3.6 Variable Declarations

With `if` guards, multiple generators, matching patterns, the `for` loop seems like a dream, but wait! There's more! You can define variables inside of the `for` loop. This is helpful for situations like we had earlier where we do not want to have to calculate the magnitude twice for each iteration.

```
for ((x,y) <- points; dist = magnitude(x,y); if dist<1.0) yield dist
```

In this sample, the magnitude is calculated and stored in dist. That value is then used in the two different locations. This way we get slightly shorter code with the real benefit that we only have to calculate the magnitude once.

The generators, `if` guards, and value declarations can be combined in any way given that a generator comes first. This provides a significant amount of power in a single construct. Just do not abuse it to make code that no one can understand.

For Comprehensions

In reality, the `for` loop in Scala is just a form of "syntactic sugar" (syntactic sugar refers to syntax within a programming language that is designed to make things easier to read or write – it makes the language "sweeter" for people to use). When you write

---

a `for` loop, it is converted to appropriate calls to `foreach`, `map`, `flatMap`, and `filter`. In this way, the implementation can be optimal for the collection type in question.

Also, because it is common to have multiple generators, `if` guards, and variable declarations in `for` loops, Scala allows you to leave out semicolons and use newlines instead if you enclose them all in curly braces instead of parentheses.

8.3.7 Multidimensional Sequences and `for` Loops

In chapter 7 we saw that the power of the type system allowed us to do things like make `Arrays` of `Arrays`. At this point you might wonder how you could produce such things with `for` loops. It is tempting to think that using multiple generators might do that, but the example of the grid of points shown above make it clear that doesn't happen. If you want to have a construct with higher dimensions, you need to have multiple nested `for` loops. As an example of this, we will use `for` loops to build the multiplication table that we built in chapter 6 with tabulate.

```scala
val multTable = for (i <- 0 until 10) yield {
  for (j <- 0 until 10) yield i*j
}
```

If you execute this code in the REPL you get the following result.

```
multTable: scala.collection.immutable.IndexedSeq[scala.collection.immutable.
IndexedSeq[Int]] = Vector(Vector(0, 0, 0, 0, 0, 0, 0, 0, 0, 0),
Vector(0, 1, 2, 3, 4, 5, 6, 7, 8, 9), Vector(0, 2, 4, 6, 8, 10, 12,
14, 16, 18), Vector(0, 3, 6, 9, 12, 15, 18, 21, 24, 27),
Vector(0, 4, 8, 12, 16, 20, 24, 28, 32, 36), Vector(0, 5, 10, 15,
20, 25, 30, 35, 40, 45), Vector(0, 6, 12, 18, 24, 30, 36, 42, 48, 54),
Vector(0, 7, 14, 21, 28, 35, 42, 49, 56, 63), Vector(0, 8, 16, 24,
32, 40, 48, 56, 64, 72), Vector(0, 9, 18, 27, 36, 45, 54, 63, 72, 81))
```

This has the same values as the `Array[Array[Int]]` that we made earlier, and we can use it much the same way despite the fact that it is technically a `Vector[Vector[Int]]`. We just can't assign values into it.

The key here is that each `for` loop with a `yield` produces a sequence. If you want to have sequences of sequences, you need to actually put `for` loops inside of `for` loops.

Parallel `for` Loops

Modern processors have the ability to do multiple things at the same time through a process called multithreading. This is a topic that is increasingly important to computer science and which we cover in detail in *Object-Orientation and Abstraction Using Scala* [1]. Here we can provide a brief introduction to a simple way to include multithreading in your programs through parallel collections.

One of the primary motivations for multithreaded parallelism is to more fully utilize the hardware of modern processors and speed things up. So, if you have a loop with a body that does a lot of work, you can make it parallel so that the processor does work on different parts at the same time. You can make an `Array` or a `Range` parallel by calling the **par** method. (Note that you can also call the **par** method on a List, but it is not efficient because it has to convert the List into a different structure that is efficient to work with in parallel.) Many calls to these parallel collections will get split up across multiple threads, as will **for** loops. To see the impact of this, consider the following.

```scala
scala> def fact(n:BigInt):BigInt = if (n<2) 1 else n*fact(n-1)
fact: (n: BigInt)BigInt

scala> for (i <- 30 to 15 by -1) println(fact(i))
265252859812191058636308480000000
8841761993739701954543616000000
304888344611713860501504000000
10888869450418352160768000000
403291461126605635584000000
15511210043330985984000000
620448401733239439360000
25852016738884976640000
1124000727777607680000
51090942171709440000
2432902008176640000
121645100408832000
6402373705728000
355687428096000
20922789888000
1307674368000

scala> for (i <- (30 to 15 by -1).par) println(fact(i))
1124000727777607680000
51090942171709440000
2432902008176640000
265252859812191058636308480000000
121645100408832000
6402373705728000
403291461126605635584000000
355687428096000
15511210043330985984000000
20922789888000
620448401733239439360000
304888344611713860501504000000
25852016738884976640000
10888869450418352160768000000
1307674368000
8841761993739701954543616000000
```

The factorial on **BigInt** is used in part because it is fairly slow. The first version everything runs in the way that you expect and the values of 30! down to 15! are printed in order. With the addition of **par**, the values are no longer printed in order. Instead, the work is broken up into different threads and each value prints after it has been calculated. The biggest values take longer to calculate; so, they are not the first ones to print.

While you can use this to speed things up, it has to be done with care. To get some idea of why this is, consider the following few lines typed into the REPL.

```
scala> var count = 0
count: Int = 0

scala> for (i <- (1 to 1000000).par) count +=1

scala> count
res0: Int = 930420
```

At the end of this, `count` should be 1000000, but it is not. It is about 70000 shy of that. Were you to do this on your own machine, you would certainly get a different value. Doing it on the same machine a second time will even produce a different value. This code has what is called a race condition. While the details of race conditions and how to deal with them is beyond what we will cover here, you can consider using `par` when you are not mutating values. That also means that if it is part of a `for` loop, you should probably be using `yield`.

8.4 Testing

We have now gotten to the point where you can write programs of reasonable complexity. You know most of the constructs that exist in the Scala programming language. As soon as you start writing larger programs, there are some new elements of the programming process that becomes more significant such as testing and debugging.

Just because you have written the code and it compiles and runs does not mean that it is correct. To determine if it actually does what you want it to, you need to test it. This means that you need to run the code with a variety of different inputs to make sure that they all work, and then fix the problems when they do not.

The first part of this, running the program with different inputs is called testing. The challenge in testing is trying to figure out what inputs make good tests. When you are testing code you are actually looking for inputs that will break what you have written. You want to give it some things that you know it will work on, but you also want to give it some things that you think might break it. In addition, a good set of tests will go through all of the functionality of the code.

Thinking of things that will break the code often involves looking for boundary conditions. Things that are right at the edge of valid input. For example, if you have a function that takes an `Array` or a `List`, what happens if the `List` or `Array` you pass in is empty? You want to check different small number input sets as well as large ones. If the code takes numeric values, does it work for both positive and negative? What about zero? Giving you back answers that are wrong or do not make sense can be worse than crashing. If the input is a `String`, what happens when the `String` is empty?

There are some situations where you will be certain that the input has a particular structure. In other situations the input will be coming from a user who might give you something that is not quite what you expected. Even when the input is coming from another part of the program and something that you think will have the right format, there can be cases that you have not considered. Good testing will help you find these situations.

If parts of your code require that certain things be true, you can use the `require` function in Scala to force the program to terminate if a condition is violated. You can call `require` with just a `Boolean` argument. If the `Boolean` is `false`, the program will terminate. The termination can be made more informative by providing a second argument that is a message to give the user if the requirement fails. The following shows how this might be used.

```
def weightedAverage(values:Array[Double],weights:Array[Double]):Double = {
  require(values.length == weights.length,
     "Must have same number of values and weights.")
  require(weights.length > 0,"Average of zero elements not defined.")
  require(weights.sum != 0.0,"Sum of weights can't be zero.")
  (for ((v,w) <- values.zip(weights)) yield v*w).sum/weights.sum
}
```

This function is intended to take the weighted sum of a set of values. There are a number of requirements on the values passed into it. There are three `require` calls that make sure that each of these is `true` before it calculates the value. This might seem like a lot of extra typing, but if you put calls to `require` in your code whenever there really is a requirement, you will find that it makes the debugging process a lot easier.

The other part of testing is coverage. Showing that the code works for one test input is not sufficient to show that it is really correct. How many tests do you have to write to feel confident that your code works? One of the challenges of Computer Science is that you cannot, in general, prove that programs are correct. This was one of the earliest results of Computer Science and is still a fundamental aspect of the theory of the field.[2] Certainly, some programs can be proved to be correct, but generally the best we achieve is to show that our programs work across a broad range of inputs.

There are some criteria, beyond looking for boundary cases, you can use to determine if you have enough tests. The metric to determine this is called code coverage. You want to know what fraction of your code has been executed by the tests. There are a number of different code coverage metrics that can be used.

- Function coverage - Has every function been called?

- Statement coverage - Has every statement been executed?

- Decision coverage - Has every option in branching structures (if and match) been executed?

- Condition coverage - Has every Boolean sub-expression been evaluated as both `true` and `false`?

- Condition/decision coverage - Combination of the two above.

- Loop coverage - Has every loop been executed zero, one, and more than one times?

- Path coverage - Has every path through part of the code been executed?

---

[2]The proof itself was due to Alan Turing showing that you cannot write a program that will take any program and an input and determine if the program terminates when run on that input. This is called the "Halting Problem". The implication is that you cannot, in a completely general way, even show that your program will terminate, much less give the right answer assuming it does stop. There are ways of writing things that avoid errors, but no systematic way of demonstrating correctness. It is worth noting that one nice thing about `for` loops is that they do always terminate as long as they are run on finite collections.

It should also be mentioned that while there is no completely systematic way to prove programs correct, there is a significant amount of work that has gone into proofs of correctness. Unfortunately, proving a program or algorithm correct is often challenging; so, it is only done for small algorithms or when it is absolutely required. Making this more applicable to general programming could be a significant boost to a world that is increasingly dependent on the proper functioning of programs.

The more complete the coverage your test set has, the more confident you are that your code is correct. The levels of coverage higher in this list are basically minimal standards. If your tests have not gone to every function or every statement, then there are parts of the program that you simply have not tested. Going beyond those you start looking at different ways for things to happen. There are often several different places from which a function can be called. Covering decisions will make sure that you have called them from different locations. Covering conditions makes sure that all the possibilities for why different parts of the code might be reached have been exercised.

If you stop and think about it, you will probably realize that getting condition/decision coverage requires quite a few test cases. Even these options potentially leave a lot of possibilities unchecked as they do not force loops to happen different numbers of times. The ultimate form of coverage, path coverage, is generally unattainable for any program of even modest size. Having path coverage implies that you have tested every possible path that the execution could take through the code. Consider a simple function with three `if` statements one after the other. One path through the code would have all three evaluate to `true`. Another path might have the first two `true` and the last `false`. There are actually eight different paths through such a function. If you add another `if`, the number of paths doubles to 16. Path coverage requires exponentially many different cases be tested as conditional statements are added. If that was not bad enough, a single `while` loop or recursion generally creates an infinite number of different paths as the loop could execute zero, one, two, or more times. Each one is a different path through the code. As such, path coverage is typically viewed as an unattainable ideal for anything beyond fairly simple functions.

Due to the challenges of getting good coverage on large collections of code, it is common to do testing on small blocks at a time. This process is called UNIT TESTING. Each different unit of the code has a test suite written for it that checks it's functionality independent of other parts of the code. These test suites are run over and over again during development to make sure that no new changes break code that was written earlier.

The real advantage of Unit testing is that in a small unit, one can hope to get fairly complete path coverage. However, it is not sufficient to only do Unit tests. As the different units are assembled, they have to be put through integration tests that test how the pieces work together. It is very possible for two units of code to work perfectly in isolation and fail miserably when they are put together.

Views (Advanced Topic)

The collection methods that we learned about in chapter 6 provide you with the ability to write concise expressions that have remarkable power. Unfortunately, if you string together many of these methods, the result can be inefficient code. Consider the following for some `List` of `Ints`.

```
numList.filter(_>70).map(_/10-6).sum
```

This expression makes two `List`s and runs through `List`s a total of three times. It first runs through `numList` with the `filter` and produces a new `List` of the elements that pass the `filter`. It then runs through that `List` and maps the elements to create another `List`. Finally it runs through that `List` and takes the sum of the elements. The multiple intermediate `List`s and the iteration through them is inefficient.

All of this extra work happens because the `List` type is a STRICT TYPE. That means that whatever it contains is truly kept in memory as a real structure. For expressions like this we would like the ability to have a non-strict representation of the List. In Scala such things are called Views. Most operations on a View accrue the operation without actually performing it. Later on, the operations can be forced which will cause them to actually happen and produce a strict representation.

To get a View, call the view method of a collection. Operations like map and filter that are done on the View will give you a new View type that has a memory of what operations are to be performed, but the work will not have been done yet. You can force the operations to be applied to give you back a strict representation of the data with the force method. Some other methods, such as sum, which produce a final value, will also force the accrued operations to be performed. So the above could be done using Views in the following way.

```
numList.view.filter(_>70).map(_/10-6).sum
```

This code only runs through the collection once at the call to sum and does not create any intermediate Lists. If numList were particularly long, this could provide a significant benefit.

8.5 Putting It Together

Going back to the theme park, imagine that you have the job of scheduling workers to operate rides. Your scheduling needs to take into account a number of different factors. Each ride needs a minimum number of operators and, on days when there are lots of people riding, that number needs to be increased. Also, the people who are working have to be trained to operate rides. Not everyone has been trained for every ride, so you have to make sure you have enough people scheduled who can operate each ride.

You have data from multiple weeks telling you how many people ride each ride on different days of the week. That is fairly consistent so you will use averages of those values to plan for each day. It is possible to write a program that will generate optimal schedules for an entire week. We are not yet at the point where we are ready to write such a program. Instead, we will write a program that outputs potential schedules for each day of the week. This will help you to build schedules, but will not complete the entire task for you.

The script needs to start by reading in all the data on rides and employees. There will need to be a fair bit of this; so, this is a script that should probably be run using input redirection and having the contents of a file put in as the standard input. The input will start by telling you how many rides there are followed by information for each ride. That information will include a name, the number of operators needed on a slow day, and the number of riders that qualifies as a heavy day. We will assume that on heavy days, one extra operator is needed. That will be followed by the number of employees. For each employee there will be a name, a number of rides they are trained on, and the names of those rides.

The last type of data in the input will be information on sample days. This will start by telling you how many samples there are. Each sample will have the name of a day, the name of the ride, and the total number of people who rode it that day. No assumptions will be made about the days or how many times each day appears.

Once the data has all been read in, the script should run through every day that there is data for, average the number of riders for each ride on that day, and list possible combinations of workers who can cover the rides that day. Any ride that does not have data for a given day can be assumed to be closed and does not need an operator.

The approach to finding possible groups of ride operators requires looping through the rides that are active on a given day and determining how many operators each one needs based on the average number of riders in the data. Our code will store this by having a single sequence with one entry for each operator needed on each ride. The length of that sequence tells us how many total operators are needed.

The `combinations` method is then used to pick all groupings of that many workers as our goal is to not bring in more people than we have to. For each combination, the code will run through permutations of the ride-operator list using `permutations`. It will check whether that permutation has operators who match up with rides they know how to run. If any permutation matches, that combination of operators is a possibility and it is printed. Code for doing this is shown here.

Listing 8.1: RiderSchedule.scala

```scala
 1  import io.StdIn._
 2
 3  def readRide():(String, Int, Int) = {
 4    val name = readLine()
 5    val numOps = readInt()
 6    val heavyRiders = readInt()
 7    (name, numOps, heavyRiders)
 8  }
 9
10  def readEmploy():(String, List[String]) = {
11    val name = readLine()
12    val num = readInt()
13    val rides = List.fill(num)(readLine())
14    (name, rides)
15  }
16
17  def readDay():(String, String, Int) = {
18    val day = readLine()
19    val ride = readLine()
20    val numRiders = readInt()
21    (day, ride, numRiders)
22  }
23
24  val numRides = readInt()
25  val rideInfo = Array.fill(numRides)(readRide())
26  val numEmploys = readInt()
27  val employInfo = Array.fill(numEmploys)(readEmploy())
28  val numDays = readInt()
29  val daysInfo = Array.fill(numDays)(readDay())
30
31  val days = daysInfo.map(_._1).distinct
32
33  for (day <- days) {
34    val thisDay = daysInfo.filter(_._1==day)
35    val rides = thisDay.map(_._2).distinct
36    val operatorRides = rides.flatMap(ride => {
37      val nums = thisDay.filter(_._2==ride).map(_._3)
38      val avg = nums.sum/nums.length
39      val rideData = rideInfo.find(_._1==ride).get
```

```
40      Array.fill(rideData._2+(if (avg>=rideData._3) 1 else 0))(ride)
41    })
42    val totalOps = operatorRides.length
43    for (choice <- employInfo.combinations(totalOps)) {
44      val perms = operatorRides.permutations
45      var works = false
46      while (!works && perms.hasNext) {
47        val perm = perms.next
48        if ((perm,choice).zipped.forall((r,op) => op._2.contains(r)))
49          works = true
50      }
51      if (works) {
52        println(day+" : "+choice.map(_._1).mkString(", "))
53      }
54    }
55  }
```

The top of the code defines three functions on lines 3-22 for reading information. Then lines 24-29 declare values and read in all the data. Once the data has been read in, line 31 finds the days we have data for and stores them in the value `days` using the `distinct` call to remove duplicates.

The primary functionality of the code is for `for` loop that covers lines 33-55. Inside this loop that runs through the days, the variable `thisDay` gets all the ride data for the day being considered on line 34. That is used to build the value `rides` on line 35, which contains the unique rides that we have data for on that day. The next step is to expand that so we have a sequence, called `operatorRides` with each ride duplicated a number of times equal to how many operators are needed for it. This is done using `flatMap` on lines 36-41 with a function that returns an `Array` of the proper size that is built using `fill`.

Line 42 defines `totalOps` as a short name for the total number of ride operators that are needed. Another loop on lines 43-54 goes through all combinations of employees with a length matching the number of operators needed. The selection of operators goes into `choice`. Permutations of `operatorRides` are then taken on line 44, and a check is done to see if operators match with rides in that permutation. This is done with a `while` loop on lines 46-50 so that it can exit early if any match is found.[3] If there is a match, the `choice` sequence with operator names is printed along with the day in question.

A sample input can be found at the books GitHub repository. The output from running that program on the sample input is shown here. This sample input had only four rides and ten employees, but it shows the basic functionality.

Listing 8.2: Schedules.scala

```
Fri : Mark, Lisa, Madison, Kelsey, John, Jason
Fri : Mark, Lisa, Madison, Kelsey, John, Kevin
Fri : Mark, Lisa, Kelsey, John, Jason, Kevin
Fri : Mark, Madison, Kelsey, John, Jason, Jane
Fri : Mark, Madison, Kelsey, John, Kevin, Jane
Fri : Mark, Kelsey, John, Jason, Kevin, Jane
Sat : Mark, Lisa, Madison, Amber, Kelsey, John, Jason, Kevin, Jane
Sat : Mark, Lisa, Madison, Amber, Kelsey, John, Jason, Jim, Jane
Sat : Mark, Lisa, Madison, Amber, Kelsey, John, Kevin, Jim, Jane
Sat : Mark, Lisa, Madison, Kelsey, John, Jason, Kevin, Jim, Jane
Sat : Mark, Lisa, Amber, Kelsey, John, Jason, Kevin, Jim, Jane
```

---

[3]This `while` loop could be replaced by a call to `exists` or `forall` to create a more functional solution. However, the use of a loop is more fitting with the learning objectives of this particular chapter.

```
Sun : Mark, Madison, Amber, Kelsey, John, Jason, Kevin
Sun : Mark, Madison, Amber, Kelsey, John, Jason, Jim
Sun : Mark, Madison, Amber, Kelsey, John, Kevin, Jim
Sun : Mark, Madison, Kelsey, John, Jason, Kevin, Jim
Sun : Mark, Amber, Kelsey, John, Jason, Kevin, Jim
```

One of the significant aspects of this example is the use of `combinations` and `permutations` to run through various possibilities. We will explore alternate ways of solving problems like this that can be more efficient in chapter 15. For now, these methods give us the ability to solve complex problems that would otherwise be out of our reach.

8.6   End of Chapter Material

8.6.1   Problem Solving Approach

This chapter added quite a few new constructs for you to pick from for any given line of code in the form of three different types of loops. These have been added below to what was given previously.

1. Call a function just for the side effects.

2. Declare something:

   - A variable with `val` or `var`.
   - A function with `def`. Inside of the function will be statements that can pull from anything in this list.
   - A type declaration with `type`.

3. Assign a value to a variable.

4. Write a conditional statement:

   - An `if` statement.
   - A `match` statement.

5. Write a loop statement:

   - Use a `while` loop when you do not have a collection or know how many times something will happen, nor do you need to use it as an expression.
   - Use a `do-while` loop in a situation where you could consider a `while` loop, but you know that it should always happen at least once.
   - Use a `for` loop to run through the elements of a collection or to do simple counting.

8.6.2   Summary of Concepts

- The `while` loop is a pre-test conditional loop. It will repeat the body of the loop until a condition check returns `false`. The condition is checked before the first execution of the body and then before any subsequent executions. The `while` loop is used as a statement only. It results in Unit so it cannot be used as a productive expression.

- The `do-while` loop is a post-test conditional loop and is just like the `while` loop except that the condition is checked after each execution of the body. This means that the body of a `do-while` loop will always execute at least once.

- The most commonly used loop is the `for` loop. Scala's `for` loop is a for-each loop that iterates through each member of a collection. It has many options that give it a lot of flexibility and power.

    - A generator in a `for` loop has a pattern followed by a `<-` followed by a collection that is iterated through. The `<-` symbol should be read as "in".

    - To make counting easy, there is a `Range` type that can specify ranges of numeric values. The methods `to` and `until` can produce `Range`s on numeric types. The method `by` can adjust stepping. Floating point `Range`s require a stepping.

    - The `yield` keyword can be put before the body of a `for` loop to cause it to produce a value so that it is an expression. When you have a `for` loop yield a value, it produces a collection similar to the one the generator is iterating over with the values that are produced by the expression in the body of the loop.

    - The left side of a generator in a `for` loop is a pattern. This can allow you to pull values out of the elements of the collection, such as parts of a tuple. In addition, any elements of the collection that do not match the pattern is skipped over.

    - `if` guards can be placed in `for` loops. This is particularly helpful when using `yield`, and the values that fail the conditional check will not produce an output in the result.

    - You can also place value declarations in the specification of a `for` loop. This can help make the code shorter, easier to read, and faster.

- Testing is an essential part of software development. This is where you run the program using various inputs to make certain that it does not fail or produce incorrect output. Proper testing should exercise all parts of the code. It is generally impossible to test all paths through the code, though good coverage is desirable. Challenging test cases often include boundary values.

8.6.3   Self-Directed Study

Enter the following statements into the REPL and see what they do. Some will produce errors. You should try to figure out why. Try some variations to make sure you understand what is going on.

```scala
scala> var i = 0
scala> while (i<20) {
  println(i)
  i += 2
}
scala> while (i<30) {
  println(i)
}
scala> do {
  println(i)
  i -= 1
} while (i>0)
scala> var resp = ""
```

```
scala> do {
  println("Go again? (Y/N)")
  resp = readLine()
} while (resp=="Y")
scala> 1 to 10
scala> 1 to 10 by 2
scala> 0.0 to 1.0 by 0.1
scala> for (i <- 1 to 10) println(i)
scala> for (i <- 1 to 10) yield i
scala> for (i <- 1 to 5; j <- 2 to 4) println(i+" "+j)
scala> val tups = for (i <- 1 to 5; j <- 2 to 4) yield (i,j)
scala> for ((n1,n2) <- tups) yield n1*n2
scala> val twoD = List.fill(6,4)(99)
scala> val mult = Array.tabulate(10,10)((i,j) => i*j)
scala> mult(3)(4)
scala> twoD(1)
```

8.6.4   Exercises

Many of these exercises are identical to ones that were given in chapter 5. The only difference is that those problems were to be solved with recursion and these are to be solved with loops.

1. Write the function `isPrime` that returns a `Boolean` telling if a number is prime using a loop.

2. Write a function using a loop that will print powers of two up to some value.

3. Write a function using a loop that will print powers of two up to some power.

4. Write a function using loops that will print a multiplication table up to 10s. Try to get it running first, then consider how you could make everything line up.

5. Write a function that returns a `List[Int]` of the prime factors of a number using a loop.

6. Repeat exercise 5.11 using a loop instead of recursion.

7. Write code that can take a `List[Int]` and give you back a new one where all the values have been doubled. Do this with a `while` loop, a `for` loop without a `yield`, and a `for` loop with a `yield`.

8. Ask the user to enter an integer number. Next, loop from zero to that number and count how many numbers contain the digit 3. Do this without using `toString`.

9. This problem is like 6.12 in that you are supposed to count the number of even values in an `Array[Int]`. The difference is that now you will do it once with a `while` loop and once with a `for` loop.

10. Another problem that is significant when doing real cryptography is solving linear equations under modulo arithmetic. That sounds complex, but it is really just solutions to the following:

$$ax \equiv b \mod n,$$

where we know a, b, and n and want to find x. To find the solutions to this, there can be more than one, you need to use the extended Euclid's algorithm for exercise 5.16.

You start off by calling the extended Euclid's algorithm on a and n, putting the returned values into d, x, and y. If b is not divisible by d then there is no solution. Otherwise, make $x_0 = x(b/d) \mod n$. The solutions are given by $(x_0 + i(n/d)) \mod n$ for $i \in [0, d-1]$.

11. Try to write functions to do these different things with `String`s in the following ways: with a `while` loop and an index, with a `for` loop and an index, with a `for` loop and no index, with a `Range` and higher-order methods but no loops, and with only higher-order methods.

   - Determine if a `String` is a palindrome.
   - Count the number of times a letter occurs.
   - Remove all occurrences of a letter.
   - Replace all occurrences of a letter (without using any `replace` methods).
   - Count the number of occurrences of a substring.
   - Remove all occurrences of a substring.
   - Replace all occurrences of a substring (without using any `replace` methods).
   - Count the number of vowels.
   - Remove all vowels.
   - Convert all characters to uppercase (without using `toUpper`).
   - Convert all characters to lowercase (without using `toLower`).

8.6.5 Projects

1. This project builds on top of project 6.6. For this you will fill in an entire grid of values with intersection parameters for a set of geometry. Most images on computers are made as grids of values where the values give the colors. We do not quite have the ability to introduce colors and images yet, but we are getting close.

   For now you will fill an `Array[Array[Double]]` with the t parameters for a collection of geometry. You should write a function that takes the following arguments: location of the viewer as a 3-D point, forward direction for the viewer as a 3-D vector, up direction for the viewer as a 3-D vector[4], a sequence of geometry (spheres and planes), and the number of cells across the square grid should be. You will cast one ray for each cell in the grid and see if it hits anything, and, if so, how far out it hits. Fill the grid with the values for minimum intersection parameter.

   The grid represents a plane in space that is one unit in front of the viewer position with a top left corner that is one unit up and one unit to the left. (You can find a left vector by doing the cross product of the up and forward vectors.) The grid extends to one unit to the right and one unit down. This is the basic approach for building images using ray tracing.

2. One of the useful things that you learn in calculus is that functions can be approximated. Your calculus text will mention both the MacLaurin series approximation and the Taylor series approximation. They are basically the same other than MacLaurin

---

[4]For a standard projection the up and forward directions should be perpendicular. However, the math works as long as they are not parallel. You simply get a distorted view in that case.

series are always taken about $x = 0$ and this is what we will be working with here. The definition of the MacLaurin series is

$$f(x) \sim \sum_i \frac{f^{(i)}(0)}{i!} x^i$$

So, this is the sum from $i = 0$ up to some n (or infinity if you want to be really accurate). In the sum we have x raised to the i power times the ith derivative of $f(x)$ evaluated at 0 divided by i factorial. Obviously, this is a real pain to use on functions where taking the derivative is not easy. However, for some functions where the derivatives are straightforward, performing this approximation is very easy. Examples of that would be e^x, $\sin(x)$, and $\cos(x)$.

Write a program that does a Maclaurin approximation of $\cos(x)$. That is not that hard because the derivative is $-\sin(x)$, which has a derivative of $-\cos(x)$ which goes to $\sin(x)$ then back to $\cos(x)$. Also note that you are always evaluating at $x = 0$; so, all the terms for sin go to zero.

The first few terms in this series are:

$$1 - 0 - \frac{x^2}{2!} + 0 + \frac{x^4}{4!} - 0 - \frac{x^6}{6!} + 0 + \frac{x^8}{8!} + \dots$$

For this project, you should ask the user for the x to use, as well as an accuracy. Use the `math.cos` function to get the "real" value of cosine at that value of x. Iterate until the difference between the series value and what that function gives you is less than the input accuracy. After the loop, print out the real value, the value you got from the series, and how many terms you had to sum to get that. (For an extra challenge, make your program use a Taylor series instead. This means inputing another value x_0 which the series is expanded around.)

3. Computers are used extensively for simulating physical systems, especially when the system is hard or impossible to build in a lab. For this project you will write a simple simulation of the gravitational Kepler problem. You will also explore the accuracy of what you are doing in a little bit. Imagine you have a body in orbit around a star. We will assume that the star is much larger than the other body so it stays at the origin, $(0,0)$, of our coordinate system. The other body starts at some position (x,y) with a velocity (v_x, v_y). A simple "integrator" for a system like this can be constructed by a discretization of Newton's laws (a fancy way of saying that we avoid calculus and do things in a way that is more computer friendly). Newton's second law tells us $F_1 = m_1 * a_1$ and for gravity $F = -G \frac{m_1 * m_2}{d^2}$. We are going to simplify this for our toy system and just say that $a = -\frac{1}{d^2}$. We can break this into components and get $a_x = -\frac{x}{d^3}$ and $a_y = -\frac{y}{d^3}$. Now, the trick on the computer is to say that instead of moving smoothly, the particle jumps over certain time steps, Δt. So after one time step the new position is $x = x + \Delta t * v_x$ and $y = y + \Delta t * v_y$. Similarly, $v_x = v_x + \Delta t * a_x$ and $v_y = v_y + \Delta t * a_y$. Doing this in a loop "integrates" the motion of the body. (Use the `math.sqrt` function to calculate d.)

This integrator is very simple, but far from perfect. If you start with your body at $(1,0)$ with a velocity of $(0,1)$ it should be in a nice circular orbit staying that distance forever. By measuring how that distance changes, you can get a measure of how accurate, or inaccurate, the integrator is. You can play with other positions and velocities to see what happens.

You will write a program that takes the initial x, y, v_x, and v_y as well as a time step,

Δt, as inputs. It should advance the system for a total time of 10.0 (so if $\Delta t = 0.1$ that requires 100 iterations). At the end of it you should measure the distance from the origin and print a message giving that and the original distance. Then check to see if the change is less than 1%. If it is, say that the integration was accurate enough, otherwise, say it is not. In a comment in your code you should tell how small you had to make your time step for this to be reached given the coordinate 1 0 0 1. (Note that this measure of accuracy is only good for circular orbits. We are not going to do enough physics to go beyond that, but if you happen to want to, the real objective is to conserve total energy. For an extra challenge, comparing initial and final total energies of the system.)

For fun, you can change it so it prints the x and y values during the simulation and see what is happening with a spreadsheet of using gnuplot in a manner similar to what is described in project 5.4. This can also be helpful for debugging. Such plots are shown on the website.

4. An alternate physics problem that can be solved in the same way as that for the previous project is calculating the trajectory of a projectile. If you consider air resistance, the path of a body is not a simple parabola. Using a numerical integrator that was described in the previous project, you can figure out how far a projectile will go assuming there is air resistance.

   The force of gravity near the ground can be approximated a $\overrightarrow{F}_g = -gm\hat{j}$.[5] The friction force from the air can be approximated by $F_d = \frac{1}{2}\rho v^2 C_d A$, where ρ is the density of the fluid, C_d is the drag coefficient of the shape, and A is the cross sectional surface area of the particle. The value of C_d for a smooth sphere is 0.1. The density of air is about $1.2 kg/m^3$. This force is directed in the opposite direction of the motion.

   Using a `while` loop write a script that will tell you how far a ball will go before it hits the ground with the user specifying the height from which it is thrown/launched, its initial speed, its initial angle, its radius, and its density. If you want a bit of extra challenge, allow the user to input a wind speed.

5. For this problem you will do some string parsing that has relationships to chemical formulas in chemistry. We are going to keep things fairly simple for this. The basic idea is that the user types in a string that is a chemical formula, and your program should parse that string and tell how many of each type of element are on each side of the equation. This is the first step in balancing a chemical equation. A later project will have you go through the process of doing the actual balancing.

   The format of the chemical equation will look something like this: CH4+O2=H2O+CO2. This is a reaction for burning/oxidizing methane. Note that it is not well balanced as there need to be coefficients in front of each term. Your program will assume a coefficient on each term in the equation as a lower case letter starting with 'a' in alphabetical order from left to right and output how many of each element there are. So for this input the output would be:

```
C: a*1=d*1
H: a*4=c*2
O: b*2=c*1+d*2
```

   or if you want to clean it up a bit,

---

[5] If you are not familiar with the notation, $\hat{i}$ and $\hat{j}$ represent unit vectors in the x and y directions, respectively.

```
C: a=d
H: a*4=c*2
O: b*2=c+d*2
```

This gives us three linear equations that we could try to solve (actually we have 3 equations for 4 unknowns so the system is under-determined, but that is often the case, so we will find the solution where a, b, c, and d have the smallest values possible and are all integers but you do not have to worry about that now). We will not be solving it in this project.

To be more specific about the input, it has a sequence of terms that are separated by + or =. The reagents are in terms on the left hand side of the = and the products are on the right hand side of the =. Each term can have one or more element names, each followed by the number of that element in the given molecule. The element names will all be one character long and capitalized. Also, the number of elements will be just one digit. If no number is present you assume there is only one. Allowing elements with more than one letter (uppercase followed by lowercase) or numbers with more than one digit makes a nice project for anyone looking for an extra challenge.

The output should have a separate line for each element that was present in the equation. It should list the symbol for the element followed by a colon and then the equation that tells what the coefficients have to satisfy for that element to be balanced on both sides of the equation. You can choose either format above.

6. You are on the Planning and Development Commission for your city and need to have an idea of how the city's population is going to grow over the years. Create a script that asks the user to enter the birth rate, death rate, current population, and the number of years. The script should then calculate and print the estimated population after that number of years has elapsed. Your script should include a function that calculates the population growth rate and a function that calculates the estimated population after a certain number of years. Your script should not accept a negative birth rate, negative death rate, or a population less than two.

If P is the population on the first day of the year, B is the birth rate, and D is the death rate, the estimated population at the end of the year is given by the formula:

$$P + (B * P)/100(D * P)/100$$

The population growth rate is given by the formula:

$$B - D$$

7. The Luhn algorithm is a simple checksum algorithm used to validate a variety of identification numbers, such as credit card numbers. All credit cards have between 13 and 16 digits and those digits follow certain patterns. For example, the number on American Express cards start with 37. Discover cards start with a 6. MasterCard credit cards start with a 5, and Visa credit cards start with a 4. Luhn's algorithm can be described as follows using card number 4258130280679132 as an example:

- Double every second digit from right to left. If the doubling of a digit results in a two-digit number, you then add up each digit to get a single-digit number. Thus, using our example we get the following:

  $3 * 3 = 6$

$$9 * 9 = 18 \rightarrow 1 + 8 = 9$$
$$6 * 2 = 12 \rightarrow 1 + 2 = 3$$
$$8 * 2 = 16 \rightarrow 1 + 6 = 7$$
$$0 * 2 = 0$$
$$1 * 2 = 2$$
$$5 * 2 = 10 \rightarrow 1 + 0 = 1$$
$$4 * 2 = 8$$

- Add all single-digit numbers produced in the previous step. Using our example we get the following:

  $$6 + 9 + 3 + 7 + 0 + 2 + 1 + 8 = 36$$

- Add all of the digits in the odd locations from right to left (the right-most digit is in the 1's location which is odd) in the credit card number. Using our example we get:

  $$2 + 1 + 7 + 0 + 2 + 3 + 8 + 2 = 25$$

- Sum the results from the second and third steps:

  $$36 + 25 = 61$$

  If this number is divisible by 10, the credit card number is valid; otherwise, it is invalid. Thus 4258130280679132 is invalid, but 4208130280679132 is valid. Using functions, write a script to determine whether or not a credit card number supplied by the user is valid.

8. A word find is well represented by a 2 dimensional array of characters. Enter the following into a 2D array.

```
z p i q x t p c o f r g t y r
d r y n r o r a u o t j u n x
j o h a o j r n y m f g m o a
r g a h a v c e s k x e b s v
e r t z g t e c x w s x l b d
t a y m i o a l w d i r r y e
t m e o d l m d l l u a e y r
i m n r a g z r f q p h u d k
w i i i o t k t y e x l q x o
t n i n i d e k n i l h v l s
j g i z a n y k t c v p v h i
e c l f v p a h e z t x f g a
l a c h e r z k v e z g w m l
j p p m e g k x q b w d m l g
p g o p p j d k z s i w e l i
```

Now, write a program that searches for the following words in this word find and determine if each word is in the puzzle.

art

binary

class

function

geometry

lacher

lewis

linkedin

math

netflix

novell

odersky

pesky

programming

scala

sony

tumblr

twitter

wonderful

xerox

9. Suppose you are at a school that has 1000 lockers which are all shut and unlocked. The school also has 1000 students. The first student opens every locker. The second student then shuts every second locker. Next, the third student stops at every third locker and opens the locker if it is closed or closes the locker if it is open. After that, the forth student stops at every forth locker and opens the locker if it is closed or closes the locker if it is open. This pattern keeps continuing until all thousand students have followed the pattern with all thousand lockers. At the end, which lockers will be open and which will be closed. Write a program to figure this out.

10. For this project you can keep working with recipes. You can think of this as an extension of project 6.8, but you do not have to have completed that project to do this one. For this project you will write a text menu with the following options.

   (a) Add a pantry item.
   (b) Print pantry contents.
   (c) Check a recipe.
   (d) Cook recipe.
   (e) Quit

   If the user selects to add a pantry item you ask for the name of the item and how much they are adding then return to the menus. The option to check a recipe has them enter names and amounts until they give a name of "quit". It then tells them whether or not they can make the recipe based on what they currently have. The last option will subtract the appropriate amounts for the items in the last recipe that was successfully checked. If no recipe has been successfully checked, it should print an appropriate message.

11. For this project you can upgrade what you did for project 6.9 so that there is a text menu with options.

   (a) Add course of interest.

(b) Print current courses.

(c) Remove course of interest.

(d) Build a schedule.

(e) Quit

Adding a course should have them type in a unique `String` for the course number or description along with a numeric value for how much they want that course and an integer for the time slot.[6] When they select remove they should type in the unique ID and that course will be removed from consideration. The schedule building option should ask for how many hours they want to take that semester. It will then print out the three "best" schedules that match that number of hours and do not contain courses at conflicting times.

Additional exercises and projects, along with data files, are available on the book's web site.

---

[6] Real time slots involve days and times. That would make this problem a lot harder. You can do that if you want the challenge, but to keep things simple you could use a number for each standard time slot in the schedule. So use 1 for the first MWF slot, 2 for the next one and so on.

Chapter 9

Text Files

Most of the programs that you use on a regular basis would be far less useful if they did not have the ability to read from and write to files. Consider a word processor that could not save your files and let you load them back in later. A program where you had to print output to paper before you closed to program to have any record of what you had done. Such a program would not be very useful, and it would not be used much. That is true even if it provided you with a very full set of other capabilities.

The reason for this is that all the work you do without a file sits in the temporary memory of the computer that is given to the application when it runs. When the application is stopped, for any reason, all of that memory is lost. It might be given over to other applications or used by the operating system. Whatever happens to it, you no longer have the ability to get to it in any useful way.

Files give you the ability to store values from one run of a program to the next. Files can also be used to store information that you do not want to have to write into the source code or just to store amounts of information larger than what will fit in the memory of the machine. Disk drives are much slower than memory so this last part has to be done with a level of care.

9.1   I/O Redirection

In a certain way, you have already had the ability to make your program deal with files through I/O redirection. When you run the program from the command line as a script, you can have standard input come from a file using "<". You can also make the standard output go to a file using ">". This works very nicely if you are going to be entering the same values multiple times or want to preserve the output so that you can look at it.

The down side of using I/O redirection is that it is rather limited. You only get to read from one file or write to one file. What is more, if you decide to read from a file, you cannot also have the user provide input from the keyboard, or if you decide to have the output go to a file, your user will not see anything that is printed on the screen. These limitations make this approach impractical for most applications. As such, it is important for us to learn other ways to get data into a program.

9.2   Packages and `import` Statements

Before we can learn about reading from files and writing to files, we need to learn a little about how code is organized in large projects, and specifically in the standard libraries of Scala. Almost everything that we have used so far was available in Scala without us having to specify where to look for it. The basic parts of the language are simply available by default. The one exception to this was the various `read` methods. To use those without giving long names, we have put `import io.StdIn._` at the top of scripts and in the REPL. Similarly, file handling is not available by default. We will have to tell Scala where to go looking in the libraries for these things.

The Scala language uses packages to organize code. To understand the reason for packages, consider the `List` type that we learned to use in chapter 6. This type is defined by a class written in the Scala libraries that is available to you by default because it is so commonly used. However, the word `List` is not an uncommon one. Without packages there could only be one type called `List`. If anyone tried to create another one, it would conflict with the first one. Packages allow you to have multiple types that all have the same base name, like `List`, but which are differentiated by being in different packages. In the case of `List`, the one you have been using is technically a `scala.collection.immutable.List`. However, the name `List` could also refer to `java.util.List` or `java.awt.List` from the Java libraries.

These longer names are the fully specified names of the types. They specify both a package name and the specific type name. Packages are typically named in all lower case and the dots separate subpackages in a hierarchy going from broadest to most specific. So the `List` that you have been using sits in the top level package `scala`, which has a subpackage called `collection`, which has a package inside of it called `immutable`. The `List` type is inside of that subpackage. We will not worry about creating packages at this point. Right now all you need to understand is that they exist and how to deal with code that is in packages.

To illustrate what you need to know, we will consider one of the types we will use for the topic of this chapter. To help us read from files we will use instances of the `scala.io.Source` type. While it is useful to know the fully specified name of this type as well as that of others, these full names are longer than what we want to type in all the time. Imagine if you had to type in `scala.collection.immutable.List` any time you wanted a `List` in your code. This is the reason that the `import` statement exists. An `import` statement gives Scala directions on how to find things. If it sees the name of a type or value that is not declared locally, it will look in things that have been imported to see if it can find it there. The basic syntax of an `import` statement is to have the keyword `import` followed by what you want to import. To make it easy to use `scala.io.Source` you can do the following.

```
import scala.io.Source
```

After putting in this line, you can just use the name `Source` instead of the fully specified name and Scala will know what you mean.

So what about the `import io.StdIn._` statement that we have been putting in our code so far? It is actually short for `import scala.io.StdIn._`. The `import` statement in Scala understands nesting, and everything in the package called `scala` is imported by default, so that can be left off. `StdIn` is an object in the `io` package, and the underscore is used as a wildcard to signify that everything in that object should be imported so that we can use one import statement to get all the `read` methods. You could also manually import each method on separate lines like this.

```
import io.StdIn.readInt
import io.StdIn.readLine
import io.StdIn.readDouble
```

You might notice that the `StdIn` object and the `Source` type happen to be in the same package, `io`. This makes sense as both deal with input/output, commonly abbreviated as I/O. In this way, packages do not just prevent name conflicts, they also provide organization and structure to large code bases.

The `import` statement in Scala is very flexible and powerful. You can use `import` anywhere you want, and it will be in effect from that point down to the end of the current code block. So, `import`s have the same scope as other declarations. If it appears inside of curly braces, it will only be in effect inside those curly braces. When you want to import everything in a package or object, you can do so by using an underscore as a wild card. For example, this line would bring everything in the `scala.io` package into scope.

```
import scala.io._
```

Note that this `import` statement does not make the `read` methods visible, it would allow you to refer to them by names such as `StdIn.readInt`, but it only brings the `StdIn` object into scope, not the members of that object. You can also import several things with a single line by grouping what you want to import in curly braces. If you wanted to have the `Source` class and the `BufferedSource` class from the `scala.io` package, but no other classes, you could use a line like this.

```
import scala.io.{Source,BufferedSource}
```

There are a number of other possibilities for `import` in Scala that we will not cover at this time.

By default, Scala always includes three `import`s for every file.

```
import java.lang._
import scala._
import Predef._
```

The first one beings all the standard classes from Java into scope. It also imports the basic `scala` package and then all the contents of the object `scala.Predef`. You can see the full structure of packages for Scala by looking at the API.

9.3   Reading from Files

Now it is time to actually use `scala.io.Source` to help us read information from sources other than standard input. There are other ways that we could read from files that use the Java libraries, but for now we will use `Source` because it provides us with sufficient capabilities and ties in well with the Scala collections that we have already discussed. Any code in this section will assume that you have done an `import` of `scala.io.Source` so that we can refer to it as just `Source`.

There is an object called `Source` that has methods we can call to get instances of type `Source` that we will use to read from things. The simplest way to read from a file is to use the `fromFile` method of the `Source` object.

```scala
scala> val fileSource = Source.fromFile("sampleFile.txt")
fileSource: scala.io.BufferedSource = non-empty iterator
```

As you can see from this output, the object we got back was specifically of the type `scala.io.BufferedSource`. The `BufferedSource` type will provide better performance for files which is why we were given that. It happens that reading from a hard disk is one of the slowest things you can do on a computer. A big part of the slowdown is the result of having to move to the right part of the disk. As a result, it is much faster to read a bunch of data at once, even if you do not need it all yet, than it is to read each little piece of data one byte at a time. The `BufferedSource` does exactly this. It reads in a relatively large amount of data into a buffer and then gives you that data as you request it. When the buffer is emptied, it will read again.

When you are done with the `Source` object, you should call close on it. In this case that would look like the following.

```
fileSource.close
```

There are many reasons you should remember to do this, but two really stand out. The first is that a program is only allowed to have a certain number of files open at a given time. If you fail to close files for an extended period of time, you can run out of available file descriptors and then when you try to open another file it will fail. It also signals to Scala that you are done with the file so that it can clean up things like the memory that was used for the buffer.

FileNotFoundException and IOException

File handling activities include a lot of operations where things can go wrong. In the case of `fromFile` you could have mistyped the file name or the file that is specified might be one that you cannot read from. Either of these situations would result in a `java.io.FileNotFoundException`. When reading from files, you could also attempt to do something like read beyond the end of a file. This and other mishaps would result in a more general `java.io.IOException`. If we want to make your code deal with these situations gracefully, you will need to use the `try`/`catch` expression that was mentioned in section 5.6.

In this situation you might have code like the following.

```scala
def readStuffFromFile(fileName:String):SomeType = {
  try {
    val fileSource = Source.fromFile(fileName)
    // ...
    fileSource.close()
  } catch {
    case e:java.io.FileNotFoundException =>
      // An expression to build an empty SomeType
  }
}
```

9.3.1 Iterators

When the value of the `BufferedSource` is printed all it shows is "`non-empty iterator`". This is because the `BufferedSource`, and the normal `Source`, are both subtypes of a more general type called an `Iterator`. An `Iterator` is much like the `Array` and `List` types that we saw previously. In fact, virtually all the methods that we could call on the `Array` and the `List` are available on an `Iterator`. The difference is that a basic `Iterator` is consumed as you go through it, so it can only be used once. The `Array` and the `List` are said to be `Iterable` which means that they can give us multiple `Iterator`s to use to call those methods over and over again. In the case of an `Iterator`, once you call a method that runs through the whole thing, it is spent and you cannot use it again.

```scala
scala> fileSource.mkString
res0: String =
"This is a sample text file that I have written to use for the files chapter.
There really is not all that much to this file. It is simply supposed to be
used to illustrate how we can read from files.
"

scala> fileSource.mkString
res1: String = ""

scala> fileSource
res2: scala.io.BufferedSource = empty iterator
```

The first call to `mkString` gives us back a string that has the contents of the file. The second call gives us back nothing. The reason for this is clear if we look at the value of `fileSource`. After reading through the file the first time, it has gone from a `non-empty iterator` to an `empty iterator`. There is nothing left for us to read. The `Source` type provides a method called `reset` that you do not get in most `Iterator`s. The `reset` method gives you a new `Source` that is set back to the beginning for you to read again. Note that it is a new `Source`, it does not make the original `Source` non-empty again.

At a fundamental level, the `Iterator` type is based on two methods: `hasNext` and `next`. The `hasNext` method gives a `Boolean` and tells you whether or not there is something more in the `Iterator`. The `next` method will give you that next thing. You should be able to picture how these two methods can be used in either a `while` loop or with recursion to let you run through the entire contents of the `Iterator`. Here we reset our source and call `next` on it.

```
scala> val newSource = fileSource.reset
newSource: scala.io.Source = non-empty iterator

scala> newSource.next
res3: Char = T
```

The reason the `BufferedSource` is just an iterator is not hard to understand. Remember, reading files is extremely slow. You should do it once and get the information you want that time. Reading it over and over again would be inefficient. To help force you toward efficiency, you would have to explicitly reset the file or open it back up to iterate through it again. In addition, files can be very large. An advantage of the `Iterator` is that it only needs to keep the current contents in memory, not the entire file. In contrast, the `Array` and `List` need to use memory for everything that they store.

With the `List` and `Array` types, you saw that they were parametric. We could make a `List[Int]` or a `List[String]`. This is also true of `Iterators`. The `Source` and `BufferedSource` types are specifically `Iterator[Char]`. This means that they operate naturally on individual characters. The call to `next` above illustrated this. Also, if you call functions like `map` or `filter`, or if you convert the `Iterator` to a `List` or an `Array`, the result will be a bunch of characters. We can see that explicitly here.

```
scala> newSource.toList
res4: List[Char] =
List(h, i, s, , i, s, , a, , s, a, m, p, l, e, , t, e, x, t, , f, i, l, e, , t, h,
    a, t, , I, , h, a, v, e, , w, r, i, t, t, e, n, , t, o, , u, s, e, , f, o, r,
    , t, h, e, , f, i, l, e, s, , c, h, a, p, t, e, r, .,
, T, h, e, r, e, , r, e, a, l, l, y, , i, s, , n, o, t, , a, l, l, , t, h, a, t, ,
    m, u, c, h, , t, o, , t, h, i, s, , f, i, l, e, ., , I, t, , i, s, , s, i, m,
    p, l, y, , s, u, p, p, o, s, e, d, , t, o, , b, e,
, u, s, e, d, , t, o, , i, l, l, u, s, t, r, a, t, e, , h, o, w, , w, e, , c, a,
    n, , r, e, a, d, , f, r, o, m, , f, i, l, e, s, .,
)
```

Note that the original character 'T' is missing. That is because it was consumed by the call to `next`. Converting an `Iterator` to an `Array` or `List` only takes the elements that have not yet been iterated through. It also consumes the `Iterator` in the process.

```
scala> newSource
res5: scala.io.Source = empty iterator
```

While technically all the data that you want is in the form of characters, it can be a bit difficult to do what you really want to with it in that format. For this reason, there is a method called `getLines` that will give you back a different `Iterator`. This new `Iterator` is of the type `Iterator[String]` and each element is a full line in the file without the newline character at the end.

```
scala> val lines = fileSource.reset.getLines
lines: Iterator[String] = non-empty iterator

scala> lines.next
res6: String = This is a sample text file that I have written to use for the files
    chapter.
```

9.3.2   String `split` Method

Even lines are not always all that useful because there might be multiple pieces of data on each line. There are many ways that you can split up a `String` into different pieces. For most purposes, the simplest of these is the `split` method, which was introduced in section 6.7.1. The `split` method takes a single argument that is a `String` which should be the delimiter between the pieces you want split up. It will then return an `Array[String]` with everything that was separated by that delimiter. Here is a simple example.

```scala
scala> "This is a test.".split(" ")
res7: Array[String] = Array(This, is, a, test.)
```

The `String`, "`This is a test.`", is split up using a single space as the delimiter. The result has each word from the `String` as a separate element.

Technically, the argument to split is a REGULAR EXPRESSION. We will not go into the details of regular expressions in this book. They are introduced in *Object-Orientation, Abstraction, and Data Structures using Scala*[1]. There are a few things that are worth mentioning at this point. The regular expression can have more than one character. Also, the characters '+' and '*' have special meanings. The '+' says that the character before it can occur one or more times while the '*' says that it can occur zero or more times. This is worth mentioning at this point because it is not uncommon for inputs to potentially have multiple spaces between words. To handle this, you will often call `split` with " +" instead of just a space.

Now we want to put the `split` method into action with the ability to read the file line-by-line. What we want to do is create code that will read in numeric data from a file into a 2-D data structure. In this case, we will create a 2-D `Array` because the direct access capability is useful for most of the applications we would want to use this in. We will do this in two different ways. We will break this up into two functions. The first function will work with any `Iterator[String]` and give us back the `Array[Array[Double]]` that we want. The `Iterator[String]` is helpful because that is what `getLines` will give us and that is more flexible than forcing it to be a file. The second function will take a file name and return the `Array[Array[Double]]`. It will not do much itself other than use a `Source` to read the file and pass that `Iterator` to the first function. Both functions will also take a delimiter for `split` so that the person calling them can choose what we split on.

```scala
def dataGrid(lines:Iterator[String],delim:String):Array[Array[Double]] = {
  (lines.map(s => s.split(delim).map(_.toDouble))).toArray
}

def fileToDataGrid(fileName:String,delim:String):Array[Array[Double]] = {
  dataGrid(Source.fromFile(fileName).getLines,delim)
}
```

Now we can demonstrate how this works by calling it on a file called `numbers.txt`. The file has the values for a 3x3 identity matrix with the values separated by commas and spaces as shown here.

```
1.0, 0.0, 0.0
0.0, 1.0, 0.0
0.0, 0.0, 1.0
```

The choice of format is particular because this is what you would get if you had Excel® or some other spreadsheet program write out a CSV format with only numbers. CSV stands for Comma Separated Values. It is more complex if we include non-numeric data. Anything that is not numeric is put inside of double quotes in a CSV file. We will ignore that for now and only deal with numbers.

We can read in this file and see the result in the REPL.

```
scala> fileToDataGrid("numbers.txt"," *, *")
res8: Array[Array[Double]] = Array(Array(1.0, 0.0, 0.0),
Array(0.0, 1.0, 0.0), Array(0.0, 0.0, 1.0))
```

Everything here should be pretty clear with the possible exception of the delimiter. For this file, it would have worked to use the delimiter ", ". However, that delimiter is not all that robust. If there were any extra spaces either before or after the comma it would fail. It would also fail if a space was left out after the comma. Using " *, *" as the delimiter means that this code will work as long as there is a comma with zero or more spaces either before or after it.

Even with this delimiter, the `dataGrid` method leaves some things to be desired. We are not trying to make it deal with `String`s so we do not consider it a problem if this function crashes when there is a value in the file that is not a number. More problematic is the fact that it also does not deal well with blanks. In a CSV file there are two types of blanks that can occur. One is when there are two commas with nothing between them and one is when there is a completely blank line. If either of these occurs in the file right now, our code will not respond well. We could fix this in our current version by adding some extra logic, but as long as we are good with just skipping the blanks, it is easier to do this with `for` loops and `if` guards.

```
def dataGrid(lines:Iterator[String],delim:String):Array[Array[Double]] = {
  (for (l <- lines; if !l.trim.isEmpty ) yield {
    for (n <- l.split(delim); if !n.trim.isEmpty ) yield n.trim.toDouble
  }).toArray
}
```

The `if` guards make it easy to skip things that do not fit what we want. In this case, what we want is a `String` that is not empty. Recall that the `trim` method on a `String` gives back a new `String` with all leading and trailing white space removed.

9.3.3 Reading from Other Things

One of the nice things about the `scala.io.Source` object is that it has more than just the `fromFile` method for us to use in reading files. Indeed, there are quite a few different methods that start with `from` in `scala.io.Source`. We will only consider one of them here, though you can certainly go to the API to look at others. The one you might find most interesting is the `fromURL(s:String)` method. URL stands for Uniform Resource Locator, and they are the things that you type into the address bar of a web browser. You call this method passing it a URL as a `String`, very much like what you would put in the web browser.

This makes it remarkably easy to have your program read information off of the web. For example, executing the following line will give you back the contents of my web page as a `String`.

```
Source.fromURL("http://www.cs.trinity.edu/~mlewis/").mkString
```

You can do everything with this `Source` that you would do with a `Source` from a file. Like with the file, the `Source` is an `Iterator` so you can only read through it once without calling

reset. This also makes sense because if there is anything slower on a modern computer than reading from a hard disk, it is reading information from the network.

9.3.4 Other Options (Java Based)

The advantage of using a `scala.io.Source` is that you get a Scala based `Iterator` that has all the methods you have gotten used to in working with the `List` and `Array` types. However, that does not mean that `Source` is always your best option. For some situations, you might find that the `java.util.Scanner` class is better.

The `java.util.Scanner` class has methods for reading specific types of data as well as for checking if there are specific types of data available. Here is a sampling of the methods that are available in `java.util.Scanner`.

- `hasNext():Boolean` - Check to see if there is another "word".

- `hasNextDouble():Boolean` - Check to see if there is a `Double` ready to read.

- `hasNextInt():Boolean` - Check to see if there is an `Int` ready to read.

- texttthasNextLine():Boolean - Check to see if there is a line of text ready to read.

- `next():String` - Read the next "word".

- nextDouble():Double - Read the next `Double`.

- `nextInt():Int` - Read the next `Int`.

- `nextLine():String` - Read the next line of text.

- `useDelimiter(pattern:String)` - Change the pattern used for the delimiter.

This list should be fairly self-explanatory. The only thing one might wonder about is why "word" is in quotes. The reason for this is reflected in the last method listed. When you want to read a "word" with a `Scanner`, it will read up to the next delimiter, whatever that happens to be. By default it is any grouping of white space.

In order to use a `Scanner`, you first have to make one. To do this you probably want to `import java.util.Scanner` and use the following expression.

```
new Scanner(file)
```

The `file` needs to be an expression that has the type `java.io.File` which we will also make with a new expression. You will probably want to `import` that type as well. Putting this together, you could put something like this into a program to read and print a bunch of `Int`s if you had a file with integers separated by white space. Note that this would not work on our earlier file called "numbers.txt" because of the commas and the fact that the values in that file were `Double`s.

```
import java.util.Scanner
import java.io.File

val sc = new Scanner(new File(''numbers.txt"))
while (sc.hasNextInt) {
  println(sc.nextInt())
}
sc.close()
```

The choice of printing here comes from the fact that the `Scanner` does not nicely produce a Scala collection. You can get the values into a collection, but it will typically take a bit more effort, for example you could cons it onto a var `List`. You should also remember to `close` your Scanners when you are done with them. If they link to a file, they are holding onto valuable resources.

9.4   Writing to File

Scala does not provide any functionality for writing to files in its own libraries. That is something that is already well supported by Java and adding Scala libraries has not been seen as providing a significant benefit. There are many ways to set up a file to write to in Java. The easiest approach, and the one that we will use, is the `java.io.PrintWriter`. We can make a new `PrintWriter` with new and telling it the name of the file we want to write to. So we can get a `PrintWriter` that we can use for doing output with this code.

```
import java.io.PrintWriter
val pw = new PrintWriter("output.txt")
```

Now we could call methods on `pw` that will cause things to be written to that file. You can use `print` and `println` methods to print to the file in much the same way that you have been using those functions in Scala to print to standard output.

Using this we could write the following code to print 100 random points with x and y values between 0 and 1.

```
for (i <- 1 to 100) {
  pw.println(math.random+" "+math.random)
}
pw.close()
```

The last line is critical because the `PrintWriter` is also used as a file resource with buffering, just like the `BufferedSource`. However, it holds things in memory until there is enough to make it worth writing to the disk. If you do not close the file, the text in the buffer will not have been written out. If you are not done with the file, but you want to make sure that what you have written goes into the file, you can use the `flush` method.

```
pw.flush
```

When you call this, anything in the buffer will be written out. This way if the computer crashes, loses power, or something else goes wrong, that information will be out of the volatile memory.

9.4.1   Appending to File

Creating a `PrintWriter` in the way just described has the side effect that it will either create a new file or delete an existing one and write over it. If there is an existing file and you want to append to the end of it instead of overwriting it, there is one more step that needs to be added. Instead of simply telling the `PrintWriter` what file name to use, we tell it a `java.io.FileWriter` to use. When we create a `FileWriter` we can tell it the file name and also give it a `Boolean` value that tells it whether or not to append. If the `Boolean` value is `true`, the contents of the existing file will be left in place and anything else written will go to the end of the file. If it is `false` it will behave like what we had before. So if you wanted to append to the `output.txt` file you could do the following.

```scala
import java.io.{FileWriter,PrintWriter}
val pw = new PrintWriter(new FileWriter("output.txt",true))
```

---

printf and format

If you have an application where you have to be picky about the formatting of output there are some additional functions/methods that you should be aware of. These are the `printf` function for printing to screen or file and the `format` method on `String`. These two methods provide you with a way of encoding how you want something printed or how you want a string to look. They use a style of formatting that is the default way of doing things in the C programming language.

A full description of formatting is beyond the scope of this book and can be found in the Java API under `java.util.Formatter`. You can find the Java API through a web search or going to the `java.oracle.com` web site. As of this writing, Java 8 is the current version. You would want to look at the Java SE 8 version of the API or whatever newer version might be available. A brief introduction is presented here.

Let us start by creating a few variables and printing them using `println` the way we are accustomed to doing.

```scala
scala> val g = 6.67e-11
g: Double = 6.67E-11

scala> val name = "Mark"
name: java.lang.String = Mark

scala> val classSize = 20
classSize: Int = 20

scala> println(g+" "+name+" "+classSize)
6.67E-11 Mark 20
```

This approach requires using + a lot for `String` concatenation. It also gives us very little control over how the different values are printed. The first issue could be addressed with string interpolation, but both complaints can be addressed using `printf`.

The `printf` function uses variable argument length parameters. The first argument is a format `String`. This is followed by arguments for that format. The format `String` can include format specifiers. These start with a percent sign and end with a character that specifies a type of conversion. Between these there can be some additional information that describes how the conversion should be applied. Some of the more common conversion types are:

- `d` – decimal formatting for an integer,

- `e` – scientific notation for a floating point value,

- `f` – standard decimal notation for a floating point value,

- `g` – uses either `e` or `f` formatting depending on the number, the precision, and the width,

- s – a string,

- x – hexadecimal formatting for an integer.

Here is an example using the variables above with spaces between them.

```
scala> printf("%e %s %d",g,name,classSize)
6.670000e-11 Mark 20
```

By default, the scientific notation displays a significant number of trailing zeros. It is possible to specify the width, in characters, of values by putting a number between the percent sign and the conversion. For fractional numbers, precision can also be specified by putting a dot and the number of digits of precision that are desired. Here are some examples showing that.

```
scala> printf("%.2e %s %d",g,name,classSize)
6.67e-11 Mark 20
scala> printf("%19.2e %s %d",g,name,classSize)
        6.67e-11 Mark 20
scala> printf("%19f %s %d",math.Pi,name,classSize)
        3.141593 Mark 20
```

By default, values are right-aligned when the width is larger than the number of characters displayed. This can be changed by putting a flag after the percent sign. The - flag says you want the value left-aligned. The comma flag says that you want proper regional separators put in long numbers. Examples of both of these are shown here.

```
scala> printf("%-19f %s %d",math.Pi,name,classSize)
3.141593           Mark 20
scala> printf("%,19d %s %d",Int.MaxValue,name,classSize)
    2,147,483,647 Mark 20
```

9.5   Use Case: Simple Encryption

To demonstrate the use of files, we will write some very basic encryption and decryption scripts. At this point we will not be using any serious cryptography methods. There are some end of chapter exercises and projects that build up to that. The approach we will take will probably stop others from reading your stuff, but you would not want to use them to access your bank accounts.

9.5.1   Command Line Arguments

We want to write scripts that allow us to specify all the information needed, including the names of the input and output files, on the command line. In order to do this we need to discuss how we can get the command line arguments into our programs in Scala.

When you run a program as a script in Scala, anything that appears on the command line after the name of the script file is available in the program as an element of the `Array args`, which has the type `Array[String]`. We can get the values out of this `Array` the same way we would any other `Array`. You can refer to `args(0)` to get the first argument after the file name, `args(1)` to get the second, and so on. A simple example to illustrate the use of arguments would be a script that converts two arguments to `Doubles` and adds them.

```
println(args(0).toDouble+args(1).toDouble)
```

If we put this into a file called `add.scala` we could call it with the following command on the command line.

```
scala add.scala 88 12
```

This would print 100.0 as the sum. The decimal point appears because the `Strings` were converted to `Doubles` and so the sum is a `Double` and prints as a `Double`.

9.5.2 Mapping a File

To start off with, we will write a script that uses two command line arguments for the input file name and the output file name. The main functionality is in a function that will read a file and write a modified file using a function that is passed in as a parameter to map from the character read to the one printed.

To make it so that the formatting is reasonable, the transform function will only be called for characters that are in the alphabet, whether they are uppercase or lowercase. This will leave white space and punctuation intact. Such a function could be written in the following way.

```
def mapFile(inFile:String,outFile:String,trans:Char => Char):Unit = {
  val pw = new PrintWriter(outFile)
  val in = Source.fromFile(inFile)
  for (c <- in) {
    pw.print(if (c.isLetter) trans(c) else c)
  }
  in.close
  pw.close
}
```

Make sure you do the two calls to `close` at the end to clean up things before the function exits. Both `pw` and `in` are local variables so once the program leaves this function you will not have any way to manually close them.

This function could be invoked with a simple identity function to make a copy of a file.

```
mapFile(args(0),args(1),c => c)
```

The function `c=>c` simply returns whatever is passed into it. If you put this in a file that has `imports` for `Source` and `PrintWriter`, you can call it from the command line specifying two file names, one that exists and one that does not, and the contents of the file that exists will be copied to the new file name.

9.5.3 Character Offset

A simple way to make it much harder for anyone to read what is in your file is to take the characters and offset them by a specified amount in the alphabet. We will write a script that uses three command line arguments: one for the input file name, one for the output file name, and the last for the offset amount. The simplest code for doing this might look like the following.

```
val offset = args(2).toInt
mapFile(args(0),args(1),c=>(c+offset).toChar)
```

Here we convert the third argument to an `Int` and then add it to whatever character we are encrypting. This has to be turned back into a `Char` with `toChar` because doing arithmetic with the `Char` type implicitly results in an `Int`.

This is simple enough and works well for both encoding and decoding. To decode a message, simply use an offset that is the additive inverse of what you used originally. The only thing lacking in this is that letters can be shifted out of the alphabet making them look a bit odd when the text is printed. This problem can be resolved by having the characters wrap around the end of the alphabet. So if the offset would move the letter beyond 'z', it is wrapped back to 'a'. Similarly, we need to handle the case where the offset is wrapped before 'a'. The easy way to do this is with a modulo operator. The following code will work for any offset larger than -26.

```
val offset = args(2).toInt
mapFile(args(0),args(1),c=> {
  if (c.isLower) ('a'+(c-'a'+offset+26)%26).toChar
  else ('A'+(c-'A'+offset+26)%26).toChar
})
```

This code is a bit more complex because it has to differentiate between lowercase and uppercase letters so that it knows what to subtract from the character to get a value that we can take the modulo of.

This code also does something that might look a bit odd to you at first. The function literal now includes curly braces. Remember that the curly braces just make a code block which is an expression whose value is the value of the last expression in the code block. This block only has one statement, an `if` expression, but putting it in a block helps to set it off and make it easier to read.

9.5.4   Alphabet Flip

A slightly different way to alter the letters is to flip the alphabet around. So a 'z' will become an 'a', a 'y' will become a 'b', and so on. This encoding does not need an extra argument for something like the offset. It also has an interesting side effect that the transformation is its own inverse. The code to do this might look like the following.

```
mapFile(args(0),args(1),c=> {
  if (c.isLower) ('a'+(25-(c-'a'))).toChar
  else ('A'+(25-(c-'A'))).toChar
})
```

This has the same basic structure of the normal offset version. It just does not need to get an offset, and it uses an adjustment of 25 minus the characters position in the alphabet.

9.5.5   Key Word

Both of the methods described above are reasonably easy to crack assuming someone has some idea about what you are doing. For the offset method they just need to know what offset you used, and they can probably figure that out if they get to look at a decent sample of encoded text. For the alphabet flipping model all they need to know is what you are doing, and they can decrypt any message you send.

To make things a little harder, you can use a key string to provide a variable offset. The encoding starts with the first letter in the key string and offsets the message character by the number of positions that character is above 'a'. It then uses the second letter in the key string to offset the second character in the message. This repeats until the end of the key string is reached as which point it wraps back around and starts using characters at the beginning. Code that does this is shown here.

```scala
val key = args(2)
val factor = args(3).toInt
var keyIndex = 0
mapFile(args(0),args(1),c=> {
  val offset = factor*(key(keyIndex)-'a')+26
  keyIndex = (keyIndex+1)%key.length
  if (c.isLower) ('a'+(c-'a'+offset+26)%26).toChar
  else ('A'+(c-'A'+offset+26)%26).toChar
})
```

This method needs both a `key` and a `factor`. The `factor` is required, because decoding is done by applying the same `key` with the negative of the `factor`. Most of the time the `factor` should be either 1 or -1. The way this code is written, the `key` should only include lowercase letters. After it uses two arguments to create the `key` and the `factor`, it then makes a mutable variable for the index in the `key` which begins at zero. The transforming function calculates an offset using the value of the `key` at the index and then increments the index using modulo so that the value wraps back around to zero. The code for calculating the new character is the same as that used for the first offset method.

9.5.6 Putting It Together

To make a more useful script, all three of these approaches can be put into a single script that takes an extra argument to specify the style to be used. The full contents of such a script are shown here including the `imports` and a repeat of the `mapFile` function.

Listing 9.1: mapFile.scala

```scala
import scala.io.Source
import java.io.PrintWriter

def mapFile(inFile:String,outFile:String,trans:Char=>Char):Unit = {
  val pw = new PrintWriter(outFile)
  val in = Source.fromFile(inFile)
  for (c <- in) {
    pw.print(if (c>='a' && c<='z' || c>='A' && c<='Z') trans(c) else c)
  }
  in.close
  pw.close
}

args(2) match {
  case "copy" =>
    mapFile(args(0),args(1),c => c)
  case "offset" =>
    val offset = args(3).toInt
```

```
    mapFile(args(0),args(1),c => {
      if (c.isLower) ('a'+(c-'a'+offset+26)%26).toChar
      else ('A'+(c-'A'+offset+26)%26).toChar
    })
  case "flip" =>
    mapFile(args(0),args(1),c => {
      if (c.isLower) ('a'+(25-(c-'a'))).toChar
      else ('A'+(25-(c-'A'))).toChar
    })
  case "key" =>
    val key = args(3)
    val factor = args(4).toInt
    var keyIndex = 0
    mapFile(args(0),args(1),c => {
      val offset = factor*(key(keyIndex)-'a')+26
      keyIndex = (keyIndex+1)%key.length
      if (c.isLower) ('a'+(c-'a'+offset+26)%26).toChar
      else ('A'+(c-'A'+offset+26)%26).toChar
    })
}
```

This script uses the third command line argument to tell it what type of transform function to use. Any information, like an offset or a key and factor, should be specified in the arguments after that.

9.5.7 Primes and Real Cryptography

The type of cryptography that you would want for your financial transactions involves significantly more math than what we have just covered. In particular, real cryptography makes extensive use of concepts from number theory. To understand and write cryptography algorithms does not require that you have a full and complete knowledge of number theory. There are some key concepts that you will need to understand though. We will start developing those a little here and add onto it in future chapters with the objective that you will be able to write code for the RSA public-key encryption and decryption system later in the book.

The first concept we need to cover is that of prime numbers. A prime number is a number that is only divisible by one and itself. So the sequence of primes begins as 2, 3, 5, 7, 11, 13, 17, ... All the values that were skipped in this sequence have other divisors. For example, all even numbers greater than two are divisible by two and are therefore not prime. There are an infinite number of primes, and cryptography algorithms work when using very large primes. Since we are not actually trying to secure bank transactions, we will be happy just using numbers that fit inside of our 64-bit Long type.

Now that we have had a little refresher on primes, we can write some code that deals with primes. For this section it will be sufficient to just write a function called isPrime that tells us whether or not a number is prime. Thanks to the Range type and the higher-order methods present in Scala, this can be done in one line.

```
def isPrime(n:Long):Boolean = (2L until n).forall(n%_!=0)
```

This code takes all the numbers between 2 and n-1 and checks whether n is divisible by any of them using modulo. Recall that if a divides evenly into b then b%a==0. As long as that is not true for any of the values between 2 and n-1 then the number is prime.

While this code is easy to write, it is also inefficient, particularly in the cases where **n** is prime. The reason is that we really do not need to go up to **n-1**. If there is not a divisor less than the square root of **n**, there cannot be one above that. If **n** were really large, the difference between **n** and the square root of **n** would be significant. A more efficient function can be written in the following way.

```
def isPrime2(n:Long):Boolean = {
  var i = 2
  while (i*i<=n && n%i!=0) i += 1
  n%i!=0
}
```

A **Range** type could have been used with a square root function, but the **math.sqrt** function works with **Doubles** that introduce difficulties we did not want to deal with here.

You might wonder about the use of a **while** loop here. After all, we generally prefer the **for** loop. In this case, the **while** loop not only allows us an easy way to check if we have gone above the square root of **n**, it also allows this function to stop as soon as it finds anything that divides into **n**. On the other hand, a **for** loop runs for every member of a collection. The **forall** method used above is smart enough to break out early as well.

There are more efficient ways of determining if a number is prime. You do not really need to check all the values even up to the square root of **n**. You only need to check the prime numbers up to that point. We leave it as an exercise to the reader to determine how you might write code that would use that approach to more efficiently determine if a number is prime across many calls to the function.

9.6 End of Chapter Material

9.6.1 Summary of Concepts

- Files are important to real applications as values stored in memory are lost when a program terminates. Files allow data to persist between runs. They can also be used as a source of large data sets.

- Minimal file interactions can be accomplished with I/O redirection. This approach comes with significant limitations for interactive programs.

- Large groups of code, like the libraries for Scala and Java, have to be broken into pieces. Packages are groups of code that have common functionality.

- Names that include full package specifications can be very long. **import** statements can be used to allow the programmer to use shorter names.

- One way to read from files is using **scala.io.Source**.

  - A call to **Source.fromFile** will return an instance of **BufferedSource** that pulls data from the file.

  - **Source** is an **Iterator[Char]**. The **Iterator** part implies that it is a collection that is consumed as values are pulled off it. It gives individual characters.

  - The **getLines** method returns an **Iterator[String]** with elements that are full lines.

- The `split` method on `String` is useful for breaking up lines into their constituent parts. It takes a delimiter as a `String` and returns an `Array[String]` of all the parts of the `String` that are separated by that delimiter.

- You can use the `Source.fromURL(url:String)` method to get a `Source` that can read data from a source located on the web in the form of a URL.

- For some applications it is easier to read data with a `java.util.Scanner`. This does not provide a Scala style collection, but it has methods for checking if certain types of values are present and reading them in a way that is more independent of line breaks.

- You can use the `java.io.PrintWriter` to write text data out to file. This type has print methods like those that you have already become familiar with.

9.6.2 Self-Directed Study

Enter the following statements into the REPL and see what they do. Some will produce errors. You should try to figure out why. Try some variations to make sure you understand what is going on.

```scala
scala> import java.io._
scala> import scala.io._
scala> val pw = new PrintWriter(new File("RandomMatrix.txt"))
scala> for (i <- 1 to 20) {
     | pw.println(Array.fill(20)(math.random).mkString(" "))
     | }
scala> pw.close()
scala> val nums = {
     | val src = Source.fromFile("RandomMatrix.txt")
     | val lines = src.getLines
     | val ret = lines.map(_.split(" ").map(_.toDouble)).toArray
     | src.close()
     | ret
     | }
scala> import java.util.Scanner
scala> val sc = new Scanner(new File("RandomMatrix.txt"))
scala> var nums = List[Double]()
scala> while (sc.hasNextDouble) {
     | nums ::= sc.nextDouble()
     | }
scala> nums
scala> val webPageSource = Source.fromURL("http://www.google.com")
scala> val webLines = webPageSource.getLines
scala> webLines.count(_.contains("google"))
scala> webPageSource.close()
```

9.6.3 Exercises

1. If you did project 6.7, you can now modify it so that it does not have to use input redirection. Make a function that reads in the dictionary of words and then write a script that will allow the user to input Scrabble letter sets until they enter a special value to stop the program.

2. Create a file with the letters of the alphabet on one line separated by spaces. The challenge is that you cannot manually type in a `String` with that.

3. Create a file with the letters of the alphabet with one letter on each line.

4. Write scripts that copy a file using each of the following constructs. They should take two command line arguments for the input file name and the output file name.

   - `while` loop
   - `for` loop
   - Higher order methods
   - Recursion
   - You can repeat each of the above using `getLines` if you did not the first time or without it if you used it the first time.

5. Write scripts using each of the different methods from exercise 4 that capitalizes every letter.

6. Write scripts using each of the different methods from exercise 4 that shift each vowel up one. So "a" becomes "e" and so forth.

7. Write a script that takes a number of rows and columns as command-line arguments and outputs a file with a matrix of random numbers of that size.

8. Write a script that takes a filename as a command-line argument and reads in a matrix of numbers form that file. It should print out the row and column sums and averages.

9. The Raven, by Edgar Allan Poe, is considered by some to be scary. Make this poem less scary by substituting a parrot (or some other bird of your choice) for the raven and then output the newly revised poem to a file. A text file for this poem can be found at http://www.textfiles.com/etext/AUTHORS/POE/poe-raven-702.txt.

10. Go to http://www.gutenberg.org/files/3201/files/, you will find a number of text files that you could play with. The text file that we need for this exercise is CROSSWD.TXT. Write a program that counts the number of words that do not have an "e" in them and prints each word.

9.6.4 Projects

1. There is a utility in Linux called `wc` that stands for "word count". You can run it on one or more files, and it will tell you how many lines, words, and characters there are in the files. For this project you will do the same thing, only you should also count how many times each letter occurs in each file (regardless of case). The files you should work on will be specified on the command line. Your program should read through each file and count up how many characters, words, and lines there are as well has how many times each letter occurs. You will print this information for each file followed by a grand total for all the files. Your program might be invoked as `scala wc.scala *.txt`, which would go through all the ".txt" files in the current directory.

   You will consider any string that is between white spaces to be a word. To make the counting of letters easier try doing this: `"abcde".map(c=> c-'a')` to see what it does and think about how that can help you.

2. Back in section 9.5 we went through a few simple forms of encryption. These might keep most people from reading your files, but you would not want to make changes to your bank account using them. The encryption techniques that are used for sending truly sensitive messages rely heavily on number theory. The most common systems today are what are called public-key cryptosystems. In this type of system, each user has both some information that they make public, the public-key, and some information they keep private, the private-key. For this project, you will implement a simple version of the RSA cryptosystem.

The way the RSA cryptosystem works is the following:

(a) Select two prime numbers p and q such that $p \neq q$. The level of security of the system is dependent on how big p and q are. For good security you want them to be 512-1024 bit numbers, far larger than what you can store in an Int or even a Long. Finding large primes is beyond the scope of this book so to keep this simple you should use two primes that multiply to give you a number between 128 and 256.

(b) Compute $n = pq$.

(c) Select an small odd integer, e, that is relatively prime to $(p-1)(q-1)$.

(d) Compute d, the multiplicative inverse of e, modulo $(p-1)(q-1)$. The technique for finding this is described in exercise 8.10.

(e) The pair (e, n) is the RSA public-key. You or anyone who wants to send you a message know these values.

(f) The pair (d, n) is the private-key. You need to know this to decode a message and you do not want anyone else to have it.

(g) To encode a message, M, the sender uses $P(M) = M^e \bmod n$. For this to work you need to break up the message into chunks that can be encoded into numbers, M, that are less then n.

(h) To decode the message you use $S(C) = C^d \bmod n$, where C is the encrypted message you got from the sender.

3. The proper way to represent a recipe is using a file. The same thing is true for the contents of your pantry. For this project you will extend project 8.10 to include options for building a cookbook along with file access for the pantry contents. The script should take two command-line arguments for a pantry file and a recipe file. Those files should be loaded in when the program starts. The menu should have the following options.

(a) Add a pantry item.

(b) Print pantry contents.

(c) Add a recipe.

(d) Check a recipe.

(e) Cook recipe.

(f) Quit

When the user selects the "Quit" option. The files should be written to reflect new changes. You can decide what file format to use. You simply have to make the code that writes it work with the code that reads it.

In this version, the "Add recipe" option should have the user type in a unique name of a recipe followed by item names and amounts. That should be remembered so that the "Check a recipe" option allows the user to type in just a name.

4. For this project you will extend what you did as part of project 8.11 so that the courses you are interested in will be stored in a file for reuse. The script should take a single command-line argument for the name of the storage file. The file stores the course information along with the level of interest. Menu options should be as follows.

   (a) Add course of interest.

   (b) Print current courses.

   (c) Remove course of interest.

   (d) Modify course interest.

   (e) Build a schedule.

   (f) Quit

   There is a new option to modify interest because if the program is run over a period of time, the user might become more or less interested in the course. Also, for the build option, you should add the option to have the user say they are taking one of the printed schedules and remove all those courses from the list of interests. Save the modified list of courses of interest to the file when the user quits.

5. The National Weather Service computes the windchill index using the following formula:

$$33.74 + 0.6215T - 35.75(V^{0.16}) + 0.4275T(V^{0.16})$$

   Where T is the temperature in degrees Fahrenheit, and V is the wind speed in miles per hour. This formula only applies for wind speeds over 3 miles per hour. Write a program that calculates the windchill for all the data in a file. A file that you can use for this project can be found at http://www.wunderground.com/history/. Search for your location and date, then go to the bottom of the page for a comma separated file.

6. Write a hangman game. If you go to http://www.gutenberg.org/files/3201/files/, you will find a number of text files that you could play with. The text file that we need for this project is CROSSWD.TXT. These files were placed in the public domain by Grady Ward and CROSSWD.TXT a list of words permitted in crossword games such as Scrabble®. Randomly select a word from this list and then prompt the player to guess what the word is - one letter at at time. At the beginning, display a series of asterisks '*' to represent each letter. When the player makes a correct guess, display the letter in its correct location as well as the remaining asterisks '*'. After the player has guessed the word correctly, display the number of misses and ask the user if they would like to continue to play.

7. Grimm's Fairy Tales are available in text file format in Project Gutenberg and can be found at http://www.gutenberg.org/files/2591/2591.txt. Download this file for it is one of the files that you will need for this project. You will need to look through Grimm's Fairy Tales for a series of words. These words will be found in a second file created by you. For example, you could create a file that had the words "gold", "kingdom", and "queen" in it. The goal of this project is for you to create an output file that contains each word, how many times that word was found in Grimm's Fairy Tales and a listing of all the lines that that word was found on.

8. For this project, you will need to download a DVD list from the book's web site. This is a relatively complete list of all the Region 1 DVD's that someone may have in their home theater library. The original list was obtained at

http://www.hometheaterinfo.com/dvdlist.htm, and we are very grateful for the enormous effort that it takes for the authors to maintain this list. You may find it interesting to play with this data set, but for the purposes of this project, we want you to simply search for several things: 1) Find the most expensive DVD in the list; 2) Find all the films produced by New Line™ studios; 3) Find how many titles there are from Netflix™ studios; 4) What is the title of the oldest DVD in the list; 5) Print out a list of the different Genre's in the file (listing each Genre only once).

9. The Beatles were an English rock band from the 60's. They were widely regarded as the greatest and most influential act of the rock era. The group broke up in 1970, but they are still considered one of the most successful groups of all time. A list of all of the Beatles record titles through 1974 can be found at http://textfiles.com/music/beatle.u-k. Download this file. Then ask the user for a word and list all of the albums, singles, and extended plays contain that word. Next, print out all the dates that include that word in the title.

10. Sports have a tendency to produce a lot of information in the form of statistics. You should be able to go out to the Internet and find many text files that contain statistics information for basketball teams, football teams, water polo teams, etc. For this project you will need to find a text format of box scores for a sport that interests you. One place you can look for text file statistics is at http://www.dougstats.com/ which has NBA and MLB statistics. Read in this data. Then provide a menu of relevant statistics to the user. For example, you may calculate averages for a player, or averages for a team, and you can allow the user a menu option that allows them the ability to select which average they would like to see.

Additional exercises and projects, along with data files, are available on the book's web site.

Chapter 10

Case Classes

One of the things that you should have noticed by this point is that there are times when it can be helpful to group pieces of data together. We have seen two ways that we can do this. If all the data is the same type, you could use something like an `Array` or a `List`. The down side of this is that the compiler cannot check to make sure that you have the right number of elements. For example, if you want a point in 3-D with x, y, and z coordinates then you would need an `Array[Double]` that has three elements in it. Having more or less could cause the code to break and Scala would not be able to check for that until the program was running. The alternative, which also works if the types of the values are different is to use a tuple. That same 3-D point could be represented as a `(Double, Double, Double)`. While this works reasonably well for a 3-D point, it too has some significant limitations. The main limitation is that being a `(Double, Double, Double)` does not tell you anything about what those three `Double`s mean (i.e. which is x, which is y, and which is z) or how it should be used.

To illustrate this, consider some different things that could be represented as three `Double`s. One is a point in 3-D. Another closely related one is a vector in 3-D. Either the vector or the point could be represented in Cartesian coordinates, cylindrical coordinates, or polar coordinates and the tuple would not tell you which it is. The three `Double`s could also represent the high temperature, low temperature and average temperature for a day; or the regular price, sales price, and cost of a sweater; or three sub-averages for a students grade in a class. For example, they might be the test, quiz, and assignment averages. The tuple does not tell you which *type* of data is in the tuple and this does not help you at all with keeping things straight.

What is more, the tuple lacks some flexibility and the syntax for getting things out of them is less than ideal. Calling the `_1` and `_2` methods all through your code can make it difficult to read. Imagine if instead you wanted to represent a full student in a class. Then the tuple might have a `String` and a large number of grades, all of the same type. Keeping track what numbers are associated with what grades would very quickly become problematic. To get around these limitations, we will consider the use of `case class`es for grouping data.

10.1   User Defined Types

What we really need to break away from the limitations of using tuples to deal with groupings of data is the ability to define our own *types* that have meaningful names. Tuples definitely have their place, and they are very useful in those situations, such as when you need to return 2 or 3 values from a function. However, there are many times when it would be handy to create a type specifically for a particular purpose. Then we could give the type a name that indicated its purpose and have the compiler check for us that we were using that type in the proper way.[1]

User defined types are a common feature of programming languages and have been for decades. Scala provides two constructs for creating user defined types: `class`es and `trait`s. For this chapter, in order to keep things simple, we will only consider a specific type of one of these, the `case class`.

Back in chapter 7 we defined a type as a collection of values and the operations that can be performed on them. The user defined types typically take the form of being collections of other types. This makes them fundamentally similar to just using a tuple. Where they prove to be more than just a tuple is the control they give you in determining what can be done with the types. We will remain somewhat limited in this regard for now, but even with those limitations you should find that our user defined types provide a significant boost to our programming capabilities.

10.2   `case classes`

The simplest way to start with user defined types in Scala is with `case class`es. We will start with two examples that were mentioned above: a 3-D point and a student with some grades.

```scala
case class Point3D(x:Double, y:Double, z:Double)
```

```scala
case class Student(name:String, assignments:List[Double], tests:List[Double],
  quizzes:List[Double])
```

The first one declares a type called `Point3D` that stores inside of it three different `Double`s that are named `x`, `y`, and `z`. The second declares a type called `Student` that has a `name` which is a `String` and three different `List`s of `Double` to use as grades.

There can be more to the declaration of a `case class`, but for now we will limit ourselves to this syntax that begins with the keywords "`case class`". After that is the name you want to give the type. This could be any valid Scala name, but it is customary to begin type names with uppercase letters and use camel case naming so all subsequent words also begin with uppercase letters. After that is a list of name/type pairs in parentheses. The format of these is just like the arguments to a function. The elements of this list give the names and types of the values stored in this new type. The values stored in a `class` are called MEMBERS and a `case class` is a `class`.

---

[1]Using a `type` declaration, it is possible to give a more meaningful name to a tuple, but it is still an equivalent type, so the compiler will accept any tuple of the correct structure and does not help you as much finding errors.

10.2.1 Making Objects

After you have declared a type, you need to be able to create objects of that type. With a `case class`, you can do this with an expression that has the name of the `class` followed by an argument list, just like a function call. The two classes listed above could be created and stored in variables with the following lines of code.

```
val p = Point3D(1, 2, 3)
val s = Student("Lisa", Nil, Nil, List(89))
```

The first line makes a point that has x=1, y=2, and z=3 and stores a reference to it in a variable named `p`. The next line makes a student with the name `"Lisa"` who has no assignment or test grades, but who has an 89 on one quiz and stores a reference to it in a variable named `s`.

You could insert the word `new` and a space after the equals signs so that these lines look like the following.

```
val p = new Point3D(1, 2, 3)
val s = new Student("Lisa", Nil, Nil, List(89))
```

The result of this would be exactly the same. The first syntax is shorter and works for all `case class`es so we will stick with that in our sample code.[2]

10.2.2 Accessing Members

In order to be able to use these objects, we must be able to access the different members in them. This is very simple to do, just use the dot notation to access the members. So if you want the x value in the `Point3D`, `p` that we made above, you would just do this.

```
scala> p.x
res1: Double = 1.0
```

To get the `name` of the `Student` you would do this.

```
scala> s.name
res2: String = Lisa
```

The dot notation in Scala simply means that you are using something from inside of an object. It could be a method or a value that is stored in the object. For now we will only be concerning ourselves with the values that we store in our `case class`es.

We could put this to use by writing a function to find the distance between two `Point3D`s. It might look something like this.

```
def distance(p1:Point3D, p2:Point3D):Double = {
  val dx = p1.x-p2.x
  val dy = p1.y-p2.y
  val dz = p1.z-p2.z
  math.sqrt(dx*dx + dy*dy + dz*dz)
}
```

We could also use it to calculate and average for a `Student` with code like this. Note that the minimum quiz grade is thrown out.

---

[2]In chapter 16 you will learn that normal `class`es require the use of `new` by default. To get around this requires writing some code in a companion `object`, a technique also covered in that chapter.

```
def classAverage(s:Student):Double = {
  val assnAve = if (s.assignments.isEmpty) 0.0
    else s.assignments.sum/s.assignments.length
  val quizAve = if (s.quizzes.length<2) 0.0
    else (s.quizzes.sum-s.quizzes.min)/(s.quizzes.length-1)
  val testAve = if (s.tests.isEmpty) 0.0
    else s.tests.sum/s.tests.length
  0.5*assnAve + 0.3*testAve + 0.2*quizAve
}
```

The `if` expressions here prevent us from doing division by zero.

One of the things to note about `case class`es is that the members in them are `val`s by default. As such, you cannot change what they refer to. If you try to make such a change you get something like the following.

```
scala> p.x = 99
<console>:13: error: reassignment to val
       p.x = 99
           ^
```

Whether you can change the fields in an object created from a `case class` depends on whether the things in it are mutable or not. In our two examples, all of the contents are immutable. As a result, the `case class` as a whole is immutable. Once you create a `Point3D` or a `Student`, the object you create cannot change its value in any way. However, if one or more of the fields in the `case class` were an `Array`, then the values in the `Array` would be mutable. You would not be able to change the size of the `Array` without making a new object, but you could change the values stored in it.

10.2.3 Named and Default Arguments (Advanced)

Normally, arguments are passed into a function by position to the corresponding parameter variables, thus Scala figures out which argument is which, by their order. Consider this function.

```
def evalQuadratic(a:Double, b:Double, c:Double, x:Double):Double = {
  val x2 = x*x
  a*x2 + b*x + c
}
```

If you load this into the REPL you can execute it as follows.

```
scala> evalQuadratic(2, 3, 4, 5)
res3: Double = 69.0
```

In this call, a=2, b=3, c=4, and x=5. This is because that is the order the arguments appear in both the definition of the function and the call to it. For functions where there are a significant number of arguments that are all of the same type, this can lead to confusion. To get around this, Scala has NAMED ARGUMENTS. When you call the function, you can specify the names you want the values associated with. So the call above would be like the following:

```
scala> evalQuadratic(a=2, b=3, c=4, x=5)
res4: Double = 69.0
```

In this call it is now explicit what values are going to what parameters. One advantage of this is you can enter the arguments in a different order. For example, you might inadvertently think that x was the first argument instead of the last. Without named arguments, this would lead to an error with no error message. You would simply get the wrong answer. However, if you use named arguments everything is fine because the names supersede the order.

```scala
scala> evalQuadratic(x=5, a=2, b=3, c=4)
res5: Double = 69.0
```

Here we see that even though x is first, the value we get is correct.

You can use named parameters without naming all the parameters. You can start the list with arguments that are based on position and then use names for the later ones. All the arguments after the first named one have to be named, and they cannot duplicate any that you gave using the position.

For some functions, there are some arguments that will have a particular value a lot of the time. For example, if you were creating a program to help a local bank calculate loans, it might be likely that most of their customers live in the same state. In that situation, it is nice to make it so that people calling the function do not have to provide a value for the state and let the function use a DEFAULT VALUE. When you declare the function simply follow the type with an equals sign and the value you want to have for the default. If the caller is happy with the default value, then that argument can be left out. Default arguments at the end of the list can be simply omitted. If they are in the middle then you will have to use named arguments to specify any arguments after them in the list. Consider a function to add a grade to a **Student**.

```scala
def addGrade(name:String, grade:Int = 0):Student = ...
```

Here the default **grade** is a zero. So this function can be called in two ways.

```scala
addGrade("Quinn", 95)
addGrade("Kyle")
```

The first call is like everything we have seen to this point. The second one leaves off the **grade**. As a result, Kyle gets a 0 for whatever grade this was.

10.2.4   The copy Method

The fact that you cannot mutate the values in a **case class** means that it would be helpful to have a way to make new **case class** objects that are only slightly different from existing instances. To see this, consider what happens when you want to add a new grade to a **Student**. The grades are in **Lists**, and it is easy to add to a **List**. The problem is, that does not mutate what is in the original **List**, it just gives us a new **List** that includes the new values as well as what was already there.

To help get around this problem, **case class**es come with a copy method. The copy method is intended to be used with the named arguments that were discussed in section 10.2.3. The arguments to copy have the same names as the fields in the class. Using named arguments, you only provide the ones you want to change. Anything you leave out will be copied straight over to the new object. So using the **Student** object we gave the name s above, we could use copy to do the following.

```scala
val ns = s.copy(tests = 99::s.tests)
```

This gives us back a new **Student** with the same name, assignments, and quizzes as we had in **s**, only it has a test grade of 99.

You can specify as many or as few of the fields in the **case class** as you want. Whatever fields you give the names of will be changed to the value that you specify. If you leave the parentheses empty, you will simply get a copy of the object you have originally.

10.2.5   `case class` **Patterns**

Another capability that comes with **case class**es is that you can use them in patterns. This can be used as a simple way to pull values out of an instance or to select between objects in a **match**. As an example of pulling out values, consider the following code using **Point3D**.

```
for (Point3D(x, y, z) <- points) {
  // Do stuff with x, y, and z.
}
```

This is a **for** loop that runs through a collection of points. Instead of calling each point with a name like **point**, this pulls out the values in the point and gives them the names **x**, **y**, and **z**. That can make things shorter and more clear in the body of the loop.

As an example of limiting what is considered, we can use another **for** loop that goes through a course full of students.

```
for (Student(name, _, List(t1, t2, t3), _) <- courseStudents) {
  // Processing on the students with three test grades.
}
```

This does something with patterns that we have not seen before, it nests them. You can nest patterns in any way that you want. This is part of what makes them extremely powerful. In this case, the assignment and quiz grades have been ignored and the loop is limited to only considering students with exactly three test grades. Those grades are given the names **t1**, **t2**, and **t3**. That could also have been specified with the pattern **t1::t2::t3::Nil**. Students who have more or fewer test grades will be skipped over by this loop. Note that the underscores in this pattern are used to match anything when we do not care about what it is and so we do not give it a name.

10.3   Mutable `classes`

While the functional style benefits from immutable data, and there are some distinct benefits to having all of your data be immutable, we can create **case class**es that mutate by putting the **var** keyword in front of any fields that we need to mutate.[3] So if you wanted to be able to add grades to students through mutation instead of the **copy** method, you could define Student in the following way.

```
case class Student(name:String, var assignments:List[Double], var
    tests:List[Double],
```

---

[3]It is worth noting that this approach is strongly frowned upon in many Scala style guides. If you need a **class** to be mutable, it is strongly suggested that you use the approach discussed in chapter 16. Adding **var** to a **case class** is presented here to keep thing simple, but really is not the appropriate approach.

```
var quizzes:List[Double])
```

There the fields for `assignments`, `tests`, and `quizzes` are declared as `vars`, so they can be mutated and used in a more imperative style. With that declaration, one could made a student and add grades to it in the following way.

```
scala> var s = Student("Lisa", Nil, Nil, Nil)
s: Student = Student(Lisa,List(),List(),List())

scala> s.assignments ::= 99

scala> s.tests ::= 85

scala> s.quizzes ::= 68

scala> s
res5: Student = Student(Lisa,List(99.0),List(85.0),List(68.0))
```

This might look simple, but the mutable state of `s` makes it harder to reason about what is going on and to find flaws when things are done incorrectly. It can cause even bigger headaches when you get to the point where your programs are running in parallel as we saw when briefly looking at parallel loops in chapter 8.

10.4   Putting It Together

Now we want to use a `case class` along with other things that we have learned to create a small, text based application. The application that we will write will use our first definition of `Student` along with the `classAverage` function, which are repeated here.

```
case class Student(name:String, assignments:List[Double], tests:List[Double],
  quizzes:List[Double])

def classAverage(s:Student):Double = {
  val assnAve = if (s.assignments.isEmpty) 0.0
    else s.assignments.sum/s.assignments.length
  val quizAve = if (s.quizzes.length<2) 0.0
    else (s.quizzes.sum-s.quizzes.min)/(s.quizzes.length-1)
  val testAve = if (s.tests.isEmpty) 0.0
    else s.tests.sum/s.tests.length
  0.5*assnAve + 0.3*testAve + 0.2*quizAve
}
```

We will add other functions to make a complete grade book script. This program will be run from a text menu and give us various options similar to what was done in section 8.2.

The program will also use the file handling capabilities that we have learned so that the grades of the students in the course can be saved off and then be loaded back in when we restart the program. The menu for the program will have the following options:

1. Add Test Grade

2. Add Assignment Grade

3. Add Quiz Grade

4. Print Averages

5. Save and Quit

The program will take a command line argument for the file to load in. If none is given, the user will be asked how many students are in the section and their names along with the file name to save it under. When one of the first three menu options is selected, the program will list each student's name and ask for their grade. The "Print Averages" option will print out the names of each student along with their grades in each area, their average in that area, and their total average.

There is quite a bit to this program so it is worth breaking it up into different functions and then writing each of those. To do this we can outline what will happen when we run the program and use the outline to break things down then assign function names to things.

- Startup

  - load a file (`loadSection`)
  - or create a section (`createSection`)

- Main menu (`mainMenu`)

  - print the menu (`printMenu`)
  - act on the selection
    * add a test grade (`addTest`)
    * add an assignment grade (`addAssignment`)
    * add a quiz grade (`addQuiz`)
    * print the averages (`printAverages`)

- Save when done (`saveSection`)

Now that we have figured out roughly what we need to do, we can write these functions in any order that we want. In general the process of writing functions like this can be very non-linear. You should not feel any reason why you would have to go through the functions in any particular order. Often in a real project you would do things in a certain order as you figure out how to do them.

The more experience you gain, the more comfortable you will be in writing code and then you might decide to pick certain functions because they will give you functionality that you can test. One of the advantages of having the REPL to fall back on is that we can load in our file and test functions one by one, seeing the results along the way. Without that, the `printAverages` function would prove to be extremely important to us as it would be the only way that we could see what was going on.

For our purposes we will start with `createSection` and `saveSection`. These two functions pair well together and are closely related because we have to decide how we are going to represent a section both in the memory of the computer and in the file. We will start with `createSection` and the way in which things are represented in the memory of the computer.

We have already created a `case class` called `Student` that can be used to represent one student in the section. We just need several of them. We also need to realize that they will change over time as grades are added. It would probably be sufficient to just keep an `Array[Student]`. However, there are benefits to actually wrapping the `Array` inside of a different `case class` like this.

```
case class Section(students:Array[Student])
```

In general, using a `case class` can provide greater flexibility. We might decide at some point that we want to attach data for a course name, semester, instructor, etc. to each `Section`. Those things cannot be added to a simple `Array[Student]`. However, they could easily be added to the `case class`. It also has the advantage of providing extra meaning. This is not just a random collection of `Students`, it represents a section of a class.

Now that we know this, we can write the `createSection` function. This function will prompt the user for the information that is needed to create the `Section`. For now that is a file name to save it to, the number of students, and the names of the students. The function will return the file name and the `Section`.

```
def createSection:(String,Section) = {
  println("What file would you like to save this as?")
  val fileName = readLine()
  println("How many students are in the class?")
  val numStudents = readInt()
  println("Enter the student names, one per line.")
  (fileName,Section(Array.
    fill(numStudents)(Student(readLine(),Nil,Nil,Nil))))
}
```

The first five lines of this function are fairly self-explanatory with prompts being printed and values being read. After that is the return tuple which includes a call to `Array.fill` that has a `readLine` in the pass-by-name parameter. This means that it not only makes the return value, it also includes the input of the names.

Now that we have created a new `Section`, we can consider what it will look like in a file. There are many different ways that the file could be formatted. For obvious reasons, it is nice if the format that we use to write the file is also convenient for reading it back in. The manner that we will pick here starts with the number of students in the class on a line. After that there are four lines for each student. They are the student's name followed by a line each with assignment, test, and quiz grades. This `saveSection` function can be written as follows.

```
def saveSection(fileName:String, section:Section):Unit = {
  val pw = new PrintWriter(fileName)
  pw.println(section.students.length)
  for (s <- section.students) {
    pw.println(s.name)
    pw.println(s.assignments.mkString(" "))
    pw.println(s.tests.mkString(" "))
    pw.println(s.quizzes.mkString(" "))
  }
  pw.close()
}
```

This function takes the file name and the `Section`. It then makes a `PrintWriter` with the `fileName`, which is closed at the end of the function, and prints the needed information. The use of `mkString` on the different `Lists` makes the code for doing this much shorter.

As you are writing these functions, you need to test them. One way to do that is to load them into the REPL and call them. Another way is to end the script with calls to them. At this point, the end of the script might look something like the following.

```
val (fileName, section) = createSection
```

```
saveSection(fileName, section)
```

This comes after the definition of both the `case class`es and the different functions. If you run the script with this in it, you should be prompted for the information on the `Section` and after you enter that the script should stop. You can then look at the file that you told it to save as and make sure it looks like what you would expect.

We will hold the `loadSection` function until the end and go into the main functionality with `mainMenu` and `printMenu`. You can write them in the following way.

```
def printMenu:Unit = {
  println("""Select an option:
1. Add Test Grade
2. Add Assignment Grade
3. Add Quiz Grade
4. PrintAverages
5. Save and Quit""")
}
```

```
def mainMenu(section:Section):Unit = {
  var option = 0
  do {
    printMenu
    option = readInt()
    option match {
      case 1 => addTest(section)
      case 2 => addAssignment(section)
      case 3 => addQuiz(section)
      case 4 => printAverages(section)
      case 5 => println("Goodbye!")
      case _ => println("Invalid option. Try again.")
    }
  } while (option!=5)
}
```

You cannot test this code yet because `mainMenu` calls four other functions that have not been written yet. Once we have those written, we can put a call to `mainMenu` at the end of the script right before the call to `saveSection`.

The three different add functions will all look pretty much the same. We will only show the `addTest` function and let you figure out the others. It is worth thinking a bit about how that function will work. The `Student` type is immutable. All the fields in the `case class` are `val`s so they cannot be changed. The `String` and the three different `List[Int]` values are all immutable so once a `Student` is created, it is set forever. However, the `Section` type stores the `Student`s in an `Array`. This means we can change what `Student` objects are being referred to. We can use the `copy` capabilities of the `case class` to make new instances that are almost the same except for small variations. Using this, the `addTest` function could be written in the following way.

```
def addTest(section:Section):Unit = {
  for (i <- 0 until section.students.length) {
    println("Enter the grade for "+section.students(i).name+".")
    section.students(i) = section.students(i).
      copy(tests=readInt()::section.students(i).tests)
  }
}
```

This code works just fine, but it is a bit verbose because we have to type in `section.students(i)` so many times. We have to have the index because we need to be able to do the assignment to an element of the `Array`. The `section.students(i)` before the equal sign in the assignment is hard to get rid of because we have to mutate that value in the design of this code. The code could be shortened with appropriate use of `imports`, but there is another, more interesting solution.

```
def addTest(section:Section):Unit = {
  for ((s,i) <- section.students.zipWithIndex) {
    println("Enter the grade for "+s.name+".")
    section.students(i) = s.copy(tests=readInt()::s.tests)
  }
}
```

This version uses `zipWithIndex` and a pattern on the tuple to give us both a short name for the student, `s`, and an index into the array, `i`. Both of these are equally correct so use the one that makes more sense to you and duplicate it for assignments and quizzes.

The next function in the menu is `printAverages`. A very basic implementation of this would just print student names and the course average. However, it could be helpful to see all the grades and the partial averages as well. That is what is done in this version.

```
def printAverages(section:Section):Unit = {
  for (s <- section.students) {
    println(s.name)
    val assnAve = if (s.assignments.isEmpty) 0.0
      else s.assignments.sum/s.assignments.length
    println(s.assignments.mkString("Assignments:",", "," = "+assnAve))
    val quizAve = if (s.quizzes.length<2) 0.0
      else (s.quizzes.sum-s.quizzes.min)/(s.quizzes.length-1)
    println(s.quizzes.mkString("Quizzes:",", "," = "+quizAve))
    val testAve = if (s.tests.isEmpty) 0.0
      else s.tests.sum/s.tests.length
    println(s.tests.mkString("Tests:",", "," = "+testAve))
    println("Average = "+(0.5*assnAve+0.3*testAve+0.2*quizAve))
  }
}
```

This function uses the code from the earlier `classAverage` function and inserts some print statements. The only thing in here that might seem odd is the use of a `mkString` method that takes three arguments instead of just one. With this longer version, the first string goes before all the elements and the third one goes after all the elements. The argument in the middle is the delimiter as it has been in previous usage.

```
def loadSection(fileName:String):(String,Section) = {
  val src = Source.fromFile(fileName)
  val lines = src.getLines
  val section = Section(Array.fill(lines.next().toInt)(Student(
    lines.next(),
    lines.next().split(" ").filter(_.length>0).map(_.toDouble).toList,
    lines.next().split(" ").filter(_.length>0).map(_.toDouble).toList,
    lines.next().split(" ").filter(_.length>0).map(_.toDouble).toList
  )))
  src.close
  (fileName,section)
}
```

This function includes three lines for handling the file. The most important part of the function is in the declaration of the `section` variable which calls `lines.next()` anytime that it needs a new line from the input file. The first time is to read how many students are in the section for building the `Array`. Each student pulls in four lines for the name and three different grade types. The lines of grades are `split`, `filtered`, and them `mapped` to `Doubles` before they are converted to a `List`. The `filter` is required for the situation where you have not entered any grades of a particular type.

You might wonder why the return type of this function includes the `fileName` that was passed in. Technically this is not required, but it makes this function integrate much more nicely at the bottom of the script.

```scala
val (fileName, section) = if (args.length<1) createSection
  else loadSection(args(0))
mainMenu(section)
saveSection(fileName, section)
```

Having `createSection` and `loadSection` return the same information greatly simplifies this part of the code as they can be called together in a simple `if` expression.

That is everything. You now have a full little application that could be used to store a grade book for some course. Try putting this code in and playing with it a while. Here is a complete version with everything together, along with `import` statements.

Listing 10.1: GradeScript.scala

```scala
import io.StdIn._
import io.Source
import java.io.PrintWriter

case class Student(name:String, assignments:List[Double], tests:List[Double],
  quizzes:List[Double])

case class Section(students:Array[Student])

def classAverage(s:Student):Double = {
  val assnAve = if (s.assignments.isEmpty) 0.0
    else s.assignments.sum/s.assignments.length
  val quizAve = if (s.quizzes.length<2) 0.0
    else (s.quizzes.sum-s.quizzes.min)/(s.quizzes.length-1)
  val testAve = if (s.tests.isEmpty) 0.0
    else s.tests.sum/s.tests.length
  0.5*assnAve + 0.3*testAve + 0.2*quizAve
}

def createSection:(String,Section) = {
  println("What file would you like to save this as?")
  val fileName = readLine()
  println("How many students are in the class?")
  val numStudents = readInt()
  println("Enter the student names, one per line.")
  (fileName,Section(Array.
    fill(numStudents)(Student(readLine(),Nil,Nil,Nil))))
}

def saveSection(fileName:String, section:Section):Unit = {
  val pw = new PrintWriter(fileName)
```

```
    pw.println(section.students.length)
    for (s <- section.students) {
      pw.println(s.name)
      pw.println(s.assignments.mkString(" "))
      pw.println(s.tests.mkString(" "))
      pw.println(s.quizzes.mkString(" "))
    }
    pw.close()
  }

  def loadSection(fileName:String):(String,Section) = {
    val src = Source.fromFile(fileName)
    val lines = src.getLines
    val section = Section(Array.fill(lines.next().toInt)(Student(
      lines.next(),
      lines.next().split(" ").filter(_.length>0).map(_.toDouble).toList,
      lines.next().split(" ").filter(_.length>0).map(_.toDouble).toList,
      lines.next().split(" ").filter(_.length>0).map(_.toDouble).toList
    )))
    src.close
    (fileName,section)
  }

  def addTest(section:Section):Unit = {
    for (i <- 0 until section.students.length) {
      println("Enter the grade for "+section.students(i).name+".")
      section.students(i) = section.students(i).
        copy(tests=readInt()::section.students(i).tests)
    }
  }

  def addAssignment(section:Section):Unit = {
    for (i <- 0 until section.students.length) {
      println("Enter the grade for "+section.students(i).name+".")
      section.students(i) = section.students(i).
        copy(assignments=readInt()::section.students(i).assignments)
    }
  }

  def addQuiz(section:Section):Unit = {
    for (i <- 0 until section.students.length) {
      println("Enter the grade for "+section.students(i).name+".")
      section.students(i) = section.students(i).
        copy(quizzes=readInt()::section.students(i).quizzes)
    }
  }

  def printAverages(section:Section):Unit = {
    for (s <- section.students) {
      println(s.name)
      val assnAve = if (s.assignments.isEmpty) 0.0
        else s.assignments.sum/s.assignments.length
      println(s.assignments.mkString("Assignments:"," "," = "+assnAve))
      val quizAve = if (s.quizzes.length<2) 0.0
        else (s.quizzes.sum-s.quizzes.min)/(s.quizzes.length-1)
```

```scala
      println(s.quizzes.mkString("Quizzes:",", ","," = "+quizAve))
      val testAve = if (s.tests.isEmpty) 0.0
        else s.tests.sum/s.tests.length
      println(s.tests.mkString("Tests:",", ","," = "+testAve))
      println("Average = "+(0.5*assnAve+0.3*testAve+0.2*quizAve))
  }
}

def printMenu:Unit = {
  println("""Select an option:
1. Add Test Grade
2. Add Assignment Grade
3. Add Quiz Grade
4. PrintAverages
5. Save and Quit""")
}

def mainMenu(section:Section):Unit = {
  var option = 0
  do {
    printMenu
    option = readInt()
    option match {
      case 1 => addTest(section)
      case 2 => addAssignment(section)
      case 3 => addQuiz(section)
      case 4 => printAverages(section)
      case 5 => println("Goodbye!")
      case _ => println("Invalid option. Try again.")
    }
  } while (option!=5)
}

val (fileName, section) = if (args.length<1) createSection
  else loadSection(args(0))
mainMenu(section)
saveSection(fileName, section)
```

As a final exercise, consider how this might be altered if the `Section` type used a `List`, or we decided not to mutate the `Array`. In this situation, the functions like `addTest` have to be altered so that they return a new `Section` instead of mutating the one that was passed in. The `mainMenu` function also needs to be changed. The ideal implementation would be recursive. Here are modified implementations of `addTest` and `mainMenu` that support this approach.

```scala
def addTest(section:Section):Section = {
  section.copy(students = section.students.map(s => {
    println("Enter the grade for "+s.name+".")
    s.copy(tests=readInt()::s.tests)
  }))
}

def mainMenu(section:Section):Unit = {
  printMenu
  val option = readInt()
```

```
val newSection = option match {
  case 1 => addTest(section)
  case 2 => addAssignment(section)
  case 3 => addQuiz(section)
  case 4 => printAverages(section); section
  case 5 => println("Goodbye!"); section
  case _ => println("Invalid option. Try again."); section
}
if (option!=5) mainMenu(newSection)
}
```

The addTest function for this version is actually simpler than the non-functional approaches because we use map to produce the new set of students with updated grades. The other thing to note is that in mainMenu, the match statement is converted to an expression. To make the types work, all the cases have to give back a session.

Tuple Zipped Type (Advanced)

Something that you need to do fairly frequently is to run through two collections at the same time, pulling items from the same location of each. One way to do this is to use the zip method to zip the collections together into a new collection of tuples. While this works well if you have two collections, especially if you use a for loop, it does not work as well for three or more collections, and it is fundamentally inefficient because the zip method will go through the effort of creating a real collection with a bunch of tuples in it.

To get around these limitations, the types for tuples of length 2 and 3 have a type associated with them called Zipped. The sole purpose of the Zipped type is to let you get the benefits of running through a zipped collection without actually doing the zipping. To get an instance of the Zipped type, simply make a tuple that has all the collections you want in it and call the zipped method. The Zipped type has some of the main higher order methods that you have been using on collections: exists, filter, flatMap, forall, foreach, and map. The difference is that in the Zipped type they take multiple arguments. Specifically, they take as many arguments as there are elements in the tuple. This is significant because if you call a function like map that is mapping a collection of tuples the function has to take one argument and go through some effort to pull the elements out of the tuple. With the Zipped type you do not have to do that as the functions that are passed in are supposed to take multiple arguments instead of a single tuple with the multiple values.

A comparison of the two approaches is shown here.

```
val l1=List(1,2,3)
val l2=List(4,5,6)
l1.zip(l2).map(t => t._1*t._2)
(l1,l2).zipped.map((v1,v2) => v1*v2)
// or
(l1,l2).zipped.map(_*_)
```

For this example the first one is a bit shorter unless we use the underscore notation, but that typically will not be the case. Note that the underscore notation is not an option for the version that uses normal `zip`. More importantly, the first one relies on the `_1` and `_2` methods which will make the code hard to read and understand for anything with more logic. To get the benefit of easy to read names using `zip` you would have to do the following.

```
l1.zip(l2).map(t => {
  val (v1,v2)=t
  v1*v2
})
```

It remains an exercise for the reader to see what happens if you want to iterate over three collections using `zip`. Consider the `Zipped` type when you need to iterate over two or three collections at the same time.

10.5   End of Chapter Material

10.5.1   Summary of Concepts

- The act of grouping together data is very useful in programming. We have been doing this with tuples. The problem with tuples is that they do not provide meaning and their syntax can make code difficult to read and understand.

- User defined types let you create your own types that have meaning related to the problem you are solving.

- One way of making user defined types is with `case class`es. We will use these to group values together and give them useful, easy to read names.

  - To create a `case class`, follow those keywords with an argument list like that for a function with names and types separated by commas. Names for types typically start with a capital letter.

  - You create an instance of a `case class` give the name of the type followed by an argument list of the values it should store.

  - When you want to access the members of a case class, use the dot notation we have been using for other objects.

  - The members of a `case class` are all `val`s by default. As a result, instances of `case class`es tend to be immutable. The only way that will not be true is if a member is mutable.

  - To make new instances of `case class`es that are slightly different from old ones, use the `copy` method. This method is called with named arguments for any members that you want to have changed in the copy.

  - Another useful capability of `case class`es is that they can be used as patterns.

  - You can make `case class`es mutable by putting `var` in front of the fields you wish to mutate, but as with `var` declarations, this approach is discouraged.

10.5.2 Self-Directed Study

Enter the following statements into the REPL and see what they do. Some will produce errors. You should try to figure out why. Try some variations to make sure you understand what is going on.

```scala
scala> case class Accident(dlNumber1:String,dlNumber2:String)
scala> case class
    Driver(name:String,dlNumber:String,dob:String,history:List[Accident])
scala> def wreck(d1:Driver,d2:Driver):(Driver,Driver,Accident) = {
    | val accident = Accident(d1.dlNumber,d2.dlNumber)
    | (d1.copy(history = accident::d1.history),
    |  d2.copy(history = accident::d2.history),
    |  accident)
    | }
scala> var me = Driver("Mark","12345","long ago",Nil)
scala> var otherPerson = Driver("John Doe","87654","01/01/1990",Nil)
scala> val (newMe,newOther,acc) = wreck(me,otherPerson)
scala> me = newMe
scala> otherPerson = newOther
scala> println(me.name)
scala> println(otherPerson.dlNumber)
scala> println(me.history.length)
scala> otherPerson.name = "Jane Doe"
scala> case class Vect2D(x:Double,y:Double)
scala> def magnitude(v:Vect2D):Double = {
    | math.sqrt(v.x*v.x+v.y*v.y)
    | }
scala> def dot(v1:Vect2D,v2:Vect2D):Double = v1.x*v2.x+v1.y*v2.y
scala> def makeUnit(angle:Double):Vect2D = {
    | Vect2D(math.cos(angle),math.sin(angle))
    | }
scala> def scale(v:Vect2D,s:Double):Vect2D = Vect2D(v.x*s,v.y*s)
scala> val a = makeUnit(math.Pi/4)
scala> val b = makeUnit(3*math.Pi/4)
scala> dot(a,b)
scala> magnitude(a)
scala> magnitude(b)
scala> magnitude(scale(a,3))
```

10.5.3 Exercises

1. Write a `case class` to represent a student transcript.

2. Using your answer to the previous exercise, define a function that adds one semester of grades to the transcript.

3. Using your answer to 1, write a function that will return the student's GPA.

4. Write a `case class` to represent a recipe.

5. Using your answer to 4, write a function that take a recipe and the name of an ingredient and returns how much of that ingredient is needed.

6. Write a `case class` to represent the information needed for a house in a Realtor posting.

7. Rewrite the grade book program in a completely functional way so that it has neither `Arrays`, `vars`, or other mutable objects.

8. Play with `zip` using 3 collections. Compare it to using `zipped`.

9. Pick a favorite sport and make a `case class` that can be used to store player information.

10. Extend what you did on the previous exercise so you have a `case class` that stores the information for a team.

10.5.4   Projects

1. This is an extension on project 8.3. You will use `Arrays` and `class`es to take the Keplerian simulator a bit further. In that program all you had was one body in motion about a "central mass" that was not moving at all. Now you can store and work with the positions and velocities of many bodies because you can store their component data in `Arrays` or `case class`es. That is to say you can have an `Array` for x positions as well as y, v_x, and v_y, or an `Array` of some `case class` that stores those values. This allows you to simulate the motions of many particles at the same time which is much more fun. Earlier you only had to calculate the acceleration due to the central particle. Now you want to calculate accelerations due to all interactions between all the particles. You can also make the central particle one of the particles in the array or try not even having a central particle.

   With multiple particles, you need to have a nested loop (or a `for` loop with two generators) that calculates the accelerations on each particle from all the others and adds them all up. Keep in mind that if particle i pulls on particle j then particle j pulls back on i just as hard, but in the opposite direction. That does not mean the accelerations are the same though. The acceleration is proportional to the mass of the puller because the mass of the pullee is canceled out by its inertia. Earlier we had $a_x = -\frac{x}{d^3}$ and $a_y = -\frac{y}{d^3}$. When the particle doing the pulling is not at the origin, d is the distance between the particles, and x and y are the distances between them in x and y directions. We also need a factor of m for the mass of the puller. You want to add up the accelerations from all other particles on each one and store that into arrays so

$$a_x(i) = -\sum_j \frac{(x_j - x_i) * m_j}{d_{ij}^3}$$

   There is a similar formula for y. The d_{ij} value is the distance between particle i and particle j. Also note that given a value, c, the best way to cube it is `c*c*c`, not `math.pow(c,3)`.

   When you write your formula in the code, think a bit about it. This formula causes a problem if we really use d because particles can get too close to one another. It is suggested that you make $d = \sqrt{d_x^2 + d_y^2} + \epsilon$, where *epsilon* is a small value. You can play with how small you want it to be because that depends on the scale of your problem. It is also recommended that you have your integrator not use the normal Euler method which calculates accelerations, then updates positions, then velocities. Make sure that it does the accelerations on velocities before applying the velocities to the positions. Keep in mind that you want to break this up into functions that fit together nicely and are helpful. It will hurt your brain more if you do not.

The input for the program will have a first line that is the number of bodies, the timestep, the stopping time, and the number of steps between outputs. This will be followed by lines with the x, y, v_x, v_y, and *mass* values for each particle. A sample input file can be found on the book's website on the page for this chapter. Note that the central mass in that file has a mass of 1 and all the others are much smaller.

As output, you should write out the positions of all the particles in your simulation to a file once for every n steps, where n is a value given on the first line of the input. If you do this then you can run a spreadsheet of **gnuplot** to plot it. If you use **gnuplot** and give it the command **plot 'output'** it will make a little scatter plot showing you the paths of the particles in the simulation.

2. This project builds on top of project 8.1. You have likely been using tuples or separate values to represent the geometry elements in your ray tracer. This is information that is much more naturally represented as a case class. For this project you should go through and edit your existing code so that it includes three different **case classes**, one for spheres, one for planes, and one for a scene which has a **List[Sphere]** and a **List[Plane]**.

3. This is the first installment for you building your own text adventure. Your program will read in from a map file that you will write by hand and let the user run around in that map by using commands like "north" to move from one room to another. The map file will have a fairly simple format right now and you will create your own map file using vi. Make sure when you turn in this program you turn in both the script and the map file so it can be tested with your map.

The format of the map file should start with a line telling the number of rooms then have something like the following. You can change this if you want to use a slightly different format:

```
room_number
room_name
long line of room description
number_of_links
direction1
destination1
direction2
destination2
...
```

This is repeated over and over. (The number of rooms at the top is helpful for storing things in an **Array** so that you know how big to make it.) Each room should have a unique room number, and they should start at 0. The reason is that you will be putting all the room information into **Arrays**. There is a link on the book's website to a sample map file, but you do not have to stick exactly to that format if you do not want to. You might deviate if you are thinking about other options you will add in later. Odds are good you will be refactoring your code for later projects.

The interface for your program is quite simple. When you run the program it should read in the map file and keep all the map information stored in an **Array[Room]** where Room is a **case class** you have made to keep the significant information for each room.[4] You will start the user in room 0 and print the description of that room

---

[4]Many implementations also make a **case class** to represent an **Exit** from a room.

and the different directions they can go as well as where those exits lead to, then follow that with a prompt. You could just use > as the prompt to start with. It might get more complex later on when you have real game functionality. So when the user starts the game it might look something like this if you read in the sample map file.

```
Halsell 228
You are standing in a room full of computers and comfy chairs with
a giant air conditioning unit hanging from the ceiling. While the
surroundings are serene enough, you cannot help feeling a certain amount
of dread. This isn't just a fear that the air conditioning unit is
going to fall either. Something in you tells you that this room is
regularly used for strange rituals involving torture. You can only
wonder what happens here and why there isn't blood all over the place.
Your uneasiness makes you want to leave quickly.
The hallway is east.
>
```

The user must type in either a direction to move or "quit". If anything else is entered you should print an appropriate error message. The only goal of this project is to allow the user to move around the map. Collection methods such as `find`, `indexWhere`, `filter`, or `partition` can be extremely helpful for this project.

4. Convert the work you did for project 9.3 to use **case class**es to bind data. So you should have a **case class** for an item with an amount, one for a recipe, one for pantry contents, etc.

5. Convert the work you did for project 9.4 to use **case class**es. With this addition you can put more information into each course because you now have a better way to store it all together.

6. Convert the work you did for project 9.5 to use **case class**es to bind data. You should have a **case class** that includes CDT Time, Temperature, Dew Point, Humidity, Sea Level Pressure In, Visibility MPH, Wind Direction, Wind Speed MPH, Gust Speed MPH, Precipitation In, Events, Conditions, Wind Direction Degrees, UTC Date.

7. Convert the work you did for project 9.8 to use **case class**es to bind data. You should have a **case class** that includes a DVD title, studio, date released, status, Sound, versions, price, rating, year, genre, aspect, UPC, DVD release date, ID, and timestamp.

8. Convert the work you did for project 9.9 to use **case class**es to bind data. You should have a **case class** that includes Media Type, Song Title, and Date Released.

9. Sports have a tendency to produce a lot of information in the form of statistics. A **case class** is a good way to represent this information. For this project, convert the work you did for project 9.10 to use **case class**es to bind data. Put multiple box scores into a text file and read them in. Have menu options for calculating averages for players and teams in the different relevant statistics.

Additional exercises and projects, along with data files, are available on the book's web site.

Chapter 11

GUIs

So far, all of the programming that we have done has been part of a text interface. The programs print information to a terminal window and to give them input we enter text from the keyboard. These types of programs have their place in computing, but these days very few people actually use such programs. They typically run in the background instead, doing things that people do not see. The programs that you likely work with most are more graphical. They open up windows and you can interact with them through mouse clicks as well as typing.

This type of program is called a GUI (Graphical User Interface). In this chapter we will look at how we can write GUIs in Scala and get them to interact with the user.

11.1 GUI Libraries and History

There are reasons why we began doing our programming with the console. There are certain complications involved in programming GUIs. Your Scala programs run on top of the Java Virtual Machine and relies on the underlying virtual machine and the libraries for it to do a lot of the work in a GUI. Scala adds things on top when it is possible to improve them in a significant way by doing so. In the implementation of Scala that we are using, this means that there is a dependence on the Java GUI libraries.

When Java was originally released in 1995, there was one GUI library called the Abstract Windowing Toolkit, AWT. The AWT library makes direct calls to the operating system or windowing environment for building elements of the GUI. This gave the AWT a significant benefit in speed. However, it also provided a significant limitation. Due to the cross-platform nature of Java, they only included elements in the AWT that were present on all platforms. They are the most fundamental parts of GUI construction like windows, buttons, text boxes, etc. These elements were drawn by the underlying platform and always looked like the underlying platform.

Both the restriction in appearance and the limited number of component types led to the creation of a second GUI library for the Java standard. This second library was called Swing, and it was a "pure" Java library. That means that all the code in Swing was written in Java. It uses calls to AWT to make things happen, but does not directly call anything that is not in Java itself. The Swing library included a lot of different GUI elements that were more powerful and flexible than what was provided in AWT. It also included the ability to have elements of the GUI rendered in different ways so they do not have to look like the underlying system. Originally this flexibility came at a significant cost in speed. Over time various optimizations were made and now Swing runs perfectly fine for most applications.

In 2007, a new GUI and graphics framework called JavaFX was introduced to fix perceived problems with Swing. The primary goals were to make something that was faster and which made it easier for programmers to create rich and visually pleasing interfaces. JavaFX started its life as a separate scripting language, but later became just a library for Java. With the release of Java 8 in 2014, JavaFX became the standard that developers are supposed to use for graphical interfaces on the JVM. It has an even broader variety of options than Swing, and runs faster as well.

There is a Scala library that is a wrapper[1] around JavaFX called ScalaFX. This is the library that will be used in this book for writing GUIs. It makes calls to JavaFX, but provides a syntax that takes full advantage of the features of the Scala programming language. Unfortunately, this library is not part of the standard Scala installation. For that reason, you might go to `http://www.scalafx.org/` to get the most recent version of ScalaFX. A JAR file[2] for a reasonably recent version can be found at the book's GitHub repository.

All the types for ScalaFX are organized in packages that start with `scalafx`. The ScalaFX web site has a link to the API on the right side. At the time of this writing, the current version was 8.0. We will introduce each of the subpackages as they are needed. As with the standard Scala API, the ScalaFX API is a useful tool that you will want to familiarize yourself with. Unlike previous chapters, this chapter and the following one can truly only skim the top of what is in ScalaFX. The API is a good resource for those who want to go further.

11.2   First Steps

To get us started, we will write a little program that does nothing more than pop up a window with nothing inside of it. The code for that is shown here.

Listing 11.1: FirstGUI.scala

```
import scalafx.application.JFXApp
```

---

[1]ScalaFX provides types that allow you to make calls to JavaFX using a more Scala like style.

[2]Java Archive File which is a compressed file used by the Java platform to distribute code.

```
val app = new JFXApp {
  stage = new JFXApp.PrimaryStage {
    title = "First GUI"
    width = 500
    height = 500
  }
}
```

```
app.main(args)
```

At the top, you can see there is an import of the `JFXApp` type from the `scalafx.application` package. We then declare a `val` called `app` of that type. Inside of our new `JFXApp` object we set the `stage` to be a new `JFXApp.PrimaryStage`. If you look at `scalafx.application.JFXApp` in the API, you will see that it has a `var` called `stage`, and that is what we are setting here.

The `Stage` type has members for `title`, `width`, and `height` that we set inside of that new `JFXApp.PrimaryStage` object. The meanings of these should be fairly obvious.

The script ends with a call to the `main` method of `app` passing it the command line arguments. That call is what makes things happens and should pop up a window.

Now you have to run this script. We assume that you are running your scripts from the command line as described in Appendix A. If you try that, you will get a number of errors telling you that `scalafx.application.JFXApp` cannot be found, along with other information that is basically caused by the same issue. The problem is that ScalaFX is not part of the standard Scala libraries. For that reason, we need to tell Scala where to find the JAR file for ScalaFX.[3] This is done by specifying what is called a classpath. There are several ways that you can do this. The simplest is to use the `-cp` option at the command line. On my system, I do that with the following command.

```
scala -cp /home/mlewis/scalafx.jar FirstGUI.scala
```

The `-cp` is followed by the path to the JAR file. On your machine that path will definitely be different and the JAR file might have a longer name that includes version details.

Another approach is to specify a place that the JVM should always go to look for extra libraries. This can be done by making a `CLASSPATH` environment variable. The details of how you do this vary greatly between operating systems, versions, and other details that make it impossible to cover here. You can find such information on the web. On one of the author's Windows machines a CLASSPATH environment variable was created with a variable value of `"C:\Users\llben_000\scalafx.jar;."` (where C:\Users\llben_000 is the location of the scalafx.jar file. Do not include the double quotes.). Under the Bash shell for Linux, adding the line `export CLASSPATH=.:/home/mlewis/scalafx.jar` to your `.bashrc` file works well. Again, the exact path and file name could be different on your machine.

On a Linux machine, a third approach would be to make an alias for the command that includes the classpath specification. That might look like `alias scalafx="scala -cp /home/mlewis/scalafx.jar"`. Once you have this defined, you could execute `scalafx FirstGUI.scala` and you do not need to worry about the classpath variable.

---

[3]There are several ways you can get this JAR file. You could build it from the main repository through `http://www.scalafx.org` (there are several steps involved in this, but the advantage of this approach is that it will always match your current version of Scala) or download it from the book's GitHub repository. If you are taking a course, your instructor might tell you where you can find one.

However you go about doing it, when you run this little script, it will pop up a window that is 500 pixels by 500 pixels with a title that says "First GUI". Closing the window will end the script running.

11.3   Stages and Scenes

The `Stage` type that was created in our first little application is what defined the window that popped up. To make the GUI more interesting, we need to put things on our `Stage`. This is done by adding a `Scene` to the `Stage`. That `Scene` can then have contents that will show up in the window. The following example demonstrates adding a `Scene` with one thing inside of it to our GUI.

Listing 11.2: JustButton.scala

```scala
import scalafx.application.JFXApp
import scalafx.scene.Scene
import scalafx.scene.paint.Color
import scalafx.scene.control.Button

val app = new JFXApp {
  stage = new JFXApp.PrimaryStage {
    title = "First GUI"
    scene = new Scene(300,200) {
      fill = Color.Coral
      val button = new Button("Click me!")
      content = button
    }
  }
}

app.main(args)
```

There are three additional `imports` added here to bring in the `Scene`, the `Color` type, and the `Button` type, which is what we added to the scene. The `String` "Click me!" is the text that appears on the button. You can see the window that results from running this code in figure 11.1.

Note that the button is located in the top left corner. This probably is not the ideal location for the button, but it is where things go in the scene by default. The scene has a coordinate system using the x and y coordinates. The top left corner is at location $(0,0)$ in our scene. However, 2D graphics coordinates are slightly different than what you are used to from math. In math you are used to x growing as you move right and y growing as you move up. The scene uses typical graphics coordinates where x still grows from left to right, but y grows from top down. The coordinates that are typically used for 2D graphics are like this. The x-coordinate value, increases as it goes right and the y-coordinate increases as it goes down. The following code moves the button to be roughly centered on the window.

Listing 11.3: MoveButton.scala

```scala
import scalafx.application.JFXApp
import scalafx.scene.Scene
import scalafx.scene.paint.Color
```

FIGURE 11.1: This is the window that pops up running JustButton.scala. There is a single button in a 300 by 200 pixel scene. Note that the button is, by default, in the top left corner.

```scala
import scalafx.scene.control.Button

val app = new JFXApp {
  stage = new JFXApp.PrimaryStage {
    title = "First GUI"
    scene = new Scene(300,200) {
      fill = Color.Coral
      val button = new Button("Click me!")
      content = button
      button.layoutX = 115
      button.layoutY = 70
    }
  }
}

app.main(args)
```

This code sets the values of `layoutX` and `layoutY` on the button. Those values are specified in pixels on the graphics coordinates, so the top left corner of the button is 115 pixels to the right and 70 pixels below the top of the scene. Figure 11.2 shows the results of running this program.

Having a window with only one thing in it is a bit limiting. As it happens, the contents of the scene can also be set to a sequence, and all the elements of the sequence are placed in the scene. This is shown here in the program called SeveralFiles.scala.

Listing 11.4: SeveralItems.scala

```scala
import scalafx.application.JFXApp
import scalafx.scene.Scene
import scalafx.scene.paint.Color
import scalafx.scene.control._
import scalafx.scene.shape._

val app = new JFXApp {
  stage = new JFXApp.PrimaryStage {
    title = "First GUI"
    scene = new Scene(500,300) {
```

FIGURE 11.2: This is the window that pops up when you run MovedButton.scala. Notice that the button has been moved so that it is roughly centered in our 300 by 200 pixel window.

```scala
    fill = Color.Coral

    // Create and place elements
    val button = new Button("Click me!")
    button.layoutX = 75
    button.layoutY = 45
    val rectangle = Rectangle(200,150)
    rectangle.layoutX = 10
    rectangle.layoutY = 140
    val label = new Label("A label")
    label.layoutX = 250
    label.layoutY = 10
    val checkBox = new CheckBox("Would you like to play a game?")
    checkBox.layoutX = 250
    checkBox.layoutY = 40
    val comboBox = new ComboBox(List("Scala","Java","C++"))
    comboBox.layoutX = 250
    comboBox.layoutY = 70
    val listView = new ListView("AWT Swing JavaFX ScalaFX".split(" "))
    listView.layoutX = 250
    listView.layoutY = 100
    listView.prefHeight = 190

    // Add elements to contents
    content = List(button, rectangle, label, checkBox, comboBox, listView)
    }
  }
}

app.main(args)
```

This code uses the underscore for some `imports` as we are using multiple types from those packages. It then goes through and creates several different `Node`s to place in the scene. Each element that is placed on the screen is a `Node`. In addition to a `Button`, it includes a `Rectangle`, a `Label`, a `CheckBox`, a `ComboBox`, and a `ListView`. The end of the block for the scene sets the `contents` equal to a `List` that includes all of those elements. The `layoutX` and `layoutY` of each one are set so that they do not overlap. To prevent the `ListView` from going below the bottom of the window, we set its preferred size. The `ComboBox` and `ListView` take

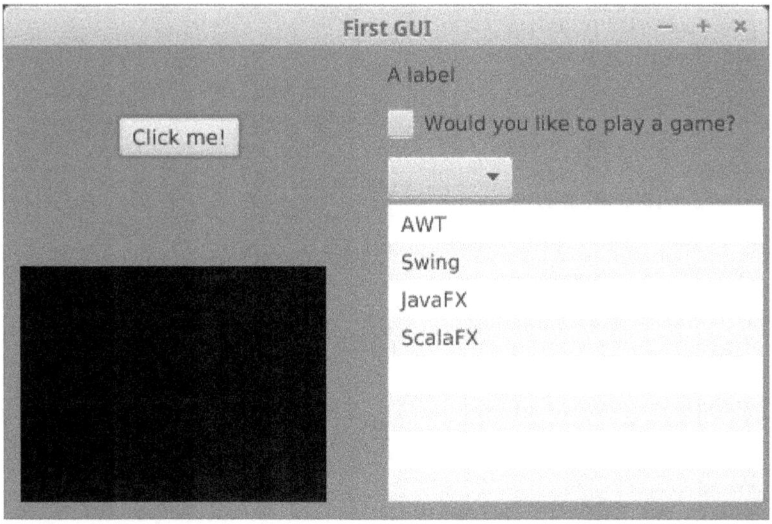

FIGURE 11.3: This is the window created when you run SeveralItems.scala. It has sequence containing several `Node`s added to the scene and moved to different locations.

arguments that are sequences of the things that they are supposed to display. To illustrate how it works, the `ComboBox` was created with a `List[String]`, while the `ListView` was created with an `Array[String]` that is built using `split`. Running this program produces the window shown in figure 11.3

11.4 Events and Handlers

The next thing that we need to do is learn how to allow the user to interact with a GUI. Before learning about GUIs, we did this using functions like `readInt` and `readLine`. Things are more complex with the GUI as there are many ways that you can interact with them. There are some elements that accept text/keystrokes, but many others interact with the mouse. Even those that do interact with keys cannot use something like `readLine` because there could be many of elements in a window, which one gets the key strokes?

The way that GUIs typically deal with user interaction is through events and handlers. When the user does something, it produces an event. In the code, you have to specify a handler that is supposed to deal with that event. The handler is basically a function that is passed the event and which then completes the desired action, potentially using information from that event.

To help illustrate this, we modify the GUI from the last example and put different types of interactivity on the various elements. Here is the modified code.

Listing 11.5: InteractiveItems.scala

```
import scalafx.Includes._
import scalafx.application.JFXApp
import scalafx.scene.Scene
import scalafx.scene.paint.Color
import scalafx.scene.control._
import scalafx.scene.shape._
```

```scala
import scalafx.event.ActionEvent
import scalafx.scene.input._

val app = new JFXApp {
  stage = new JFXApp.PrimaryStage {
    title = "First GUI"
    scene = new Scene(500,300) {
      fill = Color.Coral

      // Make and place elements
      val button = new Button("Remove Item")
      button.layoutX = 75
      button.layoutY = 45
      val rectangle = Rectangle(30,30)
      rectangle.layoutX = 10
      rectangle.layoutY = 140
      val label = new Label(s"Location is ${rectangle.layoutX()},
          ${rectangle.layoutY()}.")
      label.layoutX = 250
      label.layoutY = 10
      val checkBox = new CheckBox("Would you like to play a game?")
      checkBox.layoutX = 250
      checkBox.layoutY = 40
      val comboBox = new ComboBox(List("Scala","Java","C++"))
      comboBox.layoutX = 250
      comboBox.layoutY = 70
      val listView = new ListView("AWT Swing JavaFX ScalaFX".split(" "))
      listView.layoutX = 250
      listView.layoutY = 100
      listView.prefHeight = 190

      // Add contents
      content = List(button, rectangle, label, checkBox, comboBox, listView)

      // Add event handlers
      button.onAction = (e:ActionEvent) => {
        val selected = listView.selectionModel.value.selectedItems
        listView.items = listView.items.value.diff(selected)
      }
      comboBox.onAction = (e:ActionEvent) => {
        listView.items.value += comboBox.selectionModel.value.selectedItem.value
      }
      onMouseClicked = (e:MouseEvent) => {
        if (checkBox.selected.value) {
          rectangle.layoutX = e.x
          rectangle.layoutY = e.y
          label.text = s"Location is ${rectangle.layoutX.value},
              ${rectangle.layoutY.value}."
        }
      }
    }
  }
}

app.main(args)
```

Other than altering the text in the `Button` and the `Label`, the primary changes come near the end of the script where the event handlers are added to the different elements. There is also a new import of `scalafx.Includes._` added to the top which is needed so that Scala will do some conversions for the handlers. The handlers are added by setting values that start with "on", such as `onAction` and `onMouseClicked`. They are set to be functions that take an appropriate type of event as input. When the user does the specified thing, the function is called. These functions are just like any other lambda expressions with a variable name, a rocket (=>), and the body of the function. The one thing to note is that you generally have to specify the type of the argument for these to work. That is why the argument is given as `(e:ActionEvent)` for the first two. As usual, the variable name `e` could be anything you choose. The letter `e` is chosen here to represent an event.

In this example, handlers are added to the button and the combo box. For both of those it is the `onAction` value that is set. This function is called when the most common action for that element occurs, such as clicking on the button.[4] The code in the button click removes all the selected items from the list view. It does this by first getting all the items that are selected in the list view and then setting the list view items to be the original list of items with the selected ones removed. For the combo box, the handler adds whatever item was selected in the combo box to the end of the list view.

This code also sets the `onMouseClick` handler of the whole scene. Note that the `onAction` follows an object and a dot. The `onMouseClicked` does not. Because the code is nested inside the `scene`, it uses the `scene`'s version of `onMouseClicked`. This usage is just like setting the `title` of the `Stage` or the `fill` on the `Scene`. The handler that is created checks if the check box has been selected. If it has not, nothing else is done. If it is has been selected, then the rectangle is moved to the location of the click, and the label is updated with the appropriate text. Note that because this is part of the `Scene`, only clicks on the background invoke it. If you click on one of the other components in the GUI, this code is not called, so the rectangle cannot be placed directly on top of the other elements.

Running this code will bring up a window that looks like that shown in figure 11.4.

11.5 Controls

ScalaFX has a long list of different elements that can be added into a scene, a few of which were shown in the previous examples, but not really explained. All of these elements that can be added into a Scene are subtypes of the type called `Node`. Looking at the whole picture, for a ScalaFX GUI, we have the `JFXApp` at the top level. Inside of this we have a `Stage` that represents the primary window for the application. The `Stage` needs to contain a `Scene`, and finally the `Scene` can contain the different `Node`s that appear in it.

There are 120 subtypes of `scalafx.scene.Node` in the ScalaFX API, a fact that should give you an indication of how much you can do with this library. Only a few of the key ones will be covered in this book. This section focuses on some of the GUI controls, which are found in the `scalafx.scene.control` and `scalafx.scene.shape` packages.

The term Control is used in JavaFX to denote basic GUI elements that the user can interact with. We break these into different categories for discussion and show examples of each. The sample code shows not only how to create each element and add it into a `Scene`, but also how to code basic interactions with that element.

---

[4]There are many types of events that can occur for each element, but the Scala developers have determined which is the most common event action for each GUI element.

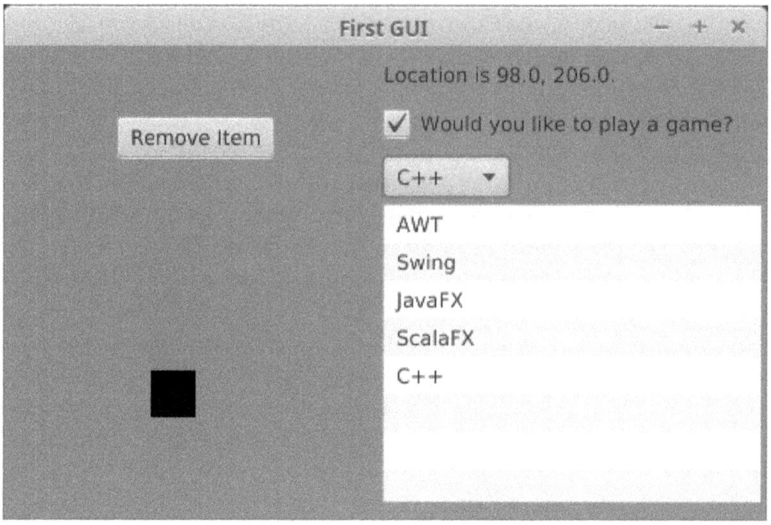

FIGURE 11.4: This figure shows the window that results from running InteractiveItems.scala after the user has done a few things.

11.5.1 Text Controls

The first set of controls that we want to look at are those that deal with text. This includes the non-interactive `Label` as well at the `TextField`, `TextArea`, and `PasswordField`. The last three are used to allow users to enter text into a GUI. The `TextField` is for shorter input that takes a single line. The `TextArea` is used for multi-line inputs. The `PasswordField` is like a `TextField`, but it hides what the user types into the field. This sample code shows the use of these four different controls.

Listing 11.6: TextControls.scala

```scala
import scalafx.Includes._
import scalafx.scene.Scene
import scalafx.application.JFXApp
import scalafx.scene.control.Label
import scalafx.scene.control.TextField
import scalafx.scene.control.TextArea
import scalafx.scene.control.PasswordField
import scalafx.event.ActionEvent

val app = new JFXApp {
  stage = new JFXApp.PrimaryStage {
    title = "Text Controls"
    scene = new Scene(500,300) {
      // Label
      val label = new Label("Generally non-interactive text.")
      label.layoutX = 20
      label.layoutY = 20

      // TextField
      val textField = new TextField()
      textField.layoutX = 20
      textField.layoutY = 50
```

```
    textField.promptText = "Single-line field"

    // TextArea
    val textArea = new TextArea()
    textArea.layoutX = 20
    textArea.layoutY = 80
    textArea.promptText = "Multi-line field"
    textArea.prefWidth = 460
    textArea.prefHeight = 120

    // Password Field
    val passwordField = new PasswordField()
    passwordField.layoutX = 20
    passwordField.layoutY = 205
    passwordField.promptText = "Password field"

    content = List(label, textField, textArea, passwordField)

    textField.onAction = (e:ActionEvent) => {
      label.text = "Field action : "+textField.text.value
    }
    textField.focused.onChange {
      if (!textField.focused.value) label.text =
        "Field focus : "+textField.text.value
    }

    textArea.focused.onChange {
      if (!textArea.focused.value) label.text =
        "Area focus : "+textArea.text.value
    }

    passwordField.onAction = (e:ActionEvent) => {
      label.text = "Password action : "+passwordField.text.value
    }
    passwordField.focused.onChange {
      if (!passwordField.focused.value) label.text =
        "Password focus : "+passwordField.text.value
    }
  }
 }
}


app.main(args)
```

Figure 11.5 shows what you see when you run this code. In addition to placing the elements, this code sets the `promptText` on each element. The `promptText` is the greyed out text that you see in each field that tells the user what they should enter in that field. It is text that appears when the element does not have focus and the user has not yet input anything. Readers might not be familiar with the term "focus". When writing simple console applications, where there is only one place for keystrokes to go, they are taken as standard input that can be read by commands like `readInt` and `readLine`. In a GUI though, keystrokes could go to any of the different elements. The one that the user selected is said to have focus. The field that has focus will be the field that receives the users actions. This is significant for code that wants to do something after a user has altered text.

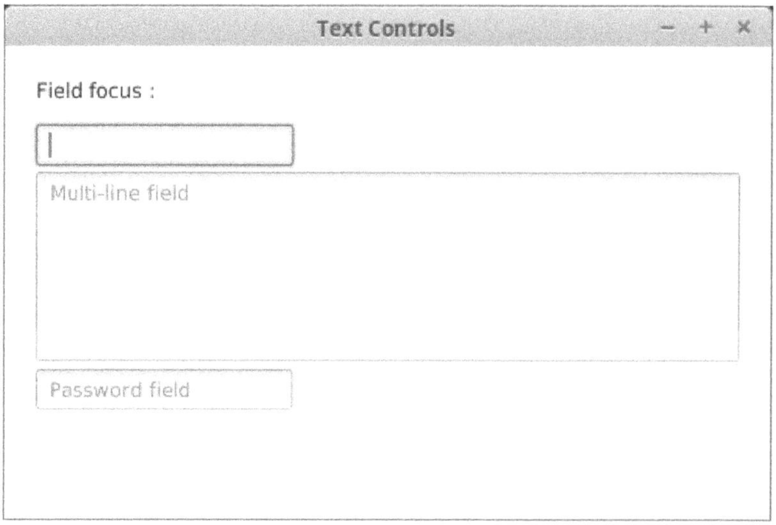

FIGURE 11.5: This figure shows the window that results from running TextControls.scala.

There are two types of handlers in this code: the `onAction` handler and the `onChange` handler. The `TextField` and `PasswordField` set the `onAction` handler. This handler is called when the user hits the Enter key on that field. However, users do not always hit Enter on a `TextField`. That type of handler does not even exist for the `TextArea` because hitting Enter on a `TextArea` simply goes to the next line. To know that the user has finished editing text, we can also detect when the element loses focus. That is what is being done with the calls to `focused.onChange`. The code passed to that method is passed by-name, and it is invoked anytime that the element gains or loses focus. In this case, we only care about when it loses focus. The `focused` member is a property, so we have to call the `value` method to get an actual `Boolean` value.[5]

11.5.2 Button-like Controls

There are four button-like controls in ScalaFX. In addition to the `Button` and `CheckBox`, which we saw earlier, there are also `RadioButtons` and `ToggleButtons` that can both be grouped together into `ToggleGroups`. The `RadioButton` appears as a circle next to some text and can be marked as selected or not. The `ToggleButton` looks like a standard `Button`, but when selected, it stays in the "down" state which generally appears in a darker color. Other than appearance, the main difference between the two is that if you click the selected `ToggleButton`, it will be unselected. This is not true of `RadioButtons`. Once a `RadioButton` from a group has been selected, you cannot get back to a state where nothing is selected. The following code sample shows a GUI with these elements and how to handle events that happen with them.

Listing 11.7: ButtonControls.scala

```
import scalafx.Includes._
import scalafx.scene.Scene
import scalafx.application.JFXApp
import scalafx.scene.control.Label
import scalafx.scene.control.Button
```

---

[5]You can also call the `apply` method to get the same result.

```scala
import scalafx.scene.control.CheckBox
import scalafx.scene.control.RadioButton
import scalafx.scene.control.ToggleButton
import scalafx.scene.control.ToggleGroup
import scalafx.event.ActionEvent

val app = new JFXApp {
  stage = new JFXApp.PrimaryStage {
    title = "Button Controls"
    scene = new Scene(300,340) {
      // Label
      val label = new Label("Just used for feedback.")
      label.layoutX = 20
      label.layoutY = 20

      // Button
      val button = new Button("Button")
      button.layoutX = 20
      button.layoutY = 50

      // CheckBoxes
      val cb1 = new CheckBox("Check Box 1")
      cb1.layoutX = 20
      cb1.layoutY = 80
      val cb2 = new CheckBox("Check Box 2")
      cb2.layoutX = 20
      cb2.layoutY = 110

      // RadioButtons
      val rb1 = new RadioButton("Radio Button 1")
      rb1.layoutX = 20
      rb1.layoutY = 140
      val rb2 = new RadioButton("Radio Button 2")
      rb2.layoutX = 20
      rb2.layoutY = 170
      val rb3 = new RadioButton("Radio Button 3")
      rb3.layoutX = 20
      rb3.layoutY = 200
      val group1 = new ToggleGroup()
      group1.toggles = List(rb1, rb2, rb3)

      // Toggle Buttons
      val tb1 = new ToggleButton("Toggle Button 1")
      tb1.layoutX = 20
      tb1.layoutY = 230
      val tb2 = new ToggleButton("Toggle Button 2")
      tb2.layoutX = 20
      tb2.layoutY = 260
      val tb3 = new ToggleButton("Toggle Button 3")
      tb3.layoutX = 20
      tb3.layoutY = 290
      val group2 = new ToggleGroup()
      group2.toggles = List(tb1, tb2, tb3)

      content = List(label, button, cb1, cb2, rb1, rb2, rb3, tb1, tb2, tb3)
```

```
    button.onAction = (e:ActionEvent) => {
      label.text = "Button clicked"
    }

    cb1.onAction = (e:ActionEvent) => {
      label.text = "Check Box 1 is " + cb1.selected.value
    }
    cb2.onAction = (e:ActionEvent) => {
      label.text = "Check Box 2 is " + cb2.selected.value
    }

    rb1.onAction = (e:ActionEvent) => {
      label.text = "Radio Button 1 is " + rb1.selected.value
    }
    rb2.onAction = (e:ActionEvent) => {
      label.text = "Radio Button 2 is " + rb2.selected.value
    }
    rb3.onAction = (e:ActionEvent) => {
      label.text = "Radio Button 3 is " + rb3.selected.value
    }
    rb1.selected.onChange(println("Radio button 1 is " + rb1.selected.value))

    tb1.onAction = (e:ActionEvent) => {
      label.text = "Toggle Button 1 is " + tb1.selected.value
    }
    tb2.onAction = (e:ActionEvent) => {
      label.text = "Toggle Button 2 is " + tb2.selected.value
    }
    tb3.onAction = (e:ActionEvent) => {
      label.text = "Toggle Button 3 is " + tb3.selected.value
    }
    tb1.selected.onChange(println("Toggle button 1 is " + tb1.selected.value))
  }
 }
}

app.main(args)
```

All of the button-like controls allow you to define `onAction`, which is called when the user clicks on that button. For the `RadioButton` and the `ToggleButton`, the state of selection can also change when the user selects one of the other elements. This does not fire the action handler, but you can set up code to be called when that type of change occurs by setting the `onChange` value of the `selected` member, as shown in the code. Code to do this was only added for the first `RadioButton` and `ToggleButton`.

11.5.3   Selection Controls

There are three controls that allow the user to select different items from a designated selection. The `ComboBox` and `ListView` appeared in earlier examples. There is also a `ChoiceBox`, which is very similar to the `ComboBox` in functionality and appearance, but can allow multiple selections. Here is code that adds these three components to a `Scene` and handles the user making selections on them.

FIGURE 11.6: This figure shows the window that results from running ButtonControls.scala.

Listing 11.8: SelectionControls.scala

```scala
import scalafx.Includes._
import scalafx.scene.Scene
import scalafx.application.JFXApp
import scalafx.scene.control.Label
import scalafx.scene.control.ChoiceBox
import scalafx.scene.control.ComboBox
import scalafx.scene.control.ListView
import scalafx.collections.ObservableBuffer
import scalafx.event.ActionEvent

val app = new JFXApp {
  stage = new JFXApp.PrimaryStage {
    title = "Selection Controls"
    scene = new Scene(667,200) {
      // Label
      val label = new Label("For display purposes.")
      label.layoutX = 20
      label.layoutY = 20

      // ChoiceBox
      val choiceBox = new ChoiceBox(ObservableBuffer("Choice 1", "Choice 2",
          "Choice 3"))
      choiceBox.layoutX = 20
      choiceBox.layoutY = 50

      // ComboBox
      val comboBox = new ComboBox(List("Combo 1", "Combo 2", "Combo 3"))
      comboBox.layoutX = 20
      comboBox.layoutY = 80
```

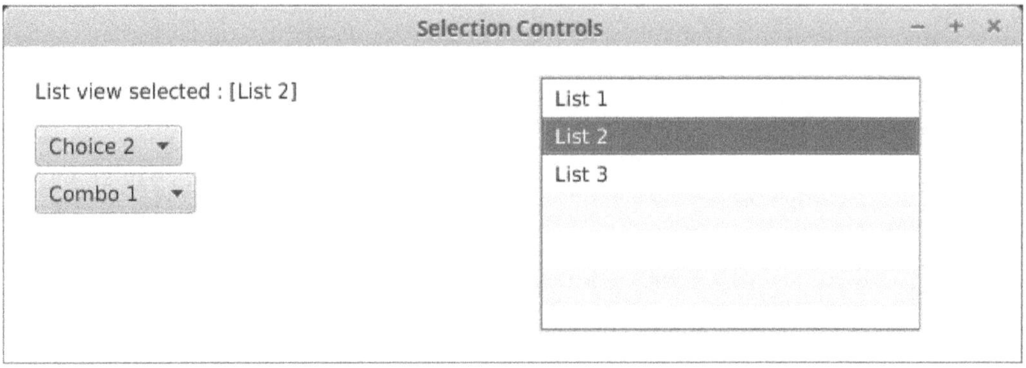

FIGURE 11.7: This figure shows the window that results from running SelectionControls.scala after the user has selected some items.

```scala
// ListView
val listView = new ListView(List("List 1", "List 2", "List 3"))
listView.layoutX = 353
listView.layoutY = 20
listView.prefHeight = 160

content = List(label, choiceBox, comboBox, listView)

choiceBox.value.onChange {
  label.text = "Choice selected : " + choiceBox.value.value
}

comboBox.onAction = (e:ActionEvent) => {
  label.text = "Combo box selected : " + comboBox.value.value
}

listView.selectionModel.value.selectedItem.onChange {
  label.text = "List view selected : " +
      listView.selectionModel.value.selectedItem.value
}
    }
  }
}

app.main(args)
```

The ChoiceBox requires that the elements be added using an ObservableBuffer instead of just a List. Also note that the three controls have different ways in which you add code that is executed when a selection is made. The ComboBox uses the standard onAction method. For the ChoiceBox, you have to use onChange with the value property. The ListView is the most complex, as it requires calling onChange, but on the selectedItem under the selectionModel.

11.5.4 Pickers

There are some types of values that the program needs to work with frequently, that require more complex interfaces for them to be easier to use. These include things like colors and dates. ScalaFX includes `ColorPicker` and `DatePicker` controls for just this purpose. Each of these appears normally as something like a `ComboBox`, but when the user clicks on them, they are presented with controls that allow them to select from various colors or a calendar that allows them to pick dates. Here is some code that uses these two controls.

Listing 11.9: PickerControls.scala

```scala
import scalafx.Includes._
import scalafx.scene.Scene
import scalafx.application.JFXApp
import scalafx.scene.control.Label
import scalafx.scene.control.ColorPicker
import scalafx.scene.control.DatePicker
import scalafx.scene.paint.Color
import scalafx.event.ActionEvent
import java.time.LocalDate

val app = new JFXApp {
  stage = new JFXApp.PrimaryStage {
    title = "Picker Controls"
    scene = new Scene(250,130) {
      // Label
      val label = new Label("Shows date once selected.")
      label.layoutX = 20
      label.layoutY = 20

      // ColorPicker
      val colorPicker = new ColorPicker(Color.White)
      colorPicker.layoutX = 20
      colorPicker.layoutY = 50

      // DatePicker
      val datePicker = new DatePicker(LocalDate.now)
      datePicker.layoutX = 20
      datePicker.layoutY = 80

      content = List(label, colorPicker, datePicker)

      colorPicker.onAction = (e:ActionEvent) => {
        fill = colorPicker.value.value
      }

      datePicker.onAction = (e:ActionEvent) => {
        label.text = "Date is : " + datePicker.value.value
      }
    }
  }
}

app.main(args)
```

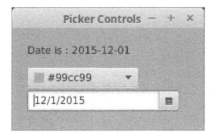

FIGURE 11.8: This figure shows the window that results from running PickerControls.scala after the user has selected some items.

Note that when make one of these types, you have to give it an initial value. For this reason, there are two extra imports at the top of the file for `scalafx.scene.paint.Color` and `java.time.LocalDate`. Both of these controls use `onAction` to set what will happen when their value is changed. In this example code, the `ColorPicker` value is used to set the fill color of the scene, so you can control what color is used for the background. The selected date is shown in the `Label`, as has been done in most of the other examples. Running this program and making some selections produced the window shown in figure 11.8.

11.5.5   TableView

One of the more complex controls that you have available to you in ScalaFX is the `TableView`. As the name implies, this control displays a table of information. It is very powerful, and includes options for editing data, but we will only look at its usage for displaying data in a table format. The code below shows a simple example.

Listing 11.10: TableControl.scala

```scala
import scalafx.Includes._
import scalafx.scene.Scene
import scalafx.application.JFXApp
import scalafx.scene.control.TableView
import scalafx.scene.control.TableColumn
import scalafx.event.ActionEvent
import scalafx.collections.ObservableBuffer
import scalafx.beans.property.StringProperty
import scalafx.beans.property.ObjectProperty

case class Student(name:String,test1:Int,test2:Int)

val app = new JFXApp {
  stage = new JFXApp.PrimaryStage {
    title = "Table Control"
    scene = new Scene(500, 300) {
      val data = ObservableBuffer(
        Student("Mark Smith", 76, 89),
        Student("Lisa Doe", 97, 96),
        Student("Bob Builder", 20, 54)
      )

      val table = new TableView(data)
      val col1 = new TableColumn[Student,String]("Name")
```

```
col1.cellValueFactory = cdf => StringProperty(cdf.value.name)
val col2 = new TableColumn[Student,Int]("Test 1")
col2.cellValueFactory = cdf => ObjectProperty(cdf.value.test1)
val col3 = new TableColumn[Student,Int]("Test 2")
col3.cellValueFactory = cdf => ObjectProperty(cdf.value.test2)
val col4 = new TableColumn[Student,Double]("Average")
col4.cellValueFactory = cdf =>
    ObjectProperty((cdf.value.test1+cdf.value.test2)/2.0)
table.columns ++= List(col1, col2, col3, col4)

root = table

table.selectionModel.value.selectedItem.onChange(
  println("Selected " + table.selectionModel.value.selectedItem.value)
)
      }
    }
  }

app.main(args)
```

In a table, the data is a sequence of rows that you make as an `ObservableBuffer`. Each column in the table has code that pulls out some value from that data. In this example, we created a `Student` type that stores a name and two grades. The table is created and passed data that contains three students. We then create four columns for the table that display the name, two grades, and average. The `TableColumn` type needs to have two type parameters. The first one is the type of data that is in the rows. The second one is the type of value being displayed in that column. It also takes an argument for the text you want to appear at the top of that column.

In order to display the data, we have to tell each column how it gets the proper value from the student. This is done by setting the `cellValueFactory` for that column. We set it to a function that takes an object of type `CellDataFeatures` and returns an observable[6] of the value to display. In this code, we have used the name `cdf` for the `CellDataFeatures`, and the only member of that class that we use is the value. The functions return either a `StringProperty` or `ObjectProperty` in this code.

Another thing that has been changed from previous examples is that instead of setting the `content` member of the `Scene`, we set the `root` value. Using `content` makes it easy to add multiple items to a `Scene`, but those items do not change in size when the window changes in size. In this program, we only have a single `Node` being added, and there is an advantage to having that `Node` conform to the size of the window. For that reason, we use `root` instead. We will see this again in section 11.7.

The last element in the code shows how to handle a user selecting a row from the table. This is done in a manner very similar to the `ListView`, where we set the `onChange` behavior for the selected item. Running this program produces a window like that shown in figure 11.9.

11.5.6 TreeView

Another more complex control that is provided in ScalaFX is the `TreeView`. This displays values that can be nested and allows the user to expand or collapse elements that have things

---

[6]Observables and properties are discussed in detail in a few sections.

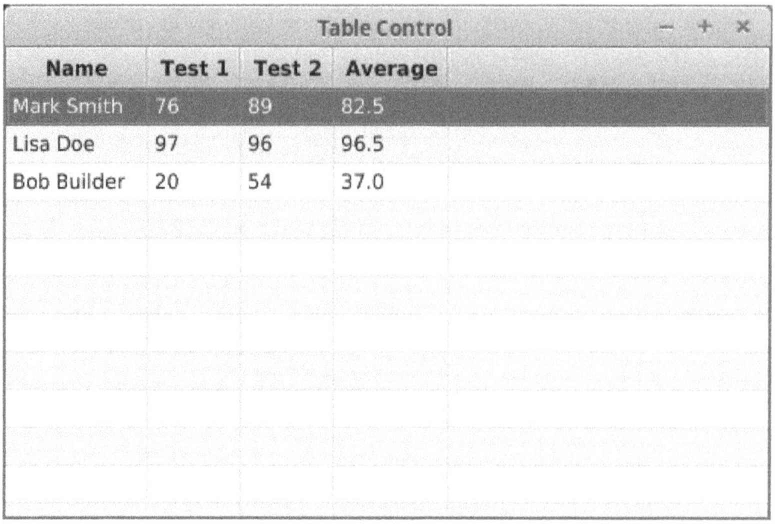

FIGURE 11.9: This figure shows the window that results from running TableControl.scala.

nested inside of them. It is likely that you have seen this type of display to show things like directory structures. Simple example code for a `TreeView` is shown here.

Listing 11.11: TreeControl.scala

```
import scalafx.Includes._
import scalafx.scene.Scene
import scalafx.application.JFXApp
import scalafx.scene.control.TreeView
import scalafx.scene.control.TreeItem
import scalafx.event.ActionEvent

val app = new JFXApp {
  stage = new JFXApp.PrimaryStage {
    title = "Tree Control"
    scene = new Scene(500, 300) {
      val imperative = new TreeItem("Imperative")
      imperative.children = List(new TreeItem("Java"), new TreeItem("C"), new
          TreeItem("C++"))
      val functional = new TreeItem("Functional")
      functional.children = List(new TreeItem("Lisp"), new TreeItem("Haskell"), new
          TreeItem("Scala"))
      val treeRoot = new TreeItem("Languages")
      treeRoot.children = List(imperative, functional)

      val tree = new TreeView(treeRoot)
      tree.layoutX = 20
      tree.layoutY = 20
      tree.prefWidth = 460
      tree.prefHeight = 260

      content = tree

      tree.selectionModel.value.selectedItem.onChange(
```

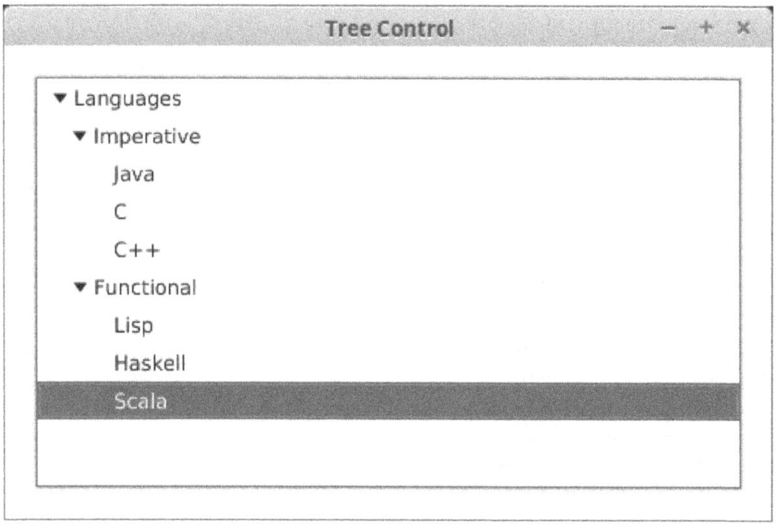

FIGURE 11.10: This figure shows the window that results from running TreeControl.scala after the items are expanded and one is selected.

```
    println("Selected " + tree.selectionModel.value.selectedItem.value)
  )
 }
 }
}
```

```
app.main(args)
```

The `TreeView` displays `TreeItems`, and each `TreeItem` may or may not have children associated with it. In this example, the top item in the tree, typically called the "root", is for programming languages. That item has two children for imperative and functional languages. Each of those is given three children that are created at the point where they are added as children.

The script ends by showing how you can register code to be run when an item is selected. The code resembles that used for the `TableView` and the `ListView`. Note that when a user expands or collapses an item, that does not register as a change in the selection. The window produced by running this code, expanding all the elements, and selecting "Scala" is shown in figure 11.10.

11.5.7 Menus and `FileChooser`

Menus typically provide one of the main ways for users to interact with GUIs, and ScalaFX includes a number of different options for creating and interacting with menus. The following code runs through and sets up a little program that makes use of most of the different options for menus. It also shows how a `FileChooser` can be utilized to allow the user to easily select files for opening or saving.

Listing 11.12: MenuControls.scala

```
import scalafx.Includes._
import scalafx.scene.Scene
import scalafx.application.JFXApp
```

```scala
import scalafx.scene.control.Label
import scalafx.scene.control.MenuBar
import scalafx.scene.control.Menu
import scalafx.scene.control.MenuItem
import scalafx.scene.control.MenuButton
import scalafx.scene.control.SeparatorMenuItem
import scalafx.scene.control.CheckMenuItem
import scalafx.scene.control.ContextMenu
import scalafx.scene.control.RadioMenuItem
import scalafx.scene.control.SplitMenuButton
import scalafx.scene.control.ToggleGroup
import scalafx.scene.layout.Priority
import scalafx.stage.FileChooser
import scalafx.event.ActionEvent
import scalafx.scene.input.KeyCode
import scalafx.scene.input.KeyCombination
import scalafx.scene.input.KeyCodeCombination

val app = new JFXApp {
  stage = new JFXApp.PrimaryStage {
    title = "Menu Controls"
    scene = new Scene(500, 200) {
      val menuBar = new MenuBar()
      val fileMenu = new Menu("File")
      val newItem = new MenuItem("New")
      newItem.accelerator = new KeyCodeCombination(KeyCode.N,
          KeyCombination.ControlDown)
      val openItem = new MenuItem("Open")
      openItem.accelerator = new KeyCodeCombination(KeyCode.O,
          KeyCombination.ControlDown)
      val saveItem = new MenuItem("Save")
      saveItem.accelerator = new KeyCodeCombination(KeyCode.S,
          KeyCombination.ControlDown)
      val exitItem = new MenuItem("Exit")
      exitItem.accelerator = new KeyCodeCombination(KeyCode.X,
          KeyCombination.ControlDown)
      fileMenu.items = List(newItem, openItem, saveItem, new SeparatorMenuItem,
          exitItem)

      val checkMenu = new Menu("Checks")
      val check1 = new CheckMenuItem("Check 1")
      val check2 = new CheckMenuItem("Check 2")
      checkMenu.items = List(check1, check2)

      val radioMenu = new Menu("Radios")
      val radio1 = new RadioMenuItem("Radio 1")
      val radio2 = new RadioMenuItem("Radio 2")
      val radio3 = new RadioMenuItem("Radio 3")
      val group = new ToggleGroup
      group.toggles = List(radio1, radio2, radio3)
      radioMenu.items = List(radio1, radio2, radio3)

      val typesMenu = new Menu("Types")
      typesMenu.items = List(checkMenu, radioMenu)
      menuBar.menus = List(fileMenu, typesMenu)
```

```
menuBar.prefWidth = 500

val menuButton = new MenuButton("Menu Button")
menuButton.items = List(new MenuItem("Button 1"), new MenuItem("Button 2"))
menuButton.layoutX = 20
menuButton.layoutY = 50

val splitMenuButton = new SplitMenuButton(new MenuItem("Split Button 1"), new
    MenuItem("Split Button 2"))
splitMenuButton.text.value = "Split Menu Button"
splitMenuButton.layoutX = 20
splitMenuButton.layoutY = 100

val contextMenu = new ContextMenu(new MenuItem("Context 1"), new
    MenuItem("Context 2"))
val label = new Label("Right click this to get a context menu.")
label.layoutX = 20
label.layoutY = 150
label.contextMenu = contextMenu

content = List(menuBar, menuButton, splitMenuButton, label)

exitItem.onAction = (e:ActionEvent) => {
  sys.exit(0)
}
saveItem.onAction = (e:ActionEvent) => {
  val fileChooser = new FileChooser
  val selectedFile = fileChooser.showSaveDialog(stage)
  label.text = "Save to : "+selectedFile
}
openItem.onAction = (e:ActionEvent) => {
  val fileChooser = new FileChooser
  val selectedFile = fileChooser.showOpenDialog(stage)
  label.text = "Open : "+selectedFile
}
    }
  }
}

app.main(args)
```

The menus at the top of the Scene are created as part of a MenuBar. That MenuBar includes different Menus that can be set with the menu property. In this example we add a "File" menu and a "Types" menu. Each menu can be populated by a combination of MenuItems, CheckMenuItems, RadioMenuItems, or other Menus as well as SeparatorMenuItems that are intended to help with layout.

The "File" menu in this example has several MenuItems that one might expect to see in such a menu, with a separator to give some offset for "Exit". To see this, you should pull the code down from the book's GitHub repository and run it. These options also have accelerator keys (also known as short-cut keys) set for them. Actions have been attached to "Open", "Save", and "Exit". Setting an action handler for a MenuItem is similar to what was done for the Button. You do this by setting the onAction value. The action for "Exit" is to call sys.exit(0), which terminates the program and tells whatever called it that it was a normal termination. The "Open" and "Save" options have actions that create FileChoosers and

FIGURE 11.11: This figure shows the window that results from running MenuControls.scala.

then call the `showOpenDialog` or `showSaveDialog` methods. When you do this, a standard dialog box for file selection is displayed. There are many other options you can provide for `FileChooser`s, but this gives you the basics.

The different options in the "File" menu also have keyboard accelerator values set for them. This allows the user to quit the program simply by pressing Ctrl-X on their keyboard or invoke the "Save" option with Ctrl-S. Having shortcuts like this is never required for a program, but it can make things much faster for users who like using the keyboard.

The "Types" menu illustrates nesting sub-menus. The first sub-menu shows the use of `CheckMenuItems`, which can be checked or unchecked and remember their state, much like a `CheckBox`. The second sub-menu demonstrates the use of `RadioMenuItems`. Like `RadioButtons`, these can be added to a `ToggleGroup` so that only one option can be selected at a time from within that group. You can have multiple `ToggleGroup`s in your `Scene`.

ScalaFX also includes three other ways to utilize menus. There is a `MenuButton`, that will pop up a menu when the button is clicked as well as a `SplitMenuButton` that has a left section that operates as a plain button, and a right section that will cause a menu to be displayed. You can see what these look like in figure 11.11. So in our example, the side of the button that has the text "Split Menu Button" is just a regular button, and the drop-down arrow to the left displays a menu when clicked.

Lastly, this code demonstrates the use of a `ContextMenu`. This is a menu that you can bring up by right clicking on some element of the GUI. A `Label` is added to our `Scene` and its `contextMenu` is set to be a little `ContextMenu`. If you run the program and right click on the label, you will see it pop up.

11.5.8 Other Stuff

There are a number of other controls that do not fit into the categories of the previous sections, so they are presented here. These include the `ProgressBar`, `ScrollBar`, `Separator`, `Slider`, and `ToolBar`. The following code shows examples of using these and how you can interact with them.

Listing 11.13: OtherControls.scala

```scala
import scalafx.Includes._
import scalafx.scene.Scene
import scalafx.application.JFXApp
import scalafx.scene.control.Button
import scalafx.scene.control.Label
import scalafx.scene.control.ProgressBar
import scalafx.scene.control.ScrollBar
import scalafx.scene.control.Separator
import scalafx.scene.control.Slider
import scalafx.scene.control.ToolBar
import scalafx.event.ActionEvent
import scalafx.geometry.Orientation

val app = new JFXApp {
  stage = new JFXApp.PrimaryStage {
    title = "Other Controls"
    scene = new Scene(500, 190) {
      val toolBar = new ToolBar
      toolBar.prefWidth = 500

      val advButton = new Button("Advance")
      val decButton = new Button("Decrement")
      toolBar.items = List(advButton, decButton, new Separator,
          new Button("tool 1"), new Button("tool 2"))

      val progress = new ProgressBar
      progress.layoutX = 20
      progress.layoutY = 70
      progress.prefWidth = 210

      val scroll = new ScrollBar
      scroll.layoutX = 20
      scroll.layoutY = 100
      scroll.min = 0
      scroll.max = 100
      scroll.prefWidth = 210

      val scrollLabel = new Label("Scroll bar value")
      scrollLabel.layoutX = 20
      scrollLabel.layoutY = 140

      val separator = new Separator
      separator.layoutX = 250
      separator.layoutY = 0
      separator.orientation = Orientation.Vertical
      separator.prefHeight = 300

      val slider = new Slider(0,10,0)
      slider.layoutX = 270
      slider.layoutY = 70
      slider.prefWidth = 210

      val sliderLabel = new Label("Slider value")
      sliderLabel.layoutX = 270
```

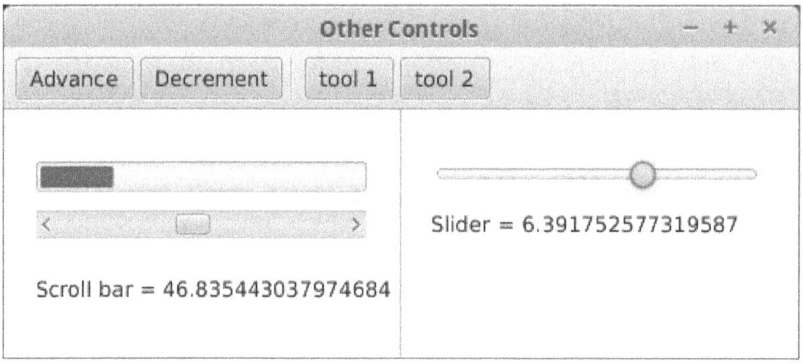

FIGURE 11.12: This figure shows the window that results from running OtherControls.scala after adjusting the various elements.

```scala
    sliderLabel.layoutY = 100

    content = List(progress, scroll, scrollLabel, separator, slider, toolBar,
        sliderLabel)

    advButton.onAction = (e:ActionEvent) => {
      progress.progress = progress.progress.value + 0.05 min 1.0 max 0.0
    }

    decButton.onAction = (e:ActionEvent) => {
      progress.progress = progress.progress.value - 0.05 min 1.0 max 0.0
    }

    scroll.value.onChange {
      scrollLabel.text = "Scroll bar = " + scroll.value.value
    }

    slider.value.onChange {
      sliderLabel.text = "Slider = " + slider.value.value
    }
  }
 }
}

app.main(args)
```

The top of the window has the `ToolBar` with four buttons. There is a `Separator` between the first two buttons and the second two buttons. The main area is divided in half by another `Separator`.

On the left side of the main area is a `ProgressBar` and a `ScrollBar`. The first two buttons in the `ToolBar` control the `ProgressBar`. By default, the `ProgressBar` has a negative value which causes it to be animated. For example, at the time of this writing shows a blue bar that moves back and forth across the field. The buttons can be used to increase or decrease the value. When these are clicked, you can see the indicator move. Figure 11.12 shows after the "Advance" button has been clicked several times.

The `ScrollBar` is set to go from 0 to 100, and when the value changes, the `Label` below the `ScrollBar` changes to tell what the value is. By default the `ScrollBar` is horizontal. You can change it to a vertical orientation if you want.

The right side of the main area has the `Slider`, which also has a label below it, and that label is updated when the value of the `Slider` is changed.

11.6   Observables, Properties, and Bindings

A number of examples above have used things with names that included the words "Observable" or "Property". For example, both the `ChoiceBox` and the `TableView` used an `ObservableBuffer` to hold the data that was placed into them. The columns for the `TableView` also explicitly made objects with the types `StringProperty` and `ObjectProperty`. Though it was not explicit in the code, the `text` and `focused` members of the `TextField` are of type `StringProperty` and `ReadOnlyBooleanProperty`, respectively. It turns out that properties are all over the place in the code that we have written. The calls to `onChange` typically followed a `value`, which was a `Property`. In addition to the ability to set the `onChange` behavior, these properties have capabilities that we have not yet explored. The most important of those is the ability to do bindings.

In the ScalaFX library, an observable is, as the name implies, something that we can observe in the code. What this means is that you can set it up so that other parts of the code are notified when the observed value is changed. The reason for the `ObservableBuffer` in the `TableView` is that the table is observing the data. So if you change the data, the table will be automatically updated. Note that this type of behavior does not happen by default normally. If you make a `var` and set a `TextField` to display that value, doing an assignment to the `var` does not alter the `TextField`. It happens that the `text` in the `TextField` is a property, so there are other ways to get it to update that are automatic.

The way to make things update automatically in the GUI is with bindings. You can bind a property to an observable of the proper type, and when the observable changes, the property will automatically be updated. You can also create bidirectional bindings between two properties, so that if either one is changed, the other will be updated as well. To see how this works, we start off by modifying a few of the earlier examples so that they use bindings instead of the `onAction` method.

We begin with the code for the `ColorPicker` and `DatePicker`. In the original version, both set `onAction` to do something. The following code shows how these can be replaced by appropriate bindings.

Listing 11.14: PickerBindings.scala

```
import scalafx.Includes._
import scalafx.scene.Scene
import scalafx.application.JFXApp
import scalafx.scene.control.Label
import scalafx.scene.control.ColorPicker
import scalafx.scene.control.DatePicker
import scalafx.scene.paint.Color
import scalafx.beans.property.StringProperty
import java.time.LocalDate

val app = new JFXApp {
  stage = new JFXApp.PrimaryStage {
    title = "Picker Bindings"
    scene = new Scene(250,130) {
      // Label
```

```scala
    val label = new Label("Shows date once selected.")
    label.layoutX = 20
    label.layoutY = 20

    // ColorPicker
    val colorPicker = new ColorPicker(Color.White)
    colorPicker.layoutX = 20
    colorPicker.layoutY = 50

    // DatePicker
    val datePicker = new DatePicker(LocalDate.now)
    datePicker.layoutX = 20
    datePicker.layoutY = 80

    content = List(label, colorPicker, datePicker)

    fill <== colorPicker.value

    label.text <== StringProperty("Date is : ") + datePicker.value.asString
    }
  }
}
```

```scala
app.main(args)
```

The only changes are at the end of the file, where the handling code had been. The colorPicker object had been set so that it would alter the fill property of the Scene when a selection was made. Now that functionality is achieved with fill <== colorPicker.value. The <== operator creates a unidirectional binding. In this case, fill is bound to the value of colorPicker.value. So when the value of the colorPicker is changed, the background color of the scene is automatically updated to reflect it.

Updating the text on the Label is a bit more complex, but only a bit. This is because we have a type difference, and we also want to bind to a String that is more than just the date itself. The code that is used is label.text <== StringProperty("Date is : ") + datePicker.value.asString. Again, the <== operator is used to actually create the binding. What we bind to needs to be a property, and more specifically a StringProperty, because the text is a StringProperty. Calling asString on the datePicker.value gives us a StringProperty for that, but we also want to prepend "Date is :" for the label. To do that, we make a StringProperty with that string literal and use + to concatenate them. This produces a new StringProperty that is itself bound to the value of the DatePicker.

11.6.1 Numeric Properties and Bindings

To illustrate what we can do with numeric bindings, we will use a new example. This example has three Sliders that control the location and size of a rectangle. The first one controls the x-location (horizontal location), the second one controls the y-location (vertical location), and the third one controls the size. There is also a ScrollBar that is bidirectionally bound to one of the Sliders to illustrate how one of the elements change when you change the other element. Lastly, there is a Label that is set to be centered on the window using bindings.

Listing 11.15: NumericBindings.scala

```scala
import scalafx.Includes._
import scalafx.scene.Scene
import scalafx.application.JFXApp
import scalafx.scene.control.Label
import scalafx.scene.control.ScrollBar
import scalafx.scene.control.Slider
import scalafx.scene.shape.Rectangle

val app = new JFXApp {
  stage = new JFXApp.PrimaryStage {
    title = "Numeric Bindings"
    scene = new Scene(600, 250) {
      val label = new Label("This stays centered.")

      val xSlider = new Slider(0,100,0)
      xSlider.layoutX = 10
      xSlider.layoutY = 10
      xSlider.prefWidth = 180

      val ySlider = new Slider(0,100,0)
      ySlider.layoutX = 210
      ySlider.layoutY = 10
      ySlider.prefWidth = 180

      val sizeSlider = new Slider(0,100,0)
      sizeSlider.layoutX = 410
      sizeSlider.layoutY = 10
      sizeSlider.prefWidth = 180

      val scroll = new ScrollBar
      scroll.layoutX = 210
      scroll.layoutY = 220
      scroll.min = 0
      scroll.max = 100
      scroll.prefWidth = 180

      val rectangle = Rectangle(10,10)
      rectangle.layoutX = 0
      rectangle.layoutY = 40

      content = List(label, xSlider, ySlider, sizeSlider, scroll, rectangle)

      label.layoutX <== (width-label.width)/2
      label.layoutY <== (height-label.height)/2
      rectangle.layoutX <== xSlider.value*6
      rectangle.layoutY <== ySlider.value+40
      rectangle.width <== (sizeSlider.value*2)+10
      rectangle.height <== (sizeSlider.value*2)+10
      scroll.value <==> ySlider.value
    }
  }
}

app.main(args)
```

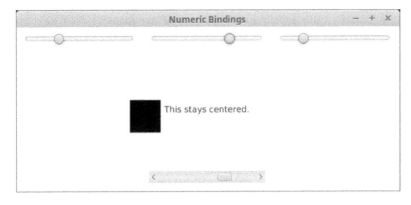

FIGURE 11.13: This figure shows the window that results from running NumericBindings.scala after the sliders have been adjusted some.

The setup of the GUI is fairly standard, based on what we have seen previously. The interesting part is the bindings at the bottom. All of these bindings are for numeric values. In particular, they have the type `DoubleProperty`. As you can see in the code, we are able to do math with these values, and the result is another `DoubleProperty`.

If you have played much with the previous GUIs, you probably noted one significant shortcoming, they do not adjust to the user changing the window size. In all of the GUIs so far, we have manually set the values of `layoutX` and `layoutY` as well as `prefWidth` and `prefHeight` when needed. This approach is simple, but it is not really flexible. We could have added the line `resizable = false` to the stage in any of these GUIs to make it so that the user is not allowed to change the size of the window, but that does not really fix the underlying problem. The proper way to do with this is to use layouts and panes, which are described in section 11.7. However, bindings can also produce the desired results in simple situations, or when you want to do something that an existing layout or pane will not do.

In this example, the `label` is always displayed in the center of the window, even when you resize the window.[7] This behavior is produced by binding the layout values to simple calculations based on the width and height of the full scene and the label. You can use this type of binding to put things in arbitrary locations that are based on the size of the window, but it is less efficient than using layouts and panes when they work.

The last binding is an example of a bidirectional binding. It uses the `<==>` operator to link the values of the `ScrollBar` and the middle `Slider` that is bound to the y-value of the rectangle. If you move either one, the other also moves, and the rectangle's vertical position also shifts. In figure 11.13 you can see that the `ScrollBar` and the `Slider` above it are in the same position.

11.6.2 Conditional Bindings

It is also possible to make bindings conditional. This is done using `when`, `choose`, and `otherwise`. To illustrate this, we have a simple script that includes a `Slider` and a `Label`. The location of the `Label` is determined by the `Slider`. Unlike the previous example, where the position moved smoothly, the position here is set in a discontinuous way so it jumps

---

[7]In the build of ScalaFX used at the time this was written, this produces a warning as the binding happens before the scene has a size. This does not have a negative performance on the behavior of the script, so we will ignore it.

from the left side to the right when the `Slider` crosses the central value. The color of the background is also varies based on whether the mouse is over the `Label`.

Listing 11.16: ConditionalBindings.scala

```scala
import scalafx.Includes._
import scalafx.scene.Scene
import scalafx.application.JFXApp
import scalafx.scene.control.Label
import scalafx.scene.control.Slider
import scalafx.scene.paint.Color

val app = new JFXApp {
  stage = new JFXApp.PrimaryStage {
    title = "Conditional Bindings"
    scene = new Scene(500, 250) {
      val label = new Label("Hover to change background.")

      val slider = new Slider(0,100,0)
      slider.layoutX = 10
      slider.layoutY = 10
      slider.prefWidth = 180

      content = List(label, slider)

      label.layoutX <== when (slider.value < 50) choose 0 otherwise
          width-label.width
      label.layoutY <== (height-label.height)/2
      fill <== when (label.hover) choose Color.Red otherwise Color.White
    }
  }
}


app.main(args)
```

At the end of the code you can see how the conditional bindings are created. The `when` is followed by an observable Boolean, such as a `BooleanProperty`. That is followed by `choose` and the value/property that should be used if the condition is true. That is followed by `otherwise` and the value/property to use if the condition is false. Unlike `if` and `else`, `when`, `choose`, and `otherwise` are not keywords in the Scala language, they are part of the ScalaFX library. The `import scalafx.Includes._` makes `when` visible, and the others are actually methods being used in the infix style.[8]

Figure 11.14 shows the window that one gets from this script with the `Slider` moved above the middle value and the mouse, not shown, hovering over the `Label`.

---

[8]Recall that the infix style is when a method with one argument is put between the object and the operand without a dot or parentheses. We typically use this with symbolic operators. Another common example was `min` and `max` when we would do `a min b` instead of `a.min(b)`.

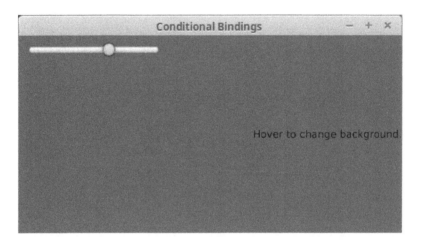

FIGURE 11.14: This figure shows the window that results from running Conditional-Bindings.scala after the slider has been adjusted with the mouse hovering over the label.

11.7   Layout and Panes

While you can use bindings to make things move around when you change the size of a window, this really is not the way you want to build complex user interfaces. The proper way to make complex interfaces in ScalaFX is with layouts and panes. Most of these are in the `scalafx.scene.layout` package, but some are part of `scalafx.scene.control`. The ones in `layout` produce fairly simple layouts that do not interact with the user. The ones in `control` allow some form of user interaction. Using either of these, you can create complex displays by nesting them inside of one another. The panes themselves are `Nodes`, so you can add them to one another in the same way that you might add a `Button` into one. Each type has its own rules for how the elements inside of it change size when the whole thing changes size.

11.7.1   `scalafx.scene.layout` Panes

The following list briefly describes the types provided in the `scalafx.scene.layout` package in alphabetical order.

- `AnchorPane` - This pane allows you to set the location of contents inside of it relative to the location of this pane. You can call methods like `AnchorPane.setBottomAnchor(child:Node, value:Double)` to specify where a child should appear in the `AnchorPane`. Items can be anchored to more than one edge.

- `BorderPane` - Arranges elements in five different regions: top, bottom, left, right, and center. The elements at the top and the bottom span the full width of the pane and take as much space as they need in height. The left and right elements take the remaining space in height and as much as they need in width. The center element gets whatever is left over.

- `FlowPane` - This pane places the contents one after another with either a horizontal or vertical orientation, giving each one as much space as it wants. When it runs out of space it starts a new row or column, based on the orientation.

- `GridPane` - Arranges its elements in a flexible grid. The elements can be placed at any row and column, and they are allowed to cover multiple rows and/or columns. The rows and columns are sized to accommodate the elements that are in them.

- `HBox` - Arranges elements horizontally next to one another.

- `StackPane` - Stacks elements on top of one another in the same area of the screen so that they overlap.

- `TilePane` - Arranges the elements in a grid of uniform "tiles". The contents are arranged going across each row first. Tile sizes are determined by the contents by default.

- `VBox` - Arranges the elements vertically each one below the one before it.

To help illustrate how these work, the following code creates a GUI that has a `BorderPane` as the primary contents, with a `HBox` on top, a `VBox` on the left, a `FlowPane` on bottom, a `StackPane` on the right, and a `TilePane` in the center. In each one of these we add a `Button`, a `Label`, a `TextField`, a `TextArea`, a `ComboBox`, and a `Slider`.

Listing 11.17: Layouts.scala

```scala
import scalafx.Includes._
import scalafx.scene.Scene
import scalafx.application.JFXApp
import scalafx.scene.control._
import scalafx.scene.layout.BorderPane
import scalafx.scene.layout.FlowPane
import scalafx.scene.layout.HBox
import scalafx.scene.layout.Pane
import scalafx.scene.layout.StackPane
import scalafx.scene.layout.TilePane
import scalafx.scene.layout.VBox
import scalafx.scene.layout.Background
import scalafx.scene.layout.BackgroundFill
import scalafx.scene.paint.Color

def addControlsToPane(pane:Pane, fill:Color):Pane = {
  val button = new Button("Click Me")
  val label = new Label("Plain label")
  val field = new TextField
  field.text = "Text Field"
  val area = new TextArea
  area.text = "Text Area\nMultiple\nLines"
  area.prefWidth = 100
  area.prefHeight = 100
  val combo = new ComboBox(List("Alpha", "Beta", "Gamma"))
  val slider = new Slider
  pane.children = List(button, label, field, area, combo, slider)
  pane.background = new Background(Array(new BackgroundFill(fill, null, null)))
  pane
}

val app = new JFXApp {
  stage = new JFXApp.PrimaryStage {
    title = "Layouts"
    scene = new Scene(750,600) {
      val borderPane = new BorderPane
      borderPane.top = addControlsToPane(new HBox(10), Color.gray(0.25))
      borderPane.left = addControlsToPane(new VBox(10), Color.gray(0.75))
      borderPane.bottom = addControlsToPane(new FlowPane(10, 10), Color.gray(0.5))
      borderPane.right = addControlsToPane(new StackPane, Color.gray(1.0))
      val tilePane = new TilePane
      tilePane.prefRows = 2
      borderPane.center = addControlsToPane(tilePane, Color.gray(0.0))
      tilePane.children.foreach(_.managed = true)
      root = borderPane
    }
  }
}

app.main(args)
```

The result of running this program is shown in figure 11.15. The `import` of the controls was changed to use an underscore as the details of those were discussed above. The `addControlsToPane` function was written because the same controls are being added to

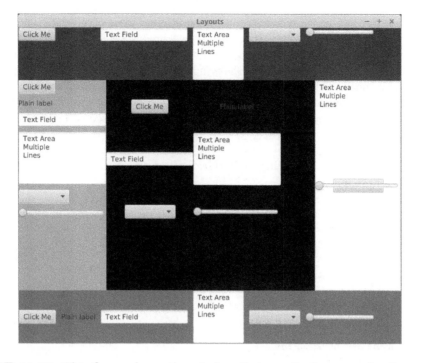

FIGURE 11.15: This figure shows the window that results from running Layouts.scala after making the window slightly smaller.

every one of the different panes and we do not want to have to duplicate that code. Making it take a `Pane` allows it to work with all of the different panes. The preferred size of the `TextArea` was set so that it would not consume too much space in the GUI. All the other elements were left at their default sizes. The `addControlsToPane` function also takes a second argument that gives the background color for that `Pane`. This is used at the end of the function when `pane.background` is set. This is done so that you can tell the different panes apart in the resulting GUI. There are many options for setting the background of `Panes`, and we will not be going into the details in this book.

In the top you can see that the `HBox`, which has a dark gray background color, positions the controls horizontally with all of them aligned to the top by default. Note that this can be changed by setting the `alignment` property. On the left is the `VBox` with a light gray background color, which behaves much like the `HBox`, but stacks the elements vertically and aligned to the left. At the bottom is a `FlowPane` with a medium gray background, which looks much like the `HBox` as far as what it produces, but there are two significant differences to note. The controls are all centered vertically by default, and when the window gets too small for them to fit across a single row, the last one is shifted down to a new row. If the window is made smaller, the `Slider` would be cut off in the `HBox`. The two arguments passed in to make the `FlowPane` are spacing values. The spacing values represent the number of pixels that are put between items vertically and horizontally. Spacing can be added to all the elements using `Insets`, but we will not go into details about those.

On the right there is a `StackPane`. Its white background is not seen as the `TextArea` here covers the whole thing and conceals not only the background, but the elements added before the `TextArea`. This produces very different results from the other panes as the controls are literally stacked one on top of the other. In the center is a `TilePane` with a black background. All the controls are given a space that is equal to the largest width and largest height of the contents. They are then laid out across rows, forming new rows as needed. The `TilePane` works much like the `FlowPane` other than how things are sized.

Note that the `BorderPane` is set to be the `root` of the `Scene`. As was mentioned before, this causes it to be bound to the size of the window, so that the contents adjust when the window size is changed. This will not happen by default if you use `content`.

The `GridPane` is a bit more complex than the panes in the previous example, so we give it its own. The following code shows how a `GridPane` could be used to lay out a GUI for a basic calculator.

Listing 11.18: GridPane.scala

```scala
import scalafx.Includes._
import scalafx.scene.Scene
import scalafx.application.JFXApp
import scalafx.scene.control._
import scalafx.scene.layout.GridPane
import scalafx.scene.layout.Priority

val app = new JFXApp {
  stage = new JFXApp.PrimaryStage {
    title = "GridPane"
    scene = new Scene(300,300) {
      val gridPane = new GridPane
      for (i <- 1 to 9) {
        val button = new Button(i.toString)
        gridPane.children += button
        GridPane.setColumnIndex(button, (i-1)%3)
        GridPane.setRowIndex(button, (i-1)/3+1)
        GridPane.setHgrow(button, Priority.Always)
        GridPane.setVgrow(button, Priority.Always)
        button.maxWidth = Int.MaxValue
        button.maxHeight = Int.MaxValue
      }
      val textField = new TextField
      gridPane.children += textField
      GridPane.setConstraints(textField,0,0,4,1)
      val zeroButton = new Button("0")
      gridPane.children += zeroButton
      GridPane.setConstraints(zeroButton,0,4,2,1)
      GridPane.setVgrow(zeroButton, Priority.Always)
      zeroButton.maxWidth = Int.MaxValue
      zeroButton.maxHeight = Int.MaxValue
      val periodButton = new Button(".")
      gridPane.children += periodButton
      GridPane.setConstraints(periodButton,2,4)
      GridPane.setVgrow(periodButton, Priority.Always)
      periodButton.maxWidth = Int.MaxValue
      periodButton.maxHeight = Int.MaxValue
      val plusButton = new Button("+")
      gridPane.children += plusButton
      GridPane.setConstraints(plusButton,3,1)
      GridPane.setHgrow(plusButton, Priority.Always)
      plusButton.maxWidth = Int.MaxValue
      plusButton.maxHeight = Int.MaxValue
      val minusButton = new Button("-")
```

```
gridPane.children += minusButton
GridPane.setConstraints(minusButton,3,2)
minusButton.maxWidth = Int.MaxValue
minusButton.maxHeight = Int.MaxValue
val multButton = new Button("*")
gridPane.children += multButton
GridPane.setConstraints(multButton,3,3)
multButton.maxWidth = Int.MaxValue
multButton.maxHeight = Int.MaxValue
val divButton = new Button("/")
gridPane.children += divButton
GridPane.setConstraints(divButton,3,4)
divButton.maxWidth = Int.MaxValue
divButton.maxHeight = Int.MaxValue

    root = gridPane
  }
 }
}
```

```
app.main(args)
```

The result of this code can be seen in figure 11.16. To add the main buttons, we use a **for** loop to prevent us from having to duplicate the same code nine times. After each **Button** is created and added, we set constraints on it. This code sets the column and row values with separate calls, then sets the **Hgrow** and **Vgrow** values. Without this, the rows and columns will not grow to fill in the full space of the window, and all the elements will remain packed in the top left corner. Note that the constraints are set by calling methods of the **GridPane** object and passing the **Node** as the first argument. That loop ends with calls to set the **maxWidth** and **maxHeight** of the buttons to the maximum **Int** value. This causes the buttons to fill the entire space available to them. Without this, there will be small buttons spaced out through the window.

The other controls are added after the loop. The constraints for these are set using **setConstraints**, which takes the **Node** followed by the row, column, row span, and column span. The span values allow us to have a single element cover more than one row or column. This is used for the **TextField** and the "0" button.

11.7.2   scalafx.scene.control **Panes**

The **scalafx.scene.control** package provides these additional panes. Each of these includes user interaction of some form or another.

- **Accordion** - This is a group of **TitledPane**s stacked vertically on top of one another, where only one can be open at a time.

- **ScrollPane** - This pane embeds a single **Node** and will display scroll bars if the child wants more space than what the **ScrollPane** has to display.

- **SplitPane** - Arranges children horizontally or vertically with a bar between them that the user can adjust.

- **TabPane** - Displays a row of tabs and one child. Selecting a tab brings up the associated child. The **Tab** type is used to describe those tabs.

FIGURE 11.16: This figure shows the window that results from running GridPane.scala.

- **TitledPane** - This element shows up as a title that can be expanded to display its contents and collapsed back down to just the title. It is primarily used in an **Accordion**.

These can be used with the other panes from the **layout** package. The example code shown here demonstrates their usage in isolation.

To demonstrate the usage of these panes, we put a **TabPane** at the top level with one tab that contains the other panes and a second tab with a single large **TextArea**. The tab with the panes has a **SplitPane** at the top level with an **Accordion** embedded in a **ScrollPane** on the left and another **TextArea** on the right.

Listing 11.19: ControlPanes.scala

```scala
import scalafx.Includes._
import scalafx.scene.Scene
import scalafx.application.JFXApp
import scalafx.scene.control._

val app = new JFXApp {
  stage = new JFXApp.PrimaryStage {
    title = "Control Panes"
    scene = new Scene(500,250) {
      val tabPane = new TabPane
      val splitPane = new SplitPane
      val tabArea = new TextArea
      val panesTab = new Tab
      panesTab.text = "Control Panes"
      panesTab.content = splitPane
      val areaTab = new Tab
      areaTab.text = "Text Area"
      areaTab.content = tabArea
      tabPane.tabs = List(panesTab, areaTab)
      val scrollPane = new ScrollPane
```

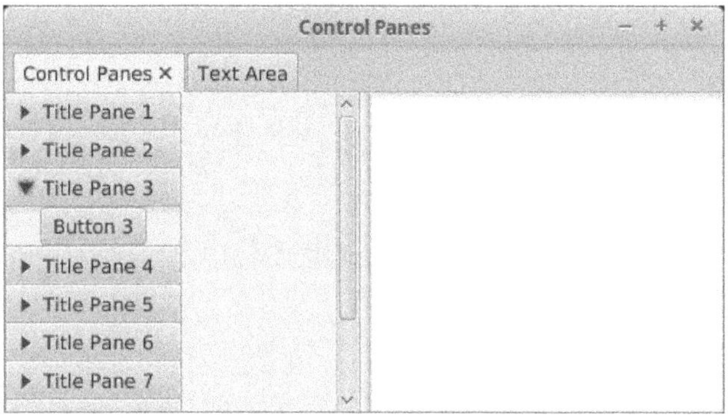

FIGURE 11.17: This figure shows the window that results from running Control-Panes.scala after expanding one of the `TitledPane`s in the `Accordion`.

```
val rightArea = new TextArea
splitPane.items += scrollPane
splitPane.items += rightArea
val accordion = new Accordion
for (i <- 1 to 10) {
  val titledPane = new TitledPane
  titledPane.text = "Title Pane "+i
  titledPane.content = new Button("Button "+i)
  accordion.panes += titledPane
}
scrollPane.content = accordion

      root = tabPane
    }
  }
}

app.main(args)
```

Figure 11.17 shows what you would see if you ran this script and expanded "Title Pane 3". The tabs appear at the top. The central bar for the `SplitPane` can be moved left or right by the user. Note that there is a scroll bar present that allows you to view the later elements in the `Accordion`. This was the reason for including a loop that added ten `TitledPane`s, so they would be long enough to need a scroll bar. If you drag the split location to the left so that it is smaller than the `Accordion`, a scroll bar will appear at the bottom as well.

Perhaps the biggest challenge in using these controls is remembering the name each one uses for where the children/contents should be placed. The names are fairly logical for each one, but there is not a single name that works for all of them. The `TabPane` uses `tabs`, the `SplitPane` uses `items`, the `ScrollPane` uses `content`, and the `Accordion` uses `panes`. You either have to remember these or go look in the API or this book when you need that information.

FIGURE 11.18: This shows a possible layout for the GUI that we want to build for editing recipes.

11.8 Putting It Together

Now that we have briefly explored the different components of ScalaFX GUIs and how we interact with them, we can try to put a number of them together in a larger program. The program is going to be something like a digital recipe book. It should be able to store multiple recipes. For each recipe, it should allow the user to specify ingredients with amounts as well as have a written description of the cooking directions. The GUI should give the user the ability to add and remove recipes. They should be able to select recipes and edit the various information that we have on them.

At this point in the analysis you would probably want to sketch out a little picture of what you want the GUI to look like. That is done for this GUI in figure 11.18. The left side is taken up by a `ListView` that shows all the recipes. When the user clicks on a recipe, the information for it should be shown on the right. That includes a list of all the ingredients and the directions. The ingredients can also be edited and there need to be some buttons to help with this. The options for adding and removing recipes will be in the menus.

The question is how we should lay this out. Any GUI can be built in multiple different ways using the different pane options that were discussed above. For example, the separation between the recipe list and the information about the selected recipe could be created by a `BorderPane`, a `HBox`, or a `SplitPane`. Which of these you pick will normally be determined by the way in which the panes would size things or whether you want to give the user the control of a `SplitPane`. We will use a `BorderPane` for the top level, which will hold the

MenuBar, the recipe list, ingredient information and tools, and the recipe directions. The
BorderPane will allow us to resize the elements if the user resizes the window, and it will
give us a top, left, and center that will be used to further organize the content. We can
place the MenuBar in the top, the recipe list in the left, and everything else in the center.
We will break up the center and use a SplitPane to separate the recipe directions from
the ingredient information. We will use a GridPanel for the buttons, ingredient list, and
ingredient settings. The following code shows how we could solve this adding "traditional"
handlers to the code instead of bindings.

Listing 11.20: Recipe.scala

```scala
import scalafx.Includes._
import scalafx.scene.Scene
import scalafx.application.JFXApp
import scalafx.scene.control._
import scalafx.scene.layout._
import scalafx.event.ActionEvent
import scalafx.scene.input.KeyCode
import scalafx.scene.input.KeyCombination
import scalafx.scene.input.KeyCodeCombination
import scalafx.stage.FileChooser
import scalafx.collections.ObservableBuffer
import java.io.PrintWriter

case class Ingredient(name:String, amount:String)
case class Recipe(name:String, ingredients:List[Ingredient], directions:String)

val app = new JFXApp {
  stage = new JFXApp.PrimaryStage {
    title = "Recipes"
    scene = new Scene(600,400) {
      // Menus
      val menuBar = new MenuBar
      val fileMenu = new Menu("File")
      val openItem = new MenuItem("Open")
      openItem.accelerator = new KeyCodeCombination(KeyCode.O,
          KeyCombination.ControlDown)
      openItem.onAction = (e:ActionEvent) => { openFile }
      val saveItem = new MenuItem("Save")
      saveItem.accelerator = new KeyCodeCombination(KeyCode.S,
          KeyCombination.ControlDown)
      saveItem.onAction = (e:ActionEvent) => { saveFile }
      val exitItem = new MenuItem("Exit")
      exitItem.accelerator = new KeyCodeCombination(KeyCode.X,
          KeyCombination.ControlDown)
      exitItem.onAction = (e:ActionEvent) => { System.exit(0) }
      fileMenu.items = List(openItem, saveItem, new SeparatorMenuItem, exitItem)

      val recipeMenu = new Menu("Recipe")
      val addItem = new MenuItem("Add")
      addItem.onAction = (e:ActionEvent) => { addRecipe }
      val removeItem = new MenuItem("Remove")
      removeItem.onAction = (e:ActionEvent) => { removeRecipe }
      recipeMenu.items = List(addItem, removeItem)
```

```scala
42        menuBar.menus = List(fileMenu, recipeMenu)
43
44        // Recipe List
45        var recipes = Array(Recipe("Pop Tarts",List(Ingredient("Pop Tart",
46          "1 packet")), "Toast the poptarts ...\nor don't."))
47
48        val recipeList = new ListView(recipes.map(_.name))
49        recipeList.selectionModel.value.selectedIndex.onChange {
50          val index = recipeList.selectionModel.value.selectedIndex.value
51          if (index>=0) setFields(recipes(index))
52        }
53
54        // Ingredients stuff
55        val addButton = new Button("Add")
56        addButton.onAction = (ae:ActionEvent) => addIngredient
57        val removeButton = new Button("Remove")
58        removeButton.onAction = (ae:ActionEvent) => removeIngredient
59        val ingredientsList = new ListView[String]()
60        val ingredientNameField = new TextField
61        val amountField = new TextField
62        val ingredientsGrid = new GridPane
63        ingredientsGrid.children += addButton
64        GridPane.setConstraints(addButton,0,0)
65        ingredientsGrid.children += removeButton
66        GridPane.setConstraints(removeButton,1,0)
67        ingredientsGrid.children += ingredientsList
68        GridPane.setConstraints(ingredientsList,0,1,2,3)
69        val nameLabel = new Label("Name:")
70        ingredientsGrid.children += nameLabel
71        GridPane.setConstraints(nameLabel,3,0)
72        val amountLabel = new Label("Amount:")
73        ingredientsGrid.children += amountLabel
74        GridPane.setConstraints(amountLabel,3,2)
75        ingredientsGrid.children += ingredientNameField
76        GridPane.setConstraints(ingredientNameField,4,0)
77        ingredientsGrid.children += amountField
78        GridPane.setConstraints(amountField,4,2)
79        ingredientsList.selectionModel.value.selectedItem.onChange {
80          val recipeIndex = recipeList.selectionModel.value.selectedIndex.value
81          val ingredientIndex =
                  ingredientsList.selectionModel.value.selectedIndex.value
82          if (recipeIndex>=0 && ingredientIndex>=0) {
83            ingredientNameField.text =
                    recipes(recipeIndex).ingredients(ingredientIndex).name
84            amountField.text =
                    recipes(recipeIndex).ingredients(ingredientIndex).amount
85          }
86        }
87        ingredientNameField.text.onChange {
88          val newName = ingredientNameField.text.value
89          alterSelectedIngredient(i => i.copy(name = newName))
90          val ingredientIndex =
                  ingredientsList.selectionModel.value.selectedIndex.value
91          if (ingredientIndex>=0) ingredientsList.items.value(ingredientIndex) =
                  newName
```

```
92      }
93      amountField.text.onChange {
94        alterSelectedIngredient(i => i.copy(amount = amountField.text.value))
95      }
96
97      // Directions
98      val directionsArea = new TextArea
99      directionsArea.text.onChange {
100       val recipeIndex = recipeList.selectionModel.value.selectedIndex.value
101       if (recipeIndex>=0) {
102         recipes(recipeIndex) = recipes(recipeIndex).copy(directions =
                  directionsArea.text.value)
103       }
104     }
105
106     val splitPane = new SplitPane
107     splitPane.orientation = scalafx.geometry.Orientation.Vertical
108     splitPane.items += ingredientsGrid
109     splitPane.items += directionsArea
110
111     // Top level layout
112     val topBorderPane = new BorderPane
113     topBorderPane.top = menuBar
114     topBorderPane.left = recipeList
115     topBorderPane.center = splitPane
116
117     root = topBorderPane
118
119     def openFile:Unit = {
120       val chooser = new FileChooser
121       val selected = chooser.showOpenDialog(stage)
122       if (selected!=null) {
123         val src = io.Source.fromFile(selected)
124         val lines = src.getLines
125         recipes = Array.fill(lines.next.toInt)(Recipe(
126             lines.next,
127             List.fill(lines.next.toInt)(Ingredient(lines.next,lines.next)),
128             {
129               var dir = ""
130               var line = lines.next
131               while (line!=".") {
132                 dir += (if (dir.isEmpty) "" else "\n")+line
133                 line = lines.next
134               }
135               dir
136             }
137         ))
138         src.close()
139         recipeList.items = ObservableBuffer(recipes.map(_.name):_*)
140         recipeList.selectionModel.value.selectFirst
141         setFields(recipes.head)
142       }
143     }
144
145     def saveFile:Unit = {
```

```scala
146          val chooser = new FileChooser
147          val selected = chooser.showSaveDialog(stage)
148          if (selected!=null) {
149            val pw = new PrintWriter(selected)
150            pw.println(recipes.length)
151            for (r <- recipes) {
152              pw.println(r.name)
153              pw.println(r.ingredients.length)
154              for (ing <- r.ingredients) {
155                pw.println(ing.name)
156                pw.println(ing.amount)
157              }
158              pw.println(r.directions)
159              pw.println(".")
160            }
161            pw.close()
162          }
163        }
164
165        def addRecipe:Unit = {
166          val dialog = new TextInputDialog
167          dialog.title = "Recipe Name"
168          dialog.headerText = "Question?"
169          dialog.contentText = "What is the name of the new recipe?"
170          dialog.showAndWait().foreach { name =>
171            recipes = recipes :+ Recipe(name,
172              List(Ingredient("ingredient","amount")),"Directions")
173            recipeList.items = ObservableBuffer(recipes.map(_.name):_*)
174            recipeList.selectionModel.value.clearAndSelect(recipes.length-1)
175            setFields(recipes.last)
176          }
177        }
178
179        def removeRecipe:Unit = {
180          if (!recipeList.selectionModel.value.selectedItems.isEmpty) {
181            recipes = recipes.patch(recipeList.selectionModel.value.selectedIndex.
                 value,Nil,1)
182            if (recipes.isEmpty) {
183              recipes = Array(Recipe("New recipe",
184                List(Ingredient("ingredient","amount")),"Directions"))
185            }
186            recipeList.items = ObservableBuffer(recipes.map(_.name):_*)
187            recipeList.selectionModel.value.clearAndSelect(0)
188            setFields(recipes.head)
189          }
190        }
191
192        def addIngredient:Unit = {
193          val recipeIndex = recipeList.selectionModel.value.selectedIndex.value
194          if (recipeIndex>=0) {
195            val newIngr = Ingredient("Stuff", "Some")
196            recipes(recipeIndex) = recipes(recipeIndex).copy(
197              ingredients = recipes(recipeIndex).ingredients :+ newIngr)
198            ingredientsList.items.value += newIngr.name
199          }
```

```
200        }
201
202        def removeIngredient:Unit = {
203          val recipeIndex = recipeList.selectionModel.value.selectedIndex.value
204          val ingredientIndex =
               ingredientsList.selectionModel.value.selectedIndex.value
205          if (recipeIndex>=0 && ingredientIndex>=0) {
206            recipes(recipeIndex) = recipes(recipeIndex).copy(ingredients =
207              recipes(recipeIndex).ingredients.patch(ingredientIndex,Nil,1))
208            setFields(recipes(recipeIndex))
209          }
210        }
211
212        def setFields(r:Recipe):Unit = {
213          ingredientsList.items.value.clear
214          ingredientsList.items.value ++= r.ingredients.map(_.name)
215          directionsArea.text = r.directions
216          ingredientNameField.text = ""
217          amountField.text = ""
218        }
219
220        def alterSelectedIngredient(diff: Ingredient => Ingredient):Unit = {
221          val recipeIndex = recipeList.selectionModel.value.selectedIndex.value
222          val ingredientIndex =
               ingredientsList.selectionModel.value.selectedIndex.value
223          if (recipeIndex>=0 && ingredientIndex>=0) {
224            val newIngredient =
                 diff(recipes(recipeIndex).ingredients(ingredientIndex))
225            val newRecipe = recipes(recipeIndex).copy(ingredients =
226              recipes(recipeIndex).ingredients.updated(ingredientIndex,
227              newIngredient))
228            recipes(recipeIndex) = newRecipe
229          }
230        }
231      }
232    }
233  }
234
235  app.main(args)
```

The GUI created by running this code can be seen in figure 11.19. This is a rather long program, so we will consider the pieces individually. Lines 1-12 have the imports that we need, mostly from the ScalaFX libraries, but also includes java.io.PrintWriter for saving the recipes out to file. Lines 14 and 15 have the case classes that are used to store the recipe book.

The GUI setup happens in lines 21-117. There are comments to offset the different sections of code that handle different parts of the GUI. The sections for the menus and the recipe list are fairly straightforward. Note that line 45 declares and initializes a var Array that stores the recipes in memory. This is used extensively by the other elements of the GUI.

The section of the GUI that shows the ingredient information uses a GridPane and is a bit longer. The interactions with the user include changing the TextFields when the user selects a different ingredient and changing the data when they edit the fields. Note that unlike previous examples, this code reacts to changes in the actual text of the TextField

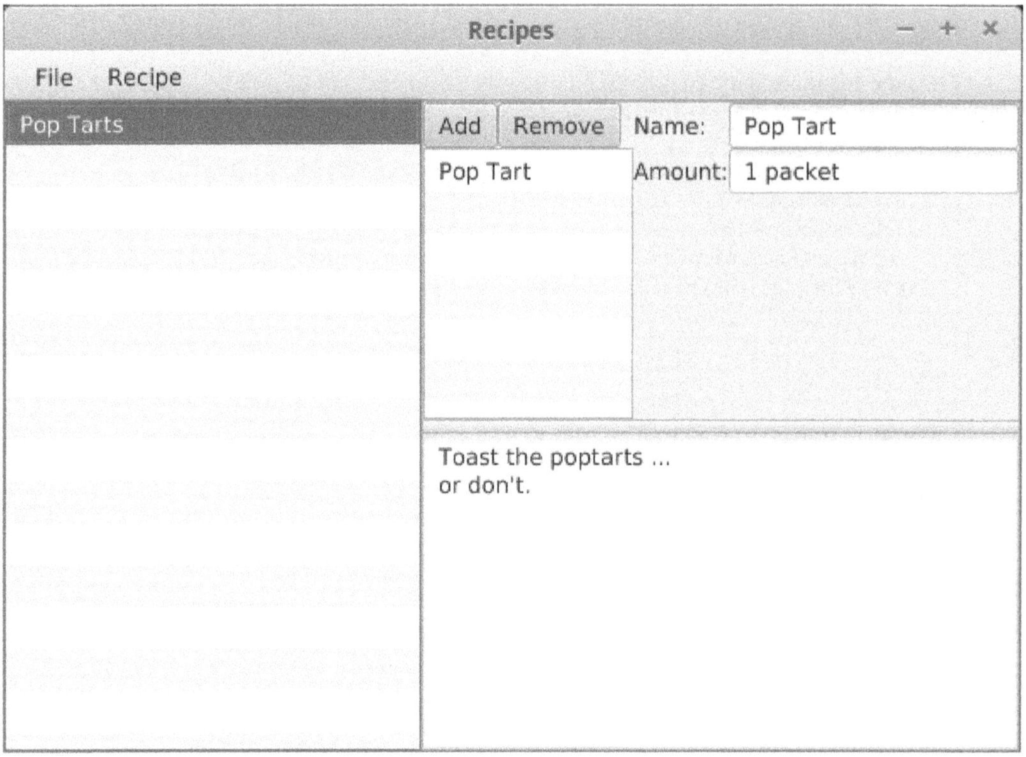

FIGURE 11.19: This shows the window that is produced when you run Recipe.scala.

and the `TextArea` below it. This is because the code that deals with the changes refers to the indexes selected in the `ListView`s. If focus is lost because the user clicks on a different option in one of the two `ListView`s, those indexes are updated before the change in focus is processed. As a result, all changes are lost. Listening to every change in the `text` is less efficient, but it prevents any changes from being lost.

The `onChange` code for the fields uses a function below called `alterSelectedIngredient` to prevent code duplication. This function takes a single argument which is a function of `Ingredient => Ingredient`. This function is passed the current value for the ingredient and should return a modified ingredient to store back into the list. If you go to the definition of `alterSelectedIngredient` on lines 220-230, you will see that on line 226 we use a method of `List` that we have not seen before called `updated`. The behavior of this method is to replace one element of a `List` at a particular index with a specified value. This is used to modify the selected recipe by replacing the old ingredient with the one built by the function that is passed in.

The GUI setup ends by making the directions `TextArea` and then placing all the pieces together using a `SplitPane` and a `BorderPane`. The `BorderPane` is set as the `root` of the `Scene`, so that it will size itself with the window.

After the GUI setup are a number of functions that are called when things change in the GUI or for buttons and menu items. These include functions for saving and loading in text files that store the recipes, as well as functions for adding and removing recipes and ingredients.

The `addRecipe` function on lines 165-177 uses another feature that we have not seen previously. It needs to ask the user for the name of the new recipe, as we do not provide any way to edit that in the GUI. To do this, it uses a type called `TextInputDialog` from the `control` package. As the name implies, this class allows us to pop up a dialog box that reads text from the user. After creating the dialog box, we set the `title`, `headerText`, and `contentText`. We then call the method `showAndWait`, which waits until the user has hit either "OK" or "Cancel". The result of that method is an `Option` which will either be some value or None. Since we do not want to do anything if the user selects "Cancel", a call to `foreach` is used to handle their input. A `foreach` will ignore Nones.

There is also a `ChoiceDialog` that can be used to allow the user to select from a set number of options. Readers can look in the API for details of that class.

We can reduce the length of this code a bit by utilizing bindings. The following code produces the same GUI, and it functions in the same way. However, instead of having handlers in the `TextField`s that set the values in our recipe objects, in this version the recipe objects store their values using the `StringProperty` type, and binds the `TextField`'s `text` properties to those values when they are selected.

Listing 11.21: Recipe2.scala

```scala
1   import scalafx.Includes._
2   import scalafx.scene.Scene
3   import scalafx.application.JFXApp
4   import scalafx.scene.control._
5   import scalafx.scene.layout._
6   import scalafx.event.ActionEvent
7   import scalafx.scene.input.KeyCode
8   import scalafx.scene.input.KeyCombination
9   import scalafx.scene.input.KeyCodeCombination
10  import scalafx.stage.FileChooser
11  import scalafx.collections.ObservableBuffer
12  import scalafx.beans.property.StringProperty
13  import java.io.PrintWriter
14
15  case class Ingredient(name:StringProperty, amount:StringProperty)
16  case class Recipe(name:StringProperty, ingredients:List[Ingredient],
        directions:StringProperty)
17
18  val app = new JFXApp {
19    stage = new JFXApp.PrimaryStage {
20      title = "Recipes"
21      scene = new Scene(600,400) {
22        // Menus
23        val menuBar = new MenuBar
24        val fileMenu = new Menu("File")
25        val openItem = new MenuItem("Open")
26        openItem.accelerator = new KeyCodeCombination(KeyCode.O,
            KeyCombination.ControlDown)
27        openItem.onAction = (e:ActionEvent) => { openFile }
28        val saveItem = new MenuItem("Save")
29        saveItem.accelerator = new KeyCodeCombination(KeyCode.S,
            KeyCombination.ControlDown)
30        saveItem.onAction = (e:ActionEvent) => { saveFile }
31        val exitItem = new MenuItem("Exit")
32        exitItem.accelerator = new KeyCodeCombination(KeyCode.X,
            KeyCombination.ControlDown)
33        exitItem.onAction = (e:ActionEvent) => { System.exit(0) }
34        fileMenu.items = List(openItem, saveItem, new SeparatorMenuItem, exitItem)
35
```

```scala
36    val recipeMenu = new Menu("Recipe")
37    val addItem = new MenuItem("Add")
38    addItem.onAction = (e:ActionEvent) => { addRecipe }
39    val removeItem = new MenuItem("Remove")
40    removeItem.onAction = (e:ActionEvent) => { removeRecipe }
41    recipeMenu.items = List(addItem, removeItem)
42
43    menuBar.menus = List(fileMenu, recipeMenu)
44
45    // Recipe List
46    var recipes = Array(Recipe(StringProperty("Pop Tarts"),List(Ingredient(
47      StringProperty("Pop Tart"), StringProperty("1 packet"))),
48      StringProperty("Toast the poptarts ...\nor don't.")))
49
50    val recipeList = new ListView(recipes.map(_.name.value))
51    var selectedRecipe:Option[Recipe] = None
52    recipeList.selectionModel.value.selectedIndex.onChange {
53      val index = recipeList.selectionModel.value.selectedIndex.value
54      if (index>=0) bindRecipeFields(recipes(index))
55    }
56
57    // Ingredients stuff
58    var selectedIngr:Option[Ingredient] = None
59    val addButton = new Button("Add")
60    addButton.onAction = (ae:ActionEvent) => addIngredient
61    val removeButton = new Button("Remove")
62    removeButton.onAction = (ae:ActionEvent) => removeIngredient
63    val ingredientsList = new ListView[String]()
64    val ingredientNameField = new TextField
65    val amountField = new TextField
66    val ingredientsGrid = new GridPane
67    ingredientsGrid.children += addButton
68    GridPane.setConstraints(addButton,0,0)
69    ingredientsGrid.children += removeButton
70    GridPane.setConstraints(removeButton,1,0)
71    ingredientsGrid.children += ingredientsList
72    GridPane.setConstraints(ingredientsList,0,1,2,3)
73    val nameLabel = new Label("Name:")
74    ingredientsGrid.children += nameLabel
75    GridPane.setConstraints(nameLabel,3,0)
76    val amountLabel = new Label("Amount:")
77    ingredientsGrid.children += amountLabel
78    GridPane.setConstraints(amountLabel,3,2)
79    ingredientsGrid.children += ingredientNameField
80    GridPane.setConstraints(ingredientNameField,4,0)
81    ingredientsGrid.children += amountField
82    GridPane.setConstraints(amountField,4,2)
83    ingredientsList.selectionModel.value.selectedItem.onChange {
84      val recipeIndex = recipeList.selectionModel.value.selectedIndex.value
85      val ingredientIndex =
              ingredientsList.selectionModel.value.selectedIndex.value
86      if (recipeIndex>=0 && ingredientIndex>=0) {
87        bindIngredientFields(recipes(recipeIndex).ingredients(ingredientIndex))
88      }
89    }
```

```scala
90      ingredientNameField.text.onChange {
91        val newName = ingredientNameField.text.value
92        val ingredientIndex =
             ingredientsList.selectionModel.value.selectedIndex.value
93        if (ingredientIndex>=0) ingredientsList.items.value(ingredientIndex) =
             newName
94      }
95
96      // Directions
97      val directionsArea = new TextArea
98
99      val splitPane = new SplitPane
100     splitPane.orientation = scalafx.geometry.Orientation.Vertical
101     splitPane.items += ingredientsGrid
102     splitPane.items += directionsArea
103
104     // Top level layout
105     val topBorderPane = new BorderPane
106     topBorderPane.top = menuBar
107     topBorderPane.left = recipeList
108     topBorderPane.center = splitPane
109
110     root = topBorderPane
111
112     def openFile:Unit = {
113       val chooser = new FileChooser
114       val selected = chooser.showOpenDialog(stage)
115       if (selected!=null) {
116         val src = io.Source.fromFile(selected)
117         val lines = src.getLines
118         recipes = Array.fill(lines.next.toInt)(Recipe(
119             StringProperty(lines.next),
120             List.fill(lines.next.toInt)(Ingredient(StringProperty(lines.next),
                 StringProperty(lines.next))),
121             {
122               var dir = ""
123               var line = lines.next
124               while (line!=".") {
125                 dir += (if (dir.isEmpty) "" else "\n")+line
126                 line = lines.next
127               }
128               StringProperty(dir)
129             }
130         ))
131         src.close()
132         recipeList.items = ObservableBuffer(recipes.map(_.name.value):_*)
133         recipeList.selectionModel.value.selectFirst
134         bindRecipeFields(recipes.head)
135       }
136     }
137
138     def saveFile:Unit = {
139       val chooser = new FileChooser
140       val selected = chooser.showSaveDialog(stage)
141       if (selected!=null) {
```

```scala
142          val pw = new PrintWriter(selected)
143          pw.println(recipes.length)
144          for (r <- recipes) {
145            pw.println(r.name)
146            pw.println(r.ingredients.length)
147            for (ing <- r.ingredients) {
148              pw.println(ing.name)
149              pw.println(ing.amount)
150            }
151            pw.println(r.directions)
152            pw.println(".")
153          }
154          pw.close()
155        }
156      }
157
158      def addRecipe:Unit = {
159        val dialog = new TextInputDialog
160        dialog.title = "Recipe Name"
161        dialog.headerText = "Question?"
162        dialog.contentText = "What is the name of the new recipe?"
163        dialog.showAndWait().foreach{ name =>
164          recipes = recipes :+ Recipe(StringProperty(name),
165            List(Ingredient(StringProperty("ingredient"),
166                StringProperty("amount"))), StringProperty("Directions"))
166          recipeList.items = ObservableBuffer(recipes.map(_.name.value):_*)
167          recipeList.selectionModel.value.clearAndSelect(recipes.length-1)
168        }
169      }
170
171      def removeRecipe:Unit = {
172        if (!recipeList.selectionModel.value.selectedItems.isEmpty) {
173          recipes = recipes.patch(recipeList.selectionModel.value.selectedIndex.
174              value,Nil,1)
174          recipeList.items = ObservableBuffer(recipes.map(_.name.value):_*)
175          recipeList.selectionModel.value.clearAndSelect(0)
176        }
177      }
178
179      def addIngredient:Unit = {
180        val recipeIndex = recipeList.selectionModel.value.selectedIndex.value
181        if (recipeIndex>=0) {
182          val newIngr = Ingredient(StringProperty("Stuff"), StringProperty("Some"))
183          recipes(recipeIndex) = recipes(recipeIndex).copy(
184            ingredients = recipes(recipeIndex).ingredients :+ newIngr)
185          ingredientsList.items.value += newIngr.name.value
186        }
187      }
188
189      def removeIngredient:Unit = {
190        val recipeIndex = recipeList.selectionModel.value.selectedIndex.value
191        val ingredientIndex =
192            ingredientsList.selectionModel.value.selectedIndex.value
192        if (recipeIndex>=0 && ingredientIndex>=0) {
193          recipes(recipeIndex) = recipes(recipeIndex).copy(ingredients =
```

```
194          recipes(recipeIndex).ingredients.patch(ingredientIndex,Nil,1))
195        bindRecipeFields(recipes(recipeIndex))
196      }
197    }
198
199    def bindRecipeFields(r:Recipe):Unit = {
200      ingredientsList.items = ObservableBuffer(r.ingredients.map(_.name.value):_*)
201      selectedRecipe.foreach(_.directions.unbind)
202      directionsArea.text = r.directions.value
203      r.directions <== directionsArea.text
204      selectedRecipe = Some(r)
205    }
206
207    def bindIngredientFields(ingr:Ingredient):Unit = {
208      selectedIngr.foreach { si =>
209        si.name.unbind
210        si.amount.unbind
211      }
212      ingredientNameField.text = ingr.name.value
213      ingr.name <== ingredientNameField.text
214      amountField.text = ingr.amount.value
215      ingr.amount <== amountField.text
216      recipes.foreach(println)
217      selectedIngr = Some(ingr)
218    }
219    }
220  }
221 }
222
223 app.main(args)
```

You can see that we were able to eliminate the `AlterSelectedIngredient` function in the above code, but we did add another function to call when binding ingredients in addition to the one for recipes. The bindings simplify the code in some ways, but they do introduce one notable complexity. When we switch from one recipe to another or one ingredient to another, we have to unbind the values that are no longer selected. This requires introducing two `vars` that store what we have bound to. The `Option` type is used for these as nothing has been bound when the program is first started.

The approach of using bindings typically works best if there is minimal changing of what values are being bound, hence, this is not the ideal application for them. In your own GUIs, you are encouraged to use whichever approach makes the most sense to you.

11.9   End of Chapter Material

11.9.1   Summary of Concepts

- Most of the programs that people interact with these days have Graphical User Interfaces, GUIs.

- We can write GUIs in Scala using types from the ScalaFX. This library wraps around the JavaFX library, though this fact is generally not obvious in our code.

- The elements of a GUI are called `Nodes`. There are a number of different active and passive `Nodes` in the ScalaFX library. You should consult the API to see everything that is available.

  - To represent windows you can use `Stage` or `Dialog` types.
  - There many different active `Nodes` ranging from the simple `Button` to the complex `TableView`.
  - Complex GUIs are built using `Panes`. These `Panes` can hold other `Nodes`, including other panes. Different panes have different rules for how their contents are placed and sized.
  - `MenuBars` hold `Menus`. These can hold several types including `MenuItems` and other `Menus`.

- User interactivity for GUIs is different from terminal applications which have a single point of input and can block when waiting for information.

  - Depending on what you want to do in the GUI, you can set one of the "on" members of a `Node` or you might provide `onChange` code that will be executed when a `Property` changes.
  - Bindings provide another mechanism for having changes to one value alter another one.

- The `FileChooser` type provides a standard file selection dialog box.

11.9.2   Self-Directed Study

ScalaFX GUIs cannot be done well in a REPL.

11.9.3   Exercises

1. Using three `ComboBoxes` and a `TextField` set up a little GUI where the user makes simple math problems. The `ComboBoxes` should be in a row with the first and third having numbers and the middle on having math operators. When the user picks a different value or operation, the `TextField` should update with the proper answer.

2. Write a GUI for a simple text editor. Use a `TextArea` for the main editing field and put in menu options to open and save files.

3. Write a GUI that has a `TableView` that can display the contents of a CSV file. You can take a normal spreadsheet and save to CSV format. You want to be able to load it in and have the values appear in your `TableView` in the appropriate locations.

4. Write tic-tac-toe with a 3x3 grid of buttons. Have something print when each is clicked.

5. Make a GUI with one of each of the `Node` types mentioned in this chapter, excluding panes. You can decide what, if anything, it does.

6. Add menu options to the last table example for removing a row or a column and implement the code to carry out those operations.

7. Make a GUI that has one `Slider` and one `TextField`. When the user moves the `Slider`, the value it is moved to should appear in the `TextField`.

8. Write a GUI that has two `ListView`s with two `Button`s in between them. The `Button`s should say "Add" and "Remove".

9. Write a GUI that has three `CheckBox`es and a `ListView`. The `CheckBox`es should be labeled "Uppercase", "Lowercase", and "Digits" for uppercase letters, lowercase letters, and the numeric digits. When the state of a `CheckBox` is altered, the `ListView` should be changed to show only the values for the things that are selected.

10. Repeat the above exercise except use `RadioButton`s and have them in a `ToggleGroup` so that only one option can be selected at a time.

11.9.4 Projects

1. Write a functional calculator in a GUI. You need to have at least the basic four math operations. Feel free to put in other functionality as well. You can get a basic idea of how an old four function calculator works by going to http://www.online-calculator.com/.

2. If you did project 8.5 on parsing chemical equations, you might consider doing this problem as well. The chemical parser is not required for this, so you can start here if you want. In that problem, the chemical equation was converted to a system of linear equations. Systems of linear equations can be represented as matrix equations. For this problem you will build a GUI that represents matrices as tables then lets you edit them and do basic operations of addition and multiplication on them.

   To keep things easy, the matrices will be square. You should have a menu option where the user can set the size of the matrices. They should be at least 2x2. The GUI will display three matrices that we will call `A`, `B`, and `C`. The user gets to enter values into `A` and `B`. The value of C is set when the user selects menu options for add or multiply.

3. Write a GUI to play a basic minesweeper game. You can use a `GridPane` or a `TilePane` of `Button`s for the display and the user interaction. The challenge in this problem is deciding how to store the mines and associate them with `Button`s. There are many ways to do this. Students should think about different approaches instead just rushing into one.

4. This project continues the line of ray tracing options. You should build a GUI that lets you edit a scene. You should have `ListView`s of spheres, planes, and light sources. Each one should have the geometric settings as well as color information. You need to have menu options to save to a file or load from a file.

5. Editing rooms for the map traversal in project 10.3 using a text editor in a basic text file can be challenging and is error prone. For this reason, it could be helpful to have a GUI that lets you edit the rooms. It can display a `ListView` of rooms that you can select from and then other options for setting values on the selected room. You need to have menu options for loading a file, saving a file, and adding a room. For a little extra challenge consider putting in the option to remove a room. Note that when you do this, the indices of all the rooms after that one change.

6. You can extend project 10.8. For this project you will write a GUI that lets users rank their favorite songs. You should provide a GUI element that allows the user to rank each song on a 5-star rating scale. You also should have GUI elements that let the user pick songs they like best and put them into a top-10 `ListView`. They should

also be able to remove songs from that top-10 `ListView`. Give the user menu options to save and load newly ranked file.

7. Make a program that displays a music database. You want to store significant information for different songs including the name of the song, the artist, the album, the year released, and anything else you find significant. The program should be able to display the information for all the songs as well as allow the user to narrow it down to certain artists, albums, years, etc.

8. For this project, we want to turn the menu driven script of project 10.4 into a GUI driven program. You can use a `ListView` or a `TableView` to display information related to the pantry and recipe items. You need to have menu options for saving and loading. Like the menu based version, users should be able to add items to their pantries, add recipes to the recipe books, check if they can make a recipe, and tell the program that they have made a recipe and have the proper amount reduced from their pantry contents. The details of how this looks and whether you use `Buttons` or `MenuItems` for most of the functionality is up to you.

9. This chapter did not use the theme park example so it appears in this project idea instead. You should write a GUI for the theme park functionality using different tabs in a `TabPane` for different areas of the park. The GUI should interact with files for saving activity in the park. You should keep at least the following three tabs, "Ticket Sales", "Food Sales", and "Ride Operation".

   For the first two, you should have `Buttons` that the user can click on to add items to a purchase. On the ticket tab those would include people of different ages, whether they were going to the water park, and various numbers of coolers. For the food purchase there are different types of food and their sizes. You can find food prices on page 66. All purchases should be added to a `ListView`. There should also be a `Button` to remove selected items.

   On the ride operation tab you need to have a `ComboBox` for selecting the ride, a `ListView` with all the possible operator names for the user to select from, a `TextField` for entering how many people rode in the last period of time, and a `Button` for submitting the data.

10. For this project you will convert the functionality of the text menu based program in project 10.5 to a GUI. You can decide the exact format of the GUI, but you should have a `ListView` of the courses of interest with the ability to add and remove entries as well as the ability to edit values for entries. The courses should have at least a course number, name, interest level, and time slot information associated with them.

    When the user chooses to generate schedules, you should show the different schedules in some format and allow the user the ability to select one that they will take. All courses from that selected schedule should be removed from the list of courses of interest. The course of interest information should be saved in a file for use from one run to another.

11. For this project you will extend the work you did for project 10.9 so that there is a GUI. The GUI should include a `TableView` for displaying the statistics from a single game as well as a `ListView` that shows the different games that are known. A second part of the GUI, perhaps a second tab in a `TabPane`, should have a `ListView` with all the games and give the user the ability to select multiple options and display average stats for what is selected.

Note that the exact details of how this works will depend significantly on the sport that you have chosen. If your sport does not match this description well, make modifications to get a better fit.

12. You can extend project 10.9 by displaying the information from a box score in a GUI. Depending on the sport and what you want to display, this can be done with a `TableView` or using a complex layout including a `GridPane` or `TilePane` with `Labels` for the different information. The user should also get a `ListView` of the different box scores that are available and clicking on an option should display the proper box score. In a separate section of the GUI you should display the average values that you have been calculating in the earlier project.

Additional exercises and projects, along with data files, are available on the book's web site.

Chapter 12

Graphics and Advanced ScalaFX

Not everything that you might ever want to put into a GUI is part of the standard GUI library. They have only added elements that are used fairly frequently. There are also times when you want to display your own graphics, something completely unique to your program, in a GUI. In order to do this, you have to be able to write graphics code.

ScalaFX includes a number of different ways to add graphics in a GUI. You can either add special `Nodes` directly to the `Scene`, or you can draw to a `Canvas` node. This chapter also presents a number of other features of the ScalaFX scene graph and `Nodes` that provide features that are more graphical in nature.

For graphical elements, you can think of the `Scene` as being like a magnetic board that you can stick various things on. This is true for the non-graphical elements as well, but the view is a bit more complex when using `Panes` that automatically move things around and resize them. We will see that you can make various graphical shapes and stick them to the board, then move them around and change how they are drawn in various ways.

Note that this is a long chapter that has fairly complete coverage of many topics in ScalaFX. The topics have been put in order of importance as seen by the authors. The first five sections have the most essential information. Sections beyond that include features that many readers might find interesting, but which are less central to understanding graphics in general.

12.1   Shapes

Some of the examples in chapter 11 drew a rectangle by putting a `Rectangle` in the `Scene`. The `Rectangle` type is one of a number of different types in the `scalafx.scene.shape` package that you can use to draw things in a scene. In this section we will look at the different options that are part of that package. These are not the only way to draw things in a ScalaFX GUI. We will look at another option, the `Canvas` node, in section 12.6.

The `Rectangle` type is one of several subtypes of `Shape` in `scalafx.scene.shape`. Here is a full list of the subtypes of `Shape`. Each has a brief description and a possible syntax for how you would create one in your code. Note that the code you want to use for instantiating these objects often does not involve `new`.

- `Arc` - Creates an arc that covers a specified angle and can be filled in. You create one with `Arc(centerX, centerY, radiusX, radiusY, startAngle, length)`, where each argument is a `Double`. The `startAngle` and `length` are measured in degrees. The way in which it is drawn and filled in depends on the `type` property which can be one of the following: `ArcType.Chord`, `ArcType.Open`, or `ArcType.Round`.[1]

- `Circle` - As the name implies, this makes a `Circle`. You can create it with one of the following options: `Circle(fill:Paint)`, `Circle(radius:Double, fill:Paint)`, `Circle(centerX:Double, centerY:Double, radius:Double)`, or `Circle(centerX: Double, centerY:Double, radius:Double, fill:Paint)`.

- `CubicCurve` - Draws a cubic Bèzier curve. You create one with `new CubicCurve`. Then you have to set eight values. Those are `startX`, `startY`, `controlX1`, `controlY1`, `controlX2`, `controlY2`, `endX`, and `endY`.

- `Ellipse` - Draws an ellipse with a specified center and radii for x and y. You can build one with the center at $(0, 0)$ using `Ellipse(radiusX, radiusY)` or specify the center using `Ellipse(centerX, centerY, radiusX, radiusY)`. All arguments have type `Double`.

- `Line` - Draws a line between two points. Create with `Line(startX, startY, endX, endY)`. All arguments have type `Double`.

- `Path` - Paths are formed by combining lines, arcs, and curves. The Path type allows you to create a complex path by putting together different elements. You create one using `new Path`. The details of the path elements are described in section 12.1.1.

- `Polygon` - Creates a general polygon with the syntax `Polygon(points:Double*)`. Note that this is a variable length argument list. The arguments are taken in pairs as x, y coordinates, so a triangle would have six values, a quadrilateral would have 8 arguments, etc. This shape is automatically closed if the last point is not the same as the first one.

- `Polyline` - Creates a shape from a sequence of line segments with the syntax `Polyline(points:Double*)`. Note that this is a variable length argument list. The

---

[1]Setting the `type` property requires a little extra effort because "type" is a keyword in Scala, used for type declarations. In order to use keywords as identifiers in Scala, they have to be enclosed in backticks. This is the character on the top left key of US keyboards. So if you want to set this, you cannot use `arc.type = ArcType.Open`, you have to use `arc.`type`= ArcType.Open`.

arguments are taken in pairs as x, y coordinates, so a triangle would have six values, a quadrilateral would have 8 arguments, etc. This shape is only an outline, and it is not closed unless you make the last point equal to the first one.

- QuadCurve - Draws a quadratic Bèzier curve. You create one with QuadCurve(startX, startY, controlX, controlY, endX, endY). All the arguments have type Double.

- Rectangle - As we have seen, this creates a basic rectangle. There are three ways to construct one: Rectangle(width:Double, height:Double, fill:Paint), Rectangle(x:Double, y:Double, width:Double, height:Double) and Rectangle (width:Double, height:Double). The x and y coordinates for the rectangle are at the top-left corner, the minimum in both x and y.

- SVGPath - Creates a path by parsing a String that is formatted with SVG[2] commands. You create one with new SVGPath, then set the content to a String like the following path.content = "M410,110 L410,190 C490,190 490,110 410,110". This moves to a location, draws a line from there, then has a cubic curve that goes back around to the points moved to create a shape like the letter "D".

- Text - This type is found in the scalafx.scene.text package instead of scalafx.scene.shape. It can be used to add general text to a scene. You can build this with either new Text(x:Double, y:Double, t:String) or new Text(t:String).

The use of all of these types with the exception of Path is shown in the following code.

Listing 12.1: Shapes.scala

```scala
import scalafx.Includes._
import scalafx.application.JFXApp
import scalafx.scene.Scene
import scalafx.scene.paint.Color
import scalafx.scene.shape._
import scalafx.scene.text.Text

val app = new JFXApp {
  stage = new JFXApp.PrimaryStage {
    title = "Shapes"
    scene = new Scene(500, 200) {
      val arc = Arc(50, 50, 40, 30, 0, 270)
      arc.fill = Color.Black
      arc.`type` = ArcType.Open

      val circle = Circle(150, 50, 40, Color.Red)

      val cubic = new CubicCurve
      cubic.startX = 210
      cubic.startY = 10
      cubic.controlX1 = 210
      cubic.controlY1 = 90
      cubic.controlX2 = 290
      cubic.controlY2 = 10
      cubic.endX = 290
      cubic.endY = 90
```

---

[2]Simple Vector Graphics

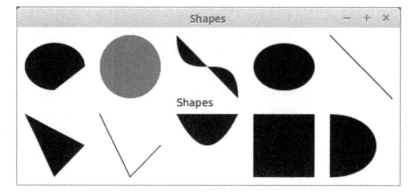

FIGURE 12.1: This window is produced by running Shapes.scala and the shapes are as follows: arc, circle, cubic curve, ellipse, line, polygon, polyline, quad curve, rectangle, and SVG path. It demonstrates all the different subtypes of **Shape** with the exception of **Path**.

```scala
    val ellipse = Ellipse(350, 50, 40, 30)

    val line = Line(410, 10, 490, 90)

    val polygon = Polygon(10, 110, 50, 190, 90, 150)

    val polyline = Polyline(110, 110, 150, 190, 190, 150)

    val quad = QuadCurve(210, 110, 250, 190, 290, 110)

    val rectangle = Rectangle(310, 110, 80, 80)

    val path = new SVGPath
    path.content = "M410,110 L410,190 C490,190 490,110 410,110"

    val text = new Text(210, 100, "Shapes")

    content = List(arc, circle, cubic, ellipse, line, polygon, polyline, quad,
        rectangle, path, text)
  }
 }
}

app.main(args)
```

The output from this program is shown in figure 12.1. Note that the `CubicCurve` and `QuadCurve` are both filled in. Only the `Line` and `Polyline` types produce outlines in this example.

12.1.1　Path Elements

The `Path` type allows you to build up more complex paths by adding various subtypes of `PathElement`. This allows you to make a single `Shape` that includes some of the line and curve types that were shown above. To make a `Path` you call `new Path`, then add the items you want into the `elements` member. You generally begin with a `MoveTo` that positions the starting location of the `Path`. Subsequent elements begin from the current element and specify where they will end up being placed.

Here are the different `PathElement` subtypes that can be added to a `Path`, along with the code you would use to create them.

- `ArcTo` - Adds an arc from the current location to some x, y location. An arc is essentially a piece of an ellipse. You can specify the radius of the arc in both x direction and y direction as well as the angle of rotation of the ellipse the arc is taken from. There are also options for whether you want the arc to cover the longest distance along the ellipse or the shortest distance along the ellipse to create the arc, as well as which direction the arc should bend. You create an arc using `ArcTo(radiusX: Double, radiusY: Double, xAxisRotation: Double, x: Double, y: Double, largeArcFlag: Boolean, sweepFlag: Boolean)`.

- `ClosePath` - As the name implies, this closes off a path. When it closes off the path, it draws a line back to a part of the path. You create this with `new ClosePath`. Note that you have to use `new` here. This is because there are no arguments for this element.

- `CubicCurveTo` - Adds a cubic Bèzier curve to the path starting at the current location and using the specified control points and end point. You create one with `CubicCurveTo(controlX1: Double, controlY1: Double, controlX2: Double, controlY2: Double, x: Double, y: Double)`.

- `HLineTo` - Adds a horizontal line to the path that moves across from the current position to the specified x-coordinate. Create one with `HLineTo(x:Double)`.

- `LineTo` - Adds a line to the path that connects the current position to a specified end point. Create one of these with `Line(x:Double, y:Double)`.

- `MoveTo` - This moves the current position to a new location without drawing anything. You can imagine this as picking up your pen so that you can start drawing somewhere else. You generally want to start a `Path` off with one of these. Create one of these with `MoveTo(x:Double, y:Double)`.

- `QuadCurveTo` - Adds a quadratic Bèzier curve to the path that goes from the current position to the end position using the specified control point. Create one of these with `QuadCurveTo(controlX: Double, controlY: Double, x: Double, y: Double)`.

- `VLineTo` - Adds a vertical line to the path from the current position to the specified y-coordinate. Create one of these with `VLineTo(y:Double)`.

The following sample program demonstrates all of the different `PathElement`s by building a long path that includes all of them. It also has two shorter paths that are used to demonstrate the meaning of the `fillRule` setting on a path.

Listing 12.2: Path.scala

```
import scalafx.Includes._
import scalafx.application.JFXApp
import scalafx.scene.Scene
import scalafx.scene.shape._
import scalafx.scene.paint.Color

val app = new JFXApp {
  stage = new JFXApp.PrimaryStage {
    title = "Path"
```

```scala
scene = new Scene(500, 300) {
  val path = new Path
  path.elements += MoveTo(50, 50)
  path.elements += VLineTo(250)
  path.elements += HLineTo(250)
  path.elements += LineTo(300, 200)
  path.elements += CubicCurveTo(500, 300, 500, 0, 400, 100)
  path.elements += QuadCurveTo(250,0,200,100)
  path.elements += ArcTo(100, 50, 45, 100, 50,false, true)
  path.elements += new ClosePath

  val path2 = new Path
  path2.elements += MoveTo(75,75)
  path2.elements += VLineTo(175)
  path2.elements += HLineTo(175)
  path2.elements += VLineTo(75)
  path2.elements += CubicCurveTo(0,175,250,175,75,75)
  path2.fillRule = FillRule.EvenOdd
  path2.fill = Color.Blue

  val path3 = new Path
  path3.elements += MoveTo(275,75)
  path3.elements += VLineTo(175)
  path3.elements += HLineTo(375)
  path3.elements += VLineTo(75)
  path3.elements += CubicCurveTo(200,175,450,175,275,75)
  path3.fillRule = FillRule.NonZero
  path3.fill = Color.Cyan

  content = List(path, path2, path3)

  }
 }
}

app.main(args)
```

You can see the window produced by this code in figure 12.2. The variable named `path` sets up the long line that goes roughly around the perimeter and includes each of the different types of elements. The smaller paths just have a box where one edge is a cubic curve that forms a loop. Having the loop allows us to illustrate that there are two different methods that can be used for filling in paths. There are `FillRule.EvenOdd` and `FillRule.NonZero`. The darker shape on the left uses the `EvenOdd` rule. With this rule, whether a region is filled in depends on if you cross the path an odd or even number of times to get to it. The central loop is not filled in because getting into there requires crossing two parts of the path, an even number. Getting to the parts that are filled in requires crossing the path one or three times. The lighter shape on the right uses the `NonZero` rule. This fills in the central loop as everything that would require going over the path more than zero times gets filled in. Note that you can also specify the `fillRule` of a `SVGPath`.

12.1.2   Paint and Stroke

The way in which different shapes are drawn is determined by different settings. When you create a `Circle` or a `Rectangle`, you can pass in a color to use for the `fill` property.

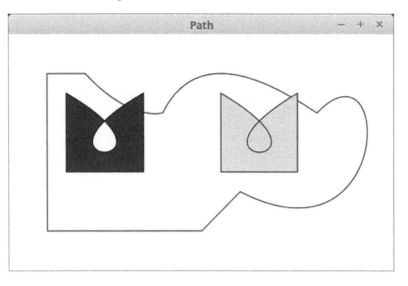

FIGURE 12.2: The window that is produced by running Path.scala. The long curve demonstrates each of the different `PathElements`. The smaller two show the different fill rules for paths.

For any other shape, you need to make the shape first and then set the `fill` property. So far we have only looked at using `Colors` for the fill, but there are two other subtypes of the general `Paint` type, `LinearGradient` and `RadialGradient`. All of these types are found in the `scalafx.scene.paint` package.

You can also change the way that lines are drawn using settings for the stroke. You can change the paint style, the dashing, the way lines end, the way they are joined, where the line is relative to the shape, and the width of the line. The code below demonstrates both paint and stroke settings.

Listing 12.3: ShapeSettings.scala

```
import scalafx.Includes._
import scalafx.application.JFXApp
import scalafx.scene.Scene
import scalafx.scene.paint._
import scalafx.scene.shape._

val app = new JFXApp {
  stage = new JFXApp.PrimaryStage {
    title = "Shape Settings"
    scene = new Scene(500, 300) {
      val polygon1 = Polygon(10, 10, 100, 190, 190, 100)
      polygon1.fill = Color.Green
      polygon1.stroke = Color.Black
      polygon1.strokeWidth = 5
      polygon1.strokeType = StrokeType.Centered
      polygon1.strokeLineJoin = StrokeLineJoin.Bevel

      val polygon2 = Polygon(160, 10, 250, 190, 340, 100)
      polygon2.fill = LinearGradient(160, 10, 340, 100, false, CycleMethod.NoCycle,
          Stop(0.0, Color.White), Stop(0.3, Color.Cyan), Stop(0.7,Color.Blue),
          Stop(1.0, Color.Black))
```

```
    polygon2.stroke = Color.Black
    polygon2.strokeWidth = 5
    polygon2.strokeType = StrokeType.Inside
    polygon2.strokeLineJoin = StrokeLineJoin.Miter

    val polygon3 = Polygon(310, 10, 400, 190, 490, 100)
    polygon3.fill = RadialGradient(45, 0.5, 400, 100, 50, false,
        CycleMethod.Reflect, Stop(0.0, Color.White), Stop(0.5, Color.Red),
        Stop(1.0, Color.Black))
    polygon3.stroke = Color.Black
    polygon3.strokeWidth = 5
    polygon3.strokeType = StrokeType.Outside
    polygon3.strokeLineJoin = StrokeLineJoin.Round

    val line1 = Line(50, 220, 450, 220)
    line1.stroke = Color.Black
    line1.strokeWidth = 5
    line1.strokeLineCap = StrokeLineCap.Butt
    line1.strokeDashArray = List(30, 10, 20, 15)

    val line2 = Line(50, 250, 450, 250)
    line2.stroke = Color.Black
    line2.strokeWidth = 5
    line2.strokeLineCap = StrokeLineCap.Round
    line2.strokeDashArray = List(30, 10, 20, 15)

    val line3 = Line(50, 280, 450, 280)
    line3.stroke = Color.Black
    line3.strokeWidth = 5
    line3.strokeLineCap = StrokeLineCap.Square
    line3.strokeDashArray = List(30, 10, 20, 15)

    content = List(polygon1, polygon2, polygon3, line1, line2, line3)
  }
 }
}

app.main(args)
```

You can see the output of running this code in figure 12.3. The first polygon uses a basic Color for the fill. The second polygon uses a linear gradient. This type of paint has a starting point and an ending point, and you can specify the various colors that appear along the gradient and where they appear. These are called Stops. You can also tell if the positions for the Stops are proportional or not. If not, then the values for them go from 0 to 1. If it is proportional, then they should use units comparable to the pixel size you want the gradient to vary over. The last thing you can specify is a cycle method that determines what color is used beyond the start and end. There are three settings for this, CycleMethod.NoCycle, CycleMethod.Reflect, and CycleMethod.Repeat. With NoCycle, anything before the start position is the first color and anything after the end is the last color. With Reflect, the sequence repeats over and over in alternating order. With Repeat, the same sequence is done repeatedly.

You can construct a LinearGradient with either LinearGradient(startX: Double, startY: Double, endX: Double, endY: Double, proportional: Boolean, cycleMethod: CycleMethod, stops: Stop*) or LinearGradient(startX: Double,

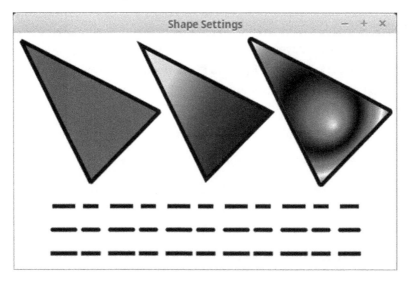

FIGURE 12.3: The window that is produced by running ShapeSettings.scala. The three triangles at the top demonstrate the three different types of Paint that are available as well as the three styles of joins. The lines at the bottom demonstrate dashing and the cap styles.

startY: Double, endX: Double, endY: Double, proportional: Boolean, cycleMethod: CycleMethod, stops: List[Stop]). The only difference between the two is the last argument. The sample code uses the first option and NoCycle. Note that the Stops are created using Stop(offset: Double, color: Color).

The third polygon uses a RadialGradient for the fill. This creates a circular or elliptical pattern of colors. You can provide an orientation and length for the ellipse as well as a center and radius for the pattern. Like the LinearGradient, you then specify the CycleMethod and the Stops. You can construct a RadialGradient using RadialGradient(focusAngle: Double, focusDistance: Double, centerX: Double, centerY: Double, radius: Double, proportional: Boolean, cycleMethod: CycleMethod, stops: Stop*) or RadialGradient (focusAngle: Double, focusDistance: Double, centerX: Double, centerY: Double, radius: Double, proportional: Boolean, cycleMethod: CycleMethod, stops: List[Stop]). The code uses the first option. For the CycleMethod it uses Reflect.

There are more options for the stroke than there are for the fill. The properties that you can set as well as their types and possible values are shown here.

- stroke: Paint - The color of the lines. Gradients are allowed.

- strokeDashArray: ObservableBuffer[Double] - Lengths in pixels for how long a section should be drawn or not drawn.

- strokeDashOffset: DoubleProperty - An offset for where the dashing begins.

- strokeLineCap: StrokeLineCap

  - StrokeLineCap.Butt - Ends the line at the end with no cap.

  - StrokeLineCap.Round - Puts a round, half-circle, cap on the end of the line.

  - StrokeLineCap.Square - Put a square cap with a size equal to the width at the end of the line.

- strokeLineJoin: StrokeLineJoin

  - StrokeLineJoin.Bevel - Flattens off the join between the segments.
  - StrokeLineJoin.Miter - Extends the segments out to meet in a point.
  - StrokeLineJoin.Round - Rounds out the transition from one segment to the next.

- strokeMiterLimit: Double - Specifies how far a miter join can extend.

- strokeType: StrokeType - Tells where the line is drawn for a shape.

  - StrokeType.Centered - Draws centered on the perimeter.
  - StrokeType.Inside - Draws inside of the perimeter.
  - StrokeType.Outside - Draws outside of the perimeter.

- strokeWidth: Double - Specifies how wide the stroke should be.

The polygons demonstrate the StrokeTypes and StrokeJoins. The styles used are centered, inside, and outside going from left to right. The joins are bevel, miter, and round going from left to right. The dashing and cap types are illustrated in the three lines at the bottom. All three have the same dashing structure so that you can see the impact of the different cap types. The lines use cap styles of Butt, Miter, and Square going from top to bottom.

12.2   Basic Keyboard, Mouse, and Touch Input

In chapter 11 we saw how we could add in code to allow the user to interact with the various GUI elements. It is also possible to interact with the user at a lower level by getting individual actions from the keyboard and mouse as well as touch. This can be done using many different properties that begin with on, each of which specifies a function that is called when something happens. We will look at these separately for the different types of interactions.

For the keyboard there are only three properties of significance: onKeyPressed, onKeyReleased, and onKeyTyped. The onKeyPressed function is called when a key goes down. The onKeyReleased is called when a key is let back up.[3] The onKeyTyped is called for the full typing of a character and is only called for keys that wind up being printed. This means that onKeyTyped does not register things like arrow keys, Shift, or Ctrl while onKeyPressed and onKeyReleased do.

These handlers all take an argument of a scalafx.scene.input.KeyEvent. This provides you with the information about what key was part of the event as well as other things that you may need to know. The KeyEvent object contains members such as character for typed values, code to tell you which key was hit (including non-printable keys), as well as altDown, controlDown, metaDown, and shiftDown. The code member has the type scalafx.scene.input.KeyCode. You can compare those codes to values in KeyCode such as KeyCode.A, KeyCode.F1, KeyCode.Left, or many others that are included to represent whatever keys might be found on your keyboard.

In section 11.5.1 we discussed the concept of focus as it relates to text controls. This

---

[3]Note that keyboards have a "repeat" rate so that if you press a key and hold it, multiple events will wind up being fired.

same concept applies to handling keyboard events. If you have multiple elements in your GUI, and you want to deal with the keyboard events for certain ones, you will have to request focus on the `Node` that is handling the events. It can be helpful to put in a handler for `onMouseClicked` along with the keyboard handling so that you can call `requestFocus` on that `Node` when it is clicked.

The `onMouseClicked` handler is just one of several event handlers that you can register for dealing with user interactions involving the mouse. The following list shows the different options, when those handlers are called, and the type of event that gets passed into them.

- `onDragDetected` - Takes a `MouseEvent` and is called when a drag is detected.

- `onDragDone` - Takes a `DragEvent` and is called when a drag-and-drop event ends after the release and the current `Node` was the source of the drag.

- `onDragDropped` - Takes a `DragEvent` and is called when a mouse button is released at the end of a drag-and-drop on the current `Node`.

- `onDragEntered` - Takes a `DragEvent` and is called when a drag-and-drop motion enters this `Node`.

- `onDragExited` - Takes a `DragEvent` and is called when a drag-and-drop event exits the current `Node`.

- `onDragOver` - Takes a `DragEvent` and is called when a drag-and-drop event moved over the current `Node`.

- `onMouseClicked` - Takes a `MouseEvent` when the mouse is clicked over this `Node`.

- `onMouseDragEntered` - Takes a `MouseDragEvent` and is called when the mouse enters the `Node` while the mouse button is down.

- `onMouseDragExited` - Takes a `MouseDragEvent` and is called when the mouse leaves the `Node` while the mouse button is down.

- `onMouseDragOver` - Takes a `MouseDragEvent` and is called when a full mouse pressed-drag-release gesture happens on this `Node`.

- `onMouseDragReleased` - Takes a `MouseDragEvent` and is called when a mouse press-drag-release ends on this `Node`.

- `onMouseDragged` - Takes a `MouseDragEvent` when and is called when the mouse is pressed on this `Node` then dragged.

- `onMouseEntered` - Takes a `MouseEvent` when the mouse enters this `Node`.

- `onMouseExited` - Takes a `MouseEvent` when the mouse exits this `Node`.

- `onMouseMoved` - Takes a `MouseEvent` when the mouse moves within this `Node` and no buttons are pressed.

- `onMousePressed` - Takes a `MouseEvent` when a mouse button goes down over this `Node`.

- `onMouseReleased` - Takes a `MouseEvent` when a mouse buttons is released over this `Node`.

- `onScroll` - Takes a `ScrollEvent` and when the user performs a scroll action.

- `onScrollFinished` - Takes a `ScrollEvent` and when the user finises a scroll action.

- `onScrollStarted` - Takes a `ScrollEvent` and when the user begins a scroll action.

The `MouseEvent` object has all the information that you need to know about the state of the mouse and other information for the particular event. The following list shows significant members of the `MouseEvent` type that you might need to use to interpret what to do with that action.

- `altDown: Boolean` - Tells if the Alt key was down during this event.

- `button: MouseButton` - Gives a value that can be compared to `MouseButton.Middle`, `MouseButton.None`, `MouseButton.Primary`, or `MouseButton.Secondary`.

- `clickCount: Int` - Tells you how many times the mouse has been clicked in a row so the program can respond to double clicks, triple clicks, etc.

- `controlDown: Boolean` - Tells you if the Control key was down during this event.

- `metaDown: Boolean` - Tells you if the Meta key was down during this event.

- `middleButtonDown: Boolean` - Tells you if the middle mouse button was down for this event.

- `pickResult: PickResult` - Used for 3D graphics discussed in section 12.11.

- `primaryButtonDown: Boolean` - Tells you if the primary mouse button was down during this event.

- `sceneX: Double` - Tells you the x-coordinate of the mouse in the `Scene`.

- `sceneY: Double` - Tells you the y-coordinate of the mouse in the `Scene`.

- `screenX: Double` - Tells you the x-coordinate of the mouse on the full screen.

- `screenY: Double` - Tells you the y-coordinate of the mouse on the full screen.

- `secondaryButtonDown: Boolean` - Tells you if the secondary mouse button was down during this event.

- `shiftDown: Boolean` - Tells you if the Shift buttons was down during this event.

- `shortcutDown: Boolean` - Tells you if the shortcut modifier for your platform was down during this event.

- `source: AnyRef` - Tells you the object on which the event initially occurred.

- `stillSincePress: Boolean` - Tells you if the mouse pointer has stayed in the same location since the last pressed event occurred.

- `x: Double` - Gives you the x-coordinate of the mouse relative to the origin of the event's source.

- `y: Double` - Gives you the y-coordinate of the mouse relative to the origin of the event's source.

- `z: Double` - Gives you the z-coordinate (the depth position) of the mouse relative to the origin of the event's source.

The MouseDragEvent is a subtype of MouseEvent that only adds a method called gestureSource which can be used to determine the object that the drag started with. The DragEvent removed the members that tell you about buttons and instead has a method called dragboard that returns an object of type Dragboard that is used for transferring data. Full details and usage of drag-and-drop are beyond the scope of this book. The ScrollEvent has many of the same methods as a MouseEvent related to keyboard keys and position, but it adds a few others specific to scrolling.

- deltaX: Double - Gets the amount of horizontal scroll.

- deltaY: Double - Gets the amount of vertical scroll.

- textDeltaX: Double - Gets the amount of horizontal text-based.

- textDeltaY: Double - Gets the amount of vertical text-based.

Mouse and keyboard interactions are demonstrated in the following program which allows the user to draw paths with the mouse then use the arrow keys to move a little circle around. The circle is not allowed to cross the lines.

Listing 12.4: DrawMaze.scala

```scala
import scalafx.Includes._
import scalafx.application.JFXApp
import scalafx.scene.Scene
import scalafx.scene.paint.Color
import scalafx.scene.shape._
import scalafx.scene.input.MouseEvent
import scalafx.scene.input.KeyEvent
import scalafx.scene.input.KeyCode

val app = new JFXApp {
  stage = new JFXApp.PrimaryStage {
    title = "Draw Maze"
    scene = new Scene(500,500) {
      val polyline = Polyline(0, 0, 500, 0, 500, 500, 0, 500, 0, 0)
      var walls = List(polyline)

      val ball = Circle(21, 21, 20)

      content = List(polyline, ball)

      onKeyPressed = (e:KeyEvent) => {
                        println("Pressed")
        val oldX = ball.centerX.value
        val oldY = ball.centerY.value
        if (e.code == KeyCode.Up) ball.centerY = ball.centerY.value - 2
        if (e.code == KeyCode.Down) ball.centerY = ball.centerY.value + 2
        if (e.code == KeyCode.Left) ball.centerX = ball.centerX.value - 2
        if (e.code == KeyCode.Right) ball.centerX = ball.centerX.value + 2

        // Collision detection with walls
        val clear = walls.forall(shape => {
          Shape.intersect(ball,shape).boundsInLocal.value.isEmpty
        })
```

```scala
      // If it collided, go back to old location
      if (!clear) {
        ball.centerX = oldX
        ball.centerY = oldY
      }
    }

    onMousePressed = (e:MouseEvent) => {
      walls ::= Polyline()
      content += walls.head
      walls.head.points ++= List(e.x,e.y)
    }
    onMouseDragged = (e:MouseEvent) => {
      walls.head.points ++= List(e.x,e.y)
    }
    onMouseReleased = (e:MouseEvent) => {
      walls.head.points ++= List(e.x,e.y)
    }
  }
 }
}
```

```scala
app.main(args)
```

The code begins by creating a scene that has a `Polyline` that goes around the scene as well as a `Circle` that begins in the top left corner. Note that the perimeter line is added to a `var List` called `walls`. This is used to keep track of all the shapes that the circle cannot go through.

The `onKeyPressed` handler is set to code that moves the ball. To do this, it remembers the current location of the ball, then moves it two pixels in the proper direction. It then uses a call to `forall` on `walls` to see if this new location would overlap any walls. This is done using a call to `Shape.intersect`. The results of that call is a new `Shape` object that represents the intersection. To tell if this shape is empty, we call `boundsInLocal`, which gives us a `Property`. To get the actual bounds from the property we call `value`, then finally ask the bounds if it is empty. If the new position does intersect something, the circle is set back to its initial position.

The three mouse handlers are all rather simple. When the mouse is pressed, a new `Polyline` is prepended to the front of `walls` using cons. That new shape is added to the contents of the scene, and the current point of the mouse is added onto its set of points. The handlers for dragging and releasing the mouse simply append the mouse location to the `Polyline` at the head of `walls`, adding the most recent segment to that line.

Figure 12.4 shows the result of running this program after the user has drawn some walls and moved the ball around a bit using the keys.

One of the features that has been added to JavaFX, and hence ScalaFX, is the ability to handle touch interactions. Earlier GUI libraries for Java were written before the widespread use of touch interfaces. JavaFX came out after touch interfaces on phones and computers become more common. As such, it includes the ability to detect those types of events and gives you the ability to write code that uses them. To do this, you set a handler much as you would for the mouse or keyboard. The handler is passed an event type that gives the needed information. The following list gives the different properties that you can set to be handlers as well as the types of events that they take as inputs.

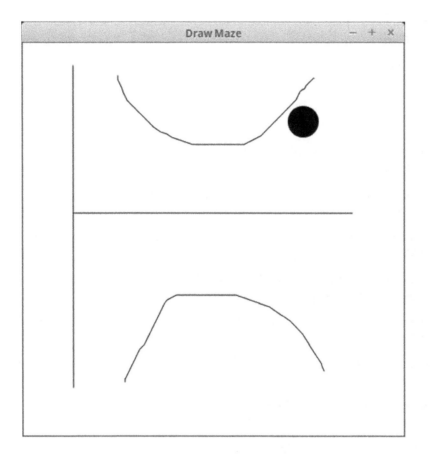

FIGURE 12.4: The window that is produced by running DrawMaze.scala after the user has drawn in a few lines and moved around with the keys.

- onRotate - This is called when the user performs a rotate action. It is passed a RotateEvent.

- onRotationFinished - This is called at the end of a rotate action. It is passed a RotateEvent.

- onRotationStarted - This is called when the user begins a rotate action. It is passed a RotateEvent.

- onSwipeDown - This is called when a swipe down gesture is started. It is passed a SwipeEvent.

- onSwipeLeft - This is called when a swipe left gesture is started. It is passed a SwipeEvent.

- onSwipeRight - This is called when a swipe right gesture is started. It is passed a SwipeEvent.

- onSwipeUp - This is called when a swipe up gesture is started. It is passed a SwipeEvent.

- onTouchMoved - This is called when the user performs a move gesture. It is passed a TouchEvent.

- onTouchPressed - This is called when the user does a touch press. It is passed a TouchEvent.

- onTouchReleased - This is called when the user does a touch release. It is passed a TouchEvent.

- onTouchStationary - This is called when the user does a stationary touch gesture. It is passed a TouchEvent.

- onZoom - This is called when the user performs a zooming action. It is passed a ZoomEvent.

- onZoomFinished - This is called when the user finishes a zooming action. It is passed a ZoomEvent.

- onZoomStarted - This is called when the user starts a zooming action. It is passed a ZoomEvent.

As you can see, there are five different types of events for touch gestures. Each one has methods that specifically fit that type of action. For example, the RotateEvent has angle and totalAngle methods that tell you how far the rotation has gone. The SwipeEvent has a touchCount: Int that tells you the number of touch points for this gesture. The TouchEvent also has touchCount, but also has touchPoint and touchPoints which give back instances of TouchPoint that contains information on a single point where the user is touching the screen. Lastly, ZoomEvent has a totalZoomFactor and zoomFactor that both give Doubles for the amount of zoom.

If you do not have a touch screen on your computer, you will not be able to take advantage of these handlers. However, there are projects that are designed to compile JavaFX programs for use on iOS and Android mobile devices. While you cannot use these with the Scala scripts we are currently writing, this is something you could consider doing after you have learned how to write Scala applications and use a number of other tools, such as appropriate build tools.

12.3 Images

Most applications that do graphics cannot do what they need just with shapes. Another critical component for most applications is images. Most of the types that you need to work with images are provided in the `scalafx.scene.image` package. The most fundamental of these types is the `Image` type, which represents an image. These can be created using any of the following.

- `new Image(url: String, requestedWidth: Double, requestedHeight: Double, preserveRatio: Boolean, smooth: Boolean, backgroundLoading: Boolean)`

- `new Image(url: String, requestedWidth: Double, requestedHeight: Double, preserveRatio: Boolean, smooth: Boolean)`

- `new Image(url: String, backgroundLoading: Boolean)`

- `new Image(url: String)`

The `url` argument specifies the source of the image data. These can start with "file:" to pull the information from the local disk. The `requestedWidth` and `requestedHeight` arguments allow you to specify the size of the image if you want one that is different from the natural size of the image. The `preserveRatio` argument, if true, will cause the aspect ratio to be preserved despite the scaling. The `smooth` argument tells ScalaFX whether it should use a higher quality filtering algorithm or a faster one. Lastly, the `backgroundLoading` argument, if true, will let the program continue to execute while this image loads in the background.

The `Image` type has `height` and `width` properties. The most significant method on the `Image` type is probably `pixelReader: Option[PixelReader]`. This will provide a `PixelReader` assuming that the image type in question allows for reading pixels. The details of the `PixelReader` type are discussed later.

If your primary goal is to get an image to display in your GUI, the `ImageView` type is a subtype of `Node` that serves this purpose. You can create one of these with `new ImageView(url: String)` or `new ImageView(image: Image)`. After you have created it, you can set the `fitHeight` and `fitWidth` to alter the bounding box that the image is fit inside.

For many applications, it is helpful to be able to find out the color of individual pixels in an image. For these purposes there is the `PixelReader` type, which allows you to get this type of information. At this point, the two that are significant are `getArgb(x: Int, y: Int): Int` and `getColor(x: Int, y: Int): Color`. Given the x and y coordinate values of a pixel, these methods return either the numeric ARGB value for the color or a `Color` value for it.

There is a subtype of `Image` called `WritableImage`. As the name implies, this type of allows you to write into it. Specifically, you can get a `PixelWriter`, the couterpart to the `PixelReader` and set the value of individual pixels. To create a `WritableImage`, use one of the following.

- `new WritableImage(reader: PixelReader, x: Int, y: Int, width: Int, height: Int)`

- `new WritableImage(reader: PixelReader, width: Int, height: Int)`

- `new WritableImage(width: Int, height: Int)`

The two versions that take a `PixelReader` make a copy of the image that `PixelReader` is attached to, potentially with an offset and a different scale. The last version creates a blank, transparent image of the specified width and height.

Once you have a `WritableImage` you can call the `pixelWriter` method to get a `PixelWriter`. There are three methods in `PixelWriter` that are significant to us here. They are `setArgb(x: Int, y: Int, argb: Int): Unit`, `setColor(x: Int, y: Int, c: Color): Unit`, and `setPixels(dstx: Int, dsty: Int, w: Int, h: Int, reader: PixelReader, srcx: Int, srcy: Int): Unit`. The first two parallel methods in `PixelReader`, in that they set a single pixel in the image to a particular ARGB numeric value or a specific `Color`. The last method sets a rectangular region of pixels at a location and size determined by the first four arguments and uses values taken from the specified location in a `PixelReader`.

To illustrate the user of these different types, the following example loads in an image that is specified at command line and scales it down to be only 200 pixels across. It then tries to gets a `PixelReader` for that image. If it can, it makes three `WritableImages` and their associated `PixelWriters` and copies over the red, green, and blue components to those images. They are then put in `ImageViews` and stored in the variables `red`, `green`, and `blue`. If the reader cannot be acquired those variables simply get `Labels` that say such. All four images are then added to a `TilePane` to display.

Listing 12.5: ImageView.scala

```scala
import scalafx.Includes._
import scalafx.application.JFXApp
import scalafx.scene.Scene
import scalafx.scene.image.{Image, ImageView, WritableImage, PixelReader,
    PixelWriter}
import scalafx.scene.control.Label
import scalafx.scene.layout.TilePane
import scalafx.scene.paint.Color

if (args.length < 1) {
  println("You must provide an argument with the name of a file to load.")
  sys.exit(0)
}

val app = new JFXApp {
  stage = new JFXApp.PrimaryStage {
    scene = new Scene(800,200) {
      val original = new Image("file:"+args(0), 200, 200, true, true)
      val (red, green, blue) = original.pixelReader match {
        case Some(reader) =>
          val rimg = new WritableImage(original.width.value.toInt,
              original.height.value.toInt)
          val rwriter = rimg.pixelWriter
          val gimg = new WritableImage(original.width.value.toInt,
              original.height.value.toInt)
          val gwriter = gimg.pixelWriter
          val bimg = new WritableImage(original.width.value.toInt,
              original.height.value.toInt)
          val bwriter = bimg.pixelWriter
          for (i <- 0 until original.width.value.toInt;
              j <- 0 until original.height.value.toInt) {
            val c = reader.getColor(i, j)
```

FIGURE 12.5: The window that is produced by running ImageView.scala with a sample image.

```
29          rwriter.setColor(i, j, Color(c.red, 0, 0, 1.0))
30          gwriter.setColor(i, j, Color(0, c.green, 0, 1.0))
31          bwriter.setColor(i, j, Color(0, 0, c.blue, 1.0))
32        }
33        (new ImageView(rimg), new ImageView(gimg), new ImageView(bimg))
34      case None =>
35        (new Label("No Reader"), new Label("No Reader"), new Label("No Reader"))
36    }
37    val tilePane = new TilePane
38    tilePane.children = List(new ImageView(original), red, green, blue)
39
40    root = tilePane
41    }
42  }
43 }
44
45 app.main(args)
```

A possible output of this program can be seen in figure 12.5.[4]

It is interesting to note that you can get images onto `Button`s by using the code `new Button(text: String, graphic: Node)` and providing an appropriate `ImageView` as the graphic.

12.3.1   Writing Images to File

While JavaFX and therefore ScalaFX support a remarkable number of features, there is one rather significant one that they did not add direct support for, writing images to file. In order to do this, we have to go through some older Java based libraries. The functionality to read and write images in Java is provided by the `javax.imageio.ImageIO` type. This type has a method called `write` that can write out image files in various formats. The challenge is that `ImageIO` works with the `BufferedImage` that was part of the older libraries, not the `Image` type from JavaFX. Fortunately, there are ways to convert the `Image` type that we use to the `BufferedImage` type using the `scalafx.embed.swing.SwingFXUtils` object. The following line of code will write an `Image` called `img` out to a file specified by the `java.io.File` named `file` as a PNG file.

```
ImageIO.write(SwingFXUtils.fromFXImage(img, null), "png", file)
```

You might find that you want to save off your whole `Scene` or a particular `Node` as an image. In order to do this, you can use some **snapshot** methods in order to get what is

---

[4]This image provided by Quinn Bender.

displayed on the screen into a `WritableImage`. To take a snapshot of the full `Scene` you can use one of the following methods.

- `snapshot(callback:(SnapshotResult) => Unit, image:WritableImage):Unit`

- `snapshot(image: WritableImage): WritableImage`

For both of them you pass in a `WritableImage` that is supposed to be used. In the first one, you also provide a callback function that takes a `SnapshotResult` and returns `Unit`. This version will do the rendering separately, and call the callback function when it is done. The `SnapshotResult` type has a method called `Image` that will give you the `WritableImage` with the snapshot in it. The second version will pause until the image has been written and give it back to you. Either way, you can then use the code above to write the image out to a file.

12.4   Transformations

One of the most fundamental capabilities common to nearly all graphical displays is the ability to transform the things that are being drawn. Transforms include combinations of rotations, translations, scaling, and shearing. You can see an example of what each of these looks like in figure 12.6. ScalaFX takes this to the extreme of allowing you to apply transforms to every `Node` in a `Scene`. This means you can not only transform the `Shape` types, but controls, panes, and everything else. The types of transforms that are allowed are AFFINE TRANSFORMS. An affine transform is one where lines that are parallel before the transform are still parallel after the transform. This is true for rotations, translations, scaling, and shear. Any combination of these transforms is also an affine transform.

You can apply transforms using a number of different methods of the `Node` class given in the following list.

- `rotate` - Can be set to specify the amount of rotation in degrees.

- `rotationAxis` - Can specify a `Point3D` that is the location to rotate about.

- `scaleX` - Can be set to a `Double` that is the amount the `Node` is scaled in the x-coordinate.

- `scaleY` - Can be set to a `Double` that is the amount the `Node` is scaled in the y-coordinate.

- `scaleZ` - Can be set to a `Double` that is the amount the `Node` is scaled in the z-coordinate.[5]

- `transforms` - Allows you to specify a sequence of transforms that are all applied to this `Node`.

- `translateX` - Can be set to a `Double` that specifies the amount to translate in the x-coordinate.

- `translateY` - Can be set to a `Double` that specifies the amount to translate in the y-coordinate.

- **translateZ** - Can be set to a `Double` that specifies the amount to translate in the z-coordinate.[5]

Most of these are straightforward other than `transforms`. To make the these you use

```
Transform.rotate(angle: Double, pivotX: Double, pivotY: Double): Rotate,
Transform.scale(x: Double, y: Double, pivotX: Double, pivotY: Double): Scale,
Transform.scale(x: Double, y: Double): Scale,
Transform.shear(x: Double, y: Double, pivotX: Double, pivotY: Double): Shear,
Transform.shear(x: Double, y: Double): Shear,
```

and

```
Transform.translate(x: Double, y: Double): Translate.
```

This approach is shown in the following code segment. These transforms are represented in the computer using matrices. That details of how that works goes beyond this text, but you should know that you can build your own transforms using the methods

```
Transform.affine(mxx: Double, mxy: Double, mxz: Double, tx: Double, myx: Double,
    myy: Double, myz: Double, ty: Double, mzx: Double, mzy: Double, mzz: Double,
    tz: Double): Affine
```

or

```
Transform.affine(mxx: Double, myx: Double, mxy: Double, myy: Double, tx: Double,
    ty: Double): Affine.
```

Listing 12.6: SimpleTransforms.scala

```scala
import scalafx.Includes._
import scalafx.application.JFXApp
import scalafx.scene.{Scene, Group}
import scalafx.scene.shape.Line
import scalafx.scene.transform.Transform

def makeParallelLines:Group = {
  new Group(Line(-50, -50, -50, 50), Line(50, -50, 50, 50))
}

val app = new JFXApp {
  stage = new JFXApp.PrimaryStage {
    title = "Transforms"
    scene = new Scene(400,400) {
      val lines1 = makeParallelLines
      lines1.transforms = List(Transform.translate(100,100))

      val lines2 = makeParallelLines
      lines2.transforms = List(Transform.translate(300,100), Transform.rotate(45,
          0, 0))

      val lines3 = makeParallelLines
      lines3.transforms = List(Transform.translate(100,300), Transform.scale(0.5,
          0.5))
```

---

[5]Only used for 3D graphics discussed in section 12.11.

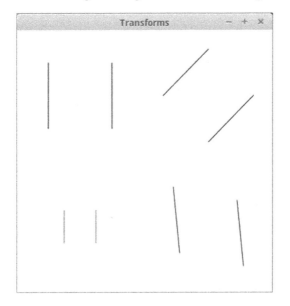

FIGURE 12.6: The window that is produced by running SimpleTransforms.scala.

```
val lines4 = makeParallelLines
lines4.transforms = List(Transform.translate(300,300),
    Transform.shear(0.1,0.2))

content = List(lines1, lines2, lines3, lines4)
    }
  }
}
```

```
app.main(args)
```

The output of this code is shown in figure 12.6.

This example demonstrates something else that we have not seen before, using Groups to group together Nodes in the scene graph. The parallel lines are separate Line objects, and we want to provide the same transformations to each of them. This is achieved by putting the two lines in a Group, then applying the transforms to that Group and adding the Group into the Scene. You can nest Groups inside of Groups to create more complicated structures as well. This capability is we say that ScalaFX uses a scene graph, and you should take advantage of it if you intend to build more complex graphics using the geometry types in Scalafx.

12.5   Animation

Up to this point, any changes that happened in our GUIs had to be completely motivated by user interactions with our programs. That is to say that we could change aspects of the GUI when the user clicked a mouse or hit a key. Consider the DrawMaze.scala example where the mouse cuased lines to be drawn and the keys caused a circle to move. Having things happen between user interactions, or possibly even without the user doing anything at all is essential for a number of different types of applications. This is the realm of animations. ScalaFX provides a `scalafx.animation` package, which has a number of types that help with the creation of animations, including the `Animation` type. There are two main subtypes of `Animation` called `Transition` and `Timeline` that can be used to produce different types of animations. There is also an `AnimationTimer` that gives you the ability to do whatever you want at fairly regular intervals in an application. These three approaches are covered in the following subsections.

There are a number of useful members and methods provided by the `Animation` type that are common the both `Transitions` and the `Timeline`. Here is a partial list of those.

- `autoReverse` - This is a `BooleanProperty` that tells if the animation should reverse when it gets to the end.

- `currentRate` - A `ReadOnlyDoubleProperty` that tells the current direction and speed that the animation is being played at.

- `currentTime` - A `ReadOnlyObjectProperty[Duration]` that tells the current position in playing the animation. The `scalafx.util.Duration` type is used to represent lengths of time and is covered in more detail below.

- `cycleCount` - This `IntegerProperty` defines how many times the animation should cycle.

- `cycleDuration` - This `ReadOnlyObjectProperty[Duration]` gives how long it takes to run through the full animation.

- `delay` - This `ObjectProperty[Duration]` can be used to get or set a delay to wait before the animation begins.

- `jumpTo(cuePoint: String)` - Jumps to a predefined location in the animation.

- `jumpTo(time: Duration)` - Jumps to a specified point in time in the animation.

- `onFinished` - This `ActionEvent` handler can be given code that should happen when the animation is finished. You can use this to schedule subsequent animations.

- `pause()` - Pauses play of the animation.

- `play()` - Causes the animation to begin playing.

- `playFrom(cuePoint: String)` - Begins playing the animation from a predefined point.

- `playFrom(time: Duration)` - Begins playing the animation from a specified time.

- `playFromStart()` - Begins playing the animation from its beginning.

- `rate` - This `DoubleProperty` can be used to set the speed and direction of play for the animation.

- `status` - This `ReadOnlyObjectProperty[Animation.Status]` allows you to determine the current state of the animation. The possible statuses are `Animation.Status.Paused`, `Animation.Status.Running`, and `Animation.Status.Stopped`.

- `stop()` - Stops play of the animation and sets it back to the beginning.

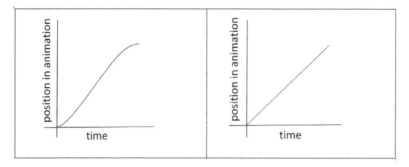

FIGURE 12.7: On the left you will see an example of `Interpolator.EaseBoth`. On the right you see an example of `Interpolator.Linear`.

- `targetFramerate: Double` - Tells you the maximum frame rate that this animation is expected to be played.

- `totalDuration` - This `ReadOnlyObjectProperty[Duration]` tells you how long the full animation takes to play, including any repeating.

The fact that animations take time to execute leads to several of the methods listed above working with the `scalafx.util.Duration` type. Instances of this type represent time durations in milliseconds. You can make one by calling `new Duration(millis: Double)`. You can perform standard math operations and comparisons between `Duration`s. There are also some special values defined at `Duration.Indefinite`, `Duration.One`, `Duration.Unknown`, and `Duration.Zero`.

12.5.1   Transitions

The first subtype of `Animation` that we want to consider is the `Transition` type. The `Transition` type has all the members of `Animation` listed above, but adds an `ObjectProperty[Interpolator]` called `interpolator` that governs the way in which the animation goes from beginning to end.

You can imagine an animation as taking time from 0.0 to 1.0 and the position in the animation going from 0.0 to 1.0. An `Interpolator` is a function that takes a fractional time and gives back a fractional position. The simplest approach to this gives back the same value that is passed in. This can look a bit unnatural for some actions though as there are abrupt stops at the ends. For this reason, there are alternative functions as well as mechanisms that you can use to define your own interpolation. Here are the options defined in the `Interpolator` object.

- `Discrete` - Gives back 1.0 for the input 1.0 and 0.0 otherwise.

- `EaseBoth` - Slows the transition at both the beginning and the end and is linear in between.

- `EaseIn` - Slows the transition at the beginning, but is linear in the middle and end.

- `EaseOut` - Slows the transition at the end, but is linear at the beginning and middle.

- `Linear` - This is the basic interpolator that gives back the input value.

- `Spline(x1:Double, y1:Double, x2:Double, y2:Double): Interpolator` - This creates a cubic function that uses the two specified control points.

- `Tangent(t1:Duration, v1:Double, t2:Duration, v2:Double): Interpolator` - This creates an interpolator with a specified tangent value coming in and out for a particular duration. This should be used with a `Timeline` that is described in the next section.

- `Tangent(t: Duration, v: Double): Interpolator` - This creates an interpolator that uses the same tangent and duration at both the start and the end.

There are ten subtypes of `Transition` that are defined in ScalaFX. Two of them are used for combining other transitions, but the other eight give control over specific things that can be transitioned for `Nodes`. Here is a list of the different types, how you create them, and extra members that are useful in each.

- `FadeTransition` - This causes a `Node` to fade in or out by altering the opacity. You can create them with `new FadeTransition(duration: Duration, node: Node)` or `new FadeTransition(duration: Duration)`. If you use the second option, you have to set the `Node` property. The `fromValue` and `toValue` properties allow you to specify the opacity at the beginning and end of the transition.

- `FillTransition` - This causes the `fill` of a `Shape` to transition from one color to another. Instances can be created with

  ```
  new FillTransition(duration: Duration),
  new FillTransition(duration: Duration, shape: Shape),
  new FillTransition(duration: Duration, fromValue: Color, toValue: Color),
  ```

  or

  ```
  new FillTransition (duration: Duration, shape: Shape, fromValue: Color,
      toValue: Color).
  ```

  There are properties with the names `duration`, `shape`, `fromValue`, and `toValue` that can be set if you do not use the option that specifies all of these.

- `ParallelTransition` - This transition is used to combine other transitions making them happen at the same time. You can create them with

  ```
  new ParallelTransition(node: Node, children: Seq[Animation]),
  new ParallelTransition(node: Node),
  ```

  or

  ```
  new ParallelTransition(children: Seq[Animation]).
  ```

  You will have to set the `node` or the `children` if you use an option that does not specify one of them.

- `PathTransition` - Causes a `Node` to follow a path. This is done by altering the `translateX` and `translateY` values of the `Node` as well as the `rotate` value if the `orientation` has the value `OrthogonalToTangent`. Instances can be created using `new PathTransition(duration: Duration, path: Shape)` or `new PathTransition(duration: Duration, path: Shape, node: Node)`. You can also set the `orientation`, which can be either `PathTransition.OrientationType.None` or `PathTransition.OrientationType.OrthogonalToTangent`.

- **PauseTransition** - This does not do anything but call the `onFinished` code when it is completed. You can create one with `new PauseTransition(duration: Duration)`.

- **RotateTransition** - Rotates a `Node` during the transition. You can make one of these with `new RotateTransition(duration: Duration, node: Node)` or `new RotateTransition(duration: Duration)`. You should specify the `axis`, which is a `scalafx.geometry.Point3D`. You should also set the `fromAngle` property and either the `toAngle` or `byAngle` property.

- **ScaleTransition** - Scales a given `Node` through a transition. You can create instances using

  `new ScaleTransition(duration: Duration)`

  or

  `new ScaleTransition(duration: Duration, node: Node)`.

  You should specify the `fromX`, `fromY`, and `fromZ` properties. Then you can either specify `toX`, `toY`, and `toZ` or `byX`, `byY`, and `byZ`.

- **SequentialTransition** - This is the other approach to combining transitions. The transitions put in a `SequentialTransition` will happen one after another. You can create instances of this using `new SequentialTransition (children: Seq[Animation])`, `new SequentialTransition (node: Node)`, or `new SequentialTransition (node: Node, children: Seq[Animation])`. You will need to specify the `node` or the `children` properties if you use an option that does not give both.

- **StrokeTransition** - This transitions the color of a stroke on a `Shape`. You can create instances using

  `new StrokeTransition(duration: Duration)`,
  `new StrokeTransition (duration: Duration, shape: Shape)`,
  `new StrokeTransition (duration: Duration, fromValue: Color, toValue: Color)`,

  or

  `new StrokeTransition (duration: Duration, shape: Shape, fromValue: Color, toValue: Color)`.

  If you use an option that does not set all the properties, you should set them on the created object.

- **TranslateTransition** - Moves/translates the `Node` through a transition. You can create instances using `new TranslateTransition(duration: Duration, node: Node)` or `new TranslateTransition(duration: Duration)`. You should specify `fromX`, `fromY`, and `fromZ`, then also give either `toX`, `toY`, and `toZ` or `byX`, `byY`, and `byZ`.

Nearly all of these transitions are demonstrated in the following example. The scene has two rectangles. The top one has a number of sequential transitions that are controlled by the user with some buttons. The bottom one is doing a number of transitions in parallel.

Listing 12.7: Transitions.scala

```scala
1   import scalafx.Includes._
2   import scalafx.application.JFXApp
3   import scalafx.scene.Scene
4   import scalafx.event.ActionEvent
5   import scalafx.animation._
6   import scalafx.scene.control.Button
7   import scalafx.scene.layout.FlowPane
8   import scalafx.scene.paint.Color
9   import scalafx.scene.shape._
10  import scalafx.util.Duration
11
12  val app = new JFXApp {
13    stage = new JFXApp.PrimaryStage {
14      title = "Transitions"
15      scene = new Scene(600, 400) {
16        val rect1 = Rectangle(20, 50, 100, 100)
17        rect1.strokeWidth = 10
18        val fadeTrans = new FadeTransition(new Duration(1000))
19        fadeTrans.fromValue = 1.0
20        fadeTrans.toValue = 0.5
21        fadeTrans.autoReverse = true
22        val path = new CubicCurve
23        path.startX = 70
24        path.startY = 100
25        path.controlX1 = 600
26        path.controlY1 = 400
27        path.controlX2 = 0
28        path.controlY2 = 400
29        path.endX = 530
30        path.endY = 100
31        val pathTrans = new PathTransition(new Duration(1000), path)
32        val strokeTrans = new StrokeTransition(new Duration(1000), Color.Black,
              Color.Cyan)
33        val seqTrans = new SequentialTransition(rect1, List(fadeTrans, pathTrans,
              strokeTrans))
34        seqTrans.autoReverse = true
35        seqTrans.cycleCount = 4
36        val startButton = new Button("Start")
37        startButton.onAction = (e:ActionEvent) => seqTrans.play
38        val pauseButton = new Button("Pause")
39        pauseButton.onAction = (e:ActionEvent) => seqTrans.pause
40        val stopButton = new Button("Stop")
41        stopButton.onAction = (e:ActionEvent) => seqTrans.stop
42        val flowPane = new FlowPane
43        flowPane.children = List(startButton, pauseButton, stopButton)
44        val rect2 = Rectangle(20, 280, 100, 100)
45        val transTrans = new TranslateTransition(new Duration(2000))
46        transTrans.fromX = 0
47        transTrans.toX = 460
48        val fillTrans = new FillTransition(new Duration(2000), Color.Black,
              Color.Green)
49        val rotTrans = new RotateTransition(new Duration(2000))
50        rotTrans.fromAngle = 0
51        rotTrans.toAngle = 360
```

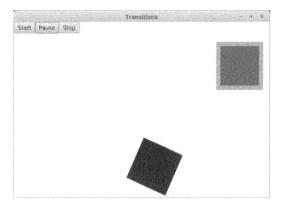

FIGURE 12.8: The window that is produced by running Transitions.scala.

```scala
52    val parallel = new ParallelTransition(rect2, List(transTrans, fillTrans,
          rotTrans))
53    parallel.autoReverse = true
54    parallel.cycleCount = 1000000
55
56    // Uncomment to see non-eased motion
57    // transTrans.interpolator = Interpolator.LINEAR
58    // rotTrans.interpolator = Interpolator.LINEAR
59
60    content = List(rect1, rect2, flowPane)
61
62    parallel.play
63    }
64   }
65 }
66
67 app.main(args)
```

You can see a sample of what this program looks like in figure 12.8. Lines 16 to 35 set up the first rectangle with a `FadeTransition`, a `PathTransition`, and a `StrokeTransition`. Those three are put together in a `SequentialTransition` that is applied to the rectangle. Lines 36 to 41 set up buttons to control that `SequentialTransition`. Lines 44 to 54 set up the second rectangle with a `TranslateTransition`, a `FillTransition`, and a `RotateTransition`. These are all combined in a `ParallelTransition`, so the rectangle at the bottom spins from side to side of the window while changing color from black to green and back.

All of the transitions use the `EASE_BOTH` interpolation by default. If you run this code and watch it, you should be able to tell that the movement accelerates smoothly at the beginning and decelerates smoothly at the end. If you uncomment lines 57 and 58 and rerun the script, you will be able to see what the `LINEAR` interpolation looks like by comparison.

12.5.2   Timelines

Another approach to doing animation is with the `Timeline` type. Using a `Timeline` you are able to specify given times as `KeyFrames`. Each `KeyFrame` has different `KeyValues` associated with it. The `KeyValue` is a part of a pair which consists of a writeable value or property and the value it should have when the animation reaches that `KeyFrame`. The

KeyValue can optionally specify an interpolator for how the change from the previous KeyFrame to the current one should be handled. The default interpolator for KeyValues is Linear.

You can build a Timeline with Timeline(keyFrames: Seq[KeyFrame]). To make a single KeyFrame, you would use the following.

```
KeyFrame(time: Duration, name: String = null, onFinished:
    EventHandler[ActionEvent] = null, values: Set[KeyValue[\_, \_]] = Set.empty)
```

Note that all the parameters other than time have default values, so this can be called with just a single argument of type Duration. Most likely, you will want to specify the Set[KeyValue[_,_]]. You can do so by name if you do not want to provide a name or an onFinish handler that is called when this KeyFrame is reached. The individual KeyFrames are created with KeyValue(target, endValue, interpolator) or KeyValue(target, endValue). The target and endValue can have many different types, but they need to match. So if the target value is a DoubleProperty then the endValue needs to be a Double. In general, the types need to be things that can be interpolated, you will find it works best with numeric types and Boolean.

The following example shows the use of a basic Timeline with three different KeyFrames. It begins with a rectangle that is in the top left corner. When the user presses the "Start" button, the rectangle slides along the top and rotates one full time around. After getting half way across the top, it stops spinning and slides down to the bottom while still moving across.

Listing 12.8: Timelines.scala

```
 1  import scalafx.Includes._
 2  import scalafx.application.JFXApp
 3  import scalafx.scene.Scene
 4  import scalafx.event.ActionEvent
 5  import scalafx.animation._
 6  import scalafx.scene.control.Button
 7  import scalafx.scene.layout.FlowPane
 8  import scalafx.scene.shape._
 9  import scalafx.util.Duration
10
11  val app = new JFXApp {
12    stage = new JFXApp.PrimaryStage {
13      title = "Transitions"
14      scene = new Scene(600, 400) {
15        val rect1 = Rectangle(20, 50, 100, 100)
16
17        val kv11 = KeyValue(rect1.translateX, 0)
18        val kv12 = KeyValue(rect1.translateY, 0)
19        val kv13 = KeyValue(rect1.rotate, 0)
20        val kf1 = KeyFrame(new Duration(0), values = Set(kv11, kv12, kv13))
21        val kv21 = KeyValue(rect1.translateY, 0)
22        val kv22 = KeyValue(rect1.rotate, 360)
23        val kf2 = KeyFrame(new Duration(1000), values = Set(kv21, kv22))
24        val kv31 = KeyValue(rect1.translateX, 460)
25        val kv32 = KeyValue(rect1.translateY, 230)
26        val kf3 = KeyFrame(new Duration(2000), values = Set(kv31, kv32))
27
28        val timeline = Timeline(List(kf1, kf2, kf3))
29
```

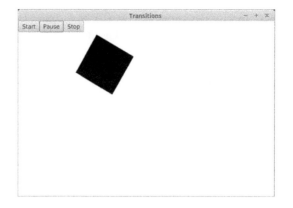

FIGURE 12.9: The window that is produced by running Timelines.scala.

```scala
30        val startButton = new Button("Start")
31        startButton.onAction = (e:ActionEvent) => timeline.play
32        val pauseButton = new Button("Pause")
33        pauseButton.onAction = (e:ActionEvent) => timeline.pause
34        val stopButton = new Button("Stop")
35        stopButton.onAction = (e:ActionEvent) => timeline.stop
36        val flowPane = new FlowPane
37        flowPane.children = List(startButton, pauseButton, stopButton)
38
39
40        content = List(rect1, flowPane)
41      }
42    }
43  }
44
45  app.main(args)
```

A KeyFrame is added with a time of 0 that sets translateX, translateY, and rotate all to zero. The second KeyFrame is at one second into the animation, and it keeps the translateY value at 0 and has the rotate value go to 360. This produces the spinning motion moving across the top. If the translateY value were not set to zero here, the rectangle would begin moving down immediately at the beginning of the animation. The third KeyFrame is at two seconds, and it has the translateX and translateY values set to 460 and 230, respectively, to put the rectangle at the bottom of the window. The output of this script after pressing "Start" is shown in figure 12.9.

This script stopped at three KeyFrames to keep the example reasonably short. You can add far more than this and create arbitrarily complex animations using this approach.

12.5.3   AnimationTimer

Transitions are good for doing fairly short animated actions. Timelines allow for longer sequences of behaviors that alter numeric properties of Nodes. Both of these are best for scripted types of behaviors that the programmer can lay out in advance and that have less dependence on what the user is doing. If you want to simply have actions that occur at fairly regular intervals where you write logic to do things based on the state of the program, including user actions, the AnimationTimer is probably a better choice. An example of the type of program that could benefit from an AnimationTimer is a game. In a game, the user typically interacts with the program through keyboard or mouse events, and the behavior of elements in the game needs to vary depending on where things are and what has happened. This is challenging to set up using `Transitions` or `Timelines`. It works best if you can simply write code that is called every so often to update the state. This is what an `AnimationTimer` does.

You can create an `AnimationTimer` using `AnimationTimer(handler: (Long) => Unit)`. The `handler` is a function that takes a `Long` and produces `Unit`. The `Long` represents the current time measured in nanoseconds. You can make a `var` outside of the function that is used to store the previous time. Then each time the function is invoked, you can use the difference of the last time and the current time to figure out how much time has passed and update things in an appropriate manner. The handler code can do whatever it is that you want it to do, including altering the properties of `Nodes` in the `Scene`. You can control the timer with `start` and `stop` methods. Remember that the timer will not do anything until you call the `start` method.

To illustrate the use of an `AnimationTimer`, the following code shows a simple implementation of the classic video game Pong. There are paddles on the left and right side of the window. The left paddle is controlled by the W and S keys while the right paddle is controlled by the up and down arrow keys. There is a circle that moves around with a particular speed and angle. The speed and angle are stored in `vars` so that they can change as the game goes on. There is also a label that is centered near the top of the window that shows the score. The individual player scores are stored in properties and bindings are used to both position the label and to set its text so that when either score is changed, the label updates and always stays centered.

Listing 12.9: Pong.scala

```scala
1   import scalafx.Includes._
2   import scalafx.application.JFXApp
3   import scalafx.scene.Scene
4   import scalafx.scene.shape._
5   import scalafx.scene.control.Label
6   import scalafx.scene.paint.Color
7   import scalafx.scene.input._
8   import scalafx.animation.AnimationTimer
9   import scalafx.beans.property._
10
11  val app = new JFXApp {
12    stage = new JFXApp.PrimaryStage {
13      title = "Pong"
14      scene = new Scene(500, 500) {
15        val paddle1 = Rectangle(0, 200, 20, 100)
16        val paddle2 = Rectangle(480, 200, 20, 100)
17        val ball = Circle(250, 250, 20)
18        var speed = 100.0
19        var theta = 0.0
20        val score1 = IntegerProperty(0)
21        val score2 = IntegerProperty(0)
22        val scoreDisplay = Label("")
23        scoreDisplay.layoutY = 30
24        scoreDisplay.layoutX <== (width-scoreDisplay.width)/2
25
26        scoreDisplay.text <== StringProperty("")+score1.asString+" : "+score2.asString
27
28        content = List(paddle1, paddle2, ball, scoreDisplay)
```

```scala
29
30        onKeyPressed = (e:KeyEvent) => {
31          if (e.code == KeyCode.W) {
32            paddle1.y = paddle1.y.value - 5
33          }
34          if (e.code == KeyCode.S) {
35            paddle1.y = paddle1.y.value + 5
36          }
37          if (e.code == KeyCode.Up) {
38            paddle2.y = paddle2.y.value - 5
39          }
40          if (e.code == KeyCode.Down) {
41            paddle2.y = paddle2.y.value + 5
42          }
43        }
44
45        var lastTime = 0L
46        val timer = AnimationTimer(t => {
47          if (lastTime!=0) {
48            val vx = speed*math.cos(theta)*(t-lastTime)/1e9
49            val vy = speed*math.sin(theta)*(t-lastTime)/1e9
50            ball.centerX = ball.centerX.value + vx
51            ball.centerY = ball.centerY.value + vy
52            speed += (t-lastTime)/1e8
53          }
54          lastTime = t
55          if (!Shape.intersect(ball,paddle1).boundsInLocal.value.isEmpty) {
56            val offset = (paddle1.y.value+paddle1.height.value/2-ball.centerY.value)
57            theta = -offset/100*math.Pi/2
58          }
59          if (!Shape.intersect(ball,paddle2).boundsInLocal.value.isEmpty) {
60            val offset = (paddle2.y.value+paddle2.height.value/2-ball.centerY.value)
61            theta = math.Pi + offset/100*math.Pi/2
62          }
63          if (ball.centerY.value < ball.radius.value && math.sin(theta) < 0.0) {
64            theta = math.atan2(-math.sin(theta),math.cos(theta))
65          }
66          if (ball.centerY.value > height.value-ball.radius.value && math.sin(theta)
                > 0.0) {
67            theta = math.atan2(-math.sin(theta),math.cos(theta))
68          }
69          if (ball.centerX.value < -ball.radius.value) {
70            speed = 100
71            score2.value = score2.value+1
72            ball.centerX = 250
73            ball.centerY = 250
74            theta = 0
75          }
76          if (ball.centerX.value > 500+ball.radius.value) {
77            speed = 100
78            score1.value = score1.value+1
79            ball.centerX = 250
80            ball.centerY = 250
81            theta = math.Pi
82          }
```

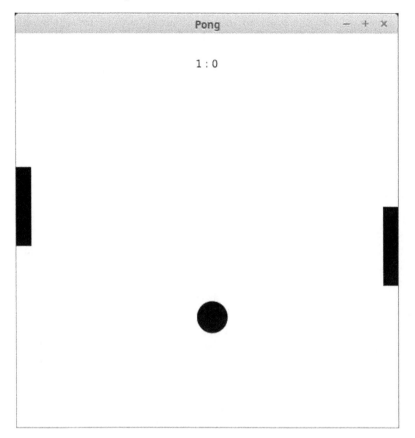

FIGURE 12.10: The window that is produced by running Pong.scala.

```
83              })
84
85          timer.start
86        }
87      }
88  }
89
90  app.main(args)
```

Figure 12.10 shows what the display might look like in the middle of a game. The new element in this code is the `AnimationTimer`, which is created on lines 46-83. All of the logic of the game occurs in here. On line 45 a `var` called `lastTime` is declared and given a value of 0L. Remember that the capital L on an integer literal forces it to be a `Long`. Without this, `lastTime` would be an `Int`, which would cause a type error on line 54 where it is updated with the new time.

Lines 48 to 52 move the ball and speed it up a bit based on the difference between the last time and the current time. The initial value of `lastTime` is not a valid time value, so that logic needs to be put in an `if` so it is skipped the first time the handler is called and does not happen until we have a valid time to use in the `t-lastTime` calculations. Lines 55 to 58 check to see if the ball intersects the left paddle. If it does, it sets the `theta` value for the ball motion to be heading to the right with some variation depending on what part of the paddle was hit. Lines 59 to 62 do the same thing for the second paddle.

Lines 63 to 82 deal with the ball and the edges. Lines 63 to 68 cause the ball to bounce off the top and bottom edges. Lines 69 to 82 deal with the ball going out the left or right sides. When that happens, the ball is reset to the center with the original speed of 100 pixels/second and the score of the appropriate player is updated. Do not worry if not all the math related to the motion of the ball makes sense to you. What is important is that you see how the code in an `AnimationTimer` can be used to control a graphical game.

12.6   Canvas

We have been creating our graphics by adding `Node`s to the scene graph that represent the elements that we want to have drawn. The `Canvas` is a `Node` that you can draw onto itself. Going back to the analogy from the beginning of the chapter of the `Scene` being a magnetic board that we can stick our shapes to, the `Canvas` is like a paper or some other drawing surface that we can stick onto the board. This paper can be moved around with transformations or treated in many ways like the other shapes we can put on the board, but it is also a surface that we can draw directly to.

There are a number of reasons why you might want to use a `Canvas` instead of placing shapes directly in the `Scene`, but the main reason would be for speed and efficiency. When you put shapes into the `Scene`, the computer stores memory for each of those shapes. This has an advantage that you can easily move them around or adjust them over time in various ways. However, if you are going to be drawing a large number of things, having to store significant information for each one can create a lot of overhead. With the `Canvas`, you simply draw stuff to it and while the result is stored, all the information that went into the individual steps of creating the drawing do not have to be. The down side of this is that if you draw a bunch of circles (or other shapes), you cannot pick one and change its location or appearance, you would need to redraw everything. For many applications though, this type of behavior is perfectly fine.

`Canvas` and the types associated with it are in the package `scalafx.scene.canvas`. You can instantiate a `Canvas` with `new Canvas(width: Double, height: Double)`. The fact that `Canvas` is a subtype of `Node` means that you can insert it into your `Scene` and do any of the other things that we have discussed for `Node`s with it. When you want to draw on the `Canvas`, you call the `graphicsContext2D` method, which returns a `GraphicsContext`. It is the `GraphicsContext` type that includes most of the functionality. Most of this functionality mirrors things that we have already discussed such as paths, filling, and strokes. The following subsections run through most of the methods of `GraphicsContext` breaking them up into different types of methods.

12.6.1   Settings

Since the settings impact everything that is drawn, we will start with those. Here is a list of the members of `GraphicsContext` that deal with the settings. The ones that do not have parentheses for the arguments are properties that you can get the value from or set using assignment. Many of these settings use types that were described in previous sections and apply in the same way here as they did before.

- `applyEffect(e: Effect): Unit` - Applies the given `scalafx.scene.effect.Effect` to the entire canvas.

- `clip(): Unit` - Sets the current path to be a clipping region. This means anything that would normally be drawn outside of this path will not appear.

- `fillRule: FillRule` - Allows you to set a `scalafx.scene.shape.FillRule`.

- `fill: Paint` - Get or set the `Paint` that is used to fill.

- `font: Font` - Get or set the font that is used for text.

- `getEffect(e: Effect): Effect` - Get the effect that is currently in use.

- `getTransform: Affine` - Get the `scalafx.scene.transform.Affine` transform that is currently in use.

- `globalAlpha: Double` - Get or set the transparency used for drawing.

- `globalBlendMode: BlendMode` - Get or set the `scalafx.scene.effect.BlendMode` that is being used.

- `lineCap: StrokeLineCap` - Get or set the `scalafx.scene.shape.StrokeLineCap` that is being used.

- `lineJoin: StrokeLineJoin` - Get or set the `scalafx.scene.shape.StrokeLineJoin` that is being used.

- `lineWidth: Double` - Get or set the width of the line for strokes.

- `miterLimit: Double` - Get or set the miter limit used with the stroke when the line join style is set to be a miter join.

- `rotate(degrees: Double): Unit` - Rotate the current transform by the specified number of degrees.

- `scale(x: Double, y: Double): Unit` - Scale the current transform by the specified amount.

- `setEffect(e: Effect): Unit` - Allows you to specify a `scalafx.scene.effect.Effect` that will be applied to subsequently drawn elements.

- `setTransform(mxx: Double, myx: Double, mxy: Double, myy: Double, mxt: Double, myt: Double):Unit` - Sets the transform to a particular matrix.

- `setTransform(xform: Affine): Unit` - Sets the transform to the provided affine transform.

- `stroke: Paint` - Sets the paint style used for subsequent strokes.

- `textAlign: TextAlignment` - Sets the `scalafx.scene.text.TextAlignment` that is used for rendering text.

- `textBaseline: VPos` - Sets the `scalafx.geometry.VPos` to be used to position text.

- `transform(mxx: Double, myx: Double, mxy: Double, myy: Double, mxt: Double, myt: Double): Unit` - Applies the specified matrix to the current transformation.

- `transform(xform: Affine): Unit` - Applies the specified affine transform to the current transformation.

- `translate(x: Double, y: Double): Unit` - Applies a translation to the current transform.

With the exception of `clip`, these methods should be fairly straightforward as they refer to the same types of settings that have been covered earlier for non-`Canvas` based drawing.

There are two other methods related to settings on the `GraphicsContext` that allow you to store and recover settings. The `save(): Unit` method will take all the current settings and store them in a structure called a stack. You can later call `restore(): Unit` to bring back earlier settings. If you call `save()` multiple times without calling `restore()`, they stack up and the next call to restore will pull off the most recently saved values. Another call will pull off the values before that and so on.

12.6.2   Basic Fills and Strokes

One way of drawing things to a `GraphicsContext` is to simply draw shapes out directly. This can be done by filling in the shape or just by drawing the outline as a stroke. The following methods provide this functionality.

- `fillArc(x: Double, y: Double, w: Double, h: Double, startAngle: Double, arcExtent: Double, closure: ArcType): Unit`

- `fillOval(x: Double, y: Double, w: Double, h: Double): Unit`

- `fillPolygon(points: Seq[(Double, Double)]): Unit`

- `fillPolygon(xPoints: Array[Double], yPoints: Array[Double], nPoints: Int): Unit`

- `fillRect(x: Double, y: Double, w: Double, h: Double): Unit`

- `fillRoundRect(x: Double, y: Double, w: Double, h: Double, arcWidth: Double, arcHeight: Double): Unit`

- `fillText(text: String, x: Double, y: Double, maxWidth: Double): Unit`

- `fillText(text: String, x: Double, y: Double): Unit`

- `strokeArc(x: Double, y: Double, w: Double, h: Double, startAngle: Double, arcExtent: Double, closure: ArcType): Unit`

- `strokeLine(x1: Double, y1: Double, x2: Double, y2: Double): Unit`

- `strokeOval(x: Double, y: Double, w: Double, h: Double): Unit`

- `strokePolygon(points: Seq[(Double, Double)]): Unit`

- `strokePolygon(xPoints: Array[Double], yPoints: Array[Double], nPoints: Int): Unit`

- `strokePolyline(points: Seq[(Double, Double)]): Unit`

- `strokePolyline(xPoints: Array[Double], yPoints: Array[Double], nPoints: Int): Unit`

- `strokeRect(x: Double, y: Double, w: Double, h: Double): Unit`

- `strokeRoundRect(x: Double, y: Double, w: Double, h: Double, arcWidth: Double, arcHeight: Double): Unit`

- strokeText(text: String, x: Double, y: Double, maxWidth: Double): Unit

- strokeText(text: String, x: Double, y: Double): Unit

Given the previous discussions of shapes, the meaning of these methods and their arguments should be fairly clear.

There is another method of GraphicsContext that is worth mentioning here, and that is clearRect(x: Double, y: Double, w: Double, h: Double): Unit. By default, the Canvas is transparent. When you draw on it, the drawing covers up anything that is below the Canvas in the Scene. The clearRect method sets a region back to being transparent so that you can see through it again to what is below.

12.6.3 Building a Path

Instead of drawing simple shapes, you can also build more complex paths, much the same way that you could with the Path type as opposed to just using things like Rectangle and Circle. To use this approach, you start off by calling beginPath(): Unit. This starts a new path for you to work with. When you are done with a path, you can call either fillPath(): Unit or strokePath(): Unit depending on how you want to add that path to your drawing.

After you have called beginPath, the following methods can be used to add to the path.

- appendSVGPath(svgpath: String): Unit

- arc(centerX: Double, centerY: Double, radiusX: Double, radiusY: Double, startAngle: Double, length: Double): Unit

- arcTo(x1: Double, y1: Double, x2: Double, y2: Double, radius: Double): Unit

- bezierCurveTo(xc1: Double, yc1: Double, xc2: Double, yc2: Double, x1: Double, y1: Double): Unit

- closePath(): Unit

- lineTo(x1: Double, y1: Double): Unit

- moveTo(x0: Double, y0: Double): Unit

- quadraticCurveTo(xc: Double, yc: Double, x1: Double, y1: Double): Unit

- rect(x: Double, y: Double, w: Double, h: Double): Unit

The behavior of these methods is similar to the ones described previously for adding PathElements to a Path.

12.6.4 Image Operations on Canvas

The GraphicsContext also has a number of methods that deal with images. Those methods are as follows.

- drawImage(img: Image, sx: Double, sy: Double, sw: Double, sh: Double, dx: Double, dy: Double, dw: Double, dh: Double): Unit - Draw a portion of the specified image to the given destination in the Canvas. The sx, sy, sw, and sh parameters represent the x, y, width, and height of the rectangular region of the image or source. The dx, dy, dw, and dh parameters give the rectangular region of the destination for drawing.

- drawImage(img: Image, x: Double, y: Double, w: Double, h: Double): Unit - Draws the specified image into the given rectangular region of the Canvas.

- drawImage(img: Image, x: Double, y: Double): Unit - Draws the specified image with the top left corner at the supplied x, y coordinates.

There is also a method called pixelWriter that provides you with a PixelWriter that can be used to write specific pixels directly to the Canvas. Note that you cannot get a PixelReader for the Canvas, so if you are writing something that needs to be able to read pixels, you will need to make a WritableImage and draw that to the Canvas using one of the drawImage methods.[6]

12.6.5 A Canvas Based Game

We finish this section with an example game that is based entirely on Canvas style graphics. In the game the player controls a little green dot with the keyboard. Little red dots try to catch the green dot and if they succeed, the player loses. The mouse is used to "draw" puddles on the screen area that slow down both the player and the enemies. That is the basic idea.

There are a few more details in the analysis. Puddles evaporate over time to become smaller and enemies only live so long. New enemies pop up at regular intervals. As the game goes on, enemies live longer. So the real objective is to see how long the player can stay alive. Here is code for this game.

Listing 12.10: EvadeGame.scala

```scala
import scalafx.Includes._
import scalafx.application.JFXApp
import scalafx.scene.Scene
import scalafx.animation.AnimationTimer
import scalafx.scene.canvas._
import scalafx.scene.control._
import scalafx.scene.input._
import scalafx.scene.layout.BorderPane
import scalafx.scene.paint.Color
import scalafx.event.ActionEvent

case class Enemy(x:Double, y:Double, time:Double)
case class Puddle(x:Double, y:Double, size:Double)
case class Player(x:Double, y:Double)

val CanvasSize = 600

var enemies = List[Enemy]()
var puddles = List[Puddle]()
var player = Player(300,300)
var leftPressed = false
var rightPressed = false
var upPressed = false
var downPressed = false
var regenDelay = 0.0
```

---

[6]There are two methods on Canvas called snapshot that can give you a WritableImage that you could read pixels from. However, this approach would be far less efficient than simply making your own WritableImage that you can read from, write to, and draw to the GraphicsContext.

```scala
26   var enemyLifespan = 10.0
27   var lastTime = 0L
28   var playerDead = false
29
30   val app = new JFXApp {
31     stage = new JFXApp.PrimaryStage {
32       title = "Puddle Dash"
33       scene = new Scene(CanvasSize, CanvasSize+30) {
34
35         val startItem = new MenuItem("Start")
36         val exitItem = new MenuItem("Exit")
37         val fileMenu = new Menu("File")
38         fileMenu.items = List(startItem, new SeparatorMenuItem, exitItem)
39         val menuBar = new MenuBar
40         menuBar.menus = List(fileMenu)
41
42         val canvas = new Canvas(CanvasSize, CanvasSize)
43         val border = new BorderPane
44         border.top = menuBar
45         border.center = canvas
46         content = border
47
48         canvas.onMousePressed = (me:MouseEvent) => { puddles ::= Puddle(me.x, me.y,
                5) }
49         canvas.onMouseDragged = (me:MouseEvent) => { puddles ::= Puddle(me.x, me.y,
                5) }
50         canvas.onKeyPressed = (ke:KeyEvent) => {
51           if (ke.code==KeyCode.Left) leftPressed = true
52           if (ke.code==KeyCode.Right) rightPressed = true
53           if (ke.code==KeyCode.Up) upPressed = true
54           if (ke.code==KeyCode.Down) downPressed = true
55         }
56         canvas.onKeyReleased = (ke:KeyEvent) => {
57           if (ke.code==KeyCode.Left) leftPressed = false
58           if (ke.code==KeyCode.Right) rightPressed = false
59           if (ke.code==KeyCode.Up) upPressed = false
60           if (ke.code==KeyCode.Down) downPressed = false
61         }
62         val gt = canvas.graphicsContext2D
63
64         def inMud(x:Double,y:Double):Boolean = {
65           puddles.exists(p => {
66             val dx = p.x-x
67             val dy = p.y-y
68             dx*dx+dy*dy<p.size*p.size
69           })
70         }
71
72         def movePlayer(dt:Double):Unit = {
73           val speed = if (inMud(player.x,player.y)) 10 else 30
74           if (leftPressed) player = player.copy(x = player.x-speed*dt)
75           if (rightPressed) player = player.copy(x = player.x+speed*dt)
76           if (upPressed) player = player.copy(y = player.y-speed*dt)
77           if (downPressed) player = player.copy(y = player.y+speed*dt)
78         }
```

```scala
 79
 80     def moveEnemies(t:Double, dt:Double):Unit = {
 81       enemies = for (e <- enemies; if e.time+enemyLifespan>t) yield {
 82         val speed = if (inMud(e.x,e.y)) 15 else 40
 83         val dx = if (e.x < player.x) speed else if (e.x > player.x) -speed else 0
 84         val dy = if (e.y < player.y) speed else if (e.y > player.y) -speed else 0
 85         e.copy(x = e.x+(dx+util.Random.nextInt(5)-2)*dt,
 86                y = e.y+(dy+util.Random.nextInt(5)-2)*dt)
 87       }
 88     }
 89
 90     def checkKill():Unit = {
 91       for (enemy <- enemies) {
 92         val dx = enemy.x-player.x
 93         val dy = enemy.y-player.y
 94         if (dx*dx+dy*dy < 25) {
 95           playerDead = true
 96           timer.stop
 97         }
 98       }
 99     }
100
101     def updatePuddles(dt:Double):Unit = {
102       puddles = puddles.filter(_.size > 1).map(p => p.copy(size =
                math.sqrt(p.size*p.size-5*dt)))
103     }
104
105     val timer = AnimationTimer(t => {
106       if (lastTime>0) {
107         val dt = (t-lastTime)/1e9
108         movePlayer(dt)
109         moveEnemies(t/1e9, dt)
110         checkKill()
111         updatePuddles(dt)
112         regenDelay -= dt
113         if (regenDelay < 0) {
114           val cx = util.Random.nextInt(2)
115           val cy = util.Random.nextInt(CanvasSize-10)
116           enemies ::= Enemy(10+cx*(CanvasSize-20), cy, t/1e9)
117           if (math.random<0.1) enemyLifespan += 1
118           regenDelay = 10.0
119         }
120       }
121       lastTime = t
122
123       // Draw stuff
124       gt.fill = Color.Black
125       gt.fillRect(0, 0, canvas.width.value, canvas.height.value)
126       for (p <- puddles) {
127         gt.fill = Color.Brown
128         gt.fillOval(p.x-p.size, p.y-p.size, p.size*2, p.size*2)
129       }
130       for (enemy <- enemies) {
131         gt.fill = Color(1f, 0f, 0f, (t/1e9-enemy.time)/enemyLifespan.toFloat)
132         gt.fillOval(enemy.x-5, enemy.y-5, 10, 10)
```

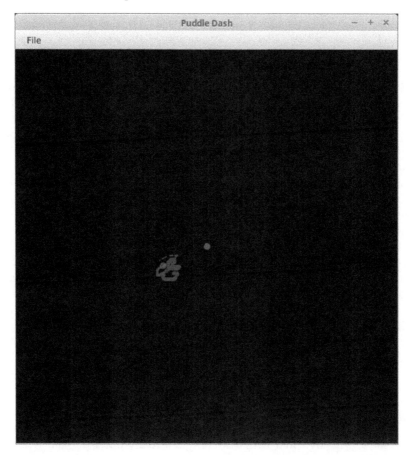

FIGURE 12.11: The window that is produced by running EvadeGame.scala after drawing some puddles with an enemy nearby.

```
133          }
134          gt.fill = Color.Green
135          gt.fillOval(player.x-5, player.y-5, 10, 10)
136          if (playerDead) {
137            gt.fill = Color.White
138            gt.fillText("You Lose!", 200, 200)
139          }
140        })
141
142        startItem.onAction = (e:ActionEvent) => {
143          canvas.requestFocus
144          timer.start
145        }
146        exitItem.onAction = (e:ActionEvent) => sys.exit(0)
147
148      }
149    }
150  }
151
152  app.main(args)
```

You can see what this game looks like in figure 12.11. The first 28 lines have various declarations. Those are followed by the declaration of the **app** with a **Stage** and **Scene**. Lines 35 to 46 build a simple layout with a **MenuBar** at the top of a **BorderPane** and a **Canvas** at the center.

Lines 48 to 61 set up the event handling on the **canvas**. Mouse presses and drags add elements to the **puddles** list. The key presses and releases change the value of some **Boolean** variables. The actual movement of the player occurs in the **movePlayer** function on lines 72 to 78. This is a different approach to player movement than was done previously. It has two main advantages. It allows the motion to be part of the normal timing of the game, so that you have more control over things like speed. It also allows the program to easily work with multiple keys being held down at the same time. That allows diagonal motion, which would not work if we moved the player in the **onKeyPressed** handler the way we did in the Pong game.

Lines 64 to 103 have some helper functions that are just used to break up the logic some and make the code easier to read and work with.

Lines 105 to 140 have the **AnimationTimer** declaration, which is where the real work is done. This includes calls to the helper functions for moving the player, the enemies, and evaporating the puddles. It also has code to create new enemies at regular intervals. After that code is the code that actually draws out the game to the **GraphicsContext** of the **Canvas** which was stored in the variable **gt** on line 62. This starts by filling a rectangle that is the size of the whole **canvas** with black, then it draws the puddles, enemies, and player. If the player has died, a text message is also drawn.

Thread Handling

One of the aspects of ScalaFX that we have not really worried about is the fact that certain types of changes should only be made in certain parts of the code. In particular, ScalaFX has a thread that is used to handle all the events and drawing that goes on. A proper discussion of threads goes beyond the scope of this book (it is covered in "Object-Orientation, Abstraction, and Data-Structures using Scala"), but there are times when you need to worry about it. The issue is that you should only change values of things in the scene graph, or draw to a **GraphicsContext** inside of that event thread. That means either inside of an event handler or in a timer, basically something that is called by ScalaFX. If you want to make changes to values or draw things outside of those chunks of code, you need to schedule it to run in the event thread. You can do this by calling **Process.runLater(op)** where **op** is a pass-by-name argument. Whatever code you put into there will be run in the ScalaFX event thread when it is free to do so. This is presented here because you might find it handy to use the **GraphicsContext** outside of the events or timer handling.

12.7 Effects

Another cool feature of ScalaFX is that it allows you to attach effects to **Nodes**. The different effects that you can add are all in the **scalafx.scene.effect** package, and they are subtypes of **Effect**. Here is a list of the different options, how to create them, and what they do. Many of these are somewhat advanced, and their descriptions will use terms that are specific to graphics and image manipulation that are not defined in this text. The interested reader is encouraged to investigate these topics further.

- `Blend` - This effect determines how colors at the same pixel are combined. By default, the pixel being drawn overwrites the one below it with transparency causing some blending. You can create one using `new Blend(mode: BlendMode)`. The different `BlendModes` are listed here. A color has red, green, and blue components with values between 0 and 1 as well as an alpha component with the same range that is the inverse of transparency.

  - `BlendMode.Add` - Adds together the color values and an alpha value.
  - `BlendMode.Blue` - Replaces only the blue component of the color.
  - `BlendMode.ColorBurn` - Inverts the bottom color components, then divides by the top color components and inverts again for the final color.
  - `BlendMode.ColorDodge` - The bottom color components are divided by the inverse of the top color components.
  - `BlendMode.Darken` - Takes the darker of the bottom and top colors.
  - `BlendMode.Difference` - Subtracts the darker of the two inputs from the lighter and displays the result.
  - `BlendMode.Exclusion` - The color components of the two inputs are multiplied and doubled. That is subtracted from the bottom input color component to give the resulting color.
  - `BlendMode.Green` - Replaces only the green component of the color.
  - `BlendMode.HardLight` - The input color components are either multiplied or screened, depending on the top input color.[7]
  - `BlendMode.Lighten` - Takes the lighter of the bottom and top colors.
  - `BlendMode.Multiply` - Multiplies the color components to get the resulting color.
  - `BlendMode.Overlay` - The input color components are either multiplied or screened, depending on the bottom input color.[7]
  - `BlendMode.Red` - Replaces only the red component of the color.
  - `BlendMode.Screen` - The color components for top and bottom are inverted, multiplied by one another and then that result is inverted to get the resulting color.
  - `BlendMode.SoftLight` - The bottom color components are either darkened or lightened depending on the top input.
  - `BlendMode.SrcAtop` - The part of the top input lying inside of the bottom input is blended with the bottom input.[7]
  - `BlendMode.SrcOver` - The top input is blended over the bottom input.[7]

- `Bloom` - This makes brighter parts of the image appear to glow. You create one using `new Bloom(threshold: Double)`.

- `BoxBlur` - Blurs an image using a box filter kernel. You create one using `new BoxBlur(width: Double, height: Double, iterations: Int)`. The `width` and `height` specify the size of the blur region and `iterations` tells how many times that blur should be processed. More times will lead to more bluring.

---

[7]Taken directly from the JavaFX API.

- ColorAdjust - This effect does per-pixel adjustments to hue, saturation, brightness, and contrast. You create one using `new ColorAdjust(hue: Double, saturation: Double, brightness: Double, contrast: Double)`.

- ColorInput - Renders a rectangular region that is filled with the given `Paint`. Create one using `new ColorInput(x: Double, y: Double, width: Double, height: Double, paint: Paint)`.

- DisplacementMap - This causes the image to be rendered with offsets specified by a `FloatMap`. This allows you to do things like make a `Node` appear wavy. Create one using `new DisplacementMap(mapData: FloatMap, offsetX: Double, offsetY: Double, scaleX: Double, scaleY: Double)` or `new DisplacementMap (mapData: FloatMap)`. The `FloatMap` can be created using `new FloatMap(width: Int, height: Int)` and the values in it are set using the `def setSamples(x: Int, y: Int, s0: Float, s1: Float): Unit` method.[8]

- DropShadow - Renders a shadow behind the given content with a specified radius, offset, and color. You can create one using `new DropShadow(radius: Double, offsetX: Double, offsetY: Double, color: Color)`, `new DropShadow(radius: Double, color: Color)`, or `new DropShadow(blurType: BlurType, color: Color, radius: Double, spread: Double, offsetX: Double, offsetY: Double)`. The `BlurType` can be one of the following four values: `BlurType.Gaussian`, `BlurType.OnePassBox`, `BlurType.ThreePassBox`, or `BlurType.TwoPassBox`.[9]

- GaussianBlur - This blurs the image using a Gaussian function with the specified radius. You can create one using `new GaussianBlur(radius: Double)`.

- Glow - Makes an input image appear to glow. You can create one with `new Glow(level: Double)`.

- ImageInput - A source effect that passes an unmodified image through to another effect. Create one with `new ImageInput(source: Image, x: Double, y: Double)` or `new ImageInput(source: Image)`.

- InnerShadow - Renders a shadow inside the edges of some content with a specified radius, offset, and color. You can create one using `new InnerShadow(radius: Double, offsetX: Double, offsetY: Double, color: Color)`, `new InnerShadow(radius: Double, color: Color)`, or `new InnerShadow(blurType: BlurType, color: Color, radius: Double, choke: Double, offsetX: Double, offsetY: Double)`. The `BlurType` is one of the same values listed above for `DropShadow`.

- Lighting - Creates the effort of a light shining on an element to give it a more realistic look. You create one with `new Lighting(light: Light)`. The `Light` type is generally created by instantiating one of three subtypes, `Light.Distant`, `Light.Point`, or `Light.Spot`. These can be made using the following calls.

  - `new Light.Distant(azimuth: Double, elevation: Double, color: Color)`
  - `new Light.Point(x: Double, y: Double, z: Double, color: Color)`

---

[8]The `FloatMap` type allows different "bands" that can be used for different things. The `DisplacementMap` uses two bands for x and y values, respectively.

[9]These are the names used at the time of writing. The developers of ScalaFX appear to be moving to a camel-case standard, so it is likely that in the future these will be `Gaussian`, `OnePassBox`, `ThreePassBox`, and `TwoPassBox` instead.

     – new `Light.Spot(x: Double, y: Double, z: Double, specularExponent: Double, color: Color)`

- `MotionBlur` - Creates the effort of blurring from motion. You can create one with `new MotionBlur(angle: Double, radius: Double)`. The angle specifies the perceived direction of motion and the radius tells it how much blurring to do.

- `PerspectiveTransform` - This effect provides a faux perspective transformation. This cannot be done with affine transforms, because when viewed in perspective, parallel lines appear to meet in the distance. You can create one with `new PerspectiveTransform(ulx: Double, uly: Double, urx: Double, ury: Double, lrx: Double, lry: Double, llx: Double, lly: Double)`, the values passed in specify the location for the corners of a quadrilateral that the `Node` is mapped to.

- `Reflection` - This renders a reflected version of the node below the actual content. You can create one using `new Reflection(topOffset: Double, fraction: Double, topOpacity: Double, bottomOpacity: Double)`. The opacities allow the reflection to have a fading appearance.

- `SepiaTone` - Applies a sepia coloration to an image. You can create one with `new SepiaTone(level: Double)`.

- `Shadow` - This creates a shadow of the `Node` it is attached to. Unlike `DropShadow`, it does not show both the original and the shadow. You can create one with `new Shadow(radius: Double, color: Color)` or `new Shadow(blurType: BlurType, color: Color, radius: Double)`. The `BlurType` has the same options as mentioned for `DropShadow`.

Usage of the majority of these is shown in the following code. In particular, this code displays the ones that work well with text. Effects that work best with images are shown in section 12.3.

Listing 12.11: Effects.scala

```scala
import scalafx.Includes._
import scalafx.application.JFXApp
import scalafx.scene._
import scalafx.scene.effect._
import scalafx.scene.paint._
import scalafx.scene.shape._
import scalafx.scene.text._

def makeText(x:Double, y:Double, s:String, effect:Effect, darkBack:Boolean):Node =
    {
  val text = new Text(x, y, s)
  text.font = Font("serif", FontWeight.Bold, 40)
  text.fill = if (darkBack) Color.White else Color.Black
  text.effect = effect
  if (darkBack) {
    val group = new Group
    val b = text.boundsInLocal.value
    val r = Rectangle(b.minX, b.minY, b.width, b.height)
    r.fill = Color.Black
    group.children = List(r, text)
    group
```

```scala
  } else {
    text
  }
}

val app = new JFXApp {
  stage = new JFXApp.PrimaryStage {
    title = "Effects"
    scene = new Scene(450,720) {
      val bloom = makeText(20, 40, "Bloom", new Bloom(0.1), true)
      val boxBlur = makeText(20, 100, "BoxBlur", new BoxBlur(7, 7, 2), false)
      val floatMap = new FloatMap(400, 40)
      for (i <- 0 until floatMap.width.value;
           offset = (0.1*math.sin(i/30.0)).toFloat;
           j <- 0 until floatMap.height.value) {
        floatMap.setSamples(i, j, 0.0f, offset)
      }
      val dMap = new DisplacementMap(floatMap)
      val displacementMap = makeText(20, 160, "DisplacementMap", dMap, false)
      val dropShadow = makeText(20, 220, "DropShadow", new DropShadow(5, 20, 20,
          Color.Black), false)
      val gaussianBlur = makeText(20, 300, "GaussianBlur", new GaussianBlur(7),
          false)
      val innerShadow = makeText(20, 360, "InnerShadow", new InnerShadow(3, 3, 3,
          Color.Green), false)
      val lightEffect = new Lighting(new Light.Distant(0, 45, Color.White))
      lightEffect.surfaceScale = 5.0
      val lighting = makeText(20, 420, "Lighting", lightEffect, false)
      val motionBlur = makeText(20, 480, "MotionBlur", new MotionBlur(15, 10),
          false)
      val perspective = makeText(20, 540, "Perspective", new
          PerspectiveTransform(20, 500, 400, 510, 400, 520, 20, 540), false)
      val reflection = makeText(20, 600, "Reflection", new Reflection(5, 0.8, 1.0,
          0.0), false)
      val shadow = makeText(20, 700, "Shadow", new Shadow(5, Color.Black), false)

      content = List(bloom, boxBlur, displacementMap, dropShadow, gaussianBlur,
          innerShadow, lighting, motionBlur, perspective, reflection, shadow)
    }
  }
}

app.main(args)
```

The output of this code is shown in figure 12.12. Each effect is applied to a string with the name of the effect. There is a helper method called `makeText` that creates an instance of `Text` and gives it a large font. If the effect needs a background to be visible, one is added by making a group that also contains a `Rectangle` that matches the bounds of the `Text`.

The above example showed the effects that worked well with text. The `ColorAdjust`, `SepiaTone`, and `Glow` effects are better suited for working with images, so we have a separate example for those. The following code allows the user to specify multiple files on the command line. It loads in those files and puts them on `MenuButtons` that have menu options of `ColorAdjust`, `SepiaTone`, and `Glow`. If the user selects a menu option, that effect is applied to the image on the button. All the buttons are put in a `TilePane`.

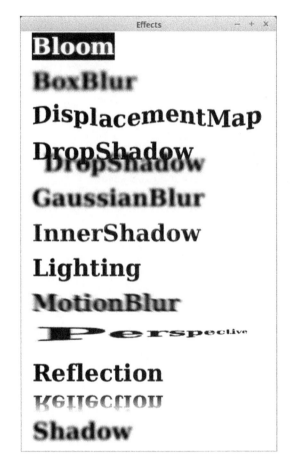

FIGURE 12.12: The window that is produced by running Effects.scala.

Listing 12.12: ImageEffects.scala

```scala
1  import scalafx.Includes._
2  import scalafx.application.JFXApp
3  import scalafx.scene.Scene
4  import scalafx.scene.image.{Image, ImageView}
5  import scalafx.scene.layout.TilePane
6  import scalafx.scene.effect._
7  import scalafx.scene.control._
8  import scalafx.event.ActionEvent
9
10 if (args.length < 1) {
11   println("You must provide arguments with the names of files to load.")
12   sys.exit(0)
13 }
14
15 val app = new JFXApp {
16   stage = new JFXApp.PrimaryStage {
17     scene = new Scene(680,420) {
18       val tilePane = new TilePane
19       for (file <- args) {
20         val img = new Image("file:"+file, 300, 200, true, true)
21         val view = new ImageView(img)
22         val colorAdjust = new MenuItem("Color Adjust")
23         colorAdjust.onAction = (e:ActionEvent) => {
24           view.effect = new ColorAdjust(-0.05, 0.2, 0.1, 0.1)
25         }
26         val glow = new MenuItem("Glow")
27         glow.onAction = (e:ActionEvent) => {
28           view.effect = new Glow(0.5)
29         }
30         val sepia = new MenuItem("Sepia")
31         sepia.onAction = (e:ActionEvent) => {
32           view.effect = new SepiaTone(0.5)
33         }
34         val button = new MenuButton("",view)
35         button.items = List(colorAdjust, glow, sepia)
36         tilePane.children += button
37       }
38
39       root = tilePane
40     }
41   }
42 }
43
44 app.main(args)
```

A possible output of this program can be seen in figure 12.13.[10]

---

[10]The images shown here were provided by Quinn Bender.

FIGURE 12.13: The window that is produced by running ImageEffects.scala with four sample images. The top-left and bottom-right images were set to the `ColorAdjust` effect, the top-right used `SepiaTone`, and the bottom-left used `Glow`.

12.8   Charts

A common task done on computers when you have data sets is to use charts to plot the data. For this reason, ScalaFX includes a `scalafx.scene.chart` package with a number of different chart types as well as other types that are required to build those charts. There are eight different chart types that are part of the library. These are as follows.

- `AreaChart` - Fills in an area from zero up to a line for one or more sequences of data numeric x-y data.

- `BarChart` - Displays data as bars for different text categories.

- `BubbleChart` - Draws bubbles of various sizes at numeric x-y coordinates.

- `LineChart` - Draws a line for a sequence of numeric x-y data.

- `PieChart` - Draws a pie chart for a sequence of labeled numeric data.

- `ScatterChart` - Draws a scatter chart for a sequence of x-y data.

- `StackedAreaChart` - Draws area charts stacked on top of one another for multiple series of x-y data.

- `StackedBarChart` - Draws bar charts stacked on top of one another for different categories.

With the exception of `PieChart`, all of the chart types inherit from a `XYChart`, which is not listed here because it is abstract. All of the subtypes of `XYChart` require two axes as well as the data to be plotted. ScalaFX provides three different subtypes of `Axis` including `CategoryAxis` for text categories and `NumberAxis` for numeric values.

To help illustrate the use of these charts in a program, the following code creates three different charts, a `PieChart`, a `BarChart`, and a `ScatterChart`, based on data about the planets in our solar system.

Listing 12.13: Charts.scala

```scala
import scalafx.Includes._
import scalafx.application.JFXApp
import scalafx.scene.Scene
import scalafx.scene.chart._
import scalafx.scene.control._
import scalafx.scene.layout._
import scalafx.collections.ObservableBuffer
import scalafx.geometry.Pos

val app = new JFXApp {
  stage = new JFXApp.PrimaryStage {
    title = "Charts"
    scene = new Scene(1000, 800) {
      case class PlanetData(name:String, mass:Double, a:Double)
      val planets = List(PlanetData("Mercury", 0.0553, 0.387),
                        PlanetData("Venus", 0.815, 0.723),
                        PlanetData("Earth", 1.0, 1.0),
                        PlanetData("Mars", 0.107, 1.524),
                        PlanetData("Jupiter", 317.83, 5.203),
```

```
20              PlanetData("Saturn", 95.159, 9.537),
21              PlanetData("Uranus", 14.536, 19.191),
22              PlanetData("Neptune", 17.147, 30.069))
23
24      val pieChart = PieChart(ObservableBuffer(planets.map(p =>
            PieChart.Data(p.name,p.mass)):_*))
25
26      val barSeries = XYChart.Series("Planetary Mass",
            ObservableBuffer(planets.map(p => XYChart.Data(p.name,
            p.mass:Number)):_*))
27      val barChart = BarChart(CategoryAxis(), NumberAxis("Mass [Earth Mass]"),
            ObservableBuffer(barSeries))
28
29      val scatterSeries = XYChart.Series("Mass vs. Orbital Distance",
            ObservableBuffer(planets.map(p => XYChart.Data[Number, Number](p.a,
            p.mass)):_*))
30      val scatterChart = ScatterChart(NumberAxis("Semimajor Axis [AU]"),
            NumberAxis("Mass [Earth Mass]"), ObservableBuffer(scatterSeries))
31
32      val tilePane = new TilePane
33      tilePane.children = List(pieChart, barChart, scatterChart)
34      tilePane.alignment = Pos.TopCenter
35
36      root = tilePane
37    }
38  }
39 }
40
41 app.main(args)
```

The result of running this code can be seen in figure 12.14. As usual, the code begins with a number of imports, this time that includes `scalafx.scene.chart._`. That is followed by the general setup of the app, the stage, and the scene. Lines 14-22 define a `case class` and a list of instances of it that provide the data to be used for the plots. This data has the planets in our Solar System along with their masses in units of Earth masses and the semimajor axes (average distance from the Sun) in AU.[11]

Line 24 creates a `PieChart` using the masses to show how much of the mass of the solar system, outside of the Sun, is in each planet. This is done by passing `PieChart` an `ObservableBuffer` filled with `PieChart.Data` instances. All the charts in ScalaFX use `ObservableBuffer`s because they update if the values are changed. This can be used to nice effect in some applications. There is also a property of the charts called `animated` that will cause such changes to be animated if it is set to true. In this case, the contents of the `ObservableBuffer` are created by mapping over the `planets` data pulling out the names and the masses. Note the use of `:_*`. Recall that this is used when a function or method expects a variable length argument list, and we want to pass it a sequence such as a `List` or `Array`.

Lines 26 and 27 set up the `BarChart`. All of the `XYChart`s can display multiple series of data. Your example only shows one, but the way it is set up should make it fairly clear how others would be added. Line 26 makes the single series. Note that it is a `XYChart.Series`. All of the subtypes of `XYChart` use the `Series` and `Data` types defined in `XYChart` instead of defining their own. After seeing the creation of the `PieChart`, the way in which this is created should be fairly straightforward. There is only one element that might seem a bit

---

[11] An AU, short of Astronomical Unit, is the average distance between the Earth and the Sun.

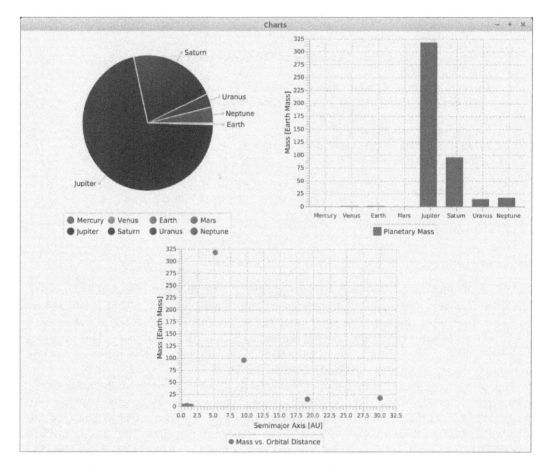

FIGURE 12.14: The window that is produced by running Charts.scala.

unusual, and that is the type specification in `p.mass:Number`. This is because the next line uses a `NumberAxis` and the type of the data has to be an exact match to the types of the axes. Using the type specification forces Scala to treat the `Double` from the `case class` as a `Number`. We could have also changed the `case class` so that the number fields were of type `Number` instead of `Double`.

On line 27 the `BarChart` is built. It needs to be given two axes. The data can be added later, or in this call. For this example, the first axis, the x-axis, is a `CategoryAxis`. The second axis, the y-axis, is a `NumberAxis` with a specified label. The data is an `ObservableBuffer[XYChart.Series]`. This code makes a buffer with our one series, but if there were multiple series of data that we wanted to display, they could easily be added to the buffer.

Lines 29 and 30 create the `ScatterChart`. The series declaration is very similar to that for the `BarChart`. One difference is that the types for the data are given as type parameters instead of putting type specifications on the arguments. That is to say that we put `[Number, Number]` after `XYChart.Data` instead of putting `:Number` after each argument. This is just used to show a different approach. You can use either style in your own code.

The `ScatterChart` itself is created by giving two labeled `NumberAxes` followed by an `ObservableBuffer` of our data series. Again, adding multiple series is as easy as adding them to the buffer of data.

The script ends by putting all three charts in a `TilePane`, setting the `TilePane` to display its children at the top-center, and setting that `TilePane` to be the `root` of the `scene`. This gives us the layout that appears in figure 12.14.

This example allows the axes to use default settings, but there are times when you want to provide more direction in how axes are displayed. To do this on a `NumberAxis`, we would create the axis on a separate line and give it a name, then change the appropriate properties. The properties you are most likely to want to change are the bounds and the tick marks. You can change the bounds by setting `lowerBound` and `upperBound`. You can make tick marks visible or not by setting `tickMarkVisible` or `minorTickVisible`. You can set the number of minor tick marks between major tick marks with `minorTickCount`. There are other settings for tick lengths and fonts that we will not describe here, that you can find in the API.

One note of caution related to the charts in ScalaFX. These types were built to be very flexible and to provide a lot of options. Using observable values for lists and data allows them to do interesting animations without putting too much burden on the programmer as well. Unfortunately, all of these things have overhead associated with them. For that reason, you probably do not want to use these charts for large data sets. If you have much more than 10,000 data elements to display, you will probably find that these charts are too slow and consume too much memory to be useful.

12.9 Media

Media, both audio and video, has become a significant part of many modern user interfaces. ScalaFX includes a number of types that we can use to add media to our applications in the `scalafx.scene.media` package. We will only cover the most significant of those here.

Short sounds that you want to play repeatedly are best represented by the `AudioClip` type. You can create an `AudioClip` using `new AudioClip(source: String)`. The `source` needs to be a URL. If you have a file in your current directory, you can simply specify `"file:filename"`. ScalaFX supports a reasonable number of audio formats including .wav and .mp3.[12]

Longer audio files and pretty much any video files should not be loaded completely into memory. The `Media` type should be used to represent such files. You create a `Media` instance using `new Media(source: String)` where the `source` should again be a URL, like for the `AudioClip`. In addition to the audio formats mentioned above, ScalaFX has fairly general support for video as well, but it is not as accepting here as for audio.[13] You are likely to find that you have to put in some effort to get video files in formats that it is happy with. The authors' experience is that the formats recorded by cell phones seem to be well supported.

In order to play a `Media` instance you use a `MediaPlayer`. You can create an instance of `MediaPlayer` using `new MediaPlayer(media: Media)`. If the `Media` is audio, this is all that you need. For video, you probably want to have it displayed in the GUI. This is done by creating a `MediaView` with `new MediaView(mediaPlayer: MediaPlayer)`. The `MediaView` is a subtype of `Node` and can be added to a GUI. Calling the `play`, `pause`, and `stop` methods on the `MediaPlayer` will control what is displayed in the `MediaView`

A program demonstrating the use of these media types is shown here. A button at the top will play an `AudioClip` of a laser.[14] Most of the display shows a video with some buttons to control it on the left.

Listing 12.14: Media.scala

```scala
1   import scalafx.Includes._
2   import scalafx.application.JFXApp
3   import scalafx.scene.Scene
4   import scalafx.scene.media._
5   import scalafx.scene.layout._
6   import scalafx.scene.control.Button
7   import scalafx.event.ActionEvent
8
9   val app = new JFXApp {
10    stage = new JFXApp.PrimaryStage {
11      title = "Media"
12      scene = new Scene(600, 350) {
13        val audioClip = new AudioClip("file:Laser_Cannon-Mike_Koenig-797224747.mp3")
14
15        val fireButton = new Button("Fire!")
16        fireButton.onAction = (ae:ActionEvent) => audioClip.play
17        fireButton.prefWidth = Int.MaxValue
```

---

[12]At the time of writing, JavaFX 8 listed their audio support as "MP3; AIFF containing uncompressed PCM; WAV containing uncompressed PCM; MPEG-4 multimedia container with Advanced Audio Coding (AAC) audio".

[13]At the time of writing, JavaFX 8 list their video support as "FLV containing VP6 video and MP3 audio; MPEG-4 multimedia container with H.264/AVC (Advanced Video Coding) video compression".

[14]The sound for the `AudioClip` was downloaded from `http://soundbible.com/1771-Laser-Cannon.html`.

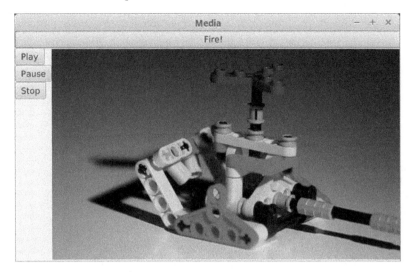

FIGURE 12.15: The window that is produced by running Media.scala.

```
18
19    val media = new Media("http://techslides.com/demos/sample-videos/small.mp4")
20    val mediaPlayer = new MediaPlayer(media)
21    val mediaView = new MediaView(mediaPlayer)
22
23    val playButton = new Button("Play")
24    playButton.onAction = (ae:ActionEvent) => mediaPlayer.play
25    val pauseButton = new Button("Pause")
26    pauseButton.onAction = (ae:ActionEvent) => mediaPlayer.pause
27    val stopButton = new Button("Stop")
28    stopButton.onAction = (ae:ActionEvent) => mediaPlayer.stop
29
30    val borderPane = new BorderPane
31    borderPane.top = fireButton
32    borderPane.left = new VBox(playButton, pauseButton, stopButton)
33    borderPane.center = mediaView
34
35    root = borderPane
36  }
37  }
38 }
39
40 app.main(args)
```

The window produced by this code is shown in figure 12.15.

12.10   Web

Just as media has become increasingly important in applications, so has the web. ScalaFX includes types that allow you to bring the web into your application in the

scalafx.scene.web package. The most obvious of these is the WebView, which is basically a fully capable web browser in a Node, so you can add it to any GUI that you want to
display Web based content. The WebView is really a graphical front end for the WebEngine
type, which does the real work associated with running web pages.

The following example uses a WebView and a TextField to make a very basic web
browser. They are put in a BorderPane with the TextField at the top and the WebView in
the center. The WebView is initially set to display the home page for Scala.

Listing 12.15: Web.scala

```scala
import scalafx.Includes._
import scalafx.application.JFXApp
import scalafx.event.ActionEvent
import scalafx.scene.Scene
import scalafx.scene.control.TextField
import scalafx.scene.layout.BorderPane
import scalafx.scene.web._

val app = new JFXApp {
  stage = new JFXApp.PrimaryStage {
    title = "Web"
    scene = new Scene(750,600) {
      val urlField = new TextField
      val webView = new WebView
      webView.engine.load("http://www.scala-lang.org")
      urlField.text = webView.location.value

      val borderPane = new BorderPane
      borderPane.top = urlField
      borderPane.center = webView

      root = borderPane

      urlField.onAction = (ae:ActionEvent) => {
        webView.engine.load(urlField.text.value)
      }
      webView.location.onChange(urlField.text = webView.location.value)
    }
  }
}

app.main(args)
```

At the end of the code, handlers are added so that when the user enters a URL in the
TextField and hits enter, that page will be displayed, and when the page updates from
something like a clicked link, the displayed text is shown. If you enter this code and run it,
take note of the fact that you have to enter a full URL, including the protocol (normally
http) for it to work.

Figure 12.16 shows what this script looks like when you run it. You can view most
content on the web with just this small amount of code. One of the main things it will not
do is play media on sites like YouTube.

The other Node type in the scalafx.scene.web package is HTMLEditor. As the name
implies, this type provides you with an editor that can be used to edit formatted text. The
user sees an editor with controls for how things are displayed. The fact that the back-end
representation uses HTML is not revealed to the user. If you want to allow users to edit

FIGURE 12.16: The window that is produced by running Web.scala.

text beyond the plain text of a `TextArea`, this is how you would want to do it in ScalaFX. The following code shows a script that uses this along with menu options for saving and opening files to produce a functional text editor with formatting that saves in HTML.

Listing 12.16: Web.scala

```scala
1  import scalafx.Includes._
2  import scalafx.application.JFXApp
3  import scalafx.event.ActionEvent
4  import scalafx.scene.Scene
5  import scalafx.scene.control._
6  import scalafx.scene.layout.BorderPane
7  import scalafx.scene.web._
8  import scalafx.stage.FileChooser
9
10 val app = new JFXApp {
11   stage = new JFXApp.PrimaryStage {
12     title = "Rich Editor"
13     scene = new Scene(600,400) {
14       val menuBar = new MenuBar
15       val fileMenu = new Menu("File")
16       val openItem = new MenuItem("Open")
17       val saveItem = new MenuItem("Save")
18       val exitItem = new MenuItem("Exit")
19       fileMenu.items = List(openItem, saveItem, new SeparatorMenuItem, exitItem)
20       menuBar.menus = List(fileMenu)
21       menuBar.prefWidth = Int.MaxValue
22
23       val editor = new HTMLEditor
```

```
24
25        val borderPane = new BorderPane
26        borderPane.top = menuBar
27        borderPane.center = editor
28
29        root = borderPane
30
31        openItem.onAction = (ae:ActionEvent) => {
32          val chooser = new FileChooser
33          chooser.showOpenDialog(stage) match {
34            case null =>
35            case file =>
36              val source = io.Source.fromFile(file)
37              editor.htmlText = source.mkString
38              source.close
39          }
40        }
41
42        saveItem.onAction = (ae:ActionEvent) => {
43          val chooser = new FileChooser
44          chooser.showSaveDialog(stage) match {
45            case null =>
46            case file =>
47              val pw = new java.io.PrintWriter(file)
48              pw.print(editor.htmlText)
49              pw.close
50          }
51        }
52
53        exitItem.onAction = (ae:ActionEvent) => {
54          sys.exit(0)
55        }
56      }
57    }
58  }
59
60  app.main(args)
```

The window created by running this and entering some text is shown in figure 12.17. You can see the setting controls that are provided by the HTMLEditor type along the top of the window under the menu bar. You probably recognize most of these as standard controls for text editors that you have worked with.

12.11   3D Graphics

The last topic of graphics in ScalaFX that we will discuss is 3D graphics. Everything that we have done to this point has been distinctly 2D. There are a number of reasons for that, the main one being that 2D graphics are a lot simpler. Despite that relative "simplicity", we have still managed to spend over 50 pages on the topic. There are a lot more details to 3D graphics than what we will cover in this section. This will be a very basic introduction to the topic that gives you enough knowledge to do some basic things with 3D graphics.

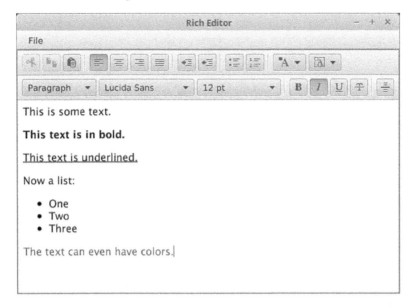

FIGURE 12.17: The window that is produced by running HTMLEditor.scala.

There are four different 3D shapes in the `scalafx.scene.shape` package that are 3D. These are `Box`, `Cylinder`, `MeshView`, and `Sphere`. With the exception of `MeshView`, these should be fairly self-explanatory as they are all basic 3D shapes. The `MeshView` covers everything that is not those other three shapes by allowing you to make complex geometries out of triangles. Proper coverage of the `MeshView is outside of the scope of our discussion`.

For 2D shapes, you needed to specify the `fill` that was used for drawing them. When you go to 3D, this changes to a material. ScalaFX provides the `scalafx.scene.paint.PhongMaterial` type for this purpose. You can instantiate this type with `new PhongMaterial(diffuseColor: Color, diffuseMap: Image, specularMap: Image, bumpMap: Image, selfIlluminationMap: Image)` or `new PhongMaterial( diffuseColor: Color)`. The first option gives you signfiicant control over how the material appears. The different "map" arguments specify differences in how things are drawn across the surface. For example, the `diffuseMap` allows you to apply a texture to the surface of the shape.

In addition to having the shapes and material, when you move to 3D you have to deal with cameras and lighting, both of which are defined in the `scalafx.scene` package. There are two types of cameras provided. By default you get a `ParallelCamera`. You can change this to be a `PerspectiveCamera`. When you use the `ParallelCamera`, object do not get smaller as they move away from the camera. In this view, parallel lines heading away from the viewer remain parallel. This is not what happens in real life, but it is sufficient for some application. The `PerspectiveCamera` more closely resembles the way we actually see, with objects shrinking as they get further away and everything converging to a point at infinite distance. You make a `ParallelCamera` with `new ParallelCamera`. To make a `PerspectiveCamera` you use `new PerspectiveCamera(fixedEyeAtCameraZero: Boolean)`. The `Stage` has a member called `camera` that you can assign to your desired camera type.[15]

There are also two types of light defined in ScalaFX. They are the `AmbientLight` and

---

[15]The ability to only have one camera for the full stage can be a bit limiting. The `scalafx.scene` package

the `PointLight`. An `AmbientLight` lights everything evenly while a `PointLight` can be moved around, and it lights items in the scene as if there is a light in that location. They are both created with **new** and require a single argument of a `Color` that is the color of the light.

The following code shows an example of these features of 3D graphics in ScalaFX. It adds one of each of the basic shapes to the scene as well as one of each type of light source. There are controls that let you pick the camera type and the color of the lights. You can also specify if you want the `Sphere` to be texture mapped or not. It starts off a flat red, but if you select that `CheckBox`, it is textured with a surface map of Mars.

Listing 12.17: 3D.scala

```scala
import scalafx.Includes._
import scalafx.application.JFXApp
import scalafx.scene.{Scene, ParallelCamera, PerspectiveCamera, AmbientLight,
    PointLight}
import scalafx.scene.image.Image
import scalafx.scene.paint.{Color, PhongMaterial}
import scalafx.scene.shape.{Box, Cylinder, MeshView, Sphere}
import scalafx.scene.control._
import scalafx.event.ActionEvent

val app = new JFXApp {
  stage = new JFXApp.PrimaryStage {
    title = "3D"
    scene = new Scene(400, 400) {
      val marsImage = new Image("file:marssurface.jpg")
      val sphere = new Sphere(50)
      sphere.translateX = 200
      sphere.translateY = 200
      val sphereMaterial = new PhongMaterial(Color.Red)
      sphere.material = sphereMaterial

      val cylinder = new Cylinder(40, 100)
      cylinder.translateX = 100
      cylinder.translateY = 100
      cylinder.material = new PhongMaterial(Color.Green)

      val box = new Box(50, 40, 60)
      box.translateX = 300
      box.translateY = 300
      box.material = new PhongMaterial(Color.Blue)

      val ambient = new AmbientLight(Color.Gray)
      val point = new PointLight(Color.White)
      point.translateX = 400
      point.translateZ = -400

      val usePerspective = new CheckBox("Perspective View")
      usePerspective.layoutY = 270
      usePerspective.onAction = (e:ActionEvent) => {
        if (usePerspective.selected.value) {
          camera = new PerspectiveCamera(false)
```

also has a type called `SubScene` that is a subtype of `Node`. Each `SubScene` gets its own camera so you can create multiple different views that way.

```
41    } else {
42      camera = new ParallelCamera
43    }
44  }
45
46  val useTexture = new CheckBox("Texture Sphere")
47  useTexture.layoutY = 300
48  useTexture.onAction = (e:ActionEvent) => {
49    if (useTexture.selected.value) {
50      sphereMaterial.diffuseMap = marsImage
51      sphereMaterial.diffuseColor = Color.White
52    } else {
53      sphereMaterial.diffuseMap = null
54      sphereMaterial.diffuseColor = Color.Red
55    }
56  }
57
58  val ambientPicker = new ColorPicker(ambient.color.value)
59  ambient.color <==> ambientPicker.value
60  ambientPicker.layoutY = 330
61
62  val pointPicker = new ColorPicker(point.color.value)
63  point.color <==> pointPicker.value
64  pointPicker.layoutY = 360
65
66  content = List(sphere, cylinder, box, ambient, point, usePerspective,
        useTexture, ambientPicker, pointPicker)
67    }
68  }
69  }
70
71  app.main(args)
```

A possible output of this program can be seen in figure 12.18.

12.12 Putting It Together

Given the breadth of this chapter, and the fact most of the sections had their own examples highlighting usage of the elements in a program, we are not going to attempt to write a single application that pulls everything together. It is not really clear that such a thing would even make sense.

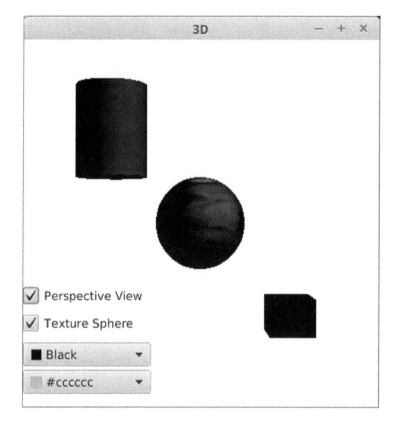

FIGURE 12.18: The window that is produced by running 3D.scala and changing some of the settings.

12.13 End of Chapter Material

12.13.1 Summary of Concepts

- If you understand how to do graphics you can go beyond the standard GUI components and draw whatever you want in a `Scene`.

- One way to add graphics is by adding shape `Nodes` directly to the scene.
  - A number of different types of shapes are included in the `scalafx.scene.shape` package.
  - The `Path` type allows you to make more complex shapes.
  - You can vary settings for filling shapes as well as the way outlines are drawn with stroke settings.

- Setting the handlers for keyboard, mouse, and touch allows you to make your graphics interact with the user.

- You can load images from files or the web using the `scalafx.scene.image.Image` type.
  - The `PixelReader` type will let you read pixel values for individual pixels as colors or numeric ARGB values.
  - The `WritableImage` type works with a `PixelWriter` to allow you to set the values on individual pixels.

- Affine transformations can be applied to any `Node` to move, rotate, scale, or shear their appearance in the display.

- ScalaFX includes three different styles of creating animations.
  - Transitions do fairly simple variations of values for certain types of values.
  - Timelines allow you to build more complex combinations of transitions for numeric properties.
  - The `AnimationTimer` gives general purpose control over things happening in the GUI. This is useful for applications like games.

- The `Canvas` type is a subtype of `Node` that you can draw on and manipulate like any other `Node`. This is done through a `GraphicsContext`. The system does not use resources storing properties of things drawn in this manner, but updating elements requires a full redraw.

- The `scalafx.scene.effect` package contains a number different effects that can be applied to nodes to produce interesting graphical results.

- The `scalafx.scene.chart` package provides functionality for plotting data with a variety of different chart styles.

- The `scalafx.scene.media` package contains types that you can use to add audio and video into your applications.

- The `scalafx.scene.web` package contains a node type for a web browser and an HTML-based rich text editor.

- ScalaFX also include basic 3D graphics capabilities. You can add 3D shapes from the `scalafx.scene.shape` package. You will also need to adjust the camera and light settings to get your desired result.

12.13.2   Exercises

1. Create a pane that draws several geometric shapes with different types of paints and strokes.

2. Use the `scalafx.scene.shape` package to create in interesting shape.

3. Write a program that displays a 10 x 10 grid.

4. Write a program that allows the user to move a ball in a scene left, right, up, and down. Check the boundaries to prevent the ball from moving out of bounds.

5. Use a gradient style paint and `Timer` to make a pattern where the colors move.

6. Use a dotted line on a wide stroke along with a `Timer` where the dotted line moves with the `Timer` ticks.

7. Write a script where you control the movement of a geometric figure with the arrows keys. Put another figure at the location of the mouse click draw them in a different color when the two intersect.

8. Polish up the evade game.

9. Write a program that creates a user interface for displaying an address book that contains a person's name, street, city, state, and zip. Your GUI should also contain buttons that would allow the user to add a new name to the book, go to the first name in the book, go to the last name in the book, go to the next name in the book, and go to the previous name in the book.

10. Write a traffic light program. This program lets the user select one of three lights: red, yellow, or green by selecting a radio button. When a color is selected, the appropriate light is turned on. Only one light can be on at a time. No light is on when the program starts.

11. Write a program that draws a circle or a square. The user selects which figure they want to draw by clicking on a radio button. You should also allow the user to select whether or not the shape is filled by selecting a check box.

12.13.3   Projects

1. If you did project 10.1 you should have noticed that looking at the numbers to figure out what is going on is quite a pain. To see if particles are moving properly it really does help to have a plot of their motion. For this project you will add a GUI with a custom drawn panel onto that program so that you can draw where particles are located.

   The only challenge in doing this is getting from the coordinates that your particles are at to coordinates that fit in a normal window. There are two ways to do this. One is to do the math yourself. Specify a minimum and maximum value for x and y and use

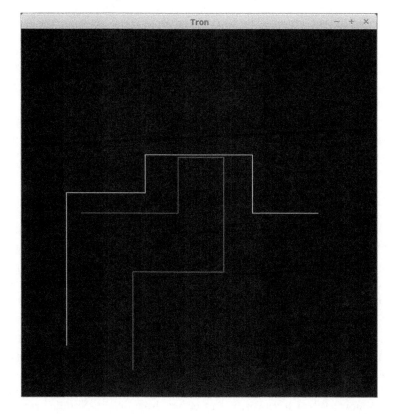

FIGURE 12.19: This figure shows what is intended as a possible output for project 3. The two players are single points that leave lines behind them. They can only turn at 90 degree angles.

linear equations to transform from the range (min, max) to $(0, width)$ or $(0, height)$. You can also do this with an `AffineTransform` using a combination of scale and translate. There is benefit to each approach and neither is significantly harder than the other.

To make the GUI more functional you could include `TextFields` that allow the user to change the values of X_{min}, X_{max}, Y_{min}, and Y_{max}. You could also include a `CheckBox` that lets the user select whether the background is cleared each time particles are drawn. Without clearing, you will see the paths of the particles as they move. With clearing you will only see whatever shapes you use to draw the particles.

2. For this project you should write a script that reads in a text file of numbers and plots the data. You can pick what style of plotting you want to enable and what format the data should be in. As with the previous project, the main challenge in this is to convert from the coordinates that the data points are in to something that appears nicely on screen.

3. The movie Tron (`http://www.imdb.com/title/tt0084827/`), released in 1982, included a number of different early arcade style games. One of these, Light Cycles, can be simplified down to a level where it is very easy to write. For this project, you will do just that to produce something like figure 12.19.

In the game, the players ride cycles that leave colored walls behind them. Running into a wall kills the player. The last player alive wins. For your implementation, the walls will be lines of pixels in a `WritableImage`, and the cycle will just be the end where the line is growing. You do not have to graphically represent the cycle. Two humans can play. One uses arrow keys and another uses letters. Only two keys are needed for each, one to turn left and one to turn right. Using an image to store the walls prevents you from having to keep that data in a separate array.

For an extra challenge, try putting in a computer controlled player. This is not really all that hard. The easiest one to write is virtually impossible for a human to beat because it has perfect reflexes. Throw in occasional random turns to make things more even.

4. A step up from the Tron Light Cycles game is the far more common Snake game. In this game a single person controls a snake that goes around eating apples or something else. Each time one is eaten, the snake grows longer. This continues until the snake runs into itself or an edge of the screen.

The reason this is more challenging than Light Cycles is that the cycles leave walls that can stay until the program is done. The walls can also be a single pixel wide. With Snake, the body parts that follow the head have to "move" because the snake does not simply get longer all the time.

5. One of the early educational programming languages, called Logo, made graphics easy to use by implementing a turtle graphics system (`http://en.wikipedia.org/wiki/Turtle_graphics`). The idea of turtle graphics is that you have a cursor, typically called the turtle, that has a position, an orientation, and pen settings. The turtle can turn and move. When it moves, it can either draw or not. Simple systems only allow the turtle to move along straight lines and that is what you will do for this project.

A simple way to encode instructions for a turtle is with a `String`. Different characters tell the turtle to do different things. An 'F' tells the turtle to move forward while drawing. An 'f' tells it to move forward without drawing. The '+' and '-' characters tell it to turn to the left and right, respectively. Other characters can be added to give the system more power and later projects will give you the opportunity to do so. The amount that the turtle moves or turns for each character is considered to be a fixed parameter. Using this system, one could draw a square by setting the angle to 90 degrees and using the string "F+F+F+F". Two squares that are separated could be made with "F+F+F+FfF+F+F+F".

Make a GUI that has a `TextField` where the user can enter a `String`. You might also have fields for segment length and turn angle. You should draw the appropriate set of lines. Simply ignore any characters that are not 'F', 'f', '+', or '-'.

6. For this project you will model the process of crystal growth. This might sound like something of a challenge, but it is not really all that hard. Crystals grow when material dissolved in a solution meets a surface of the crystal and sticks. The dissolved material moves around due to Brownian motion and is basically a random walk. You start the process by putting a seed crystal in the solution for stuff to stick to.

For our purposes, the crystal is simply represented as one color on top of a background that is another color. Use a `WritableImage` to store this so that you can get and set pixels. If the user clicks on the panel it should add a new "seed" at the click location (simply set the color at that point in the image to the crystal color).

There should be either a `Button` or a menu option to release more particles. When a particle is released, it should start at one of the edges. You just need to keep track of the x, y location of it. Using a `while` loop, have the particle move randomly around until the move would put it on top of a spot that already has crystal. At that point you change the pixel at the location the particle had been to the crystal color.

To move the particle randomly you could use `util.Random.nextInt(4)` to get a number in the 0-3 range and move either up, down, left, or right depending on that value. If a move would put a particle outside of the image, simply ignore that move. The menu option should probably run through a loop that drops 100 or so particles and lets each run until it sticks. You only need to `repaint` after all have found their place.

Note that especially early on when the crystal seed is small, it can take a long time for a particle to run around until it hits that seed.

7. If you have been doing ray tracing options in earlier projects, it is time for you to finally see something. You can use the code you wrote for project 10.2 that makes a grid of parameter values and simply set colors in a `BufferedImage` based on the t intersect parameters. To do this you will probably want to build your own colors. You can make a new `Color` object by saying `new Color(r:Int, g:Int, b:Int)` where r, g, and b are in the 0-255 range. Values outside of that range will cause an error. Simply display the image in a `Panel` that you have put in the GUI to see depth information for your ray trace.

8. Write a simple 2-D game of your choosing with simple geometry or sprites (little images for characters). Your game should have a `Timer` to make actions occur as well and take input from the user in the form of key pressed or mouse movement. At this point you know enough to write many types of games, but it is suggested that you strive to keep things simple. Have movement and some collision checking.

9. This project has you write another simple "game" that you are likely familiar with that involves sliding tiles in a 4x4 grid where one piece is missing. The tiles might have numbers or parts of a picture. Your goal is to move them around by sliding pieces into the one open space until they are in the desired order.

   Write a GUI that shows the 4x4 grid. You can choose if you want to use numbers of parts of an image. If you want to use parts of an image, you can get the small sections by loading the full image, then drawing it into smaller images with offsets so that only the part you want for a given piece appears in the smaller image.

   The user should interact with the game by click on the piece that they want to slide. In the simplest implementation, they have to click on a square immediately adjacent to the open square. For a bit of an extra challenge you can make it so that they can click on any square in the same row or column as the empty one and multiple pieces can be moved over at the same time.

10. If you did project 11.8 then you already have a GUI set up for keeping track of what a person has in their pantry and checking on recipes. However, this GUI is fairly bland. Now that you know how to put in graphics you have the ability to make it more visually appealing. For this project you should add onto your previous GUI the ability to have images associated with ingredients and recipes. You should also add in directions for how to make the recipe in the GUI.

   A basic implementation should specify a file name for images. Image files are to be copied into the directory that the script is run from. If you want a bit more challenge, make your program take a `URL` to load in an image. Then the paths to images can be given as a full URL and loaded off the web. To take this a step further, let the user type in a URL the first time, then save that locally as an image. You can save the image using the `write` method from `ImageIO`. Details are in the API.

11. An alternative upgrade to project 11.8 is to make it so that it can help you organize your shopping trip. To make this happen, you need a second data file that tells you where items are located in the store. This can be as simple as listing rows that items are in. If you want to put some information for roughly where on an isle each item is, you can.

    You need to give the user the ability to build a shopping list using the items that are in the store file. They should be able to do this while looking at the current pantry contents, or at least flip between the two views quickly. The grocery list should have the ability to specify an item and an amount. The user should be able to say that they are buying that list and have those items added to the pantry.

    The row information is used to build a simple graphical representation of the store. You can draw rectangles for rows and put labeled dots for the items in each row. The details are left up to you.

12. For this option, you can add a graphical representation to the schedule that is built in project 11.10. The goal is to have what looks like a fairly standard "week view" of the schedule. You will have a column for each day, and the courses should be displayed as labeled rectangles that span the section of the day that course would occupy. This makes it easier to see when there are overly long blocks of class time or too many small breaks spread around.

    How you specify times is left up to you. In the previous project in this line, it was suggested that each time in the schedule be given a unique integer number. To stay with that, you need to hard code the times and days for those time slots. The alternative is to add functionality so that the user can enter specific times and days. Taking that route will require more complex code in the schedule building to make sure that no two courses are scheduled for the same time.

13. The Mandelbrot set is a famous fractal in the complex plane discovered by Benoit Mandelbrot. A web search will provide you with lots of information on this beautiful structure. Part of what is so remarkably about the Mandelbrot set is that it has such a simple definition, but contains infinite complexities. The set is defined as the points, c, in the complex plane for which the equation $z_{n+1} = z_n^2 + c$, where $z_0 = 0$, gives a bounded sequence.

    In practice, if the value of $|z|$ ever goes beyond 4, the sequence will diverge. So programs to explore the Mandelbrot set count the number of iterations to reach that point and assign a color based on how many iterations it takes. Because the points in the set itself will never diverge, there has to be some maximum number of iterations at which point you simply say it is in the set.

    For this project you will write a GUI program that displays the Mandelbrot set. You have a view on the set with minimum and maximum values in the real and complex axes. Allow the user to zoom in by clicking.

14. Most applications that allow you to keep track of your music, like that written for project 11.7, will show cover art for albums. Put that feature into your solution so that when albums (or possibly songs) are selected, the cover art for the album is displayed. Cover art can be stored in the same directory as the script and data file or you can use URLs.

15. Most interactive box scores do more than just present the statistics in the way you did for project 11.12. They also use graphics to display things like team logos and photos of players. For this project you should extend what you did for project 11.12 so that it displays this information. You should also make it interactive so that the images of things like players can be changed by user actions. For example, you could listen to `MouseEntered` events on labels of their names. If the box scores involve different teams, the logo displays should change with what teams are in the current box score.

Additional exercises and projects, along with data files, are available on the book's web site.

Chapter 13

Sorting and Searching

Some of the most fundamental things that we do on computers are to order data according to certain values and to search through data for various things. These activities are known as SORTING and SEARCHING, respectively. Due to their importance in solving many different types of problems, a lot of work has gone into doing them as quickly and efficiently as possible. In this chapter we will start your exploration of the topics by working through some of the simpler approaches.

13.1   Basic Comparison Sorts

There are many different ways in which to sort data. The most general ones do not have many requirements on the nature of the data. All they require is the ability to compare elements to see if one should come before another. These comparison sorts are the ones that are used most frequently, mainly because they are so flexible. There are limits on how fast they can be because they are not customized to the data, but typically their generality trumps speed.

At this point we are not even going to consider the fastest comparison based sorts. Instead, we will start with some of the conceptually simpler sorts. Before we do that, you should take a second to think about how you sort things as a human. There are some different scenarios you could consider. When you were younger, odds are that you were given assignments in school where you had to sort a list of words into alphabetical order. What approach did you take to doing that? Could you write down a step-by-step procedure to explain to a small child who was doing it for the first time how he/she might go about it?

The procedure that you use might vary depending on what you have to work with. In the list of words, you might need to write it down as a sorted list, and you do not want to have to erase and move things around if at all possible. You might be sorting folders in a filing cabinet. The folders can slide backward and forward. They do not have any set position, and moving a folder is not all that hard to do. On the other hand, imagine the silly situation of having to sort cars in a parking lot by their license plates. The cars do not move easily. Each move takes effort, and they certainly do not slide around.

Take some time to consider how you would sort in these different scenarios. How does the type of object impact your approach? Space can also be an issue. If you are sorting folders in a big room you might pick a different approach than if you are in a small closet with a file cabinet. For the cars, you might pick one approach if you are moving the cars from one long row to a completely different row than you would if there was only one open space for you to park cars in as you move things around.

The sorts that we are going to work with here not only work with a basic comparison operation, they are also intended to work with arrays and can work IN-PLACE. This implies that they can be done using just a single array. You do not have to make a copy. Not all sorts have this property. So this is like the parking lot with only one open space.

13.1.1 Bubble Sort

The first sort that we will describe is something of a classic in computer science. It is not efficient. In fact, you would be hard pressed to convince any person to use this sort method when sorting things by hand. It works for computers because they will do whatever you tell them, even if it is extremely repetitive. It is only taught broadly because it is so simple to describe and code.

The basic idea of the BUBBLE SORT is that you want to run through the array and look at items that are next to one another. If two items are out of order, you swap them. One pass through the array will not get it sorted unless it was very close to sorted to begin with. However, if you repeat this process over and over, the array will eventually get to a situation where it is sorted. Using this description, there are a few ways in which the bubble sort can be written. They all have an inner loop that looks something like the following.

```scala
for (i <- 0 until a.length-1) {
  if (a(i) > a(i+1)) {
    val tmp = a(i)
    a(i) = a(i+1)
    a(i+1) = tmp
  }
}
```

The end index of this loop can change over different iterations to make it more efficient, but the idea is the same. The value i runs through the array comparing each element to the one after it. If the first is larger than the second, then they do a swap.

The swap code is the three lines inside of the `if`. To picture what this is doing, imagine the analogy of cars in a parking lot. The `tmp` variable is an extra space where you can move one car. So you start by moving the first car to the empty space. After that you move the second car into the space the first one had been in. To finish it off, you pull the car from the extra space into the second spot. The car analogy is not perfect. When we move a car from one spot to another, the spot that is vacated is left empty. When we do an assignment, the variable or `Array` location we are getting the value from does not lose the value. Instead, we have two memory locations that both have references to the value. That would be like having a second version of the car. Much of the time this is only a technicality, but there are times when it can be significant as those extra references can hold onto memory that you really are not using anymore.

In order to do a full sort, this loop needs to be repeated over and over again. How many times must it be repeated? To answer that we can observe that each time through, the largest unsorted value is pushed all the way to where it belongs. If there are n elements in the `Array`, then after $n - 1$ passes everything will certainly be in place.[1] Using this logic, we can write the following code.

```scala
def bubbleSort(a:Array[Double]):Unit = {
  for (j <- 0 until a.length-1) {
    for (i <- 0 until a.length-1-j) {
      if (a(i) > a(i+1)) {
        val tmp = a(i)
        a(i) = a(i+1)
        a(i+1) = tmp
      }
    }
  }
}
```

In this code, the outer loop executes $n - 1$ times as the value j counts up from 0 to $n - 2$. You will notice one other change in the inner loop. The end value for i has been changed from `a.length-1` to `a.length-1-j`. This is based on the observation that after one pass, the largest element is in the correct location at the end. After two passes, the second largest element is also in the correct place and so on. For this reason, the inner loop can stop one spot earlier each time through because there is no need to check those values that we already know are sorted. The sort would still work fine without subtracting off j, but it would do roughly twice as much work.

The other thing that this full sort includes that was not apparent from just the inner loop is that this function has been written to sort an `Array` of `Doubles`. As written, this will not work if you pass in anything other than an `Array[Double]`. If you need to sort something else, you have to write a different sort. There are many ways in which this is less than ideal, especially considering that everything except the type declaration would work equally well for `Array[Int]`, `Array[String]`, or an `Array` of anything else that works with greater than. The approach to creating code that can sort multiple different types is covered in our second semester book, *Object-Orientation, Abstraction, and Data-Structures Using Scala*[1].

Technically, this version of bubble sort might get the data sorted and then keep working on it. Smaller elements only move forward one slot on each pass. If the data is not too far out of order so that nothing has to move all that far forward, the sort might get the data in order before the $n - 1$ iterations are complete. For this reason, you might want to use

---

[1]We do not have to put n things in place because if there are only n places and $n - 1$ are in the right place, the last one must also be in the right place as it is the only one left.

a FLAGGED BUBBLE SORT. This is a small variation on the normal bubble sort that uses a `while` loop for the outer loop. It exits after $n - 1$ iterations or if it goes through a pass of the inner loop without doing any swaps as that implies that everything is already sorted.

```scala
def flaggedBubbleSort(a:Array[Double]):Unit = {
  var flip = true
  var j = 0
  while (flip && j < a.length-1) {
    flip = false
    for (i <- 0 until a.length-1-j) {
      if (a(i) > a(i+1)) {
        val tmp = a(i)
        a(i) = a(i+1)
        a(i+1) = tmp
        flip = true
      }
    }
    j += 1
  }
}
```

This code adds a few more lines to deal with the flag and the fact that we have to do the counting ourselves with a `while` loop. The flag is called `flip`, and it starts off as `true`. However, before the inner loop starts it is always set to `false` and only gets set back to `true` when a swap is done.

13.1.2 Selection Sort (Min/Max Sort)

The next sort we want to consider is one that humans are far more likely to employ, especially in a situation like sorting the cars where moving cars around takes a significant amount of effort. The sort is the SELECTION SORT, and it is called this because it runs through the values and selects one element and places it where it belongs. In order for this to work you have to know where the value should go. This is easy if the value is either the largest or the smallest in the collection. As such, the selection sort is typically implemented as either a minimum sort (MIN SORT) or a maximum sort (MAX SORT). In theory one can also write a min-max sort that picks both and moves them to where they belong.

We will write a min sort here. It is fairly trivial to convert this to a max sort. The inner loop of the min sort will run through the currently unsorted part of the collection and find the minimum element in that. If that element is not already where it should be, it is swapped into place. For an `Array`, the fact that this is a swap is significant. To understand this, think about the cars. Shifting all the cars over one is hard. Swapping two is much easier as it only requires moving cars three times. The `Array` is like the parking lot with fixed locations. Certain other data structures could act more similarly to file folders, which slide around easily and might allow shifting. If you use an `Array` and a shift instead of a swap, the code you produce will be both longer and significantly slower.

```scala
def minSort(a:Array[Double]):Unit = {
  for (j <- 0 until a.length-1) {
    var min = j
    for (i <- j+1 until a.length) {
      if (a(i) < a(min)) min = i
    }
    if (min != j) {
```

```
      val tmp = a(j)
      a(j) = a(min)
      a(min) = tmp
    }
  }
}
```

As with the standard bubble sort, the outer loop happens $n - 1$ times as it has to swap $n - 1$ elements into the locations they belong. The contents of the inner loop here are quite short, just a check that changes the `min` variable if needed. The swap itself is longer than the loop.

The selection sort is not really more efficient than bubble sort in general. However, because the swap is outside of the inner loop, this sort can be much more efficient in those situations where the swaps are expensive. This situation cannot happen for normal `Arrays` in Scala, but it can be true for `Arrays` in some other languages or if the values being sorted are contained in a file instead of the memory of the computer.

13.1.3 Insertion Sort

The last sort we will discuss in this section is another one that humans might consider using. It is called the INSERTION SORT because the way it works is to build up a sorted group of elements at the beginning of the `Array` and each new element is inserted into the proper location in that group. The advantage of this sort is that it can do less checking than the other two. Once it finds the place to put the element it can stop and move to the next element. It is particularly efficient if the `Array` is fairly close to sorted, as each element typically does not have to be moved very far.

The code for an insertion sort can be written in the following way.

```
def insertionSort(a:Array[Double]):Unit = {
  for (j <- 1 until a.length) {
    var i = j-1
    val tmp = a(j)
    while (i >= 0 && a(i) > tmp) {
      a(i+1) = a(i)
      i -= 1
    }
    a(i+1) = tmp
  }
}
```

The outer loop here still executes $n - 1$ times, but in a different way and for a different reason. Instead of starting at 0 and stopping one short of the last element, this loop starts at 1 and goes to the last element of the `Array`. The reason for this is that an `Array` of one element can be viewed as a sorted `Array`. So we take the first element as being where it should be in a group of 1. Then we start with the second element and potentially move it forward if needed.

The code inside the outer loop needs some explaining as well, as it is a bit different from what we saw with the other two sorts. It begins with the declaration of the variable i which begins at j-1. This is the index of the next element that we need to compare to the one we are moving. We also declare a temporary variable called `tmp` that stores the value in the `Array` that we are moving forward. Having the temporary variable here prevents us from doing full swaps and makes the code both shorter and faster. To understand this, consider again the example of moving cars in a parking lot. Consider you have ten cars, the first eight of which are already sorted. You need to move the ninth one forward to the place it belongs, potentially several spots down. You would not do this by swapping cars one after the other. Instead, you would pull that car out and put it in the one empty spot to the side. Then you would move the sorted cars down, one at a time, until you had cleared out the proper spot. At that point you would drive the car from the spot on the side into the correct position.

That mental image is exactly what this code is doing. The `while` loop executes as long as we have not gotten down to the beginning of the `Array` and the element we are considering is greater than the one we are inserting. As long as that is true, we move the element up and shift our attention down to the next element. When we are done, we put the temporary value where it belongs.

13.1.4 Testing and Verifying Sorts

The last three subsections presented sorts as working products. In reality, sorts are like any other code and have to be tested. Fortunately, sorts are fairly easy to test. After the sort has completed, you should be able to run through the elements and check that each element is less than or equal to the one after it. As long as this is true for all of the elements, the array is properly sorted.[2] An imperative approach to doing this might look like the following.

```scala
def isSorted(a:Array[Double]):Boolean = {
  for (i <- 0 until a.length-1) {
    if (a(i) > a(i+1)) return false
  }
  true
}
```

This code does something that we have not seen before. It uses a `return` statement. We have seen that normally, the last statement in a function should be an expression and the value of that expression is what the function returns. It is also possible to force a function to return earlier using a `return` statement as is done here. If at any point we find consecutive elements that are out of order, there is a forced return of `false`. If it gets all the way through the array it returns `true`.

This style is frowned on by some because it complicates things for those reading the code. Normally when you see a `for` loop, you know that it will run through the entire collection, and the code will exit when it gets to the end of the collection. The `return` statement forces the `for` loop to exit earlier and as such, forces the reader to look more closely to figure out what is going on here. Code that behaves in slightly unexpected ways like this is a common source of errors, especially if the code gets modified later on.

You could get around having the `return` statement by using a `while` loop. This would remove the requirement of a `return` statement, but it does not really make the code more functional or easier to read. If anything, it might make it a bit harder to read and deal with. An alternative to this is to use the `forall` method of collections. The only challenge in this is picking a way to compare sequential elements using `forall`, as that method only works on one argument at a time. One way to get around this is to use `forall` on a `Range` that is the index. This is logically very similar to the original `for` loop except that your part of the code is completely functional.

---

[2]This does not check that all the original elements are present in the sorted array. That check is harder to do, but would be required for a complete check of correctness as it is not uncommon to create bugs where some values are lost and others get duplicated.

```
def isSorted(a:Array[Double]):Boolean = {
  (0 until a.length-1).forall(i => a(i) <= a(i+1))
}
```

The fact that random access is a fast operation on an **Array** allows this code to be efficient. A second approach, which is also functional, is to use **forall** on a **zipped** tuple as is shown here.

```
def isSorted(a:Array[Double]):Boolean = {
  (a,a.view.tail).zipped.forall(_ <= _)
}
```

This code is a bit shorter, but significantly more advanced. In fact, it uses advanced topics from sections 8.4 and 10.4. The approach here is to make a tuple that contains the original elements first and everything but the first element second. If we **zipped** these two collections with the **zip** method, the result would be a sequence of tuples with each element and the one after it in a tuple and the first and last element appear only once in the first and last tuples. That **zipped** collection could then be dealt with using **forall**. Such a solution would be very inefficient, however, as it would actually construct a lot of new objects and do a lot of copying. The advanced topics of **view** and **zipped** are used to make it more efficient.

The first advanced topic is the **view** from section 8.4. A **view** is a representation of a collection that does not make a full copy unless forced to do so. In this code, taking a **tail** of a normal **Array** would produce a whole new **Array** and copy a lot of data over into it. However, the **tail** of the **view** is just a small object that will interact with code the same way the **tail** would, but through logic instead of making a real copy.

The second advanced topic is the use of the tuple **zipped** method from section 10.4. This is a method of the 2 and 3-tuple types that allows you to more efficiently run common functions that need to go through two or three collections at the same time. Instead of making a whole new **zipped** collection of tuples, this gives you back an object that has methods like **map**, **filter**, and **forall** declared such that they take as many arguments as are in the tuple. So we get a version of **forall** which needs a function of two arguments we can easily provide using the shorthand syntax.

Whichever approach you choose, you can test a particular sort with code like the following.

```
val nums = Array.fill(args(0).toInt)(math.random)
flaggedBubbleSort(nums)
assert(isSorted(nums))
```

This code creates an **Array** of random **Doubles** with a length determined by a command-line argument. It then calls a sort on that **Array** and uses the assert function to make sure it is true. Assert is a standard function in Scala that can be called with one or two arguments. The first argument is a **Boolean** that should be **true**. If it is not **true**, an error results, terminating the program. A second argument can be provided that is passed by-name and should provide a useful message if the **assert** fails.

The **assert** function can be used generally in your code. There is also a **require** function that performs similarly, but should be used for argument values. So at the top of a function, if the function will only work on certain argument values, you can use a call to **require** to make it so the code fails in an informative manner when bad arguments are passed in.

13.1.5 Sort Visualization

It can help you understand sorts if you get to see them in action. There are many different places on the web that have animations/videos that will show you how these and other sorts work. One advantage of having already studied graphics is that we can write our own code to visualize what the sorts are doing. This is part of why the sorts above were set up to work with the Double type. It makes it easy to generate with the call to math.random and to draw because the random numbers are uniformly distributed between 0.0 and 1.0.

The following code is a full listing of a program that will show you the three sorts above on random data. When you click one of the buttons, it will generate a random array and begin sorting it. While it is sorting, it draws a representation of the array using black lines. Taller lines indicate larger values in the array.

Listing 13.1: SortVis.scala

```scala
import scalafx.Includes._
import scalafx.application.{JFXApp, Platform}
import scalafx.scene.Scene
import scalafx.scene.canvas._
import scalafx.scene.control.{Button, Slider}
import scalafx.scene.layout.FlowPane
import scalafx.scene.paint.Color
import scalafx.event.ActionEvent
import scala.concurrent.Future
import scala.concurrent.ExecutionContext.Implicits.global

val numToSort = if (args.length>0) args(0).toInt else 300
val drawHeight = 300

def renderValues(gc:GraphicsContext, a:Array[Double], i:Int, min:Int):Unit = {
  gc.clearRect(0, 0, a.length, drawHeight)
  gc.stroke = Color.Black
  for (j <- a.indices) {
    gc.strokeLine(j,drawHeight*(1.0-a(j)),j,drawHeight)
  }
  gc.stroke = Color.Green
  gc.strokeLine(i, 0, i, 10)
  gc.stroke = Color.Blue
  gc.strokeLine(min, 0, min, 10)
}

def bubbleSortVis(gc:GraphicsContext, a:Array[Double], delay:Int) = {
  for (j <- 0 until a.length-1) {
    for (i <- 0 until a.length-1-j) {
      if (a(i) > a(i+1)) {
        val tmp = a(i)
        a(i) = a(i+1)
        a(i+1) = tmp
      }
      Platform.runLater(renderValues(gc, a, i, -1))
      Thread.sleep(delay)
    }
  }
}

def minSortVis(gc:GraphicsContext, a:Array[Double], delay:Int):Unit = {
```

```scala
42    for (j <- 0 until a.length-1) {
43      var min = j
44      for (i <- j+1 until a.length) {
45        if (a(i) < a(min)) min = i
46        Platform.runLater(renderValues(gc, a, i, min))
47        Thread.sleep(delay)
48      }
49      if (min != j) {
50        val tmp = a(j)
51        a(j) = a(min)
52        a(min) = tmp
53      }
54    }
55  }
56
57  def insertionSortVis(gc:GraphicsContext, a:Array[Double], delay:Int):Unit = {
58    for (j <- 1 until a.length) {
59      var i = j-1
60      val tmp = a(j)
61      while (i >= 0 && a(i) > tmp) {
62        a(i+1) = a(i)
63        i -= 1
64        Platform.runLater(renderValues(gc, a, i, -1))
65        Thread.sleep(delay)
66      }
67      a(i+1) = tmp
68    }
69  }
70
71  val app = new JFXApp {
72    stage = new JFXApp.PrimaryStage {
73      title = "Sorts"
74      scene = new Scene(numToSort, drawHeight+50) {
75        val canvas = new Canvas(numToSort, drawHeight)
76        val gc = canvas.graphicsContext2D
77
78        val slider = new Slider(1, 20, 1)
79        slider.layoutY = drawHeight+30
80        slider.prefWidth = numToSort
81
82        val bubble = new Button("Bubble Sort")
83        bubble.onAction = (e:ActionEvent) => Future {
84          bubbleSortVis(gc, Array.fill(numToSort)(math.random), slider.value.toInt)
85        }
86        val minSort = new Button("Min Sort")
87        minSort.onAction = (e:ActionEvent) => Future {
88          minSortVis(gc, Array.fill(numToSort)(math.random), slider.value.toInt)
89        }
90        val insertion = new Button("Insertion Sort")
91        insertion.onAction = (e:ActionEvent) => Future {
92          insertionSortVis(gc, Array.fill(numToSort)(math.random), slider.value.toInt)
93        }
94
95        val flow = new FlowPane
96        flow.children = List(bubble, minSort, insertion)
```

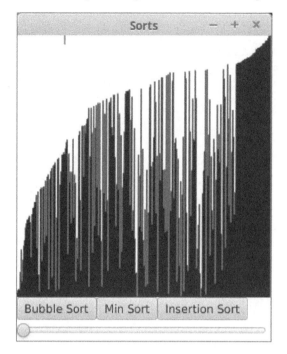

FIGURE 13.1: The window that is produced by running SortVis.scala after the "Bubble Sort" button has been clicked.

```
97          flow.layoutY = drawHeight
98
99          content = List(canvas, flow, slider)
100       }
101     }
102 }
103
104 app.main(args)
```

Figure 13.1 shows this code part way through a bubble sort.

The program allows the user to specify how many numbers are going to be sorted on the command line. If they do not give any value, 300 will be used. We also define a constant on line 13 for the height that things will be drawn so that the number 300 does not become a MAGIC NUMBER in the code. A magic number is a number that appears multiple times in the code whose meaning is not always completely clear. Magic numbers make code harder to read and understand. They also make it harder to modify and maintain. In this case, 300 appears twice with different meanings. If we did not define `drawHeight` and just used 300 instead of `numToSort`, it would be quite challenging to alter either one without messing up at least one instance of the other.

Lines 15 to 25 define the function that actually draws out our array to a `GraphicsContext`. The array values are drawn as black lines. There is also a green segment drawn at the top to show where the algorithm is working in the away. For the min sort there is a blue line that can show the location of the current minimum.

Lines 27 to 69 redefine our bubble sort, min sort, and insertion sort. They each take two different arguments that are needed for drawing. At the end of the inner loop of each of these is a call to `renderValues` followed by a call to `Thread.sleep`, which pauses the sort of rendering and allows you to adjust the speed. Note that the calls to `renderValue` are done in a call to `Platform.runLater`. That is because the sorting has to be done in a separate thread. As was mentioned in the last chapter, ScalaFX uses only a single thread for event handling and rendering. If the sort were done with that, nothing could be drawn until the sort was completed. On the other hand, the drawing needs to happen in that ScalaFX thread. The use of `Platform.runLater` makes that happen.

The actual GUI is set up on lines 71 to 102. There is a `Canvas` at the top. Below that are three buttons, one for each sort, and a `Slider`. The buttons are placed in a `FlowPane` to line them up, and the `Slider` is placed below that at the bottom of the window. When the buttons are clicked the appropriate sort is run with one nuance. Note that the call is in a block that has a call to `Future`. This is what causes the action to occur in a separate thread so it does not block the event and drawing thread. A full discussion of threads is beyond our scope here, but you should be aware that if you are going to do something that would take an extended period of time, or which needs to be animated in a button click or other event handler, this is a good way to make it happen.

13.1.6 Order Analysis

We said previously that the sorts presented above are not the most efficient sorts. These sorts are presented first because they are fairly simple and straightforward. However, it is a reasonable question to ask how "fast" they really are. This is a challenging question to answer because speed can depend on many different things including the computer you are running on and exactly what you are sorting. To address this, algorithms are generally described in terms of their "order" in a particular operation. The order is a rough functional description of how the number of operations scales with the size of the input. There are many mathematical details to proper order analysis, but for now we will use the term fairly loosely.

So the idea of order analysis is to pick one or more operations that are particularly important to the algorithm and try to find a function that models how the number of those operations changes with the size of the input. In the case of sorts, there are two operations that are particularly significant, comparisons and assignments. Most of the time we worry about comparisons and for the code we have written here, that is appropriate.

So how many comparisons does the full `bubbleSort` do on an `Array` with n elements? The first pass through the loop does $n - 1$ comparisons. The second pass does $n - 2$. The third does $n - 3$ and so on. In math terms we can write the number of comparisons in this way:

$$C = \sum_{i=1}^{n-1} i = \frac{n(n-1)}{2} = \frac{1}{2}n^2 - \frac{1}{2}n.$$

This gives the exact number of comparisons between elements that will be done. It happens to be exactly the same for the selection sort. We typically refer to this as $O(n^2)$. For order analysis we generally ignore coefficients and everything except the highest power. This does not provide a perfect picture of performance, but it works very well in general, especially when you talk about larger inputs.

The most important thing to note about an $O(n^2)$ algorithm is that if you double the number of elements, there will be four times as many comparisons done. As a result, you can generally expect the sort to take four times longer on twice as much stuff. The nature of the scaling is what we often care about the most. Different computers might be faster or slower, but just knowing this scaling and a run time for a problem of a certain size gives you a good feel for what the run time will be if you change the size of the problem.[3]

The number of comparisons in the selection sort and the bubble sort depends only on the size of the data set being sorted. It does not depend on the nature of the data. This is not the case for the flagged bubble sort or the insertion sort. These sorts can do very different numbers of comparisons depending on the data that is given to them. For example, if you give a flagged bubble sort an array that is already sorted, it will only run through it once, doing $n-1$ comparisons. We would call that $O(n)$. Granted, that is a BEST-CASE scenario. The best-case is rarely of interest. Instead, we typically worry about AVERAGE-CASE and WORST-CASE situations.[4]

So what are the average and worst-case behaviors of insertion sort? The outer loop clearly happens $n-1$ times. So we can do a summation much like what we did above. However, the inner loop happens a variable number of times. In the worst-case, which would be getting data in reverse order, the inner loop will run i times, and we get exactly the same behavior as with the bubble and selection sorts. In the best-case, which is data in the proper order, the inner loop does only one check. This gives linear performance, $O(n)$, because

$$C = \sum_{i=1}^{n-1} 1 = n - 1.$$

On average we expect something half way between these extremes, we expect the inner loop to do $i/2$ comparisons. This gives

$$C = \sum_{i=1}^{n-1} \frac{i}{2} = \frac{n(n-1)}{4} = \frac{1}{4}n^2 - \frac{1}{4}n.$$

While this is better than the behavior of the other sorts by a factor of two, it is still $O(n^2)$. This is because doubling the size of the input still makes it take four times longer.

So how good or bad is $O(n^2)$? Functions like this that grow as a polynomial function are referred to in a very general way as being tractable, because you can use them for fairly large values of n and still expect the calculation to finish in a reasonable time. We will see later on that not all algorithms have this behavior. However, $O(n^2)$ is not great if the value of n gets extremely large. We have no problem using these sorts for `Arrays` with 100 or 1000 values. However, as you continue to add zeros, these methods will prove to be too slow. Each factor of 10 in the size causes the program to do 100 times as many comparisons and generally this leads to it taking 100 times longer to complete. As such, these methods become slow if n gets bigger than about a million. Fortunately, there are alternatives.

13.1.7   Shell Sort (Diminishing Gap Sort)

The first alternative sort we will look at is the SHELL SORT, also called the diminishing gap sort. This sort was first proposed in 1959 by Donald Shell, and it was one of the first general sort algorithms developed that is faster than $O(n^2)$, though some minor changes had to be made in order to get that performance for the worst case. The basic idea of this sort is one that might seem a bit counter intuitive. It performs an insertion sort repeatedly on different subsets of the full array. To start with, the subsets are taken to be groups of

---

[3]Real hardware can break down these scaling arguments at certain critical points. For example, if the input set becomes larger than the cache of the machine, you will typically see a slowdown that does not scale quite as $O(n^2)$.

[4]The exception to this is if we happen to know that we will very often have inputs that produce best-case behavior.

elements that are widely spaced. The spacing between the elements in each subset is called the "gap". As the alternate name implies, the gap is then decreased in size for each run through the sort.

The counter intuitive aspect of this sort is that performing insertion sort multiple times should sort the array faster than doing it outright. This can work because a sort with a smaller gap size maintains the ordering at the larger gap size so work is not undone, and the insertion sort is very efficient on partially ordered data. The sorts with large gaps do very little work compared to the smaller gaps because they do not contain many elements, but once you get to the smaller gaps the data is mostly sorted so you get close to the best-case performance of the insertion sort. The sort always ends with a gap size of 1, which is doing a normal insertion sort, so you know that the result is fully sorted.

The only trick to the Shell sort is figuring out what to do about the gap. The initial suggestion was to start at half the size of the `Array` and decrease it by factors of 2. This often works well, but spacings that are factors of 2 apart will keep even and odd subsets separate until you get down to a gap of 1 and in certain cases that can actually lead to $O(n^2)$ behavior. For this reason, we use a slightly different factor where we divide by 2.2 instead of 2.

```
def shellSort(a:Array[Double]):Unit = {
  var gap = a.length/2
  while (gap >= 1) {
    for (j <- gap until a.length) {
      var i = j-gap
      val tmp = a(j)
      while (i >= 0 && a(i) > tmp) {
        a(i+gap) = a(i)
        i -= gap
      }
      a(i+gap) = tmp
    }
    gap = (gap/2.2).round.toInt
  }
}
```

This code was created by defining the `var gap` and writing the `while` loop, then cutting and pasting the insertion sort inside of the `while` loop. Once there, all the places where i was incremented or decremented by 1 were changed so the 1 was `gap`. The factor of 2.2 gives better spacings though you do need to `round` and convert back to an `Int`. You could put this into the visualization code and add the extra arguments and calls to `renderValues` to see how it works. Of the sorts we have discussed so far, this one is probably the most important to visualize to get a grasp on what it is doing and why the process of reducing the gap is significant.

The exact order of the Shell sort is harder to pin down. This is largely because it varies with the selection of gap scaling and can range from $O(n^2)$ at the poor end all the way to $O(n \log^2 n)$ at the good end. Reasonable implementations like the one shown here will tend to give $O(n^{3/2})$ performance. This might not seem like a tremendous improvement, but if n gets large, the difference can be quite dramatic.

13.2   Searching

While sorting data is something that we do a fair bit with computers, searching for data is a far more common task. A lot of the things that you do with computers on a regular basis involve searching for information either just to retrieve it or to modify information related to what is found. This makes sense, as running through data is something that computers do particularly quickly and efficiently.

13.2.1   Sequential Search (Linear Search)

The most basic form of search is the SEQUENTIAL or LINEAR SEARCH. This involves running through data one element at a time and checking each one to see if it matches what you are looking for. If it finds what you are looking for, it either returns the data or an index where it can be found. If it gets to the end without finding it, it will return something to indicate that the data was not there. The following code is a linear search through an `Array` of `Int`s.

```scala
def linearSearch(a:Array[Int], value:Int):Int = {
  var i = 0
  while (i < a.length && a(i) != value) {
    i += 1
  }
  if (i >= a.length) -1 else i
}
```

This search returns the index of the first element in the `Array` whose value is the same as what was passed in. It does this with a `while` loop so that it can stop early if it is found.

The `if` statement is needed because the common idiom when returning an index is to return -1 when the value is not found. We can get rid of this conditional at the end if we count backwards.

```scala
def linearSearchForLast(a:Array[Int], value:Int):Int = {
  var i = a.length-1
  while (i >= 0 && a(i) != value) {
    i -= 1
  }
  i
}
```

This modified version of the code starts at the end and goes backwards. It is a bit simpler, but it fundamentally alters the description of the function as we now find the last occurrence instead of the first.

There are quite a few different methods on the Scala collections that do searching. For the collections that we have learned about, these are all performing linear searches. They include the following, roughly as they are defined in `Seq[A]`:[5]

- `def find(p: (A) => Boolean): Option[A]`

- `def indexOf(elem: A, from: Int): Int`

---

[5]The exact signatures of some of the methods have been simplified so that they make sense at this point in the book.

- `def indexOf(elem: A): Int`

- `def indexOfSlice(that: Seq[A], from: Int): Int`

- `def indexOfSlice(that: Seq[A]): Int`

- `def indexWhere(p: (A) => Boolean, from: Int): Int`

- `def indexWhere(p: (A) => Boolean): Int`

- `def lastIndexOf(elem: A, end: Int): Int`

- `def lastIndexOf(elem: A): Int`

- `def lastIndexWhere(p: (A) => Boolean, end: Int): Int`

- `def lastIndexWhere(p: (A) => Boolean): Int`

Many of these methods come in pairs where one of the two takes an extra `Int` argument for an index in the collection that it should start working from. Only the first method returns an element from the collection, all the others return indices. The one that does return an element wraps it in an `Option` so that if no match is found it can return `None`. If no match is found for the methods that return an index, they will return -1.

13.2.2 Binary Search

Even though computers are very fast, linear search is far from ideal, mainly because searching is something that is done so very frequently. If the data is not ordered in any way, linear search is your only option. To understand this, imagine you are handed a normal phone book and asked to find the person who has a given phone number. Due to the fact that phone books are not ordered by phone numbers, your only recourse is to go through each and every line and check the numbers against what you are looking for. In any reasonably sized city this is something that no human would actually undertake.

If this were the only way to look through a telephone book people would not bother to keep them.[6] However, people do keep telephone books because they rarely look things up by number. Instead, people normally look things up by name and the telephone book is sorted by name. This ordering of the elements can lead to much more efficient searches. You might not be able to write a good algorithm for how you really look things up in a telephone book, but we can consider your first step and use that as direction for writing an efficient algorithm.

Given a large phone book and a name, you will open it up and look at what is on the page you open to. Odds are good that you will not get exactly the right page. However, comparing what you are looking for to what is on the page gives you a significant piece of information. If what you are looking for comes earlier in the alphabet than what is on the page you will only look at other pages before that one. You basically throw everything after that page out of your search without even looking at it. Similarly, if what you are looking for comes after the page you have opened to you will only consider pages after the current one. You will generally repeat this process in a manner that is not easy to describe in an algorithm just because your method of picking pages might be impacted by things like the binding of the book and whether one page sticks out a bit further than another.

---

[6]Thanks to rapidly changing technology and remarkable changes in computer speed and usability, it is not clear people are bothering to keep phone books anyway.

The idea of looking at a location and only considering things before or after it based on a sorted order can be used to create fast searching algorithms. The most general of which is the BINARY SEARCH. In a binary search, you keep track of a range of elements that you are considering by two integer indexes. We will call them `start` and `end`. At any given time, you know that the value you are looking for, if it is present, will be at an index in the range $i \in [\texttt{start}, \texttt{end})$.[7] So if we are searching the whole `Array`, then initially start is 0 and end is the `length` of the `Array`. We consider the midpoint of the range, `mid=(start+end)/2`,[8] and check if the element at `mid` is what you are looking for. If it is, we return `mid`. Otherwise we check if what we are looking for is greater or less than the element at `mid` and cut down our range accordingly.

To begin with, we will present an imperative version of this algorithm that uses a `while` loop and several `var` declarations.

```scala
def binarySearch(a:Array[Int], value:Int):Int = {
  var start = 0
  var end = a.length
  var mid = (end+start)/2
  while (end > start && a(mid) != value) {
    if (value < a(mid)) {
      end = mid
    } else {
      start = mid+1
    }
    mid = (end+start)/2
  }
  if (end <= start) -1 else mid
}
```

The `while` loop continues as long as the range includes at least one element and the midpoint is not the value we want. Inside the loop, a check is performed to see if the midpoint is less than or greater than what we are looking for. If it is less, we set `end=mid`. This works because `end` is exclusive, and we have just verified that the element is not at `mid`. Otherwise, we set `start=mid+1`. The `start` is inclusive so we have to move it one element beyond the `mid`. When the loop is completed we return either the value of `mid` or -1 based on whether the element was found or not.

This version of the code is fairly straightforward, but there is a simpler approach. Binary search happens to be an algorithm that lends itself very nicely to implementation as a recursive algorithm. The following code shows what this might look like.

```scala
def binarySearchRecur(a:Array[Int], value:Int, start:Int, end:Int):Int = {
  if (end <= start) -1 else {
    val mid = (start+end)/2
    if (a(mid) == value) mid
    else if (value < a(mid)) binarySearchRecur(a, value, start, mid)
    else binarySearchRecur(a, value, mid+1, end)
  }
}
```

Clearly this code is shorter than what we had before. Most people would also find this code a bit easier to read than the imperative version. The only drawback is that the function has

---

[7]As a reminder, this notation implies that the range is inclusive for `start` and exclusive for `end`.

[8]For extremely large arrays this formula for the middle value is problematic as `start+end` might overflow an `Int`. You could use `start+(end-start)/2` instead if that is an issue.

n	$\sim \log_2 n$
1,000	10
1,000,000	20
1,000,000,000	30
1,000,000,000,000	40

TABLE 13.1: Table of approximate values of $\log_2 n$ as a function of n. We use the approximation that $2^{10} \approx 10^3$. The reality is that $2^{10} = 1024$, but this approximation is rather close and is a good one to keep in your head for quick approximations.

two extra arguments. The normal way to get around that is to provide a wrapper function that only has two arguments and have it call this version. An appropriate wrapper could be written this way.

```
def binarySearch(a:Array[Int],value:Int):Int =
  binarySearchRecur(a, value, 0, a.length)
```

The logic of the recursive version is identical to the iterative version. Only the approach has changed.

Now that we have code to do a binary search, it is interesting to ask what order this function is. Again we say that the array has n elements in it. The worst case is the situation when the element is not found, and we get down to one element in the range that is not what we are looking for. So we need to figure out how many comparisons happen to narrow the range down from n to 1. After one comparison the range is cut to $n/2$. After two comparisons it is $n/4$. In general, after t comparisons, there are roughly $n/2^t$ elements left in the range.[9] We now have enough information to find the maximum number of comparisons.

$$n/2^t = 1$$
$$n = 2^t$$
$$t = \log_2 n$$

This is typically called $O(\log n)$ as the difference between logs of different bases is simply a constant multiplier. This order is generally considered to be quite fast, as it grows slowly as the input gets bigger.

To get a feel for this, let us look at a few examples of how $\log_2 n$ scales with n. A list of approximate values for this are given in table 13.1. To really put this in perspective, consider the fact that the first number is the worst case for a sequential search, and the second number is the worst case for a binary search. When n is small, the difference is not all that significant. However, as n gets large the cost savings of doing a binary search become quite apparent. The last two values of n are large enough that they pose a problem even for computers, despite their speed.

Of course, you can only use a binary search on sorted data and attempting an $O(n^2)$ sort on even a million items can be time consuming. So the real power of this $O(\log n)$ scaling is purely academic until we discuss some better ways to sort.

---

[9]This is only exact if n is a power of two. Otherwise, some rounding will occur, but that is a detail we can ignore when talking about the order.

Linear Binary Search (Advanced)

The binary search is ideal for general sorted data. However, if you happen to know that your data is fairly evenly distributed you can do even better. Instead of having `mid` be the midpoint of the range, you can place it where you would expect the `value` to be based on data being linearly distributed between the values at `start` and `end-1`. Assuming the `Array` contains `Doubles`, this could be done with a line like the following.

```
val mid = start+(((value-a(start)/(a(end-1)-a(start))*(end-1-start)).toInt
```

This picks the value of `mid` based on a linear approximation. Some care would have to be taken to insure that this does not fall into an infinite loop. In addition, if the data is not uniformly distributed, this approach can wind up begin much slower than the normal binary search. In fact, it can degrade to $O(n)$.

13.3 Sorting/Searching with `case classes`

In the sorts and searches that we have looked at, we were working with numeric types. More generally, the code that was written will work with any type that works with the comparison operators. As such, our code would have worked with `Array[Char]` or `Array[String]` if you simply altered the type that was passed into the sort because you can use < and > with the `Char` and `String` types. The `case class`es we have written do not meet this requirement. As such, we need to make other alterations to the code beyond the type if we want to sort a `case class`. Fortunately, these alterations are not all that significant.

We will work with the following case class.

```
case class Weather(id:String,year:Int,month:Int,precip:Double,tmax:Double,
  tmean:Double,tmin:Double)
```

This `case class` was created to store historical weather data. It was used to represent records for monthly data on temperature and precipitation. You would load an entire file of these records into an `Array`. If you want to see the hottest ten months you could sort the array by the high temperatures. Code for such a sort is shown here.

```
def bubbleSortWeatherHighTemps(a:Array[Weather]):Unit = {
  for (j <- 0 until a.length-1) {
    for (i <- 0 until a.length-1-j) {
      if (a(i).tmax > a(i+1).tmax) {
        val tmp = a(i)
        a(i) = a(i+1)
        a(i+1) = tmp
      }
    }
  }
}
```

A bubble sort was picked because of the simplicity of the sort. It is very easy to see what was changed. The only changes are the type in the parameter for the `Array` that is passed in and the fact that the comparison is done between fields of the `Array` elements. After applying this sort to the `Array` you can use `take` or `takeRight` to pull off the elements at one end or the other.

`case class`es also present another alternative that we did not really get with single values, the ability to have the comparison based on more than one field. For the weather data, it is likely to be stored in the data file in chronological order. Proper chronological order is a combination of both the year and month fields. If you wanted to have the ability to search for entries by time you might want to have a binary search that can look for a particular year and month. Code for doing that is listed here.

```
def binarySearchWeather(a:Array[Weather], year:Int, month:Int):Int = {
  var start = 0
  var end = a.length
  var mid = (end+start)/2
  while (end > start && (a(mid).year != year || a(mid).month != month)) {
    if (year < a(mid).year || (year == a(mid).year && month < a(mid).month)) {
      end = mid
    } else {
      start = mid+1
    }
    mid = (end+start)/2
  }
  if (end <= start) -1 else mid
}
```

Note that the comparison has become significantly more complex. It has to compare the year first, and then if there is a tie, break that tie with the month.

Unfortunately, when written in this way, we have to write a completely separate sort or search for each ordering we might want on the `case class`. For example, if you wanted wettest months instead of hottest months, you would need a separate sort. You might feel that copying a whole sort or search function only to make such small changes is not very efficient and that there should be a way to make a sort or search that works more generally. We can help improve these functions for `case class`es here. In *Object-Orientation, Abstraction, and Data Structures Using Scala*[1], we will gain the ability to abstract these ideas so that the function works with multiple types, not just different sort orders on a single type.

Back in section 5.4 we saw how we could make some recursive functions more powerful by passing functions into them. This approach is exactly what is taken by the higher order functions in the Scala collections libraries. As such, we have not had to go to it ourselves much since then. However, the desire to not write a completely new sort or search function for each and every possible ordering on a case class provides motivation to pull out these ideas again.

If you were to write a new version of bubble sort that sorts the `Weather` objects by precipitation, you would find that the only thing you change is the code related to the comparison of the elements. In order to get a sort that can sort by high temperature, precipitation, or anything else related to the `Weather` type, all we have to do is make it so that we can vary the comparison from the outside. This can be accomplished by passing in a function that does the comparison. For our sorts, we have been using less than and greater than for comparison so we just need to pass in a function that represents one of these. We pick less than here, though the code could easily be rewritten with greater than. The code for the sort after this change looks like the following.

```scala
def bubbleSortWeather(a:Array[Weather], lessThan:(Weather,Weather)=>Boolean):Unit
    = {
  for (j <- 0 until a.length-1) {
    for (i <- 0 until a.length-1-j) {
      if (lessThan(a(i+1), a(i))) {
        val tmp = (i)
        a(i) = a(i+1)
        a(i+1) = tmp
      }
    }
  }
}
```

The comparison operator is represented as a function that takes two `Weather` objects and returns a `Boolean`. A call to this function is used in the `if` statement.

Using this modified version, we could sort by the high temperatures with a call like this.

```scala
bubbleSortWeather(weather, (w1,w2)=>{w1.tmax < w2.tmax})
```

Alternately, the same method could be used to sort by precipitation with a call like this.

```scala
bubbleSortWeather(weather, (w1,w2)=>{w1.precip < w2.precip})
```

What is more, using this version of the sort, it is easy to change it so that it sorts from greatest to least instead of the standard least to greatest. In the case of precipitation this is done by changing the call in the following way.

```scala
bubbleSortWeather(weather, (w1,w2)=>{w1.precip > w2.precip})
```

All that is changed in the direction of the comparison operator. As you can see, this abstracted version that uses a function instead of a hard coded comparison operator is far more flexible.

Having seen the benefits we can get from using this in our sort, it would be nice to enable searching in the same way. If you start to write the search code with the same comparison operator you will find that there is a significant problem, the search requires more than just a less than or greater than. The search requires that we be able to tell if two things are equal. The standard way to deal with this is to have a function that returns an `Int` instead of a `Boolean`. The `Int` is negative, zero, or positive to represent less than, equal to, or greater than respectively. The other change that makes search a bit different is that the function only takes one `Weather` object because we just want to know where the thing we are looking for is relative to this value we are searching for.

Translating all of this into code gives the following.

```scala
def binarySearchWeather(a:Array[Weather],comp:(Weather)=>Int):Int = {
  var start = 0
  var end = a.length
  var mid = (end+start)/2
  var c = comp(a(mid))
  while (end > start && c != 0) {
    if (c < 0) {
      end = mid
    } else {
```

```
      start = mid+1
    }
    mid = (end+start)/2
    c = comp(a(mid))
  }
  if (end <= start) -1 else mid
}
```

The function is called `comp` and it takes a `Weather` and returns an `Int`. In order to prevent having the code call `comp` more times than it needs to, we introduce a variable named `c` that stores the result of the most recent comparison. We assign this in a code block in the condition. The loop checks to make sure it is not zero. The `if` checks if it is negative. After we calculate a new `mid` we also make a new comparison.

Now the question is, how could we use this search to find different elements in the `Array`. The first example we will give is one that duplicates the search we had above searching for a specific year and month.

```
binarySearchWeather(data, w => {
  if (w.year > 1896) -1
  else if (w.year < 1896) 1
  else if (w.month > 2) -1
  else if (w.month < 2) 1
  else 0
})
```

In this case, we are searching for February of 1896. This version uses a sequence of `if` expressions to determine if the element `w` comes before or after that time. With a little math we can make a shorter version taking advantage of the fact that there are 12 months in each year.

```
binarySearchWeather(data, w => (1896*12+2)-(w.year*12+w.month))
```

Searching for data based on date is made more complex because it depends on two values. If we use our flexible sort to reorder the `Array` by precipitation we could use the following to search for a month in which there was 1.43 inches of rain.

```
binarySearchWeather(data, w => {
  val diff = 1.43-w.precip
  if (diff > 1e-8) 1
  else if (diff < -1e-8) -1
  else 0
})
```

The logic in this can be challenging to write. Thankfully, the behavior of returning negative, zero, or positive is a common standard, and the Scala library contains code that does this with the built in numeric types. As such, the code can be expressed more simply in this way.

```
binarySearchWeather(data, 1.43 compare _.precip)
```

The `compare` call is actually a method we can call on a `Double`. The fact that a period has other meaning for numbers led to us using operator notation for the method instead of the more standard notation with dot and parentheses.

Searching for Doubles (Advanced)

Careful readers might have noticed that between the section on sorts and the section on searches, a small change was made. The sorting section used an `Array[Double]` while the searching section used `Array[Int]`. The choice of the `Double` type for sorting was motivated by the fact that they are easy to generate and visualize. However, they are not ideal for searching. This is due to the nature of the `Double` type. In our discussions we have only presented the `Double` type as the type we use for numbers when a fractional part is required. We have glossed over the details of what is happening with the `Double` type. To understand the reason that `Doubles` are not good for search algorithms requires us to dig a bit deeper into what they are.

The term `Double` stands for DOUBLE PRECISION FLOATING POINT NUMBER. The `Float` type, which we have generally ignored, is a SINGLE PRECISION FLOATING POINT NUMBER. The `Float` type can also represent fractional numbers, but it has a smaller range and lower precision. As with all numbers on a computer, floating point numbers, be they single or double precision, are really stored in binary. The best way to think about a floating point number is to think of numbers in normalized scientific notation. In decimal, you can think of a number in scientific notation as being in the following form,

$$(-1)^s * m * 10^e,$$

where $s \in [0, 1]$, $1 \le m < 10$ or $m = 0$. We call s the sign, m the mantissa, and e the exponent. Not much changes when the number goes to binary except that s, m, and e are stored in bits instead of decimal digits, and we want powers of 2 instead of 10. So scientific notation in binary would be

$$(-1)^s * m * 2^e.$$

This still does not explain why it is hard to do searching with a floating point number. The key to that comes from the fact that we only have a finite number of bits to use to store m. To understand the implication of this, consider a situation where you only have 7 decimal digits to write m. Now try to write the decimal form of the fraction 1/3. You would write $(-1)^0 * 3.333333 * 10^{-1}$. This is not exactly the same as 1/3, but it is as close as you can get with only 7 digits. The reality is that to write 1/3 in decimal you need an infinite number of digits. Lots of fractions require an infinite repeating representation in decimal.

This same thing happens in binary on the computer. Not having infinite binary digits can lead to some interesting results for factional numbers that cannot be perfectly represented. To understand this, consider the following simple example from the Scala REPL.

```scala
scala> 0.1 == 1.0 - 0.9
res0: Boolean = false
```

Mathematically you expect the expression `0.1 == 1.0 - 0.9` to be `true`, but the decimal value `0.1` is an infinite repeating sequence in binary. As such, it is truncated at some point and we get an approximation. Similarly, `0.9` cannot be represented perfectly either. The result is that subtracting `0.9` from 1 gives a value that is not exactly the same as the approximation to `0.1`. To see how different the two are we can subtract one from the other.

```
scala> 0.1 - (1.0 - 0.9)
res1: Double = 2.7755575615628914E-17
```

This is an extremely small number, but it is not zero, and because it is not zero, the two are not equal.

This is a well known challenge with floating point numbers and people who do numeric work have learned not to do checks for equality on them as the results are generally unpredictable because of the rounding that is part of even the simplest arithmetic. For our discussion, what we have seen is evidence that using == in a search algorithm on the Double type is likely to produce unexpected results. So the next question is, how do we get around that.

The basic idea behind the solution is that we generally consider floating point numbers to be equivalent as long as they are close enough to one another. So how close is close enough? That depends on whether you are working with single or double precision numbers. In math the Greek symbol ϵ, epsilon, is typically used to represent a vanishingly small value. In computer numerics it is used to describe the smallest value that you can add to one and still get a number greater than one. If you go smaller than ϵ, the value will be rounded off in the sum and all you will get back is one. Here is code that declares and calculates this for both the Double and Float types.

```
scala> val doubleEpsilon = {
     |     var eps = 1.0
     |     while (1.0+eps > 1.0) eps *= 0.5
     |     eps*2.0
     | }

doubleEpsilon: Double = 2.220446049250313E-16

scala> val floatEpsilon = {
     |     var eps = 1.0f
     |     while (1.0f+eps > 1.0f) eps *= 0.5f
     |     eps*2.0
     | }
floatEpsilon: Double = 1.1920928955078125E-7
```

Any single operation can be expected to have errors on the order of ϵ. When you string many operations together the error grows. As such, it is standard practice to consider two values equal if the relative error is less than the square root of ϵ. This means you only trust half the bits of precision in the number. As such, the following values are what you really find important.

```
scala> val sqrtDoubleEpsilon=math.sqrt(doubleEpsilon)
sqrtDoubleEpsilon: Double = 1.4901161193847656E-8

scala> val sqrtFloatEpsilon=math.sqrt(floatEpsilon)
sqrtFloatEpsilon: Double = 3.4526698300124393E-4
```

For simple order of magnitude purposes, you might remember that for Double this is about 10^{-8} and for a Float it is 10^{-4}. We will use this approximate value in the code that follows.

So how could we modify our searches to work with the Double type? We simply need to replace the check for equality with a check of the relative difference between the two. For the linear search that would look like the following.

```scala
def linearSearch(a:Array[Double],value:Double):Int = {
   var i = 0
   while (i < a.length && (a(i)-value).abs > 1e-8*value.abs) {
      i += 1
   }
   if (i > a.length) -1 else i
}
```

The second half of the condition in the while loop is performing the critical check. It takes the absolute value of the difference between the value in the **Array** and the value we are looking for. It compares to see if that is bigger than our approximate value for the square root of ϵ times the value we are looking for. You cannot simply compare to 1e-8 because if both **value** and **a(i)** have a magnitude much smaller than unity, such a comparison could be erroneous. For example, imagine if the values were positions of atoms in an object measured in meters. A separation of 10^{-8} meters apart would likely be quite significant.

13.4   Sorting Lists

The sorts discussed so far have all been very imperative, mutating value in **Array**s. These same sorts can be written in a functional manner using **List**s. The fact that the **List** type is immutable means that we can not alter the argument passed in. Instead, we have to build a new **List** that has the values in sorted order, and that sorted **List** is the result of the function. We'll look at how we can do this for our three most basic sorts in this section. Other, more efficient sorts with **List**s will be explored in chapter 15.

As before, we start off with bubble sort. The following code shows a flagged bubble sort that stops running when it goes through and there are no more swaps. It has a helper function defined inside of it that runs through and does swaps.

```scala
def bubbleSort(lst:List[Double]):List[Double] = {
  def swapper(lst:List[Double]):(Boolean, List[Double]) = lst match {
    case Nil => (false, lst)
    case h::Nil => (false, lst)
    case h1::h2::t =>
      if (h1 <= h2) {
        val (swap, rest) = swapper(h2::t)
        (swap, h1 :: rest)
      } else (true, h2 :: swapper(h1::t)._2)
  }

  val (swap, swapped) = swapper(lst)
  if (swap) bubbleSort(swapped)
  else swapped
}
```

The `swapper` function returns a `Boolean` and the `List` that results from a single pass. The `Boolean` tells us if there were any swaps done during this pass. The first two cases form something that is a common theme in these sorts because both the empty list and all lists with just one element are sorted, so they are both base cases. If there are at least two elements on the `List`, we compare them, and, if they are in the right order, we do a recursive call on all elements after the first one and give back that result with the first one consed on to the list. In that situation, this iteration was not a swap, so the value of the `Boolean` is whatever we get from further down the list. If the first element should go after the second one, the `Boolean` is true because there was a swap, and the `List` is the second element consed onto the second element of the recursive call.

The whole function works by calling `swapper` then calling again on the result if there was a swap. If there was not a swap, it returns the resulting `List`. Note that this function does not stop short of the high elements the way that our imperative version did. This code could be edited to do that, but it is a bit more challenging.

Next up is the selection sort, which is implemented as a min-sort again.

```
def minSort(lst:List[Double]):List[Double] = {
  def findAndRemoveMin(lst:List[Double]):(Double, List[Double]) = lst match {
    case Nil => throw new RuntimeException("Find and remove from Nil.")
    case h::Nil => (h, Nil)
    case h::t =>
      val (min, rest) = findAndRemoveMin(t)
      if (h < min) (h, min::rest) else (min, h::rest)
  }

  lst match {
    case Nil => lst
    case h::Nil => lst
    case h::t =>
      val (min, rest) = findAndRemoveMin(lst)
      min::minSort(rest)
  }
}
```

This code is built around a helper called `findAndRemoveMin`, which returns a tuple of the minimum value in the list and the list elements other than that minimum value. It only makes sense to call on a list with at least one element, hence the case for `Nil` which throws an exception. The way this function works is that it recurses to the end of the list, and begins returning the minimum and reduced list from the end backward. If the current element is smaller than the min after it, it becomes the new min and the old min is consed to the returned list, otherwise the min stays the same and the current element is consed to the returned list.

The primary code in the function matches the list against lists of length 0 and 1, returning simple base cases in those situations. The more complex case calls `findAndRemoveMin`, then conses that minimum value onto a recursive call of `minSort` on the whole list.

The last sort is the insertion sort, the code for which is shown here.

```
def insertionSort(lst:List[Double]):List[Double] = {
  def insert(x:Double, sorted:List[Double]):List[Double] = sorted match {
    case Nil => x::Nil
    case h::t => if (x < h) x::sorted else h::insert(x, t)
  }
```

```scala
def helper(sorted:List[Double], unsorted:List[Double]):List[Double] = unsorted
    match {
  case Nil => sorted
  case h::t => helper(insert(h, sorted), t)
}

helper(Nil, lst)
}
```

One could argue that insertion sort is the best sorted for the immutable, functional approach of the three shown here. The `insert` helper function is very simple. It assumes the input `List` is sorted, and returns a new `List` with the specified value added in the proper location. If `sorted` is empty, it returns a `List` with the new value as the only element. Otherwise, it checks if the value it insert is less than the head of `sorted`. If so, it conses it to `sorted` and if not it conses the head of `sorted` to the result of inserting the new element on the tail.

Unlike the other sorts, this one has a second short helper function that we have called `helper`. It simply builds up a sorted list, taking values one at a time from the unsorted list. The base case is when everything has been moved to the sorted list, which is returned. The function does its work by just calling the helper.

As you can certainly tell, these sorts are somewhat longer when dealing with `Lists` than they were when written for `Arrays`. The sorts shown in chapter 15 are actually often shorter when written in a functional manner than in an imperative one.

13.5   Performance and Timing

This chapter has introduced a number of different sorting and searching algorithms. We have discussed the performance of these algorithms in terms of order. This tells us how they scale as the number of inputs is changed and quite often this is all you really care about. You might not know or care about details of the hardware or the data sets you will be running on, or you know that such details will change over time. There are times though when you really do care exactly how fast an algorithm is on a particular machine, and you need to compare it to other similar algorithms on that same machine. When you do this, you need to do timing tests on the algorithm.

There are a number of different ways to get timing information on a program. Linux has a command called `time` that will let you know how much time a program consumes when it is running. For some applications this is ideal. However, for our purposes we only want to measure how long a particular part of the program takes. We do not want to measure the time taken to start things up or to initialize the `Array`. We only want to measure the time spent sorting the `Array`. You can get very detailed information like this from a PROFILER,[10] but that is overkill for what we want to do. For our purposes, we just need the ability to determine a start time and a stop time and take the difference between the two. We can do this by calling `System.nanoTime()`. This is a call to the Java libraries that returns a `Long` measuring the current time in nanoseconds.[11]

Another thing that we want to do to make the tests even is sort the same numbers for each sort. To make this happen, we really need to sort a copy of the `Array` and keep the

---

[10]You can invoke the Java profiler with the Java `-Xprof` option. To get Scala to run this option you set the `JAVA_OPTS` environment variable to include `-Xprof`.

[11]A nanosecond is 10^{-9} seconds.

original so that we can make other copies of it. We need to do this for each of the sorts; so, it is nice to put the code into a function that we can easily call with different sorts. To make it work with different sorts, we need to pass in the sort function as an argument. The following function does this for us.

```
def timeFunc(sortFunc:(Array[Double])=>Unit, a:Array[Double]):Unit = {
  val copy = Array.tabulate(a.length)(i => a(i))
  val start = System.nanoTime()
  sortFunc(copy)
  val end = System.nanoTime()
  println("Time:" + (end-start)/1e9)
  assert(isSorted(copy))
}
```

The print statement divides the time difference by `1e9`. This gives us back a value in seconds instead of nanoseconds which is much easier for us to read and deal with. We can invoke this by putting the following code at the end of our script.

```
val nums = Array.fill(args(0).toInt)(math.random)
args(1) match {
  case "bubble" => timeFunc(bubbleSort,nums)
  case "flagged" => timeFunc(flaggedBubbleSort,nums)
  case "min" => timeFunc(minSort,nums)
  case "insert" => timeFunc(insertionSort,nums)
  case "shell" => timeFunc(shellSort,nums)
}
```

Make sure that you include the sort functions that do not have calls to the rendering code when you put this together.

If you play around with this some you will notice a few things. First, you have to get up to at least 10000 numbers in the **Array** before the timing means much. If the **Array** is too small, the sort will be so fast that the resolution of the machine's clock will become a problem. Second, the amount of time spent doing the sort can vary on different invocations. For this reason, any true attempt to measure performance will run the code multiple times and take an average of the different values.

Bucket/Radix Sort (Advanced)

The sorts discussed earlier in this chapter require nothing more than the ability to compare elements. There are situations where you have more information about the values, and you can use that to sort them more efficiently. Examples of such non-general sorts include BUCKET SORT and RADIX SORT. Both use knowledge about the data being sorted to break them into groups.

To understand how this works, consider the situation where you are given a large number of folders to sort by name and you are in a big room. Many people would start off by going through the stack and breaking it into different piles. The piles would be for different parts of the alphabet. If the original stack were really large and you had a really big room, you might make one stack for each starting letter. You could continue to break things down in this way, again for each pile or switch to some other approach when the pile is smaller. This general approach is called a bucket sort and the piles would be called buckets. The bucket sort is really a general approach to breaking sorting problems into smaller, more manageable pieces.

The radix sort uses a similar approach, but it is a specific algorithm for sorting integer values. It organizes values, one digit at a time. In a counter-intuitive way, it starts from the least significant digit. This only works because each pass through preserves the order of the elements. So if A has a lower digit than B in one pass, and the next pass puts them in the same bin, A will come before B in that bin.

By convention we will implement our radix sort using decimal digits using division and modulo with powers of ten. The following code is a reasonably efficient implementation that sorts an `Array[Int]` using an `Array[List[Int]]` for the bins. Note that moving items to the bins is not an in-place operation. The values are copied back to the `Array` after binning, but it does require at least twice the memory of an in-place sort.

```scala
def radixSort(a:Array[Int]):Unit = {
  var max = a.max max a.min.abs
  var powerOf10 = 1
  while (max>0) {
    val byDigit = Array.fill(19)(List[Int]())
    for (num <- a) {
      val digit = num/powerOf10%10+9
      byDigit(digit) ::= num
    }
    var i = 0
    for (bin <- byDigit; num <- bin.reverse) {
      a(i) = num
      i += 1
    }
    powerOf10 *= 10
    max /= 10
  }
}
```

The `max` variable starts with the largest magnitude value and is divided by ten each time so that we know when to stop. The `powerOf10` variable keeps track of what digit we are currently binning. Each time through the loop, 19 empty bins are set up. You might wonder why there are 19 bins instead of just 10. The answer is that division and modulo preserves sign. If you only use 10 bins, this sort can only work on positive values. By going up to 19, it is able to handle negative numbers as well. To make that work, the bin value we get from division and modulo is incremented by 9 so that the values slide up to the domain of the `Array` index values.

The binning runs through all the numbers in the array and conses them onto the `List` for the proper digit. Consing to a `List` adds to the front. For that reason, the `for` loop that moves the values back to the `Array` has to run through each bin in reverse.

This sorting algorithm is $O(kn)$ where $k = \lceil \log_{10}(max) \rceil$. In the case of an `Int`, the value of k cannot be larger than 10. So the performance scales linearly for really large `Array`s.

13.6 Putting It Together

To illustrate concepts in this chapter and link them together with earlier chapters we will return to the theme park example. Every month the theme park picks a top employee. This is based on performance relative to the average for that month. For every day of the month you have data that tells you what operators were working each ride, and how many people went on the ride. From this, we can calculate average ridership for each ride during the month as well as how many riders rode each ride for each day that a given operator was working it. Each operator can be given an efficiency for any particular ride as the average number of people who rode on days he/she was working divided by the average for all days. Averaging these efficiencies gives an overall rating to each operator. Those can be sorted and displayed to show relative performance. Code for doing that is shown here.

Listing 13.2: EmployeeOfTheMonth.scala

```scala
import scala.io.Source

case class DailyData(ride:String, operators:Array[String], numRiders:Int)
case class RideAverage(ride:String, avNum:Double)
case class OperatorDailyData(name:String, ride:String, numRiders:Int)
case class OperatorRideAverages(name:String, rideAvs:Array[RideAverage])
case class OperatorEfficiencyFactor(name:String,factor:Double)

def parseDailyData(line:String):DailyData = {
  val parts = line.split(" *; *")
  DailyData(parts(0), parts.slice(1, parts.length-1), parts.last.toInt)
}

def readData(fileName:String):Array[DailyData] = {
  val source = Source.fromFile(fileName)
  val lines = source.getLines
  val ret = (lines.map(parseDailyData)).toArray
  source.close
  ret
}

def insertionSortByEfficiency(a:Array[OperatorEfficiencyFactor]):Unit = {
  for (j <- 1 until a.length) {
    var i=j-1
    val tmp=a(j)
    while (i>=0 && a(i).factor>tmp.factor) {
      a(i+1) = a(i)
      i -= 1
    }
    a(i+1) = tmp
  }
}

val data = readData(args(0))
val rides = data.map(_.ride).distinct
val averages = for (ride <- rides) yield {
  val days = data.filter(_.ride==ride)
  RideAverage(ride, days.map(_.numRiders).sum.toDouble/days.length)
```

```scala
39  }
40  val dataByOperator = for (day <- data; op <- day.operators) yield {
41    OperatorDailyData(op, day.ride, day.numRiders)
42  }
43  val operators = dataByOperator.map(_.name).distinct
44  val opRideAverages = for (op <- operators) yield {
45    val opDays = dataByOperator.filter(_.name == op)
46    val rideAvs = for (ride <- rides; if opDays.exists(_.ride==ride)) yield {
47      val opRides = opDays.filter(_.ride == ride)
48      RideAverage(ride, opRides.map(_.numRiders).sum.toDouble/opRides.length)
49    }
50    OperatorRideAverages(op, rideAvs)
51  }
52  val operatorFactors = for (OperatorRideAverages(op, rideAvs) <- opRideAverages)
        yield {
53    val factors = for (RideAverage(ride,av) <- rideAvs) yield {
54      av/averages.filter(_.ride==ride).head.avNum
55    }
56    OperatorEfficiencyFactor(op,factors.sum/factors.length)
57  }
58  insertionSortByEfficiency(operatorFactors)
59  operatorFactors.foreach(println)
```

This code assumes that the file has a line of data for each day that starts with the ride name, followed by operator names, with number of riders at the end. Each of these is separated by semicolons. That data is read and used to calculate the various values needed for efficiency. Once the efficiencies have been calculated, all the employees are sorted with an insertion sort and the results are printed.

13.7   End of Chapter Material

13.7.1   Summary of Concepts

- The act of ordering data according to some value is called sorting. It is a common operation on a computer as it benefits humans who view the data as well as programs when they process it. Several types of sorts were discussed in this chapter.

  - A bubble sort runs through the `Array` comparing adjacent elements and swapping them if they are out of order. This action is repeated until all the values are in proper order.

  - A selection sort picks specific elements and puts them into place. We demonstrated a `minSort`, which finds the smallest unsorted element and swaps it to the correct location. Selection sort does few swaps so it is most useful in a situation where moving data is an expensive operation.

  - An insertion sort takes each element and pushes it forward through the `Array` until it gets it into sorted order with the elements that came before it. Insertion sort is extremely efficient in situations where the data starts off close to the proper order.

  - The Shell sort is also called the diminishing gap sort. It does insertion sorts over partial data in such a way that things are moved toward proper ordering. It winds up being more efficient that any of the other three.

- When we talk about the performance of different algorithms in computer science, we typically use order analysis. This gives a rough idea of how the number of times an operation is performed will scale with the size of the input. The first three sorts above all do $O(n^2)$ comparisons.

- One of the most common tasks done on computers is looking for data, an activity we call searching.

  - If data is unorganized, the only approach is to go through all elements in a sequential/linear search. This type of search is $O(n)$.
  - When the data has a sorted order to it, you can use a binary search to find elements. A binary search is significantly faster that a sequential search because it effectively throws out half the data with each comparison. This provides $O(\log(n))$ performance.

- Sorting `case class`es requires minor alterations to the sort so that the appropriate fields are compared.

- To compare the performance of different algorithms when you really care about how quickly a program runs, you can measure how much time it takes. To measure the runtime of a whole program you can use the Linux `time` command. To measure only specific parts of code call `System.nanoTime()` before and after the section you want to time and subtract the results. Profilers can also provide more detailed information on what is taking time in a program.

- Programmers often put errors into their code by accident. These errors are commonly called bugs. The process of fixing the errors is called debugging. Bugs can be categorized into different types.

  - Sytnax errors are errors where the programmer enters something that is not valid for the language. These are found when the program is compiled. The compiler can generally give you helpful information about what is wrong and where the error is in the code.
  - Runtime errors occur when the code compiles, but crashes during the run. The crash itself can print helpful information for you. Unfortunately, runtime errors often only occur under certain situations, making them harder to track down and fix.
  - Logic errors are the term we use to describe when the code compiles and runs without crashing, but the result is inaccurate. These are generally the worst of the three as the computer does not give you pointers to where the error is occurring or what it is caused by. Instead, the programmer has to track it down. This can be done using print statements or using a debugger.[12]

- The memory of the computer can be thought of as a really big array. Each program is given a different section of memory that is divided into the heap and the stack. The stack is well organized and is where local variables are allocated. Each function that is called gets a chunk of memory called a stack frame. When the function returns, the frame is released. The heap is where all objects are allocated in Scala. A garbage collector deals with objects that are no longer in use.

---

[12]The debugger is not an option for how we are currently doing things.

13.7.2 Exercises

1. Write a `minMaxSort` that finds both the minimum and maximum values in each pass and swaps them to the proper locations.

2. Do timing tests on `Arrays` of `Ints` and `Doubles` for the sorts presented in this chapter. Note that the radix sort that is presented only works with integer types. You want to have the number of elements in the `Array` or `List` grow exponentially so that you can see variation over a large range. Recommended sizes could be 100, 300, 1000, 3000, 10000, 30000, etc. Plot the data as a log-log scatter plot.[13]

3. Following onto the timing results, do a comparison of the number of comparisons done for different sorts presented in this chapters. Plot the results in the same way.

4. While most of the work for these sorts is in comparisons, it is also interesting to look at the number of memory moves. A standard swap is three assignments. Add code to count how many assignments are done in each of the sorts and plot the results.

5. Section 7.3 showed how you can abstract functions over types. This is something that will be dealt with in detail in the second half of the book. Review that section and see if you can write a version of insertion sort and min sort that is general with regards to type.

6. Do a little web searching to find some other type of sort not described in this chapter and write it.

7. Describe what happens if you apply binary search to an unordered array.

8. Create an array of 1000 random numbers and use any sorting algorithm to sort the list. Then search the list for some items using the binary search algorithm. After that, use the binary search algorithm to search the list, switching to a sequential search when the size of the search list reduces to less than 15.

9. Print the number of comparisons performed in both searches in the previous example.

10. Rewrite the bubble sort so that it bubbles smaller numbers down instead of bigger numbers up in the array.

13.7.3 Projects

1. This project is intended for people who have been working on graphics and ray tracing, but it does not immediately link so you can do it even if you have not been doing the previous ones. For this problem you will draw polygons to a `Canvas` using a "painter's algorithm." That is where you draw things from the back to the front so that the things in front appear on top of the things behind them. Doing this properly is a challenging problem. You should base your drawing on the point in the polygon that is closest to the viewer.

   To keep things simple, the viewer will be at the origin facing out the z-axis. That way the x and y coordinates are roughly what you expect. To make it so that things that are further away are smaller you divide the actual x and y by the z value to

---

[13]This means that the N value should be the x-axis, and the time it takes should be the y-axis where both axes use a log scale. The advantage of this is that functions of the form $f(x) = x^n$ appear as straight lines in this type of plot with a slope equal to n.

get the location you would draw them at. Given a point (x,y,z) you would want it drawn on an image or panel at $((x/z+1)*size.width/2,(1-y/z)*size.height/2)$. To represent a polygon for drawing to the `GraphicsContext` you should use `moveTo` and `lineTo` to make lines. When you get to the last point you call `closePath` to close it off.

Store your polygons in a file. You can decide the exact format. In addition to having the points in each polygon, each one should have a color that it will be drawn in. Remember to sort them by z value so that the one with the smallest z value is drawn last.

2. The BASIC programming language was created to be a simple language for novice programmers. The original versions were organized by line number. Each statement of the program was a single line that has a number associated with it. The lines were ordered according to that number and flow control was implemented by allowing the program to jump to other line numbers. For this project you will create a simple GUI that lets you edit a simplified version of BASIC.

For this simplified version you have a very limited set of possible commands. The GUI will also use a "line editor" style. The allowed commands including the following: GOTO, IF-THEN, INPUT, LET, and PRINT. Each of these must be preceded by a line number that is an integer. For our purposes, variable names are single characters, and they will all be numbers. You can use a `Double`. The format of the commands is as follows.

- The `GOTO` command should be followed by an integer number that is the line number the program should execute next. (Example: 100 GOTO 50)

- The `IF-THEN` command has the following syntax: IF *comp* THEN #. The *comp* is a comparison that can have a variable or a number on either side and either = or < between them. The # is a line number that the execution will jump to if the comparison in the condition is true. If the comparison is false, the execution continues on the next line. (Example: 110 IF a<21 GOTO 50)

- The `INPUT` command should be followed by a single variable name and when it is executed the program pauses and waits for the user to input a value that is stored in that variable. (Example: 120 INPUT b)

- The `LET` command has the keyword LET followed by a variable name with an equal sign and then either a single number/variable or two number/variable operands that are separated by an operator. The operator can be +, -, *, or /. The result of the operation should be stored in the variable before the equal sign. (Example: 130 LET a=b+3)

- The `PRINT` command can be followed either by a variable name or a string in double quotes. When executed, this will print either the value of the variable or the string to the output.

Variables do not have to be declared. They come into existence when first used and they should have a value of 0 to start with if no other value was given to them.

The GUI for this program should have three main elements. The program itself is displayed in a `ListView`. This makes it simple for users to select a line to edit without letting them type random text. There should also be a `TextField` where the user can enter/edit lines. If a line is selected in the `ListView`, the text from it should appear in the `TextField`. The user can edit that line or enter anything else. The number at the beginning of the line will be used to put it in place. If a number is used that duplicates an existing line, the new one replaces the old one. Lastly there is a `TextArea` that shows the output when the program is run.

When the user hits enter on the `TextField` your program should check if what was entered is a valid command. If it is, it should put it in the program and clear. If it is not, it should leave the text there and not alter the existing program.

There should be at least four menu items for this program: "Save", "Open", "Remove", and "Run". The "Save" option saves the program to a text file. The "Open" option allows the user to select a file and open it up as a program. The "Remove" option will remove the currently selected lines in the program. Nothing happens if nothing is selected in the `ListView`. The "Run" option runs the program. It starts running at the first line and continues until execution goes beyond the last line.

3. On the book's web site, under this chapter, you will find files with historical weather data for some different cities in the US along with a link to the source of the data. You should read this data into an `Array` of some case class that you create. The data is separated by commas. Note that the second line tells you what each column of data represents. You will skip the top two lines when reading the information. Write a script that will report the months with the five warmest average temperatures and those with the five lowest average temperatures.

For a bit of an extra challenge make it so that the user can tell the program whether to report the top and bottom five months for any of the values in the file. You could do this with a lot of typing, but by passing in a function you can cut down on the length of the code greatly.

If you did any of the problems in the last chapter that included plotting points or data, you should consider sticking that functionality onto this project. That way you can bring up a GUI and show the plots of whatever field(s) the user selects.

4. If you did a game for one of the projects in chapter 12, for this one you can enhance it by adding a high scores list. That means having some way to score games. It also means saving scores to a file in a format of your choosing. Lastly, the high scores need to be sorted. Each score record needs to have at least a score and an identifying name or initials. All other details are up to the student.

5. If you have been working with the recipe projects, you now have the ability to order a shopping list according to where things are in the store. This follows most logically from project 12.11, but it does not require the graphical functionality, only a data file listing what aisles different items are in and the ability for a user to make a shopping list.

Your script should have the ability to sort the grocery list they build to go through the store in either ascending or descending order by row. The sorted list should be displayable in a `TextArea` so that the user can cut and paste it for printing.

6. This project fits in with the sequence of schedule building projects. In particular, it makes sense to do if you have already done project 11.10 where schedules were built and displayed in a GUI. That project showed the selected courses in the order the user selected them. However, it would often be helpful to have them displayed based on other criteria. For example, having them sorted by department and number or how much interest the user expressed in them could be helpful.

For a bit of extra challenge, you could include some additional information with each course indicating things like whether it is required for graduation or what graduation requirement it fulfills. Allow the user to select from multiple orderings using a `ComboBox`.

7. If you did project 12.15 or 11.12 looking at box scores you might have noticed that one significant feature that was missing was the ability to change the player listing so that it orders the players based on a particular stat. Now that you know how to sort you can fix this. How you do this depends on how you chose to display the box score, but it can be as simple as adding a few `Button`s for sorting by different statistical categories.

Additional exercises and projects, along with data files, are available on the book's web site.

Chapter 14

XML

This chapter deals with XML, short for eXtensible Markup Language. XML is technically a completely language independent topic, Scala was developed with XML in mind and makes it easy to work with.

In chapter 9, we learned how to read from and write to text files. The files that we used in that and following chapters are what are called "flat" text files. They have the data in them with nothing that tells us about the nature of the data other than formatting. The advantages flat files have are that they are fairly simple to read and write, and it can be both read and written with standard tools like a text editor, just like the one you use for your programs. The disadvantages are that it can be slow, and it lacks any inherent meaning; so, it is hard to move the information from one program to another. It is also somewhat error prone because so much information is in the formatting. XML addresses the latter disadvantage, without losing the advantages.

The eXtensible Markup Language is a standard for formatting text files to encode any type of information. It is a markup language, not a programming language. The standard simply defines a format for encoding information in a structured way. It is likely that you have heard of a different markup language called HTML, the HyperText Markup Language. HTML is used to encode web pages and has a format very similar to XML. Indeed, there is a standard called XHTML, which is basically HTML that conforms to the XML rules.

14.1   Description of XML

Everything in an XML file can be classified as either markup or content. Markup in XML is found between the symbols '<' and '>' or between '&' and ';'. The content is anything that is not markup. To help you understand XML we will look at an example XML file. This example is built on the idea of calculating grades for a course.

```
<course name="CSCI 1320">
    <student fname="Quinn" lname="Bender">
        <quiz grade="98"/>
        <quiz grade="100"/>
        <quiz grade="90"/>
        <test grade="94"/>
        <assignment grade="100">
            <!-- Feedback -->
            Code compiled and runs fine.
        </assignment>
    </student>
    <student fname="Jason" lname="Hughes">
        <quiz grade="85"/>
        <quiz grade="78"/>
        <test grade="67"/>
        <assignment grade="20">
            Code didn't compile.
        </assignment>
    </student>
</course>
```

Just reading this should tell you what information it contains. That is one of the benefits of XML, the markup can be informative. Now we want to go through the different pieces of XML to see how this was built.

14.1.1   Tags

Text between '<' and '>' characters are called TAGS. Nearly everything in our sample XML file is inside of a tag. The first word in the tag is the name of the tag, and it is the only thing that is required in the tag. There are three types of tags that can appear in an XML file.

- Start tag - Begins with '<' and ends with '>',

- End tag - Begins with '</' and ends with '>',

- Empty-element tag - Begins with '<' and ends with '/>'.

Tags can also include attributes, which come after the name and before the close of the tag. These are discussed below in section 14.1.3.

14.1.2   Elements

The tags are used to define ELEMENTs which give structure to the XML document. An element is either a start tag and an end tag with everything in between or else it is an empty element tag. The empty element tag is simply a shorter version of a start tag followed by an end tag with nothing in between.

Elements can be nested inside of one another. In our sample file, there is a `course` element that encloses everything else. This is required for XML, there needs to be one element that goes around all the other content. There are two `student` elements inside of the `course` element. Each of those includes elements for the different grades. Most of these are empty elements, but the `assignments` are not empty and have contents in them.

Any time there is a start tag, there should be a matching end tag. When elements are nested, they have to be nested properly. That is to say that if `element2` begins inside of `element1`, then `element2` must also end before the end of `element1`. In addition to other elements, you can place general text inside of elements.

14.1.3 Attributes

Additional information can be attached to both start tags and empty element tags in the form of ATTRIBUTES. An attribute is a name value pair where the value is in quotes and the two are separated by an equal sign. The example XML contains quite a few attributes. In fact, every start or empty element tag in the example has an attribute. It is not required that these tags have an attribute, but it is a good way to associate simple data with a tag. Some of the tags in the example show that you can also have multiple attributes associated with them.

14.1.4 Content

Between a start tag and an end tag you cannot only put other tags, you can put plain text. In the example XML, the `assignment` element has text in it that serves as a comment on the grade. The text that you put inside of an element is not formatted and can include anything you want, as long as it does not conflict with the XML syntax. Unlike the markup part of the XML document, there is no special formatting on content.

14.1.5 Special Characters

While the content does not have any special formatting, it is still embedded in an XML document. There are certain characters that are special in XML that you cannot include directly in the content. For example, if you put a '<' in the content it will be interpreted as the beginning of a tag. For this reason, there is a special syntax that you can use to include certain symbols in the content of an XML file. This syntax uses the other form of markup that begins with & and ends with ;. There are five standard values defined for XML.

- `&amp;` = &

- `&apos;` = '

- `&gt;` = >

- `&lt;` = <

- `&quot;` = "

There are many more defined for other specific markup languages. For example, if you have ever looked at HTML you have probably seen ` ` used to represent spaces.

14.1.6   Comments

Just like with code, it is helpful to occasionally put comments in your XML. A comment begins with '`<!--`' and ends with '`-->`'. You can put whatever text you want between these as long as it does not include the sequence to end the comment.

14.1.7   Overall Format

There are two other rules that are significant for XML. First, the entire XML document must be inside of a single element. In the example above, everything was in the `course` element. Had we wanted to have more than one `course` element, we would have had to create some higher level element to hold them.

In addition, most XML files will begin with an XML declaration that comes right before the element containing the XML document information. The declaration has a slightly different syntax and might look like the following line.

```
<?xml version="1.0" encoding="UTF-8" ?>
```

14.1.8   Comparison to Flat File

To better understand the benefit of XML, we will compare our XML file to a flat file that might be used to store basically the same information. Here is a flat file representation of the grade information from the XML above.

```
CSCI 1320
2
Quinn Bender
98 100 90
90
100
Jason Hughes
85 78
67
20
```

This is a lot shorter, but unless you happen to know what it is encoding, you cannot figure much out about it. Namely, the numbers in the file are hard to distinguish. The 2 near the top you might be able to figure out, but without additional information, it is impossible to determine which lines of grades are assignments, quizzes, or tests.

The flat file also lacks some of the information that was in the XML. In particular, the comments on the assignments in the XML format are missing in this file. It would be possible to make the flat file contain such information, but doing so would cause the code required to parse the flat file to be much more complex.

14.1.8.1   Flexibility in XML

A significant "Catch 22" of XML is that there are lots of different ways to express the same information. For example, the comment could have been given as an attribute with the name `comment` instead of as contents of the element. Similarly, if you do not want to allow comments, you could shorten the XML to be more like the flat file by changing it to the following format.

```
<course name="CSCI 1320">
    <student fname="Quinn" lname="Bender">
        <quizzes>98 100 90</quizzes>
        <tests>94</tests>
        <assignments>100</assignments>
    </student>
    <student fname="Jason" lname="Hughes">
        <quizzes>85 78</quizzes>
        <tests>67</tests>
        <assignments>20</assignments>
    </student>
</course>
```

Here all the grades of the same type have been given as contents of elements with the proper names. This makes things much shorter and does not significantly increase the difficulty of parsing the grades. It does remove the flexibility of attaching additional information with each grade such as the comments.

An alternate approach would be to not use attributes at all and store all the information in sub-elements. Doing that, one of the students might look like the following.

```
<student>
    <fname>Jason</fname>
    <lname>Hughes</lname>
    <quiz><grade>85</grade></quiz>
    <quiz><grade>78</grade></quiz>
    <test><grade>67</grade></test>
    <assignment>
        <grade>20</grade>
        <comment>Code didn't compile.</comment>
    </assignment>
</student>
```

There is no definitive answer for which of these two styles you should use. What can be said definitively is that attributes must be unique and should generally have short text. So you should not try to use attributes for things that would span multiple lines or that need multiple values that would be challenging to break apart. A possible rule of thumb to follow is that anything where you only have a single value that is short, you should represent as an attribute. Anything that has multiple values or is long should be in sub-elements.

14.2   XML in Scala

XML is not part of the normal first semester topics in computer science, but Scala makes it so easy that there is no reason not too. To see this, simply go to the REPL and type in valid XML.

```
scala> <tag>Some XML.</tag>
res0: scala.xml.Elem = <tag>Some XML.</tag>
```

The Scala language has a built in XML parser that allows you to write XML directly into your code. You can see that the type of this expression is `scala.xml.Elem`.

The `scala.xml` package contains the types related XML. We will run through some of the more significant ones here.

- `Elem` - Represents a single element. This is a subtype of `Node`.

- `Node` - Represents a more general node in the XML document. This is a subtype of `NodeSeq`.

- `NodeSeq` - Represents a sequence of `Node`s.

- `XML` - A helper object that has methods for reading and writing XML files.

There are quite a few other types that you can see in the API, but we will focus on these as they give us the functionality that we need for our purposes.

14.2.1   Loading XML

The XML object has methods we can use to either read from files or write to files. The `loadFile` method can be used to read in a file. If the first example XML that was shown is put in a file with the name "grades.xml", then the following call would load it in.

```
scala> xml.XML.loadFile("grades.xml")
res4: scala.xml.Elem =
<course name="CSCI 1320">
        <student lname="Bender" fname="Quinn">
                <quiz grade="98"></quiz>
                <quiz grade="100"></quiz>
                <quiz grade="90"></quiz>
                <test grade="94"></test>
                <assignment grade="100">


                        Code compiled and runs fine.
                </assignment>
        </student>
        <student lname="Hughes" fname="Jason">
                <quiz grade="85"></quiz>
                <quiz grade="78"></quiz>
                <test grade="67"></test>
                <assignment grade="20">
                        Code did not compile.
                </assignment>
        </student>
</course>
```

Clearly this is the content of the XML file that we had created. All that is missing is the comment, which was there for human purposes, not for other programs to worry about. In addition, the empty tags have also been converted to start and end tags with nothing in between. This illustrates that the empty tags were also just for human convenience and their meaning is the same as an empty pair of start and end tags.

14.2.2   Parsing XML

Once we have this `Elem` object stored in `res4`, the question becomes how you get the information out of it. The `NodeSeq` type, and hence the `Node` and `Elem` types which are subtypes of it, declare operators called \ and \\. These operators are used to search inside the contents of an object. Both operators take a second argument of a `String` that gives the name of what you want to look for. The difference is how far they search. The \ operator looks only for things at the top level, either `Nodes` in the current sequence if we have a true `NodeSeq`, or children of this node if we have a `Node` or `Elem`. The \\, on the other hand, finds anything that matches at any depth below the current level. To illustrate this, we will do three example searches.

```
scala> res4 \ "student"
res5: scala.xml.NodeSeq =
NodeSeq(<student lname="Bender" fname="Quinn">
             <quiz grade="98"></quiz>
             <quiz grade="100"></quiz>
             <quiz grade="90"></quiz>
             <test grade="94"></test>
             <assignment grade="100">
                   Code compiled and runs fine.
             </assignment>
      </student>, <student lname="Hughes" fname=Jason">
             <quiz grade="85"></quiz>
             <quiz grade="78"></quiz>
             <test grade="67"></test>
             <assignment grade="20">
                   Code did not compile.
             </assignment>
      </student>)

scala> res4 \ "test"
res6: scala.xml.NodeSeq = NodeSeq()

scala> res4 \\ "test"
res7: scala.xml.NodeSeq = NodeSeq(<test grade="94"></test>, <test
    grade="67"></test>)
```

The first two searches use the \ operator. The first one searches for elements that have the tag name "`student`". It finds two of them because they are at the top level and gives us back a `NodeSeq` with them in it. The second search looks for tags that have the name "`test`". This search returns an empty `NodeSeq`. This is because while there are tags with the name "`test`" in `res4`, they are nested more deeply inside of the "`student`" elements as such, are not found by the \ operator. The last example searches for the "`test`" tag name again, but does so with \\, which searches deeply, and hence gives back a `NodeSeq` with two `Nodes` inside of it.

The \ and \\ operators can also be used to get the attributes from elements. To get an attribute instead of a tag, simply put a '`@`' at the beginning of the string you are searching for. Here are three searches to illustrate this.

```
scala> res4 \ "@name"
res8: scala.xml.NodeSeq = CSCI 1320

scala> res4 \ "@grade"
res9: scala.xml.NodeSeq = NodeSeq()

scala> res4 \\ "@grade"
res10: scala.xml.NodeSeq = NodeSeq(98, 100, 90, 94, 100, 85, 78, 67, 20)
```

The first search uses \ to get the name of the top level node. Using \ to look for a @grade at the top level node does not give us anything, but using \\ will return the values of all the @grades in the document.

What you really want to do is put the information from the XML file into a structure that can be used in the program. Given what we have learned, this would mean that we want to put things into case classes. The data in this XML file corresponds very closely to the student type that was created in chapter 10. That case class looked like this.

```scala
case class Student(name:String,assignments:List[Double],tests:List[Double],
    quizzes:List[Double])
```

In that chapter we parsed a flat file into an Array of Students. Now we will demonstrate how to do the same thing using the XML. We will start with a function that takes a Node that should be a student Element and returns a Student object. Such a function might look like the following.

```scala
def studentFromXML(elem:xml.Node):Student =
  Student((elem \ "@fname")+" "+(elem \ "@lname"),
    (elem \ "assignment").map(n => (n \ "@grade").toString.toDouble).toList,
    (elem \ "test").map(n => (n \ "@grade").text.toDouble).toList,
    (elem \ "quiz").map(n => (n \ "@grade").text.toDouble).toList)
```

This function builds a Student object and passes in the four required arguments. The first is the name, which is made from the fname and lname attributes of the element. After that are three Lists of grades for the assignments, tests, and quizzes respectively. These all have a similar form. They start by doing a search for the proper tag name and mapping the result to a function that converts the value of the grade attribute to a Double. The call to text is required because the result of \ here is a Node, not a String, and the Node type does not have a toDouble method. The last part of each grade type is a call to toList. This is required because the map is working on a NodeSeq and will give back a Seq, but a List is required for the Student type.

The form of studentFromXML above is compact, but it might not be the more readable or easiest to work with. It is often helpful to introduce new variables for each value as it is parsed from the XML. The following code shows a modified version that does this.

```scala
def studentFromXML(elem:xml.Node):Student = {
  val name = (elem \ "@fname")+" "+(elem \ "@lname")
  val assignments = (elem \ "assignment").map(n => (n \
      "@grade").toString.toDouble).toList
  val tests = (elem \ "test").map(n => (n \ "@grade").text.toDouble).toList
  val quizzes = (elem \ "quiz").map(n => (n \ "@grade").text.toDouble).toList
  Student(name, assignments, tests, quizzes)
}
```

The use of map probably does not jump out to you at first. Hopefully at this point you have become quite comfortable with it and other higher order methods. However, if you think a bit you will realize that it is a bit surprising here because the thing it is being called on is not a List or an Array. Instead, it is a NodeSeq. This works because the NodeSeq is itself a subtype of Seq[Node], meaning that all the methods we have been using on other sequences work just fine on this as well.

This is useful for getting our array of students as well. The following line shows how we can use map and toArray to get the result that we want with the studentFromXML function.

```
scala> (res4 \ "student").map(studentFromXML).toArray
res15: Array[Student] = Array(Student(Quinn
    Bender,List(100.0),List(94.0),List(98.0, 100.0, 90.0)), Student(Jason
    Hughes,List(20.0),List(67.0),List(85.0, 78.0)))
```

Again, the call to **toArray** gives us back the desired **Array[Student]** instead of a more general **Seq[Student]**.

We have used the **text** method a few times in this parsing process. It is worth noting that the **text** method applied to a full **Elem** will give you all of the text that appears inside of it and all sub-elements. So calling **text** on **res4** gives the two comments along with a lot of whitespace. To get just the comment on any particular grade, you would parse down to that specific element and call the **text** method on it.

14.2.3 Building XML

So now you know how to get the contents of an XML file into a useful form in Scala. What about going the other way? Assume that the code we just wrote was used in the menu based application from chapter 10 and that changes were made and now we want to write the results back out to a file. The first step in this would be to build the **Node** that represents the data.

We saw above that we can put XML directly into a Scala program or the REPL and it will be parsed and understood. However, that alone does not give us the ability to put values from the program back into the XML file. Fortunately, this is not hard to do either. Inside of XML that is embedded in a Scala program you can embed Scala expressions using curly braces. We will start with a simple example.

```
scala> <tag>4+5 is {4+5}</tag>
res19: scala.xml.Elem = <tag>4+5 is 9</tag>
```

Here the expression 4+5 has been put in curly braces, and as you can see it evaluates to the value 9 as it should. The code you put inside of the curly braces can be far more complex and built additional XML content or tags.

We will use this to write a function that packs a **Student** object into an XML node. This code looks like the following.

```
def studentToXML(stu:Student):xml.Node = {
  val nameParts = stu.name.split(" +")
  <student fname={nameParts(0)} lname={nameParts(1)}>
    {stu.quizzes.map(q => <quiz grade={q.toString}/>)}
    {stu.tests.map(t => <test grade={t.toString}/>)}
    {stu.assignments.map(a => <assignment grade={a.toString}/>)}
  </student>
}
```

The first line **splits** the student name into pieces around spaces. It is assumed that the first element is the first name, and the second element is the last name. These are used as attribute values in the student tag. Inside of that element are three lines of code, one each for **quizzes**, **tests**, and **assignments**. Each of these **maps** the corresponding **List** to a set of elements with grade attributes.

It is worth noting two things about using code for the attribute values. First, the quotes are not written anywhere. They are automatically provided when the value is Scala code. Second, the type of the Scala expression for the value has to be **String**. This is apparent with the grade values. They are **Doubles** and have to be explicitly converted to **Strings**.

14.2.4 Writing XML to File

Once you have the `Node` you want to write, the writing process is as easy as a call to the `save` method of the `XML` object.

```
xml.XML.save("grades.xml",node)
```

The first argument is the name of the file you want to write to. The second is the `Node` that you want to write.

Validating XML (Advanced)

When XML is used for large applications, it is important to be able to verify that the contents of a file match the format that is required by the applications. This process is called validation. When the XML standard was first released, validation was done with Document Type Definition files (DTDs). A DTD is a text file that has a fairly simple format that allows you to specify what types of elements should be in a file. For each element you can say what needs to be in it or what could be in it. This ability includes attributes as well at sub-elements.

DTDs were generally considered to have two problems. First, they were a bit limited and simplistic. For example, you could say that an element must contain an attribute, but you could not put any constraints on the nature of that attribute. For example, you could not say that it had to be a number or a date. In addition, DTDs had their own syntax. The goal of XML was to be a general data storage format and some found it unfitting that you had to use some other format to specify what should go into an XML file.

For these reasons, XML schema were created. An XML schema is an XML document that uses certain tags to specify what can and cannot go into some other XML files. XML schema tend to be large and complex, but they provide great control over the format of an XML file.

There are tools for both DTDs and XML schema that will run through XML files and tell you whether or not they adhere to a given specification. There is also a package in the Scala standard libraries called `scala.xml.dtd` that can help with validation using DTDs.

14.2.5 XML Patterns

We have seen patterns in a number of places so far. With XML we can add another one. Not only can XML be written directly into the Scala language, it can be used to build patterns. The patterns look just like what you would use to build XML, except that what you put into curly braces should be names you want bound as variables. There is one significant limitation, you cannot put attributes into your patterns. This usage can be seen with the following little example in the REPL.

```
scala> val personXML = <person><name>Kyle</name><gender>M</gender></person>
personXML: scala.xml.Elem = <person><name>Kyle</name><gender>M</gender></person>

scala> val <person><name>{name}</name><gender>{sex}</gender></person> = personXML
name: scala.xml.Node = Kyle
sex: scala.xml.Node = M
```

If you decide that you like parsing XML using patterns, that might provide a motivation to use a format where you put all of your information in sub-elements instead for attributes.

14.3   Putting It Together

To illustrate the real power of XML we will make a more complete theme park program that includes the functionality of some of the earlier scripts along with editing abilities. All the information will be stored in a single XML file. This last part is something that was not highlighted before, but it implicitly comes with the ability to give meaning to data. In a flat text file, it is the position in the file that gives meaning to something. This makes it very hard to insert new information of different types. That is not a problem for XML as new tags can be added as desired. As long as the new tags have names that do not conflict with earlier tags, the earlier code will continue to work just fine.

In chapter 8, we wrote a script that would help with building schedules. In chapter 13 we wrote another script that could be used to determine the employee of the month. Both of these deal with information related to employees and rides, but there is not a 100% overlap between the required data, and the file formats are very different. We want to write a script here that will include the functionality of both of those scripts, along with the ability to add ride and employee information while keeping all of the information stored in a single XML file.

Code for this is shown below. It starts with the definition of a number of `case class`es followed by functions that can build instances of those `case class`es from XML or build XML from them. After those functions are four lines that declare the main data for the program while reading it in from an XML file specified on the command line. This is followed by slightly modified versions of the schedule builder and the employee ranker from previous chapters.

Listing 14.1: ThemeParkMenu.scala

```scala
import scala.io.Source
import scala.xml._

case class DayData(ride:String, dayOfWeek:String, operators:Array[String],
    numRiders:Int)
case class MonthData(month:Int, days:List[DayData])
case class YearData(year:Int, months:List[MonthData])
case class RideData(name:String, numberOfOperators:Int, heavyCount:Int)
case class EmployeeData(name:String, rides:List[String])

def parseDay(node:Node):DayData = {
  val ride = (node \ "@ride").text
  val dow = (node \ "@dayOfWeek").text
  val num = (node \ "@numRiders").text.toInt
  val ops = (node \ "operator").map(_.text).toArray
  DayData(ride, dow, ops, num)
}

def dayToXML(day:DayData):Node = {
  <day ride={day.ride} dayOfWeek={day.dayOfWeek} numRiders={day.numRiders.toString}>
    {day.operators.map(op => <operator>{op}</operator>)}
```

```scala
    </day>
}

def parseMonth(node:Node):MonthData = {
  val month = (node \ "@month").text.toInt
  val days = (node \ "day").map(parseDay).toList
  MonthData(month, days)
}

def monthToXML(month:MonthData):Node = {
  <month month={month.month.toString}>
    {month.days.map(dayToXML)}
  </month>
}

def parseYear(node:Node):YearData = {
  val year = (node \ "@year").text.toInt
  val months = (node \ "month").map(parseMonth).toList
  YearData(year, months)
}

def yearToXML(year:YearData):Node = {
  <year year={year.year.toString}>
    {year.months.map(monthToXML)}
  </year>
}

def parseRideData(node:Node):RideData = {
  val name = (node \ "@name").text
  val numOps = (node \ "@numberOfOperators").text.toInt
  val heavy = (node \ "@heavyCount").text.toInt
  RideData(name, numOps, heavy)
}

def rideDataToXML(rd:RideData):Node = {
  <ride name={rd.name} numberOfOperators={rd.numberOfOperators.toString}
      heavyCount={rd.heavyCount.toString}/>
}

def parseEmployeeData(node:Node):EmployeeData = {
  val name = (node \ "@name").text
  val rides = (node \ "trainedRide").map(_.text).toList
  EmployeeData(name, rides)
}

def employeeToXML(ed:EmployeeData):Node = {
  <employee name={ed.name}>
    {ed.rides.map(r => <trainedRide>{r}</trainedRide>)}
  </employee>
}

val xmlData = XML.loadFile(args(0))
var years = (xmlData \ "year").map(parseYear).toList
var rideInfo = (xmlData \ "ride").map(parseRideData).toList
var employeeInfo = (xmlData \ "employee").map(parseEmployeeData).toList
```

```scala
def buildWeeklySchedules:Unit = {
  val daysInfo = for (y <- years; m <- y.months; d <- m.days) yield d
  val days = daysInfo.map(_.dayOfWeek).distinct
  for (day <- days) {
    val thisDay = daysInfo.filter(_.dayOfWeek==day)
    val rides = thisDay.map(_.ride).distinct
    val operatorRides = rides.flatMap(ride => {
      val nums = thisDay.filter(_.ride==ride).map(_.numRiders)
      val avg = nums.sum/nums.length
      val rideData = rideInfo.find(_.name==ride).get
      Array.fill(rideData.numberOfOperators+(if (avg>=rideData.heavyCount) 1 else
          0))(ride)
    })
    val totalOps = operatorRides.length
    for (choice <- employeeInfo.combinations(totalOps)) {
      val perms = operatorRides.permutations
      var works = false
      while (!works && perms.hasNext) {
        val perm = perms.next
        if ((perm,choice).zipped.forall((r,op) => op.rides.contains(r)))
          works = true
      }
      if (works) {
        println(day+" - "+choice.map(_.name).mkString(", "))
      }
    }
  }
}


case class RideAverage(ride:String, avNum:Double)
case class OperatorDailyData(name:String, ride:String, numRiders:Int)
case class OperatorRideAverages(name:String, rideAvs:List[RideAverage])
case class OperatorEfficiencyFactor(name:String,factor:Double)

def insertionSortByEfficiency(a:Array[OperatorEfficiencyFactor]):Unit = {
  for (j <- 1 until a.length) {
    var i = j-1
    val tmp = a(j)
    while (i>=0 && a(i).factor>tmp.factor) {
      a(i+1) = a(i)
      i -= 1
    }
    a(i+1) = tmp
  }
}


def rankEmployees(data:List[DayData]):Array[OperatorEfficiencyFactor] = {
  val rides = data.map(_.ride).distinct
  val averages = for (ride <- rides) yield {
    val days = data.filter(_.ride==ride)
    RideAverage(ride, days.map(_.numRiders).sum.toDouble/days.length)
  }
  val dataByOperator = for (day <- data; op <- day.operators) yield {
    OperatorDailyData(op, day.ride, day.numRiders)
```

```scala
  }
  val operators = dataByOperator.map(_.name).distinct
  val opRideAverages = for (op <- operators) yield {
    val opDays = dataByOperator.filter(_.name == op)
    val rideAvs = for (ride <- rides; if opDays.exists(_.ride==ride)) yield {
      val opRides = opDays.filter(_.ride == ride)
      RideAverage(ride, opRides.map(_.numRiders).sum.toDouble/opRides.length)
    }
    OperatorRideAverages(op, rideAvs)
  }
  val operatorFactors = (for (OperatorRideAverages(op, rideAvs) <- opRideAverages)
      yield {
    val factors = for (RideAverage(ride,av) <- rideAvs) yield {
      av/averages.filter(_.ride==ride).head.avNum
    }
    OperatorEfficiencyFactor(op,factors.sum/factors.length)
  }).toArray
  insertionSortByEfficiency(operatorFactors)
  operatorFactors
}

def rideInput(ri:RideData):Array[String] = {
  println(ri.name)
  println(employeeInfo.filter(_.rides.contains(ri.name)).map(_.name).zipWithIndex.
      mkString(" "))
  readLine().split(" +")
}

def inputDay:List[DayData] = {
  println("What day of the week is this for?")
  val dow = readLine()
  println("For each ride displayed, enter the number of riders for the day followed
      by employee numbers from this list with spaces in between.")
  for (ri <- rideInfo;
      val input = rideInput(ri)
      if input.head.toInt>=0) yield {
    DayData(ri.name, dow, input.tail.map(_.toInt).map(employeeInfo).map(_.name),
        input.head.toInt)
  }
}

def inputRideDayData:Unit = {
  println("What month/year do you want to enter data for?")
  readLine().trim.split("/") match {
    case Array(monthText, yearText) =>
      val (month, year) = (monthText.toInt, yearText.toInt)
      if (years.exists(_.year==year)) {
        years = for (y <- years) yield {
          if (y.year==year) {
            y.copy(months = {
              if (y.months.exists(_.month==month)) {
                for (m <- y.months) yield {
                  if (m.month==month) {
                    m.copy(days = inputDay ::: m.days)
                  } else m
```

```
              }
            } else MonthData(month, inputDay) :: y.months
          })
        } else y
      }
    } else {
      years ::= YearData(year,MonthData(month, inputDay)::Nil)
    }
  case _ =>
    println("Improper format. Needs to be numeric month followed by numeric year
        with a / between them.")
  }
}

def hireEmployee:Unit = {
  println("What is the new employees name?")
  val name = readLine()
  employeeInfo ::= EmployeeData(name,Nil)
}

def trainEmployee:Unit = {
  println("Which employee is training for a new ride?")
  println(employeeInfo.map(_.name).zipWithIndex.mkString(" "))
  val empNum = readInt()
  employeeInfo = for ((e,i) <- employeeInfo.zipWithIndex) yield {
    if (i==empNum) {
      val avail = rideInfo.map(_.name).diff(e.rides)
      println("Which rides should be added? (Enter space separated numbers.)")
      println(avail.zipWithIndex.mkString(" "))
      e.copy(rides = (readLine().split(" +").map(_.toInt)).map(avail).toList ::: 
          e.rides)
    } else e
  }
}

def addRide:Unit = {
  println("What is the name of the new ride?")
  val name = readLine()
  println("How many operators does it need?")
  val ops = readInt()
  println("At what rider count should another operator be added?")
  val heavy = readInt()
  rideInfo ::= RideData(name, ops, heavy)
}

var input = 0
do {
  println("""What would you like to do?
1) Add ridership for a day.
2) Add an Employee.
3) Add training to an employee.
4) Add a ride.
5) Get schedule options for a week.
6) Rank Employees.
7) Quit.""")
```

```scala
    input = readInt()
    input match {
      case 1 => inputRideDayData
      case 2 => hireEmployee
      case 3 => trainEmployee
      case 4 => addRide
      case 5 => buildWeeklySchedules
      case 6 =>
        println("What month/year or year do you want to rank for?")
        println(readLine().trim.split("/") match {
          case Array(monthText,yearText) =>
            val year = yearText.toInt
            val month = monthText.toInt
            val y = years.filter(_.year==year)
            if (y.isEmpty) "Year not found."
            else {
              val m = y.head.months.filter(_.month==month)
              if (m.isEmpty) "Month not found."
              else {
                rankEmployees(m.head.days).mkString("\n")
              }
            }
          case Array(yearText) =>
            val year = yearText.toInt
            val y = years.filter(_.year==year)
            if (y.isEmpty) "Year not found."
            else {
              rankEmployees(y.head.months.flatMap(_.days)).mkString("\n")
            }
          case _ => "Invalid input"
        })
      case _ =>
    }
  } while (input!=7)

  XML.save(args(0), <themeParkData>
    {years.map(yearToXML)}
    {rideInfo.map(rideDataToXML)}
    {employeeInfo.map(employeeToXML)}
  </themeParkData>)
```

There is completely new code at the bottom to allow for data entry that is added to the main variables. There is also a do-while loop that handles the menu functionality. The script ends by saving the main data elements back out to the same XML file.

14.4 End of Chapter Material

14.4.1 Summary of Concepts

- XML is a text markup language that can be used to encode arbitrary data. Being text means that it is as easy to work with as flat text files, but it allows you to attach meaning to the values in the file, and the flexibility of the parsing makes it easier to extend XML files than flat text files.

- An XML file is composed of markup and content. The content is plain text. Markup has a number of options that must follow a certain format.

  - The primary markup is tags. A tag starts with < and ends with >. Each tag has a name.
  - The combination of a matching start and end tag defines an element. An element can contain content and other elements.
  - Start tags can be given attributes to store basic information.
  - Special characters that cannot appear as plain text can be specified by markup tokens that begin with a & and end with a ;.
  - You can put comments into an XML file by starting them with <!-- and ending them with -->.
  - The entire contents of an XML file must be held inside of a single element.

- The Scala language has native support for XML. This makes it significantly easier to work with XML in Scala than in most other languages. XML elements can be written directly into Scala source code.

  - An XML file can be loaded using `XML.loadFile(fileName:String):Elem`.
  - The \ and \\ operators can be used to pull things out of XML.
  - You can build XML by typing in XML literals. You can put Scala code into the XML by surrounding it with curly braces.
  - XML can be written to a file using `XML.save(fileName:String,node:Node)`.
  - Patterns can be made using XML with the limitation that attributes cannot be part of the match. Values can be bound by including names in curly braces.

14.4.2   Self-Directed Study

Enter the following statements into the REPL and see what they do. Some will produce errors. You should try to figure out why. Try some variations to make sure you understand what is going on.

```
scala> val xml1 = <tag>contents</tag>
scala> xml1.text
scala> val <tag>{str}</tag> = xml1
scala> val xml2 = <data type="simple">
     | <language>Scala</language>
     | <lesson>Programming is an art.</lesson>
     | <lesson>Software runs the world</lesson>
     | </data>
scala> xml2 \ "language"
scala> xml2 \ "lesson"
scala> val xml3 = <randPoints>
     | {(1 to 20).map(i => <point x={math.random.toString}
         y={math.random.toString}/>)}
     | </randPoints>
scala> (xml3 \ "point").map(p => {
```

```
   | val x = (p \ "@x").text.toDouble
   | val y = (p \ "@y").text.toDouble
   | math.sqrt(x*x+y*y)
   | })
scala> for (<point x={x} y={y}/> <- xml3) yield math.sqrt(x*x+y*y)
scala> val xml4 = <randPoints>
   | {(1 to 20).map(i => <point><x>{math.random}</x><y>{math.random}</y></point>)}
   | </randPoints>
scala> for (<point><x>{x}</x><y>{y}</y></point> <- xml4 \ "point") println(x+" "+y)
```

14.4.3 Exercises

1. Chapter 10 had several exercises where you were supposed to design `case class`es for a number of different types of data. For each of those, write code to convert to and from XML.

   - A transcript for a student.
   - Realtor information for a house.
   - Data for a sports team.

2. On the web site for the book there are a number of XML data files. Write `case classes` to represent that data in each one, and then write code to load the files and build objects from it.

3. Pick some other `case class` that you have written and create code to convert it to and from XML.

4. Find an RSS feed for a website you visit and save the feed as a file. Use Scala to look through the XML.

14.4.4 Projects

1. If you have been working on the different graphics options for earlier projects, the material in this chapter gives you a clear extension, store your geometry data in an XML file instead of a flat text file. You can use tags like "sphere" and "plane" to give meaning to the information. Add in a "light" tag as well in anticipation of adding lighting in a future project.

   After you have the XML format set up and some data to play with, alter the code from project 12.7 to use this data format.

2. Project 10.3 on the text adventure is very nicely extended with the use of XML data files. Using XML also makes it easier to extend the file to include additional information. For this project you should convert your map over to an XML format and have the code read in that format. In addition, you should add items. This will involve adding another `case class` for an item and putting some items into rooms in the XML data file.

   To make the items significant, you need to have it so that the `case class` for your room includes items in that room and the text description of the room lists what items are there. You should also give your player an inventory and implements commands for "get" and "drop". So if the player enters "get" followed by the name of an item in the room, that item will be taken out of the room and added to the players inventory. An "inv" command would be nice to let the player see what is in his/her inventory. If the player uses the "drop" command followed by an item in inventory, that item should be removed from inventory and placed in the current room. If "get" or "drop" are provided with an invalid item name, print an appropriate message for the user.

3. If you did project 13.4 extending your game, you had a text file that specified the high scores for players of a particular game. For this project you should modify your code and the text file to use XML instead of a flat file.

4. If you have been doing the other recipe projects, you can change the format so that it uses XML to save the data. If you do this, you need to add in instructions for the recipes as well as the ability to add comments and other information like how much certain recipes are favored. You can merge what are currently separate text files into a single XML file if you want.

   The fact that your script keeps track of recipes and pantry contents points to one other piece of functionality you should be able to add in. Allow the user to see only the recipes that they have the ingredients to cook. If you have a preference level, sort them by preference.

5. If you have been doing the scheduling options, convert the data file for courses over to XML. In doing this, you can also add the ability to include comments on courses, instructors, or other things that make sense to you, but which did not fit as well in the limited formatting of a plain text file.

6. If you did project 10.8, you can extend this project to use XML encoding for the music file. You probably want to do this by having functionality to load in a plain text record information and add it to an XML file with all the record information. To really take advantage of the XML formatting, you should allow the user to add comments to whatever elements you feel are appropriate.

7. You can extend project 12.14 on your music library by changing the format of the data file from a flat file to an XML file. Use the hierarchical nature of XML to simplify the file. You can have tags for `<artist>` at a higher level with `<album>` tags inside of those and `<song>` tags at the lowest level. With the XML you could add the ability for the user to insert notes about any element that will be displayed in a manner similar to the album cover when the appropriate item is selected in the GUI. Menu options could be used to allow editing of the notes.

8. This is a continuation of project 12.5 on turtle graphics to draw fractal shapes generated with L-systems. You can find a full description of L-systems in "The Algorithmic Beauty of Plants", which can be found on-line at `http://algorithmicbotany.org/papers/#abop`. The first chapter has all the material that will be used for this project and a later one.

   In the last turtle project you made it so that you could use a turtle to draw figures from `String`s using the characters 'F', 'f', '+', and '-'. L-systems are formal grammars that we will use to generate strings that have interesting turtle representations. An L-system is defined by an initial `String` and a set of productions. Each production maps a character to a `String`. So the production `F -> F-F++F-F` will cause any F in a `String` to be replaced by F-F++F-F. The way L-systems work is that all productions are applied at the same time to all characters. Characters that do not have productions just stay the same.

So with this example you might start with F. After one iteration you would have F-F++F-F. You would have F-F++F-F-F-F++F-F++F-F++F-F-F-F++F-F after the second iteration. The next iteration will be about five times longer than that. The `String`s in an L-system grow exponentially in length. As a result, you probably want to have the length of the turtle move for an F or f get exponentially shorter. Start with a good value like 100 pixels and divide by an appropriate value for each generation. For this example dividing by a factor of 3 is ideal. This one also works best with a turn angle of 60 degrees.

The productions for an L-system can be implemented as a `List[(Char,String)]`. You can use the `find` method on the List and combine that with `flatMap` to run through generations of your `String`. You can decide how elaborate you want your GUI to be and if users should be able to enter productions or if they will be hard coded. Look in "The Algorithmic Beauty of Plants", Chapter 1 for examples of other interesting production rules.

9. If you have done the box score options from any of the last three chapters you can extend to use XML encoding for the box score. You probably want to do this by having functionality to load in a plain text box score and add it to an XML file with all the box scores. To really take advantage of the XML formatting, you should allow the user to add comments to whatever elements you feel are appropriate.

Additional exercises and projects, along with data files, are available on the book's web site.

Chapter 15

Recursion

Back in chapter 5 we got our first introduction to recursion. At that point we used recursion to provide iteration. In chapter 8 we learned how to produce iteration using loops and have used that technique more than recursion since that point. If the only capability of recursion was to produce iteration it would not be of much interest in Computer Science because loops would be a complete substitute that have simpler syntax for most uses. However, that is not the case. Recursion allows us to express a lot more than just simple iteration, and because of this, recursion is a critical tool in writing concise solutions to many different problems.

15.1   Memory Layout

To have a proper understanding of recursion, it is helpful to have a mental model of the way that the memory of programs is laid out. The memory of the computer is basically like a huge array of bytes. It is shared between the operating system and many different programs. Different parts of memory can be allocated for different things or associated with different devices. Scala hides most of the intricacies of memory from you. It is not a language designed for doing low-level system programming. At some point in your computer science training you should learn about the details of computer memory and a language that reveals those details. For our purposes here, we will only care about the organization of memory inside of the allocation of a single program.

The memory for a program is broken into two broad pieces, the STACK and the HEAP. These terms were chosen intentionally, and the images they invoke in your mind are probably fairly accurate. A stack is orderly with one item placed on top of the previous one. A heap is much less organized with items placed almost at random. Local variables and function arguments are allocated on the stack. As was discussed in section 7.7, the memory model in Scala is such that the variables are references and they refer to objects. In this memory model, the objects are allocated on the heap. Every time a new function is called, the memory for the arguments and the local variables, along with some memory for bookkeeping, is allocated in a block that is referred to as the STACK FRAME. If that function calls another function, then another frame is allocated on top of it. When a function returns, the stack frame for that function is freed up. That same memory will be used later for another function.

This should help explain the output from a runtime error. The stack implicitly keeps track of where you are in each function when it calls the next. You can picture each of those functions stacked on top of the one that called it. That is what gets printed in the stack trace. Each line tells you what function has been called followed by the file name and line number. This memory of what functions have been called previously and the values of variables in those functions is critical to recursive functions operating properly.

The objects on the heap are allocated in free spaces. The memory for objects is freed up automatically when the object is no longer in use. The automatic freeing of heap memory is accomplished by a process called GARBAGE COLLECTION. An object can be collected if it can no longer be reached by following references that start on the stack. Not all languages include garbage collectors. In those that do not, the programmer is responsible for freeing memory that was allocated on the heap.

15.2   Power of Recursion

To understand the real power of recursion and where this power comes from, we need to revisit some code from the end of chapter 5. Early in that chapter we used a recursive function to count down from a specified number using a single argument. The code for doing that looked liked this.

```scala
def countDown(n:Int):Unit = {
  if (n>=0) {
    println(n)
    countDown(n-1)
  }
}
```

As long as the argument has not gotten below zero, this function prints the number and counts down. We also wrote code to count up using two arguments where one argument was incremented for each subsequent call. At the end of the chapter you were presented with the following code.

```scala
def count(n:Int):Unit = {
  if (n>=0) {
    count(n-1)
    println(n)
  }
}
```

You were told to enter this code and run it to see what it does. If you did so, you might have been rather surprised to see that this code counts up. That seems surprising because the argument is clearly decrementing. The reason this can count up has to do with the memory of the computer and in particular the call stack.

Call Stack

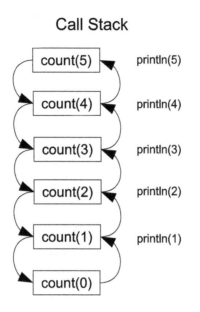

FIGURE 15.1: This shows the call stack for the function count which prints the numbers counting up. When you call the function with 5, it calls itself with 4 before doing anything else. This calls itself with 3 and so on down to 0. The call with 0 does nothing but return. It returns back into the call to count(1) that does the print and continues on back up the stack.

To understand how this works, consider figure 15.1. This shows a graphical representation for different frames on the call stack and what happens when you call count(5). This call immediately calls count(4) before it does anything else. The call to count(4) gets a new stack frame that keeps track of where it was in the call to count(5) so that when it returns it can go back to that point. The call to count(4) also goes straight into a call to count(3), which also gets a new stack frame. This continues until we get down to count(0) which does nothing at all. When the call to count(0) finishes, control returns to the call to count(1) and resumes right after the call to count(0), which is the line with the print statement. So it prints a 1. After the print, count(1) returns to count(2) which similarly does a print. Because the prints are happening as it pops back up the stack, the numbers get printed in ascending order even though the function only includes a decrement. The memory of the stack is essential for this to work because each stack frame remembers its own value of **n**.

This is a very significant point to remember about recursive calls. While the variables in the recursive calls all have the same names, they are not the same variables. It is like multiple people with the same name. Just because two or three people are named "Pat", that does not mean they are the same person. In this case, there were six different versions of **n** that were created, and they took on the values from 5 down to 0. Each was distinct from the others and occupied different parts of memory.

This example shows how the stack can come into play with recursion, but it does not really show the power of recursion. To do that, we need to have a recursive function that can call itself more than once. In the following sections we will look at several different examples of this and see problems that really require the stack and are significantly harder to convert over to loops.

15.3   Fibonacci Numbers

The classic example of recursion is Fibonacci numbers. This is a sequence of numbers where each number is defined as the sum of the previous two. So in mathematical notation this is written as $f(n) = f(n-1) + f(n-2)$. This is not a complete definition, however, because we need to know how the sequence starts. That is to say that we need a base case for the recursion. It is customary to have the first two elements be 1. So, the sequence then is 1, 1, 2, 3, 5, 8, 13, 21, ...

We can write this function in Scala with one short function definition.

```scala
def fib(n:Int):Int = if (n<3) 1 else fib(n-1)+fib(n-2)
```

So for n=1 and n=2 we get back 1. For n=3 we get back 1+1=2. For larger numbers, the process is more complex, and it is instructive to try to visualize it. We do this by considering what happens on the call stack in a manner similar to figure 15.1. Unlike the code for figure 15.1, this function calls itself twice and that makes things a bit more complex.

For example, consider a call to fib(4). This is equal to fib(3)+fib(2). In order to know what that is, recursive calls have to be made to each of those functions. By convention, we will imagine that the calls are processed from left to right. So it calls fib(3) which gets a stack frame. That in turn calls fib(2) which gets a stack frame. Unlike examples we have seen before, when fib(3) returns, there is another call to fib(2) still waiting. So immediately a new stack frame appears where the one for fib(3) had just been. This process of stack frames being removed to be replaced with new ones happens a lot in this calculation of the Fibonacci numbers as well as other recursive functions that call themselves more than once. Instead of drawing this as a vertical stack as was done in figure 15.1, it is customary to draw it in a branching structure called a tree as is drawn in figure 15.2. Each new stack frame is still below the one that called it, but different stack frames at the same depth in the stack are drawn next to one another horizontally. Arrows are used to keep track of things.

In this representation, boxes represent the stack frames. We have dispensed with putting the function names and left only the argument value. The straight arrows show function calls. The curved arrows show returns and are labeled with result values. These types of diagrams can be very helpful to you when you are trying to figure out what happens in a complex recursive function. When you draw them yourself you can dispense with the boxes and arrows. Values with lines between them will typically suffice.

The Fibonacci numbers are a standard example of recursion, but they are not a great one outside of instructional purposes. You can fairly easily write a function that calculates Fibonacci numbers that uses a loop instead of recursion, and it will be a lot faster. That is not true of the examples that follow.

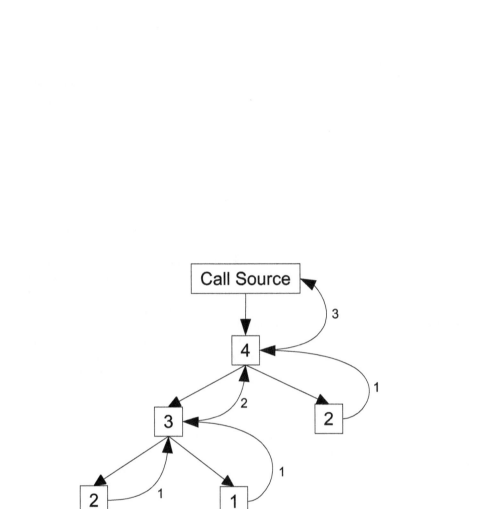

FIGURE 15.2: This figure shows a call tree for `fib(4)`. Boxes represent stack frames. Straight arrows point in the direction of function calls. Curved arrows show returns and are labeled with result values.

15.4 Towers of Hanoi

Our next example of these more powerful recursive functions is somewhat playful. You have likely seen and perhaps even played Towers of Hanoi. It is a "game" played with three pegs and a number of disks of different sizes that have holes in them; so, they can be put on the pegs. Your goal in playing is to move the disks from one peg to another following two rules:

1. You can only move one disk at a time.

2. No disk can be placed on a disk smaller than it.

We want to write a program that can solve this. To do so, we need to first analyze the problem a bit more.

If you start by picturing a peg with 7 disks on it stacked from largest at the bottom to smallest at the top and try to picture the solution, you will probably have a hard time. The first step is clearly to move the top disk to one of the other pegs, but which one? It gets more complex from there. Instead of trying to solve a hard problem, we should start with an easy problem and try to build up from there.

The easiest setup to solve is when you have only one disk. Simply move the disk from where it started to the peg you want it to finish on. That was a bit trivial, so what about two disks? First you have to move the top disk to the peg you do not want to end on, then you move the second disk to the end peg and move the smaller one over on top of it. It might not be clear at this point, but that points toward a general solution. To see this, think of the situation with three disks. We could run through every single move, but we just saw how to move two disks so we can assume we know how to do that. Using that, we move two disks to the middle peg. Note that we do not literally move two disks at a time as that would violate the rules. Instead, we use the approach we just developed to move two disks in three separate moves. Then the largest disk is moved to the end peg and the two disks are moved on top of it.

This points to a general solution. If we make the assumption that we know how to move $N-1$ disks, we can move N disks by first moving the top $N-1$ disks to the off peg, then moving the largest one to the destination, then moving the $N-1$ disks back on top of the largest one. Note that this is a recursive definition. The solution for N disks relies on the solution to $N-1$ disks. The base case is the trivial one where we move a single disk across.

We now have an approach to solving this problem. Next we need to figure out how we want to represent the problem in the memory of the computer. We need to come up with a way to represent disks and pegs. For the disks, all we really care about is the size of the disk. This can nicely be represented by an `Int`. A peg is really just a stack of disks so we need something like an `Array` or a `List` of `Int`s. To decide which, we should consider how they are used. The number of disks on a peg changes over time, but they are always added or removed at one location, the top of the stack. This points to using a `List` where the **head** of the `List` is the top disk. We can use :: to add a disk to a peg and the **tail** is what is left when a disk is removed. The script below shows code that implements the three pegs as an `Array[List[Int]]`. There is a function called `moveDisk` that moves one disk that includes a check to make sure we are not breaking the second rule.

Listing 15.1: hanoiText.scala

```
1  val num = if (args.length>0) args(0).toInt else 7
2  val pegs = Array(List.tabulate(num)(i => i+2), List[Int](), List[Int]())
```

```
 3
 4    def moveDisk(from:Int,to:Int):Unit = {
 5        require(pegs(to).isEmpty || pegs(from).head < pegs(to).head)
 6        pegs(to) = pegs(from).head :: pegs(to)
 7        pegs(from) = pegs(from).tail
 8    }
 9
10    def moveNDisks(n:Int,from:Int,to:Int):Unit = {
11        if (n==1) {
12            moveDisk(from, to)
13        } else {
14            val other = 3-from-to
15            moveNDisks(n-1, from, other)
16            moveDisk(from, to)
17            moveNDisks(n-1, other, to)
18        }
19    }
20
21    moveNDisks(pegs(0).size, 0, 2)
22
23    println(pegs.map(a => a.mkString(" ")).mkString("\n"))
```

The function `moveNDisks` is recursive and relies on the `moveDisk` function so we can feel certain the program is not cheating. The line `val other = 3-from-to` might seem a bit odd at first. The function is only passed the peg numbers to move from and to, but we have to refer to the third peg in the function as that is where the top `n-1` disks are to go first. That line calculates the number of the other peg using the simple observation that if you add the indexes of all the pegs you get 3. The script ends with a call to the `moveNDisks` function followed by a print of what is left at the end. This makes it simple to verify that the puzzle was actually solved.

Given the physical and visual nature of the Towers of Hanoi, the text print is a bit unsatisfying. This is a problem that really deserves to have a graphic display. The following script puts the earlier code inside of a ScalaFX app and uses `Rectangle`s instead of `Int`s so that you can watch the disks move around as the recursion solves the puzzle.

Listing 15.2: hanoi.scala

```
 1    import scalafx.Includes._
 2    import scalafx.application._
 3    import scalafx.event.ActionEvent
 4    import scalafx.scene.Scene
 5    import scalafx.scene.control.Button
 6    import scalafx.scene.paint._
 7    import scalafx.scene.shape._
 8    import scala.concurrent.Future
 9    import scala.concurrent.ExecutionContext.Implicits.global
10
11    val app = new JFXApp {
12      stage = new JFXApp.PrimaryStage {
13        title = "Towers of Hanoi"
14        scene = new Scene(600,300) {
15          def makeDisk(peg:Int, h:Int, size:Int):Rectangle = {
16            val s = size*10
17            val r = Rectangle(100+peg*200-s/2, height.value-h*30, s, 20)
18            r.fill = Color.Red
```

```
19       r.stroke = Color.Black
20       content += r
21       r
22     }
23
24     val num = if (args.length>0) args(0).toInt else 7
25     val pegs = Array(List.tabulate(num)(i => makeDisk(0, num-i, i+2)),
           List[Rectangle](), List[Rectangle]())
26
27     def moveDisk(from:Int,to:Int) {
28       require(pegs(to).isEmpty || pegs(from).head.width.value <
               pegs(to).head.width.value)
29       pegs(to) = pegs(from).head :: pegs(to)
30       pegs(from) = pegs(from).tail
31       Platform.runLater {
32         val p = pegs(to).head
33         p.x = 100+to*200-p.width.value/2
34         p.y = height.value-pegs(to).length*30
35       }
36       Thread.sleep(100)
37     }
38
39     def moveNDisks(n:Int,from:Int,to:Int) {
40       val other = 3-from-to
41       if (n==1) {
42         moveDisk(from, to)
43       } else {
44         moveNDisks(n-1, from, other)
45         moveDisk(from, to)
46         moveNDisks(n-1, other, to)
47       }
48     }
49
50     val button = new Button("Start")
51     content += button
52     button.onAction = (e:ActionEvent) => Future {
53       moveNDisks(pegs(0).size, 0, 2)
54     }
55   }
56 }
57 }
58
59 app.main(args)
```

To make it so that you can watch the animation, there is a Button, that when clicked makes a call to moveNDisks. This happens in a Future so that the animation can be seen. Inside of the moveDisk function, code was added that changes the location of the disk that was moved, and pauses the work for a short period so that you can see it. Right now the rectangles simply disappear from their old position and appear in the new one. It is left as an exercise for the student to use Transitions to animate this process.

If you play with this code a bit you will notice that tall stacks of disks take a very long time to move. This leads to the question of just how many moves does it take to get N disks from one peg to another? The recursive code makes this question fairly easy to answer if we are happy with a recursive function for the answer.

$$f(N) = \begin{array}{ll} 1 & N = 1 \\ 1 + 2 * f(N-1) & otherwise \end{array}$$

This can be simplified to $f(N) = 2^N - 1$. So the number of moves grows exponentially in the number of disks. Stacks of 20 or 30 disks would indeed take a very long time to complete.

15.5 Permutations

This next example is a bit more practical. We are given a `List` of values, and we want to perform some function on all permutations of this `List`. A permutation of a collection contains all the same elements, but the elements are in a different order. This type of behavior might be desired if you have different tasks and you need to find an optimal order in which to perform the tasks based on some rule.

So the next question is, how could we do this? We have seen previously that sequences in Scala have a method called `permutations` which will do for us, but we want to see how we could do it ourselves. To figure this out, we will employ a common idiom when building recursive functions. It is one that we talked about in chapter 5 with our simpler recursive functions. The idea is to take a large problem and break it down so that we solve one step and then apply the same function to solve what is left. In this case, one step is picking a first element. Unlike what we did in chapter 5, there will generally be more than one option here. Any of the elements in the current `List` could be the first element in the permutation. What follows that first element is the permutations of everything else in the `List`. That gives us a recursive definition. We want to pick an element to go first, then make a recursive call on the rest. After that returns, we pick a different element to go first and re-curse again. This should be repeated for all the elements in the `List`.

To convert this into a function we need to figure out what information has to be passed in. The obvious part is that we have to pass in the `List` of values we are permuting. We also need to pass in the elements in the permutation that has been built so far. This can be done as a second `List`. If we made the function such that it returns a `List` of `Lists` with all the permutations, these would be the only arguments we would need. However, such a function is not useful for many situations because the number of permutations grows very quickly with `List` length and the cost in memory would become prohibitive. Instead, we will pass in a function that operates on a `List` that we will call each time a complete permutation is built. That way the caller can decide what is done with the different permutations.

So we are writing a function that takes three arguments, the `List` of numbers we want to permute, a function to apply to any finished permutations, and the current `List` for this permutation.[1] This is done in the following code.

```
def permute(nums:List[Int],f:(List[Int])=>Unit,p:List[Int]):Unit = {
  if (nums.isEmpty) {
    f(p)
  } else {
    var before=List[Int]()
    var after=nums
```

---

[1]This last argument is something of an implementation detail that we could hide using a nested function if we desired. We do not do that here to keep things a bit simpler.

```
    while (!after.isEmpty) {
      val perm = after.head :: p
      permute(before ::: after.tail,f,perm)
      before ::= after.head
      after = after.tail
    }
  }
}
```

When the `List` is empty, we have a full permutation, and we call the function on it. Otherwise, we have a loop that runs through all the elements we want to permute and makes recursive calls assuming each one has been added to the head of the permutation. This is done using two variable `List`s and a `while` loop. To start off with, one `List` is empty and the other is all the elements in the input `List`. In the loop we not only make the recursive call, we also move elements from one `List` to the other. We can check to see that this function works by calling it like this.

```
permute(List(1,2,3), println, Nil)
```

When we make this call, we get the six different permutations of the list with three elements and then simply print them out. The first call should always use `Nil` for the last argument.

This function displays an interesting feature that is generally true of recursive functions. Because it is defined in terms of itself, we write it assuming that it works. When we are done writing it, it will work. This logic seems circular, but it is actually using a method called induction. First we make the function work for the simple base case. Every case above that is defined in terms of a smaller one. Eventually this gets down to the base case, which generally works in a trivial way. If the base case works, the case right above it should work. When the one above the base case works, the one above that should work. This logic progresses upward to allow us to solve problems of arbitrary size.

We mentioned earlier that we do not return a `List` of the permutations because there can be a lot of them and that would require a lot of memory. It is worth taking a second to get a slightly better understanding of what "a lot" means here. Specifically, we would like to know the order of the number of permutations. This will not only tell us how long a list would be if we built one, it will give us a measure of how long this function will take to run for any task that we might give it as it will be the number of times that the function parameter, `f`, gets called.

As is normally the case for order analysis, we will think of things in terms of the size of our input. In this case, that is the size of the input `List` which we will call n. The first call to permute makes n calls to itself, each of which is passed a `List` with $n - 1$ elements. Those then make $n - 1$ calls with `List`s of size $n - 2$. This process continues until we get down to 0 elements in the `List`. If you picture this as a tree similar to that in figure 15.2, we have a top box with n branches off it. Those lead to n boxes that each have $n - 1$ branches off of them. So at the third level we have $n * (n - 1)$ boxes. At the next level we have $n * (n - 1) * (n - 2)$. In you continue this down to 1 we get the factorial function that we played with way back in chapter 5. Indeed, the order of the permutation function is $O(n!)$. If you recall from our earlier discussion, the factorial function grows very quickly. So quickly in fact, that we had to use the `Long` or `BigInt` types if we wanted to use even modestly large inputs. Clearly you do not want to try to get all the permutations of `List`s with much more than 10 elements in them. Depending on what you are doing, even lists of 10 elements might take a while.

15.6   Mazes

Another class of problems that works well with recursion is those involving mazes. This might also seem to be more for recreation, but mazes are a simplified case of something called graphs which are very important in computer science and mathematics and are used to represent all types of different problems. The same approaches that we will use here for mazes apply to graphs as well. There are also a number of applications where doing things like finding the optimal route through some type of restricted path like a maze is significant.

To keep things simple, we are going to use a rather basic approach to building our mazes. Instead of having a grid with walls between cells, we will have a grid where complete squares can either be open or be occupied by a wall. This representation means that we can use a 2-D `Array` of values that tell us if there is a wall or not. In theory we could use an `Array[Array[Boolean]]` for this purpose, but in practice we will have use of being able to put numeric values in the "rooms" so an `Array[Array[Int]]` will be more practical.

We will define the maze with code like the following.

```
val maze=Array(Array( 0, 0, 0, 0, 0, 0, 0, 0, 0, 0),
               Array(-1,-1,-1, 0,-1,-1,-1,-1,-1,-1),
               Array( 0, 0, 0, 0,-1, 0, 0, 0,-1, 0),
               Array( 0,-1, 0,-1,-1,-1,-1, 0, 0, 0),
               Array( 0,-1, 0,-1, 0, 0,-1,-1,-1, 0),
               Array( 0,-1,-1,-1,-1, 0, 0, 0, 0, 0),
               Array( 0,-1, 0, 0,-1, 0,-1,-1,-1, 0),
               Array( 0, 0, 0,-1,-1, 0,-1, 0, 0, 0),
               Array( 0,-1, 0,-1,-1, 0,-1, 0,-1,-1),
               Array( 0,-1, 0, 0, 0, 0,-1, 0, 0, 0))
```

This builds a 2-D `Array` of `Int`s that uses a 0 to represent an open square and a -1 to represent a wall. The use of -1 for a wall is intentional so that positive numbers can be used to represent other things later. This particular maze is only 10x10 in size. This is a good size to start off with for our purposes though it does not allow for a very complex maze.

Our first task is the thing that most people would probably want to do with a maze. We want to know how to get through it in the shortest number of steps. To start with, we will ask the simple question of how many steps long is the shortest path. This has a significant advantage over trying to actually return a path. Not only is it a simple numeric value, it is also unique. There can be multiple paths all with the same length. That produces ambiguity if we ask for the optimal path. It also happens that having a function to find the length of the shortest path is sufficient to find a path, as multiple calls to this function can be used to construct the optimal path even though such a construction is less than optimal.

You can write a function to find the length of the shortest path using loops, but it is fairly complex code. We can write a simpler solution using recursion, and the recursive approach will easily adapt to solving other problems as well. As with all recursive functions, we can break the solution down into two broad parts, base cases and recursive cases. The base cases should be trivial to solve. There are two good base cases for the maze problem. One base case is when you reach the exit. If you are at the exit it takes zero steps to get out. Another base case is if you are checking a location which is out of the maze or in a wall. For that we want to return a value that could not possibly be a solution. We will come back to specifically what that value should be after discussing the recursive cases.

Picture yourself standing at a location somewhere in the maze, and you want to know how many steps it takes to get from your current location to the exit on the shortest path.

The recursive way to do this is to imagine that you had an oracle that could tell you that answer from any of the neighboring squares and figure out how you would combine those values to determine the distance from where you are. As we said earlier in the chapter, we write a recursive function by assuming we have one that works, and we write based on that assumption. Once we are done, assuming we have handled the cases properly, it will work. The oracle is our assumed working function. If you are in the maze and you can ask the oracle to tell you how many steps it takes to get out from one step to the north, south, east, and west, how would you use that information to determine the shortest distance from where you are?

To make this more concrete, picture yourself at a four way intersection. You ask the oracle about each direction and get values of 7, 13, 4, and 19. So the shortest path lies in the direction that gave you back 4. However, that was the distance from one step in that direction. The minimum distance from where you are is one greater than that, or 5 steps. It is the minimum of the four, plus one step to get to that other location. The solution of taking the minimum also points us in the proper direction for the base case of hitting a wall or going out of bounds. If the best solution is the minimum, we should return a number that cannot possibly be the minimum for a bad path. A large number such as one billion will suffice for virtually any maze. You might be tempted to use `Int.MaxValue`, but remember that we are adding 1 to it. Try doing `Int.MaxValue+1` in the REPL and you will see why that is not an ideal choice.

That defines our recursive case. There is only one other factor that needs to be considered. If you have ever been in a hedge maze or some other maze where you cannot see over the walls, it can be quite difficult to tell one location from another, and unless you do something to mark your previous locations you can wind up running in circles and never finding your way out. A rather famous solution to this problem was described by the Brothers Grimm in Hansel and Gretel who left breadcrumbs to mark their path. We will choose the same approach. Assuming you have no birds in your computer this should work better for you than it did for the children. If you do have birds in your computer, you should probably deal with that before reading on.

So the last facet that we need for our algorithm is the ability to drop down breadcrumbs to mark our path. If we ever come back upon our breadcrumbs we will treat it just like being in a wall or out of bounds. We will also sweep up our breadcrumbs before we return from a path. This was not required by Hansel and Gretel who simply wanted to be able to find their path. We, however, want to find an optimal path and leaving breadcrumbs all over the place would often cause us to miss such a path. Converting this all to code produces the following.

```scala
def shortestPath(maze:Array[Array[Int]],x:Int,y:Int,endX:Int,endY:Int):Int = {
  if (x<0 || y<0 || y>=maze.length || x>=maze(y).length || maze(y)(x)!=0) {
    1000000000
  } else if (x==endX && y==endY) {
    0
  } else {
    maze(y)(x)=1
    val dist=(shortestPath(maze, x+1, y, endX, endY) min
      shortestPath(maze, x-1, y, endX, endY) min
      shortestPath(maze, x, y+1, endX, endY) min
      shortestPath(maze, x, y-1, endX, endY)) + 1
    maze(y)(x)=0
    dist
  }
}
```

The whole function is built from if expressions. The first option is for going out of bounds, hitting a wall, or coming across one of our breadcrumbs, and it returns one billion, a number so large it cannot possibly be a valid path, but one small enough that we also will not overflow it by adding 1 for other steps through the maze. The second case is where we have reached the end. These two can be reversed and would need to be if the exit were outside of the bounds or marked by a special value in the array. The last case is the recursive case which drops a breadcrumb, makes four recursive calls and combines them with min, then picks up the breadcrumb and returns. This code could be invoked with this call.

```
println(shortestPath(maze, 0, 0, 9, 9))
```

For the maze shown above, the return value from this call is 36. For this particular maze, that is the length of the only path from 0,0 to 9,9. You should play with the maze some to make sure that the function finds the shortest path for other configurations.

This function literally tests every single path through the maze and gives us back the shortest. There are strengths and weaknesses to this approach. The weakness is that if there are a lot of paths this could take a while. We explore that more and consider ways to address it in *Object-Orientation, Abstraction, and Data Structures Using Scala*[1]. The strength of trying every path is that it takes only very minor modifications of this code to produce other similar functions such as one to find the longest path or to count up how many paths there are. These particular problems are left as exercises for the student, but we will tell you now that the only real changes are to alter the return values of the base cases and the way in which the recursive calls are combined.

15.7   Sorts

In chapter 13 we used loops to provide most our iteration. We only used recursion when working with `Lists`, and the sorts were longer than the imperative versions. We could also write those sorts recursively using `Arrays`, but then the use of recursion is only a change in presentation. It does not really alter how the sorts run or how efficient they are. The recursion in these functions is only used for iteration, nothing more. However, recursion also opens the door to some more efficient sorts. The next two subsections describe two of these.

15.7.1   Divide and Conquer Sorts

Recursion that calls itself multiple times opens up a new style of problem solving called DIVIDE AND CONQUER. The idea is exactly what the name sounds like. You take a big problem, divide it up, conquer/solve the pieces, then build a solution to the larger problem from the solutions to the pieces. The divide and conquer approach is very general and can be applied to many types of problems. In this section, we will look at two different sort algorithms that are built on this idea.

15.7.1.1   Merge Sort

The first sort we want to consider is the MERGE SORT. The general idea of a merge sort is that we are given a collection that we break into two roughly equal pieces. Each of those pieces is sorted individually and then the sorted results are merged back together.

This is a fast sort algorithm because the merge operation is $O(n)$. The reason for this is that when you want the lowest element from either collection, you do not have to look through all the elements, you only need to consider the elements at the low ends of the two collections. That means you only do one comparison. To find the next lowest element you again do one comparison. So you can make a sorted collection of n elements from two smaller sorted collections with only $n-1$ comparisons. The fact that you are repeatedly cutting the collection in half means that you get down to a single element in $\log_2(n)$ cuts. This gives an overall performance of $O(n \log(n))$. For large values of n, this is much better than the $O(n^2)$ performance of the sorts considered earlier.

There is one minor problem with the merge sort: it cannot be done in-place. Recall that being done in-place means that the sort happens in the original collection without the use of additional memory. Merge sort requires additional memory at least proportional to n to complete the sort. A well crafted merge sort can get everything done with one additional `Array` of length n. Getting the memory usage down to that level is a bit complex so it is not presented in this book.[2] In this chapter, we will use the inability to do the sort in-place as an excuse for using `Lists`, which requires extra work space anyway because they are immutable.

The first solution that we will look at is all recursive. It has a recursive `mergeSort` function that takes a `List[Int]` and returns a `List[Int]`.[3] The sort function uses a second recursive function called `merge` to put the sorted pieces together.

---

[2]Such an implementation is presented in *Object-Orientation, Abstraction, and Data Structures Using Scala*[1].

[3]Recall that unlike sorts with `Arrays`, a sort using a `List` must return a new `List` as the original one cannot be altered.

Listing 15.3: mergeSort.scala

```scala
def merge(lst1:List[Int], lst2:List[Int]):List[Int] = (lst1,lst2) match {
  case (Nil,_) => lst2
  case (_,Nil) => lst1
  case (h1::t1, h2::t2) =>
    if (h1<h2) h1 :: merge(t1, lst2)
    else h2 :: merge(lst1, t2)
}

def mergeSort(lst:List[Int]):List[Int] = lst match {
  case Nil => lst
  case h::Nil => lst
  case _ =>
    val (l1, l2) = lst.splitAt(lst.length/2)
    merge(mergeSort(l1), mergeSort(l2))
}
```

Both of the functions use pattern matching for the different cases in the recursion. This makes them nice, short, and easy to read. Unfortunately, the recursive version of `merge` has to allocate a new stack frame for every element that is merged. That means that this version of the code cannot scale up to large sizes.

To get around this limitation we need to use a more imperative `merge` that works with a `while` loop. Such a version might look like the following.

```scala
def merge(lst1:List[Int], lst2:List[Int]):List[Int] = {
  var l1 = lst1
  var l2 = lst2
  var ret = List[Int]()
  while (l1.nonEmpty && l2.nonEmpty) {
    if (l1.head<l2.head) {
      ret ::= l1.head
      l1 = l1.tail
    } else {
      ret ::= l2.head
      l2 = l2.tail
    }
  }
  if (l1.nonEmpty) ret :::= l1.reverse
  else ret :::= l2.reverse
  ret.reverse
}
```

This version is not as pretty and is about twice as long. However, it handles significantly longer `List`s making it useful if you actually have a large set of data.

One last thing to note about the merge sort is the way in which it does work. As the recursive calls go down the stack, very little happens. The input `List` gets split in half and those two halves get passed down. This repeats until we reach a single element. The real work happens as the recursion pops back up the stack. Before a function can return, it calls `merge`, which does the real work. So if you picture in your head something like 15.1, the down arrows are not associated with much work. It is the up arrows where the work is done.

15.7.1.2 Quicksort

Another divide and conquer sort is QUICKSORT. The idea of quicksort is to pick a special element called the pivot, then move it to the correct location with all the elements that are less than it before it and all those that are greater than it after it. Those elements to either side will not yet be sorted so, recursive calls are made to sort each of them. The recursion terminates when it gets down to a single element.

Unlike merge sort, quicksort can be done in-place. However, that is more complex and is not covered in this book.[4] For now we will make the quicksort like our merge sort and work on a `List`, for which the idea of being done in-place does not make sense.

The quality of a quicksort is largely dependent on the selection of a pivot. A good pivot will be in the middle of the collection to cut it in half. A really bad pivot would be the minimum or maximum element which does nothing to split up the data and only pulls out that one element from the next level of the recursion. Keeping with the idea of simplicity for the implementations in this chapter, we will use the first element as the pivot. This is not a good way to do this in general and can lead to very bad behavior on sequences that are already sorted.[5]

With these simplifications in place, we can write a version of quicksort with the following code.

Listing 15.4: quicksort.scala

```scala
def quicksort(lst:List[Double]):List[Double] = lst match {
  case Nil => lst
  case h::Nil => lst
  case _ =>
    val pivot = lst.head
    val (less, greater) = lst.tail.partition(_<pivot)
    quicksort(less) ::: (pivot :: quicksort(greater))
}
```

The `pivot` is set to the `head` of the `List`. We then use `partition` to split the rest of the elements between those that are less than the `pivot` and those that are not. Finally, we call `quicksort` on those two sublists and put the whole thing together into a result. The result is a short and simple sort function that actually works fairly well on random data.

It is worth asking what the order of this sort really is. Unlike the merge sort, which is very stable in the amount of work that it does, the performance of our quicksort can vary dramatically depending on the input. Like the merge sort, each level of the recursion does $O(n)$ work. In this case that work is the `partition` and sticking `List`s together. If the pivot is in the middle of the values, each level the data is cut in half and we get $O(\log(n))$ levels for the recursion and $O(n \log(n))$ overall performance. For random data, this is the expected behavior. However, our simple pivot selection can lead to very poor behavior and this sort is $O(n^2)$ if the input is already sorted.

The last thing to note about quicksort is that unlike merge sort, it does most of its work going down the call stack. The call to `partition` is where all the comparisons happen. This `List` version does have to merge `List`s on the way back up the call stack. A version written to work in-place with `Arrays` will not have to do even that.

---

[4]Code for an efficient quicksort is given in *Object-Orientation, Abstraction, and Data Structures Using Scala.*[1]

[5]Code for doing better pivot selection is also presented in *Object-Orientation, Abstraction, and Data Structures Using Scala*[1].

15.8   Putting It Together

To see another use of recursion we want to solve a problem that we have done before using a different approach. Twice now we have included code that makes suggestions for employee schedules based on data about ridership. The work of finding schedules was done using the **permutations** and **combinations** methods on sequences. As we saw earlier in this chapter, recursion can also be used to generate permutations. It can do combinations as well.

Instead of repeating exactly what we did before, we are going to use recursion to push it a bit further. Instead of showing possible groups of employees for each day, we want to show possible schedules for a full week that include a limit on how many days each week any given ride operator is willing/able to work.

We will not repeat the entire code from chapter 14 for the various menu options. This simply shows a function that can be used to build these more complete schedule types. For this to work, a **daysPerWeek:Int** member was added to the **EmployeeData case class**. You can also modify menu option 5 to call this function.

```scala
def recursiveBuildWeeklySchedule {
  val daysInfo = for (y <- years; m <- y.months; d <- m.days) yield d
  val days = daysInfo.map(_.dayOfWeek).distinct

  case class WorkerDays(name:String, numDays:Int)
  case class WorkerAssigns(day:String, workerRide:List[(String, String)])

  def printSchedule(schedule:List[WorkerAssigns]) {
    println("Possible Schedule:")
    println(schedule.mkString("\n"))
  }

  def recurByWorker(daysLeft:List[String], workerAvail:List[WorkerDays],
      schedule:List[WorkerAssigns], workersLeft:List[String],
      ridesNeedingOps:List[String]) {
    if (ridesNeedingOps.isEmpty) {
      recurByDay(daysLeft, workerAvail, schedule)
    } else if (workersLeft.length>=ridesNeedingOps.length) {
      val worker = employeeInfo.filter(_.name == workersLeft.head).head
      for (ride <- worker.rides) {
        ridesNeedingOps.indexOf(ride) match {
          case -1 =>
          case i =>
            val newAvail = (for (w <- workerAvail) yield {
              if (w.name == worker.name) w.copy(numDays = w.numDays-1)
              else w
            }).filter(_.numDays>0)
            val newSchedule = schedule.head.copy(workerRide = (worker.name, ride) ::
                schedule.head.workerRide) :: schedule.tail
            recurByWorker(daysLeft, newAvail, newSchedule, workersLeft.tail,
                ridesNeedingOps.patch(i,Nil,1))
        }
      }
      recurByWorker(daysLeft, workerAvail, schedule, workersLeft.tail,
          ridesNeedingOps)
```

```scala
    }
  }

  def recurByDay(daysLeft:List[String], workerAvail:List[WorkerDays],
      schedule:List[WorkerAssigns]) {
    if (daysLeft.isEmpty) {
      printSchedule(schedule)
    } else {
      val day = daysLeft.head
      val thisDay = daysInfo.filter(_.dayOfWeek==day)
      val rides = thisDay.map(_.ride).distinct
      val operatorRides = rides.flatMap(ride => {
        val nums = thisDay.filter(_.ride==ride).map(_.numRiders)
        val avg = nums.sum/nums.length
        val rideData = rideInfo.find(_.name==ride).get
        Array.fill(rideData.numberOfOperators+(if (avg>=rideData.heavyCount) 1 else
            0))(ride)
      })
      recurByWorker(daysLeft.tail, workerAvail, WorkerAssigns(day, Nil)::schedule,
          workerAvail.map(_.name), operatorRides)
    }
  }

  recurByDay(days, employeeInfo.map(e => WorkerDays(e.name, e.daysPerWeek)),
      List[WorkerAssigns]())
}
```

This is a significantly more complex recursive function than what we looked at previously. It is worth taking some time to study what is going on. The recursion is broken into two separate functions that are nested in the primary function. The primary function does little more than pull together the information on the days and then make a call to `recurByDay`.

As the name implies, the `recurByDay` function is using days as the primary recursive argument. Each level down the call stack has one fewer elements on the `daysLeft` List. The base case is when `daysLeft` is empty. That means we have a full schedule for the week, and we are ready to print it. This function borrows code from the earlier version to build up a `List` of ride names with multiple copies for the number of operators needed.

All of the information that comes into `recurByDay`, as well as other information that it figures out is passed into `recurByWorker`. The name here is again meant to give an image of what is going on. This function re-curses through the workers, considering one worker on each call. The base case is when there are no more rides that need operators. In the recursive case the code runs through all the rides that a worker is trained to operate, and if any of them still need operators, one recursive branch is taken with that worker assigned to that ride. There is also a recursive branch at the end where the current worker is not assigned to work that day.

Note that the base case of `recurByWorker` contains a call back to `recurByDay` so these are MUTUALLY RECURSIVE FUNCTIONS. It is reasonably common in more complex situations, when the recursion needs to run through a space that has several different independent parameters, to break the problem up like this into different functions that handle changes in different options and depend on one another to get a complete solution.

15.9 End of Chapter Material

15.9.1 Summary of Concepts

- Recursion can be used for much more than iteration. The memory of the stack frames on the call stack allows recursive functions to call themselves more than once and try different possibilities for solving a problem.

- One of the most common examples of a recursive function that calls itself more than once is a function to generate the Fibonacci sequence. Each number in this sequence is the sum of the previous two, so this recursive function calls itself with arguments one less than the current value and two less than the current value.

- A more interesting example is the Towers of Hanoi. Here we saw that if we knew how to solve the problem with N disks, we could extend it to $N + 1$ disks. This and a base case constitutes a complete solution for any value of N.

- Finding the shortest path through a maze works well as a recursive problem. At each point you need to find out the distance from the different options you have. Those can be combined to give you an appropriate value for the current location.

- Using a recursive divide and conquer approach to sorting leads us to two other sorts that do $O(n \log(n))$ comparisons instead of the $O(n^2)$ of our previous sorts.
    - A merge sort repeatedly breaks the collection in two going down the call stack, then merges the sorted results as it comes back up the call stack.
    - Quicksort works by picking a pivot and putting it in the right place. It then recurses on the elements that are less than the pivot as well as those that are not and sticks all the results together.

15.9.2 Exercises

1. Write functions that calculate Fibonacci numbers using the following approaches.
    - Using a loop with an `Array` or `List`.
    - Using a loop with three `var` declarations.
    - Using a recursion function that takes three arguments, but only calls itself once each time.

2. Write a function that will calculate the longest non-self-intersecting path through a maze.

3. Write a function that will count the number of non-self-intersecting paths through maze.

4. Find the size of the biggest completely empty maze on which you can do one of the recursive search functions that finishes in a minute or less.

5. The following is a function called the Ackermann function which is significant in theoretical computer science. Put this into Scala and play with it a bit.

$$f(0, n) = n + 1$$
$$f(m + 1, 0) = f(m, 1)$$
$$f(m + 1, n + 1) = f(m, f(m + 1, n))$$

6. Write a recursive function that will build the power-set of a `List`. The power-set is the set of all subsets of a given set. The fact that it is a set means that we do not care about order, unlike with permutations. For example, the power set of `List(1,2,3)` is `List(Nil, List(1), List(2), List(3), List(1,2), List(1,3), List(2,3), List(1,2,3)`.

7. Write a recursive function that builds combinations of a `List` instead of permutations. The caller should be able to specify how many elements are desired in the combination. (Hint: All combinations of all sizes are part of the power-set.)

15.9.3 Projects

1. If you have been doing the ray-tracing and graphics options, this project continues with that. Ray tracers can implement both reflections and transparency using recursion. If you have a function that tells you what color to draw for a particular ray, you can do reflection by recursively calling that function with the reflected ray and combining that color with the color of the reflecting surface. How much of each goes into the color depends on the fractional reflectivity of the object.

   For this project you will give your objects colors, but you still will not worry about lighting. So when a ray hits an object, it gets the full color of that object unless the object is reflective. If it has reflectivity R then it gets $(1-R)$ of the color of that object and R of the color from the reflected ray. To add colors you need to be able to build your own colors and get the components out of colors. The `scalafx.scene.paint.Color` class has methods called **red**, **green**, and **blue** that return the amount of each primary in a color. The values are `Doubles` in the range of 0-1 inclusive. You can also make a new `Color` object by calling `Color(red:Int,green:Int,blue:Int)`. The values passed in must be in the same range or you will get an error.

   To calculate reflection, you need the unit-normal vector for the surface you are reflecting off of. For a plane this is easy as the normal is the same everywhere. For a sphere, this is the vector from the center to the point of intersection. In both cases, you want it normalized. If you have that, the direction of the reflected ray is given by

   $$\overrightarrow{r}_{reflected} = \overrightarrow{r} - 2\left(\overrightarrow{r} \cdot \overrightarrow{n}\right)\overrightarrow{n},$$

   where $\overrightarrow{r}$ is the direction of the incoming ray, and $\overrightarrow{n}$ is the unit normal. The unit normal will also be significant for calculating lighting in a later project.

2. For this project you will write a program that solves 9x9 Sudoku puzzles. It will do this through a recursive function.

   The input will be nine lines of text, each with nine characters on it. The characters will either be a space or a single digit number. You should output a board with all the blanks filled in and put spaces between the numbers to keep it readable. It turns out that most puzzles have more than one solution. You only need to provide one.

For an extra challenge, make it so your program has a GUI. It should load puzzles from text files and use a `GridPane` to display the puzzle. Users should be able to select a solve option to have all spaces filled in or a hint option to add just a few.

3. For this project, you will write a program that recursively parses a string for an arithmetic expression and returns the value of it. Examples would be 5+7 or 9*(5+7.5)/2. Your parser should do proper order of operations (things in parentheses bind highest, * and / before + and -, and go from left to right for the same level of priority).

The approach I want you to take for solving this problem is with a divide and conquer recursive algorithm that breaks the problem into pieces starting with the lowest priority operation. You will write a function `parse(exp:String):Double`. This function will return the value of the expression in the `String exp`. First it should find the lowest priority operator (it cannot be in parentheses). If it finds one, it recurses twice with the substrings on either side of that operator (use the `substring` method of `String` or `take` and `drop`). If there is not an operator that is not in parentheses you can check if the `String` starts with '(' and pull the bounding parentheses off and recurse on what is left. If it does not start with '(' you know that part of the expression is just a number so use `toDouble` to get the value.

The user will give you a formula, that does not include any spaces, at the command line and you should simply print the value it evaluates to. So a potential invocation of your program might be as follows: `scala parser.scala 5+3*(70/5)`.

4. If you did project 14.8, you can extend it in this project. If you go further into chapter 1 of "The Algorithmic Beauty of Plants", you will find that you can use L-systems to model grasses and trees. We will not do the 3-D implementation, but you can add handling for '[' and ']' to allow branching structures. This is best done using recursive calls that start on a '[' and return on a ']'. Instead of passing in a `String`, pass in the `Iterator[Char]` that you get by calling the `iterator` method on a `String`. The advantage of the `Iterator` is that the elements are consumed when you call next so that when you return from a function call, the code will automatically be working on what remains after the ']'.

5. Determining if a particular recipe can be made with items from your pantry is not all that hard and does not require using recursion. Planning an entire meal for a large dinner party is another matter. To do this, the recipes need to be annotated with what type of dish they are: entrée, dessert, side, etc. When planning a meal, the user must be able to specify how many of each type of dish they wish to make for the meal. You can then use recursion to find the meals that can be made fitting those requirements with ingredients that are on hand.

You should attach information on how much the user likes certain recipes so that only the top 5 or so meals are shown. If you want a bit of extra challenge, consider handling the situation where there are not any meals that can be made with what is on hand and then you list top meals that need few additional ingredients that would serve as a grocery list.

6. If you have been doing the schedule building problems, you can now extend the functionality using recursion. Annotate each course with how it fits into your curriculum. Then you can specify not only how many hours you want to take, but also what requirements you want to fulfill.

This is a problem that you could solve using `permutations` and `combinations`, but as the number of course options grows, those options become less efficient. Using recursion, you can cut the recursion short for any combination if you determine that it cannot possibly work. For example, if you hit the proper number of hours before you are done with courses, you do not need to consider variations in the remaining courses, just check if you have satisfied the other requirements.

7. Having the ability to do recursion opens up a lot of possibilities for the motion of computer controlled characters in games that have barriers. Moving straight to a location is fine if the playing space is open and empty, but when there are obstacles, like in a maze, it is important to have a way to navigate them. Recursion gives you a way to do this. For this project you can implement your choice of a simple game with enemies that uses recursion to find ways around obstacles. Note that you probably do not want the enemy to have the ability to follow the shortest path to the player unless the player has a significant speed advantage. Instead, you can throw in some randomness to take choices that are somewhat less than optimal.

8. If you have been doing the text-adventure/text-map project options, you can write some utility functions that can be used to help check out the map. Recursion can let you see things like if you can get from one room to another, how many steps it takes, how many paths there are, or even the paths themselves. You can implement these as new commands when you run the program. For example a "canReach" command could be given a room number/identifier to see if the specified room can be reached from the current room.

9. If you have been working on the music library, you can throw some recursion into that as well if you give a few hints. You would need to annotate songs with hints as to what are good options for songs to follow it. You can make the recursion only follow from one song to another when the user has recommended that it is worth doing. Those recommendations can be given a "strength" as well. The user should select a starting song, and the program should find the play list with the highest total strength for all the connections.

   Note that this is a problem where you could run into problems if there are lots of songs with lots of connections. Doing this with a simple recursive algorithm could lead to extremely long run times. However, assuming that the user does not enter too many connections, recursion should work fine as long as you do not test song combos that have not been marked by the user. This is why you cannot simply consider every ordering of songs.

10. There is a common fractal structure known as Sierpinski's triangle that is created by a repeated pattern of triangles. Figure 15.3 shows the output of a program to create these. A way to view this structure is that you start with a single triangle from which you delete the triangle created by connecting the midpoints of the segments. That leaves three smaller triangles, and you then delete the middle triangles from those and repeat this process infinitely. You can also view it as replacing a triangle with three smaller triangles where the middle is missing and then repeating that process on the smaller triangles over and over. Either of these processes can be seen as a recursive method. In figure `fig:Sierpinski` the triangle on the left went down two levels of recursion while the triangle on the right went down seven levels. For this project you should write a program using recursion to generate an image like that shown in the figure. You only need to show one triangle, and you should be able to easily alter the level of detail in your output.

11. Given a regular 8x8 chess board, write a recursive program that finds and displays

FIGURE 15.3: This figure shows a Sierpinski's triangle.

all the possible ways to place 8 queens on the board in such a way that they cannot attack each other.

12. Write a program that finds all the occurrences of a word in all the files under a directory, recursively, printing out the filename and line each time one is found. Your program should take a directory and the word to search for as command line arguments. In order to do this, you need to use some additional capabilities of the `java.io.File` type, which was introduced back in chapter 9. In particular, you need to use the `isDirectory():Boolean` and `listFiles():Array[File]` methods. Use `listFiles()` to get all the files in a directory, search the contents of the non-directories, and recursively descend into the directories.

Additional exercises and projects, along with data files, are available on the book's web site.

Chapter 16

Object-Orientation

From the beginning we have said that Scala is a completely object-oriented language. You have been using objects and calling methods on them since back in chapter 2 when we were taking our first steps in the language. Despite this, we have actually done very little with object-orientation. This has been intentional. So far we have been building up your logic skills and teaching you to break problems up into different functional units. Now that your problem solving skills have reached a sufficient level we will take the next step to doing full object-oriented decomposition of problems. We finish this book by giving you a brief introduction to object-orientation and how to do it in Scala.

16.1   Basics of Object-Orientation

The basic idea of an object is that it is something that contains both data and the functionality that operates on that data. This combining of data and the functionality that works on the data is commonly referred to as ENCAPSULATION. We grouped data originally with tuples and then in chapter 10 we started using `case class`es to group data together in a more meaningful way. What we did not do with either of these methods was to bind any functionality to them directly. We wrote functions that could operate on them, but it was independent of the data itself. In the case of tuples, we cannot bind the functionality into objects. `case class`es do allow us to do this; however, we simply did not to keep things simple.

There are many different ways to think about object-orientation in programs. It is common to have beginners think of the description of the problem and the nouns in the description become objects while the verbs are methods on those objects. This is a bit restrictive, but it can be useful. A better description is that you should think of objects as having certain responsibilities, and they package together the data and functionality associated with a particular responsibility. Whichever way you want to think about it, it helps to go through some examples.

16.1.1 Analysis and Design of a Bank

We begin with a very standard example of a bank. Before we can really figure out what the code looks like, we need to know more about the problem that we are trying to solve. Without a proper description of the problem, you really do not have any chance of solving a problem. This process of properly defining the problem that you are going to solve in natural language is called analysis which was introduced back in section 6.6. Analysis is one of the first steps in any software development effort. It is something that has generally been done for you in this book.

When designing solutions to problems, programmers often need to be concerned with how user friendly their programs are for the people who will be using them. However, we are not going to worry too much here about the user interface for this system, but instead focus on the type of functionality that a simple bank system should have. We want to represent a bank where people have accounts. Our bank example only has two different types of accounts: checking and savings. Customers can make deposits to and withdraws from these accounts. Each customer can have multiple accounts. The bank itself needs to provide ways to look up customers and accounts.

A proper analysis of a real bank system would be a much more involved process and you would need to gather significantly more detailed information. What we have here is sufficient for our current purposes. We can make up details as we go. If you were writing this for a real customer, you would need to get them to actually spell out a lot more details because "making it up" would be disastrous.

Once you have a proper natural language definition of your problem, you can start thinking about how to break it up and solve it in your programming language. This is the design phase, and it is what we really want to focus on here. Our previous efforts in this area have been non-object-oriented. When we thought about the problems, we considered the grouping of data largely independent of the functions that we would write to deal with the problem. That all changes when we approach the problem from an object-oriented standpoint. Now, when we define our new types they will incorporate both data and functions. All of our functions will change into methods which will act on the data members grouped by the type.

Given the analysis from above, there are some fairly obvious things that should jump out to you for this system. They are the account, the customer, and the bank itself. We will consider each of these and try to figure out what data members and methods each one should have. The data that we will put into them will be similar to what you might have done in our previous treatment of `case class`es. The real change here is that the logic is no longer in separate functions, but is moved into the classes themselves in the form of methods. As a general rule, functions that deal closely with the data are going to move inside of our type to become methods. Any function that might mutate data for an object should definitely become a method.

Let's start by laying out the `Account` type. We will specify the data members and methods separately, giving a brief description of each.

- Account

    - Data Members:

        * **account number** - A value that uniquely identifies this account for this bank.
        * **balance** - A numeric value to keep track of how much money is in the account.
        * **account type** - Something that tells us if this is a savings or checking account.
        * **customer** - A reference to the customer who owns this account.

    - Methods:

        * **deposit** - Takes an amount of money that needs to be positive and adds it to the account balance. It also logs the transaction.
        * **withdraw** - Takes an amount of money that needs to be positive and less than or equal to the current balance and subtracts that amount from the balance. It also logs the transaction.
        * **monthly adjustment** - If appropriate for the account type, this will add interest into the account balance. It also logs any adjustment.

You could probably come up with several other pieces of data and activities to add to this example, but this provides sufficient complexity for our current purposes. The real key here is that manipulation of the data for this class is encapsulated into the methods. These methods will do safety checks to make sure that only allowed operations occur. We will also be able to prevent any outside code from altering the values. In this way, we limit the number of places in the code that could cause an error with an account.

Not only can we protect the data and make it so only certain methods are allowed to access key data (like the balance), we can also provide the ability to make sure that all changes to this key data are recorded. The descriptions of the three methods listed for the `Account` type all say that they "log" something. This has been left intentionally vague, but it is a critical element of any real financial software. If the balance were open to modification by any part of the program, it would be much harder to enforce the rule that every change has to be logged.

Next up is the `Customer` type. This type is largely just a collection of the information that is significant for a customer.

- Customer

    - Data Members:

        * **customer id** - A value that uniquely identifies this customer for this bank.
        * **first name** - The customer's given name.
        * **last name** - The customer's family name.
        * **address** - Something that stores an address for the customer. This is probably best done by defining another type for an address and having this reference an object of that type.
        * **phone number** - Something that stores the phone number of the customer. Like the address, this is probably best done by making another type, though it could start as just a `String`.
        * **accounts** - A sequence of the accounts that belong to this customer.

    - Methods:

* **change address** - Sets this customer's address to a new value. It also logs the change.
* **change phone number** - Sets this customer's phone number to a new value. It also logs the change.
* **add account** - Adds an account for this customer. It also logs the change.
* **remove account** - Removes a customer's account. It also logs the change.

The data members here should all be fairly straightforward. The only interesting aspect of them is the possibility that the address and phone number are represented with different types. You do not have to do this. Clearly an address could be represented with a few strings and a phone number could be represented by a string, but in practice, you likely want your system to do some verification to make certain that these have valid values. We could put that code here in the customer, but it is likely that addresses and phone numbers will occur in other parts of the system as this piece of software grows. When that happens, we would need to copy code from the customer into other types that deal with addresses or phone numbers. A better approach is to take the key information and functionality that these things need, and encapsulate them in their own types. That prevents code duplication and makes the code easier to work with.

If you think about this from the standpoint of objects being responsible for things, it should not be the job of the `Customer` type to validate/verify addresses and phone numbers. That gives the `Customer` type too much responsibility and dilutes the focus of what it should be doing. Creating separate types for `Address` and `PhoneNumber` allows each task to remain focused on a particular type of functionality, which can have many benefits over the lifetime of a piece of software.

Note that address and phone number were not things that we identified in our original analysis of the problem. Perhaps we would have if we have done a more complete analysis, but even if we had not, that should not prevent you from adding it to your design when you see that it makes sense. Especially as a novice programmer, you are going to miss things when you think about different problems and the code you need to write to solve them. This does not mean that you should not bother thinking about problems up front. Quite the opposite in fact. Spending the time thinking about problems before you sit down to code them will help you to gain a greater understanding of programming and problem solving. When you do realize that you missed something later in the process, you should make a mental note of that and the experience will help you the next time you solve a similar problem so that your initial approach will be more complete and you will waste less time going down paths that are not fruitful.

The last type that we will do the design for is the `Bank` type.

* Bank

    - Data Members:
        * **customers** - A sequence of `Customers`.
        * **accounts** - A sequence of `Accounts`.
    - Methods:
        * **add customer** - Creates a new customer with appropriate information and no accounts. It also logs changes.
        * **add account** - Creates a new account with a zero balance and associates it with a customer. It also logs changes.
        * **remove customer** - Closes out all accounts for the given customer and removes them from the system. It also logs changes.

      ∗ **remove account** - Closes a single account and updates the customer that had been associated with it. It also logs changes.

      ∗ **find customer** - Probably a few methods that finds customers based on various pieces information.

      ∗ **find account** - Looks up an account based on the account number.

In this design, this type is little more than a grouping of two sequences with methods for manipulating the sequences as well as methods for finding information in the sequences. Here again, encapsulation provides us with security. Not just any code should be able to manipulate the sequence of customers or accounts that we have in the system, and when it happens, those changes need to be recorded.

Clearly there are many things that have been left out of this solution. You also might have thought about other things that you would want to add in to make the solution more complete. One possibility would be the concept of branches/locations for the bank. Adding this in would illustrate why you might want to pull addresses and phone numbers out into their own types because each branch will have that information associated with it, just like the customer does.

We are not going to show how we would turn this design into actual Scala code yet. That will come a bit later. There are a fair number of details related to syntax and concepts that we have to deal with when we code this and for now we want to focus on getting you to think in a more object-oriented way without focusing too much on the syntax.

16.1.2 Analysis and Design of Pac-Man™

A very different example problem for us to break down in an object-oriented way is the classic arcade game, Pac-Man™. One reason for picking this game is that it is assumed that anyone is either already familiar with this game, or can easily find information on the basic game play.[1] For that reason, we will not explicitly describe the analysis of the problem, and instead go straight into problem design.

There are a number of different types of things that are significant in a Pac-Man™ game. Some of these are fairly obvious, such as Pac-Man™, the ghosts, and fruit. Other elements are less obvious. We definitely need a type that represents the overall playing area, we might call this the maze. Just how the maze stores things can impact what other types we will use. For example, should we have a separate type for a pellet, or should the grid cells in the maze store values that tell if the cell contains a pellet in addition to whether the cell is a wall or open space? There are pros and cons to each of these approaches. For this design, we are going to make it so that the grid stores information about whether or not pellets are present. It could also be useful to have something that represents the full game for things like score and the number of lives left. This gives us a reasonable list of types to start off with.

We will start with the `PacMan` type.

- `PacMan`

    – Data Members:

      ∗ **x** - The x-coordinate for the player likely measured in terms of the grid of the maze, not in pixels.

---

[1]For the 30th anniversary of Pac-Man™ in 2010, Google used it as the theme for their Doodle and created a playable version of the game that works in your browser. So if you are not familiar with how Pac-Man™ works, a Google search for it should be very informative.

* **y** - The y-coordinate for the player likely measured in terms of the grid of the maze, not in pixels.
* **direction moving** - The direction in which the player is moving or trying to move.
* **key direction** - The direction key the player has most recently hit.
* **animation count** - A value that is used to figure out how "open" the mouth is drawn.
* **dying** - Tells whether the player has been caught by a ghost and the program should do the dying animation.

– Methods:

* **update** - Uses the PacMan type information and the information about the maze to determine how to update the various values for this object.
* **go left** - This is called when the user does something to tell the character to move left.
* **go right** - This is called when the user does something to tell the character to move right.
* **go up** - This is called when the user does something to tell the character to move up.
* **go down** - This is called when the user does something to tell the character to move down.

This type has the responsibility of keeping track and updating everything related to the Pac-Man™ character. That includes where it is in the maze, where it is moving, and some additional information related to its movement and animation.

Perhaps the most important method there is the update method. This is to be called at regular intervals by outside code when the state of the player should be changed. This is the primary responsibility of this type, performing the key logic for the player. This will require accessing information about the maze and the locations of other elements of the game. We will see how this is done in a object-oriented fashion later.

The counterpart to the `PacMan` type is clearly the `Ghost` type.

* `Ghost`

  – Data Members:

  * **x** - The x-coordinate for the ghost likely measured in terms of the grid of the maze, not in pixels.
  * **y** - The y-coordinate for the ghost likely measured in terms of the grid of the maze, not in pixels.
  * **direction moving** - The direction in which the ghost is moving.
  * **animation count** - A value that is used to figure out aspects of the animation of the ghost.
  * **color** - Tells what color ghost this is.
  * **scared** - Tells if this ghost is running from Pac-Man™ or not.

  – Methods:

  * **update** - Uses the information stored here and the information about the maze to determine how to update the various values for this object.

Much of the data for the `Ghost` is similar to that for the `PacMan` player. The majority of the functionality also occurs in the `update` method as it does for the `PacMan` player. Since it is the responsibility of the `Ghost` type to handle all elements related to the logic of a ghost, outside code is not intended to change values in the `Ghost` directly. The changes should occur through `update`.

Up next is the `Fruit` type, which is very simple as they do not move in the original Pac-Man™ game. They only need to store their location and the type of fruit it represents. Since they do not move, they do not need an update method. Their placement and removal will be handled by other classes.

- Fruit

  − Data Members:

    * **x** - The x-coordinate for the fruit likely measured in terms of the grid of the maze, not in pixels.
    * **y** - The y-coordinate for the fruit likely measured in terms of the grid of the maze, not in pixels.
    * **fruit type** - Tells which type of fruit this is for drawing and scoring purposes.

The `Maze` type is where things get more interesting. This type is used to keep track of all the other types we have discussed as well as the layout of the maze and the pellets that the player and ghosts are moving through.

- Maze

  − Data Members:

    * **cells** - A 2D structure that tells us what state each cell is in. At this point we will probably make this an `Array[Array[Int]]` as we have not learned better ways to do this.
    * **player** - A reference to a `PacMan` that stores the one player on the board.
    * **ghosts** - A sequence of `Ghost` objects.
    * **fruit** - This should probably be an `Option[Fruit]` given that you do not always have a `Fruit` in the maze.

  − Methods:

    * **update all** - When this method is called, the player and ghosts are updated, and fruit might be added or removed. This method will also handle any other changes that need to occur to the maze.

For our purposes here, we are going to use different `Int` values to tell if we have a wall, an empty cell, a regular pellet, or a power pellet. There are better ways to do this that go beyond the scope of this book and are covered in *Object-Orientation, Abstraction, and Data Structures Using Scala*[1]. As with many of the earlier types, the functionality from the outside is handled by a single method that updates the game. This calls the update on the player as well as on the ghosts and the fruit. In some ways, this one method is largely responsible for running the logic of the game. For that reason, there is a good chance that we will break it into pieces, and have other methods that help it, but those will not be visible to outside code.

There is another class that represents the full state of the game. We are going to call it `Game`. This keeps track of score and how many lives the player has left.

- Game

– Data Members:

* **score** - An Int for how many points the player has.

* **lives** - An Int for how many lives the player has left.

– Methods:

* **main** - As we will see later, this is the method that makes the whole game run.

You might have noticed that while some of the types above dealt with some aspects of how things are drawn in the game, nothing actually did any drawing. There is a reason for this. Drawing is something of a separate responsibility from game logic, scoring, and movement. For that reason, it should be placed in a different type. This separation of responsibilities makes it easier in the future to change just how things are drawn.

• Renderer

– Methods:

* **render** - Draws everything so that we can see the game being played.

That completes the design for our Pac-Man™ game. The key thing to note is that we are putting functionality directly into the classes that have the data to deal with that data so that any mutation/alteration is limited in where it happens. In the next section, we will see how we can implement these designs in Scala.

16.2   Implementing OO in Scala

Scala is a CLASS BASED object-oriented programming language. That means that programmers create classes in their programs and those classes define types that are used to make objects. We already saw how the case classes that we made defined new types. We did not put anything in them though except some data elements.

The case keyword in front of a case class worked well for our needs in chapter 10, but it is not required to define a class. We will be writing normal classes now and only using case classes when they are called for. The way that you should think of a class is as a blueprint for an object. The class tells Scala what goes into the objects of that type and what type of functionality those objects should have. The syntax for a class is as follows.

```scala
class TypeName(arg1:Type1, arg2:Type2, ...) {
  // Methods and Members
}
```

As you can see, the only things that are different between this and the case classes we have already worked with is the lack of the keyword **case** and the curly braces with stuff in them after the **class**. These curly braces, like all curly braces in Scala, define a new scope. This scope is associated with the **class** and anything you want to put in the class goes inside of them.

16.2.1 Methods and Members

Inside of the `class` you can write code exactly the way you have anyplace else. There are small differences that will be discussed and the meaning and terminology changes a bit. When you write a variable declaration with `val` or `var` inside the body of a `class`, that is now a member declaration. It is data that is stored in the objects that are created from that `class`. When you use `def` to create a function, we call it a method because of its scope in the `class` and the fact that it is something that you will be able to call on objects created from this class. Any statements that you write that are not declarations of members (`val` or `var`) or methods (`def`) will simply be run when an object of this type is instantiated.

To see how this works, we will write a little script that includes a `class` declaration. For the `class` we will go back to the concept of grades in a course and write the `class` to represent a student. Here is a first draft of the `class` with some lines that use the class.

```scala
class Student(name:String, id:String) {
  var tests = List[Double]()
  var quizzes = List[Double]()
  var assignments = List[Double]()

  def testAverage = tests.sum/tests.size
  def quizAverage = quizzes.sum/quizzes.size
  def assignmentAverage = assignments.sum/assignments.size
  def courseAverage = testAverage*0.4 + quizAverage*0.1 + assignmentAverage*0.5
}

val john = new Student("John Doe","0123456")
john.tests ::= 78
john.tests ::= 85
println(john.testAverage)
```

The `class` begins with a `class` declaration that looks a lot like a `case class` and really is not too different from a `def`. After the keyword `class` is the name of the type, in this case `Student`. This is followed by a list of parameters to the `class`. In this case we have two `String`s called `name` and `id`. The body of the `class` is denoted with curly braces and inside of the curly braces we have three member variable declarations for grades and four method declarations for taking averages.

After the class declaration are four lines of code. The first declares a variable called `john` that is set to be a `new Student` with the name `John Doe` and a student number of `0123456`. The second and third statements add two test grades of 78 and 85 to this student. While this is simple, it is not clear that we really want the grades to be this accessible. The last statement in the script prints the average for the tests by calling the `testAverage` method.

16.2.1.1 Parameters as Members

Everything in this example should seem fairly straightforward. It really does not look much different from things that we have done before. This would change though if we try to make the print statement a bit more informative.

```scala
println(john.name+" has a "+john.testAverage+" test average.")
```

If you change the `println` to this and run the script you will get an error message that might not make much sense to you.

```
ScalaCS1/Chapters/ObjectOrientation/Code/Student1.scala:16: error: value name is
    not a member of this.Student
```

```
println(john.name+" has a "+john.testAverage+" test average.")
                ^
```

```
one error found
```

You should definitely find this surprising. This would have worked for a case class. Indeed, if you put the **case** keyword at the beginning of the **class** declaration this works fine.

The error message tells us that **name** is not a member of **Student**. This is because, by default, the arguments passed into a **class** are not turned into members. The logic behind this is that members have to be stored by each instance of the **class** that is created, so extra members consume memory. Even if they do have to be remembered, they will not be visible to any code outside of the **class**. For example, if you add the following method into the class you can call it to get the last name, but you still cannot get the name itself.

```
def lastName = name.split(" +").last
```

If you want an argument to be a member simply put **val** or **var** before the name. It will then become a member with the proper behavior for either a **val** or a **var**. In a **case class**, everything is implicitly a **val**, so we did not have to add that. For a normal **class**, we do. For our **Student class** we likely want both **name** and **id** to be constant, so we should change the class declaration to the following.

```
class Student(val name:String,val id:String) {
```

If you make this change you can use the print statement shown above without an error.

16.2.1.2   Visibility

The possibility of things being visible or not is new to us with this chapter, but it is a very important aspect of object-oriented programming because it allows you to hide details about the way an object works away from the code that uses the **class**. The value of this is that you can change those details without causing problems for the code that depends on the **class**. In the case of member data, you can also hide away details that you do not want other parts of code to have free access to. In addition, this makes it easier to find many runtime and logic errors, as certain values can only be altered in a small subset of the full program.

A great demonstration of the value of data hiding can be seen with the account **class** from the bank example we started earlier. Let us consider the simplest possible **class** that we might write to represent a bank account.

```
class Account {
  var balance = 0
}
```

This class does not take any arguments when it is constructed and, as a result, the parentheses have been left off. It has one piece of member data, an **Int** for the balance. As you might recall from chapter 2, money should be stored in **Ints** because the arithmetic of the **Double** type is inaccurate due to the limited precision. This representation is simple, but it has a significant drawback. The member balance is a **var** that can be set by any code that gets hold of an **Account** object. For a real bank program that would not be acceptable. The balance cannot be just any value. For example, negative numbers generally are not allowed. In addition, changes to the balance are typically supposed to happen in specific operations: deposits and withdraws. Those operations can enforce certain rules and, for a real bank, would log any changes to the balance so that it is possible to go back and figure out what happened. While we could make functions for those operations, there would be nothing that forces programmers to use those functions. It would be possible for a programmer to "get lazy" at some point and access the `balance` directly. The results of this in banking software could be quite extreme.

The ability to force access to data to occur through certain methods is one of the most significant capabilities of object-orientation as a way of improving software construction and controlling complexity. The real benefit of this capability is to take responsibility for proper usage away from the users of the objects. If the only ways to use the object are through methods or data that is safe for them to use, then they know they cannot mess it up and do not have to worry about that. Instead, they can focus on the logic that they are working on.

How do we accomplish this in our classes? We do it by setting the visibility of members. There are three main levels of visibility in Scala, and they are very similar to what you will find in other class based object-oriented languages.

- Public - This means that something can be accessed by any code inside or outside of the class. In Scala this is the default visibility for elements, so, there is no keyword for it.

- `private` - This means that the member can only be accessed inside the `class`. Attempts to access the member from outside of the `class` will result in a syntax error. If you prefix a declaration in a `class` with the `private` keyword, that member will be `private`.

- `protected` - This is like `private` except that `protected` members are also visible in subtypes. The details of this are beyond the scope of this book, but are dealt with in *Object-Orientation, Abstraction, and Data Structures Using Scala*[1].

In our example `Account` we really want the `balance` to be `private`. We can do this by simply adding `private` before the `var`.

```scala
class Account {
  private var balance = 0
}
```

Unfortunately, this leaves us with a `class` that is completely useless. Before making this change we could have done something like the following.

```scala
val myAccount = new Account
println(myAccount.balance)
myAccount.balance += 100
println(myAccount.balance)
```

If we do this now though we get a set of error messages.

```
ScalaCS1/Chapters/ObjectOrientation/Code/Bank.scala:6: error: variable balance in
    class Account cannot be accessed in this.Account
println(myAccount.balance)
                  ^
ScalaCS1/Chapters/ObjectOrientation/Code/Bank.scala:7: error: variable balance in
    class Account cannot be accessed in this.Account
myAccount.balance += 100
          ^
ScalaCS1/Chapters/ObjectOrientation/Code/Bank.scala:8: error: variable balance in
    class Account cannot be accessed in this.Account
println(myAccount.balance)
                  ^

three errors found
```

Note how the error message is different than what we saw previously for name in our Student class. Instead of telling us that balance is not a member, it tells us that balance cannot be accessed. This is precisely because the balance is now private, and the code is outside of the class.

To make this class useful, we would need to put some methods in it that are public that manipulate the balance in allowed ways. We had three of these in our design for the bank. We will start with the deposit and withdraw methods. Here is what the deposit method might look like. Note that this is indented because it appears inside of the Account class.

```
def deposit(amount:Int):Boolean = {
  if (amount > 0) {
    balance += amount
    true
  } else false
}
```

The method takes the amount to deposit as an Int. It returns a Boolean to tell us if the deposit went through. This method does not support logging because adding that functionality goes deeper than we want to at this point. It does check to make sure that the amount of the deposit is positive. If it is, then that amount is added to the balance and the result is true; otherwise, nothing is done and the result is false.

We can add a very similar looking withdraw method.

```
def withdraw(amount:Int):Boolean = {
  if (amount > 0 && amount <= balance) {
    balance -= amount
    true
  } else false
}
```

The only difference between deposit and withdraw is that you cannot withdraw more money than you have, and the amount is subtracted from the balance.

The last thing we need to add is a method to get the balance so that it can be checked. The proper style for this in Scala is to name the method balance and name the private member _balance. That would change our class to look like the following.

```
class Account {
  private var _balance = 0

  def deposit(amount:Int):Boolean = {
    if (amount > 0) {
      _balance += amount
      true
    } else false
  }

  def withdraw(amount:Int):Boolean = {
    if (amount > 0 && amount <= _balance) {
      _balance -= amount
      true
    } else false
  }

  def balance = _balance
}
```

The `balance` method simply gives back the value of `_balance` in the account. We can utilize this code by doing the following.

```
val myAccount = new Account
println(myAccount.balance)
myAccount.deposit(100)
println(myAccount.balance)
```

Note that because of the use of the name `balance`, the `println` statements appear the same way they did when `balance` was a public `var`. The difference is that you cannot set `balance` now unless you go through either the `deposit` method or the `withdraw` method.

Our original design included another method and several more data members. Some of these require additional elements of the bank example, so their implementations will wait until those other pieces are in place.

16.2.2 Special Methods

There are a few method names that are special when you put them into `class`es in Scala. In a way, these special methods are rules that you have to learn as special cases. This section covers the ones that are critical for you to understand to write our sample applications.

16.2.2.1 Property Assignment Methods

In the bank account example we said that in Scala when a method should simply give you the value of a member, you give it the name you want people to see for the property. In our example this was `balance`. What about when you want to alter the value? In Scala, the proper style is to have the code use the assignment method. The advantage of the assignment method is that it allows you to set a property using normal assignment syntax. To create an assignment operator, write a method with the name *prop_=*, where *prop* is the name of the property.

To help you understand this, let us assume that you did want to be able to enter something like `myAccount.balance = 700`. We had this ability with the public `var`, but we decided that was too little control. We could get that back by adding the following method.

```
def balance_=(b:Int):Unit = _balance = b
```

Thanks to the way Scala handles assignment operators, this will let all three of the following lines work.

```
myAccount.balance = 700
myAccount.balance += 40
myAccount.balance -= 50
```

So doing `+=` will work like a deposit and `-=` will work like a withdrawal. The downside is that neither returns a `Boolean` to let you know if it worked. As written, neither checks the value being deposited or withdrawn either.

This particular method leaves things too open again. We might as well have the `var` because we are doing exactly what the `var` would do. However, the advantage of an assignment method is that it can be more complex and have more logic behind it. For example, the `balance_=` method could be altered to the following.

```
def balance_=(b:Int):Unit = {
  if (b >= 0) {
```

```
    if (b < _balance) withdraw(_balance - b) else deposit(b - _balance)
  }
}
```

This version will reuse the code from `withdraw` or `deposit` and include any error checking, logging, etc. It also throws in checking so that assignments to invalid balances will not go through. If this were production code, there really should be an `else` clause on that outer `if` that would do something like throw an exception if the code attempts to set the balance to a negative value.

Note that you only include a set method for properties that are mutable so there are many situations where they are left off. If you are programming in a more functional style, you will not have these assignment methods. It is only when objects specifically need to be able to mutate that these are helpful. The real advantage, in that case, is that the code can look like a normal assignment, but will call the method that can do various types of error checking.

16.2.2.2   The `apply` Method

Another method that Scala treats in a special way is the `apply` method. This method, and how it is handled, is a major part of what makes Scala work as a functional language while remaining very object-oriented. You write `apply` just like any other method that you might put in a `class`. You can call it just like any other method as well. However, Scala will allow you to call the `apply` method without using a dot or the name `apply`. When you remove those, it makes it look like you are treating the object as a function. Indeed, this is how all functions work in Scala. As was said very early on, everything in Scala is an object. That includes functions. A function is just an object of a type that has an `apply` method that takes the appropriate arguments.

The intelligent use of `apply` is how you have been indexing into all of the collection types. Consider the following code that you could write in the REPL.

```
scala> val arr = Array(5,7,4,6,3,2,8)
arr: Array[Int] = Array(5, 7, 4, 6, 3, 2, 8)

scala> arr(3)
res0: Int = 6

scala> arr.apply(3)
res1: Int = 6
```

The call to `arr(3)` is actually just a shortcut for `arr.apply(3)`. Scala is doing nothing more here than assuming the presence of the call to `apply`.

How can we use this in our own code? In our design for the bank program, we mentioned having a `Bank class` that stores all the accounts and customers of a bank, and includes methods for looking them up. We could put an `apply` method in the `Bank class` that takes an account number and produces an `Option[Account]`. To do this, first we need to give the `Account` an ID. This is easily done by modifying the declaration to take a `val` argument.

```
class Account(val id:String) {
```

It is fine for the `id` to be public because it is a `val` of an immutable type, so people using the code cannot mess it up. Note that this assumes that account ID numbers cannot ever be changed. Doing so would basically require closing one account and opening a new one for the same customer with the same balance, but using a different ID.

With that in place, we can start our implementation of the `Bank class` giving it a collection for the accounts. We also add methods to add and find the accounts. The method to find an account is the `apply` method.

```scala
class Bank {
  private var accounts:List[Account] = Nil

  def addAccount(account:Account):Unit = accounts ::= account

  def apply(accountID:String):Option[Account] = accounts.find(_.id == accountID)
}
```

The `apply` method returns an `Option[Account]` because it is possible that there is not an account with the provided ID. In that situation, it would return `None`. After setting up an instance of `Bank` and giving it some accounts, you could search for one by ID using `bank("01234567")`, which would give back an `Option[Account]` for an account with ID "01234567".

16.2.3 `this` Keyword

By default, when you write code in a `class` that calls a method or uses a member of that `class`, the call is made on the current object or the value from the current object is used. All calls are implicitly made on the current instance. That is exactly what you want most of the time and given the scoping of variables it feels very natural. This implicit specification of an object prevents you from having to specify one. Consider the code in the `deposit` method for adding money to the balance. We wrote

```scala
_balance += amount
```

which is really short for

```scala
this._balance += amount
```

Scala simply added the `this` for you.

That is fine in general, but occasionally you will have a need to be able to put a new to the current object. For example, you might need to call a function/method that needs an instance of the object as an argument and you want to use the one that the code is currently executing on. When that situation arises you will use the `this` keyword. `this` is a name that always refers to the current instance in a `class`. In the bank example, we will us `this` in two ways when creating an account. We will use `this` in the account to pass the current account to the `Customer` who owns it to set things up. We will also use `this` in the `Customer` to check if an account belongs to the current `Customer` as it would be an error to have a `Customer` taking ownership of an account for someone else.

16.2.4 `object` Declarations

In order to complete our examples, we need to introduce another style of declaration that exists in Scala, the `object` declaration. We said earlier that you should think of a `class` as a blueprint for making objects. These objects are called instances, and the process of making them using `new` is called instantiation. On the other hand, the `object` declaration specifies how to build what is called a singleton object. As the name implies, this is an object where there is only a single instance. The syntax is very much like that of a `class` other than you cannot pass in arguments. This makes sense as you pass the arguments to a **class** when you use **new**, and you never call **new** with a singleton **object**, because the declaration creates it in the current scope. Any data members or methods you want can be declared in the **object** declaration.

Even though an **object** declaration does not create a new type, we use a capital letter as the first letter in the name to distinguish it from normal instantiated objects. To use the members or methods of an **object**, use the **object** name and call it just like you would for an instance of a class. **object** declarations can be used in any scope where you want an object, but you only want one object of that type. At the top level they are typically used as a place to organize functions, but they have far more versatility than this overall.

16.2.4.1   Applications

The fact that you can use an **object** without instantiation is very significant. The **object**s effectively exist as soon as the program starts running. You do not have to have a line that makes a new one. This is most significant when we move beyond the REPL and scripts and into applications. A top level **object** declaration defines an entry point to an APPLICATION if it contains a method called **main** that takes an argument of type **Array[String]** and results in **Unit**. Here is a simple example.

```
object FirstApp {
  def main(args:Array[String]):Unit = {
    println("My first application.")
  }
}
```

The scripts that we have been using can be converted to applications by doing nothing more than putting the code inside of a **main** that is in an **object**. If you use the variable name **args** for the command-line arguments, then references to them in a script will even work.

The main differences between a script and an application is how you organize the code in files and how you run them. By definition, a script is supposed to be a small program. All of the code for a script goes into a single file. When the amount of code that is required gets larger, it needs to be split apart and organized. The organizational structure you should follow in Scala in nearly all situations is to put each top level **class** or **object** in a separate file that is the name of the **class** or **object** followed by ".scala".

Consider the example bank application. We might use the **Account class** from above and put it in a file called "Account.scala". The **Customer class** would go in "Customer.scala", and the **Bank class** would go in "Bank.scala". Finally, we would have an **object** declaration for the application that might be called **BankBranch** and would go in "BankBranch.scala". In addition to having a **main** method, this would keep a reference to an instance of **Bank**, and handle whatever user interface we wanted, whether that was a text menu or a GUI.

Running an application like this is a two step process. First you have to compile the code. This is done with the **scalac** command, which stands for "Scala compiler". You have to tell the Scala compiler what files it is you want to compile. If you organize your code so that all the code for an application is under a directory you can execute something like **scalac *.scala**. This will compile all of the files ending in ".scala" in the current directory.

When you do this, if you have any syntax errors, they will be found and reported to you in much the same way they were when you ran programs as scripts. Once you have found and fixed all syntax errors, the compiler will produce a set of files in compiled BYTECODE that you can run. These files will end with the ".class" extension. To run the application you go back to using the **scala** command. Only now you follow it by the name of the **object** you want to run the **main** from. So in the case of our bank example we could run it by entering `scala BankBranch`. Other command-line arguments can be specified after the name of the `object`.

The application has to be an `object` because methods in a `class` can only be called on instances of that `class`, not the `class` itself. So in order to call a method you have to have executed `new` on that type. However, `main` has to be called first, before any other logic has been executed. This is not a problem with an `object` as the single instance exists without a call to `new`.

16.2.4.2 Introduction to Companion Objects

While making applications with `object` is vital, the most common use of `objects` is as COMPANION OBJECTS. A companion `object` is an `object` that has the same name as a `class`. Both the `class` and its companion `object` should be placed in the same file. The companion `object` has access to `private` members of the class it is a companion with. Similarly, the `class` can see `private` elements in the `object`.

You might have wondered why it is that when we are building objects in Scala, sometimes we use `new` and sometimes we do not. The reality is that making an object always requires invoking `new`. When you do not type it, it means that you are calling code that does. When you use the name of the type without `new` to instantiate an object, you are calling the `apply` method on the companion object, and that `apply` method is calling `new`. The first examples of this that we encountered were the simple forms for creating `Arrays` and `Lists`. When you wrote `List(1,2,3)` in your code, it actually called `List.apply(1,2,3)`, where `List` refers to the companion object.

Unfortunately, you cannot easily write `class`es with companion `object`s in the REPL or in scripts. For us to create our own companion `object`s, we have to be working in the mode of writing applications, compiling with `scalac`, and running the compiled files with `scala`.

16.3 Revisiting the API

We have already learned enough to help a bit with understanding what some things mean in the API. When you open up the API, the left side has a frame that looks like figure 16.1. Thus far we have used the generic word "type" to describe the things that are listed in this frame. Now we can be more specific. There are really three different declaration styles in Scala that appear over there and the API indicates which one you are dealing with.

The circles next to the names contain one of three letters in them. The meaning of these letters is as follows.

- c - For a `class`.

- o - For an `object`.

- t - For a `trait`, which we do not cover in this book, but are very similar to `class`es.

When you click on one of these, you are shown the methods and members that are defined in them, whether it be a `class`, an `object`, or a `trait`. The API itself is built by running a program called `scaladoc` on scala code to generate HTML descriptions of what is in the code.

When there is an "o" next to a "c" or a "t" it is a companion `object`. If you want to know if you can build objects without a direct use of `new`, look there and see if there is an `apply`

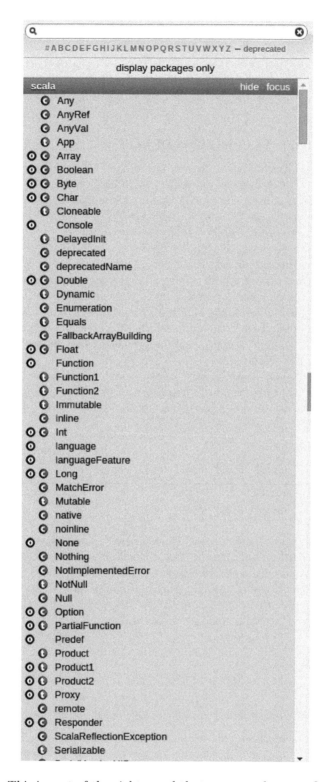

FIGURE 16.1: This is part of the right panel that you see when you first enter the API. The little circles with the letters c, o, and t indicate whether it is a `class`, `object`, or `trait`.

method defined. If there is, it generally implies that is the preferred way to make instances of that type.

16.4   Implementing the Bank Example

We now know enough about the syntax of object-oriented programming in Scala that we can implement the bank example. This is being written as an application, so each `class` is put in its own file. Note that this code does not do anything related to logging as that is beyond the scope of this text, but the code has been written such that you could add appropriate code in the methods to easily get logging in all of the places it is needed. Since we have done the most work with the `Account class` to this point, let's look at that first.

Listing 16.1: Account.scala

```scala
class Account private(val id:String, val customer:Customer, accountType:Int) {
  require(accountType == Account.Checking || accountType == Account.Savings ||
      accountType == Account.CD)
  private var _balance = 0

  customer.addAccount(this)

  def deposit(amount:Int):Boolean = {
    if (amount > 0) {
      _balance += amount
      true
    } else false
  }

  def withdraw(amount:Int):Boolean = {
    if (amount > 0 && amount <= _balance) {
      _balance -= amount
      true
    } else false
  }

  def monthlyAdjustment():Unit = {
    accountType match {
      case Account.Checking => _balance = (_balance * 1.01).toInt
      case Account.Savings => _balance = (_balance * 1.03).toInt
      case Account.CD => _balance = (_balance * 1.05).toInt
    }
  }

  def balance = _balance
  def balance_=(b:Int):Unit = {
    if (b >= 0) {
      if (b < _balance) withdraw(_balance - b) else deposit(b - _balance)
    }
  }
}
```

```
37  object Account {
38    val Checking = 0
39    val Savings = 1
40    val CD = 2
41
42    private var nextAccount = 1
43
44    def apply(customer:Customer, accountType:Int):Account = {
45      val id = "0"*(7-nextAccount.toString.length) + nextAccount
46      nextAccount += 1
47      new Account(id, customer, accountType)
48    }
49  }
```

There are quite a few things in this code that are worth taking note of. Lines 37-49 include a companion `object`. This serves two main purposes in this code. The first is a place to store the three constants that tell us the account type. If these values were put in the `class`, then every instance of the class would get their own copy. As they would all have the same values, this would be wasteful. The `object` is a singleton though, so putting them in the companion `object` means we get a single copy of each. In the `class` we refer to these constants in two locations. On line 2 there is a check to make sure that the `accountType` that is passed in matches one of the allowed values.[2] The other reference is in `monthlyAdjustment`, where interest is added based on the type of account.[3] Note that in both of these locations, the references to the constants are prefixed with the name of the `object`.

The other use of the companion `object` is to make certain that all instances of the `Account class` have unique IDs. Up on line 1 there is some syntax that we have not specifically introduced before. The `private` keyword appears in front of the argument list. This has the effect of making instantiation private so that nothing outside of this file can call `new Account(...)`. If we did not do this, other code could create new account instances with the same IDs as existing ones. Instead, we have added an `apply` method into the companion `object` and outside code must use this method to make accounts with the syntax `Account(...)`. Note that the `apply` method does not take an ID as an argument. Instead, there is a `private var` declared in the companion `object` that keeps track of the next account number to use. Inside of `apply`, this numeric value is prefixed with zeros to make a string seven characters long, and that is used as the ID in a call to `new`. The `apply` method also increments the value of `nextAccount` so that subsequent accounts will have a different ID. The combination of `apply` and making instantiation private means that outside code can never create two accounts with the same ID.

Line 3 of Account.scala shows our first usage of `this`. The `Account` and `Customer` classes have a mutual dependence. Instances of `Account` have to be constructed with a reference to a `Customer`, and the `Customer` instances keep track of the accounts they have. To deal with this, we have added line 3 to the `Account` that will automatically add the new `Account` instance to the customer it refers to. Doing this requires passing the current account, which means that we need to use `this`.

Now we can look at the `Customer class`. Like `Account`, this file includes a companion `object` with an `apply` method that helps to create unique IDs. Unlike the `Account`, it is

---

[2]As has been mentioned before, there are better ways of dealing with account types that go beyond what we have learned. Those methods would make it impossible to pass in bad values here. This approach at least makes it a runtime error that happens immediately upon creation of an `Account` with a bad type.

[3]This example is simplified in many ways, but the contents of `monthlyAdjustment` are probably the most extreme simplification. Interest rates should not be hard coded in here. Also, there should inevitably be code that makes sure this does not happen more than once each month.

possible for outside code to make instances of `Customer` using `new`. The reason for this design choice is in part to show you that it can be done, and in part based on the idea that when a customer gets a legal name change, you might want to preserve their ID, but the operation requires creating a new instance of `Customer` as the name members are not mutable.

Listing 16.2: Customer.scala

```scala
class Customer(
    val id:String,
    val firstName:String,
    val lastName:String,
    private var _address:List[String],
    private var _phone:String) {
  private var _accounts:List[Account] = Nil

  def accounts = _accounts

  def address:List[String] = _address
  def address_=(newAddr:List[String]):Unit = _address = newAddr

  def phone:String = _phone
  def phone_=(newPhone:String):Unit = _phone = newPhone

  def addAccount(account:Account):Boolean = {
    if (account.customer == this && _accounts.find(_.id == account.id) == None) {
      _accounts ::= account
      true
    } else false
  }

  def removeAccount(accountID:String):Boolean = {
    val index = _accounts.indexWhere(_.id == accountID)
    if (index < 0) false
    else {
      _accounts = _accounts.patch(index, Nil, 1)
      true
    }
  }
}

object Customer {
  private var nextID = 1

  def apply(firstName:String, lastName:String, address:List[String],
      phone:String):Customer = {
    val id = "0"*(7-nextID.toString.length) + nextID
    nextID += 1
    new Customer(id, firstName, lastName, address, phone)
  }
}
```

The `Customer` class takes quite a few arguments. When an argument list does not fit nicely on a single line, the format shown here is considered proper style, with one argument on each line following the line declaring the `class`.

The values for the address and phone number are `private vars`. Assignment methods have been added for each of them. These methods do not do anything to check the correctness of the arguments, nor do they log the changes, but that functionality could be easily added to the methods shown here. This is the advantage of using a `private var` with accessor methods, you have additional control over the values given to members even though this code does not utilize that.

The `addAccount` method is interesting because it uses `this` to help safeguard against adding invalid accounts. In order to be added, an account must refer to this instance of `Customer`, and it must have an ID that does not already appear in the list.

The last `class` that is part of the bank example is the `Bank class`. The following code adds a few methods to what we had shown previously for the `Bank` as well as a member for storing customers. Nothing in this code should seem all that unusual.

Listing 16.3: Bank.scala

```scala
class Bank {
  private var accounts:List[Account] = Nil
  private var customers:List[Customer] = Nil

  def addAccount(account:Account):Unit = accounts ::= account
  def removeAccount(accountID:String):Boolean = {
    val index = accounts.indexWhere(_.id == accountID)
    if (index < 0) false
    else {
      accounts = accounts.patch(index, Nil, 1)
      true
    }
  }

  def addCustomer(customer:Customer):Unit = customers ::= customer
  def removeCustomer(customerID:String):Boolean = {
    val index = customers.indexWhere(_.id == customerID)
    if (index < 0) false
    else {
      customers = customers.patch(index, Nil, 1)
      true
    }
  }

  def apply(accountID:String):Option[Account] = accounts.find(_.id == accountID)
  def findCustomer(customerID:String):Option[Customer] = customers.find(_.id ==
    customerID)
}
```

To demonstrate how we can put all of these things together and how to make an application, we have written an `object` called `BankBranch` that includes a `main` method.

Listing 16.4: BankBranch.scala

```scala
object BankBranch {
  def main(args:Array[String]):Unit = {
    val branch = new Bank
    val customer = Customer("Bob", "Builder",
    List("123 Broadway", "Walawala, WA"), "(123) 456-7890")
    branch.addCustomer(customer)
    val account1 = Account(customer, Account.Checking)
```

```
 8      branch.addAccount(account1)
 9      val account2 = Account(customer, Account.Savings)
10      branch.addAccount(account2)
11    }
12  }
```

The purpose of this code is to illustrate how the other `classes` and `objects` can be used. In order to run this, we must first compile it with `scalac`. If these are the only Scala files in your current directory, you can compile them using `scalac *.scala`. If you have other Scala files in the current directory, especially if you have script files that will not compile properly with `scalac`, you will have to specify the files individually. This could be done with a command like `scalac Account.scala Customer.scala BankBranch.scala`. Once you have compiled these files, you can run the main with `scala BankBranch`. Note that this command does have the `.scala` the way that we have done when running scripts.

16.5 Implementing the Pac-Man™ Example

Like the bank example, our version of Pac-Man™ is written as an application and is broken across multiple files. Not everything exactly follows the design that was laid out at the beginning of the chapter. This is to be expected. When you go to implement your designs, you often find things that were missed or realize that some pieces do not fit together the way that you expected them to. In this case, we also introduced some different Scala concepts between the design presentation and this section that have altered how we build things.

We start by looking at the `Game`, which as been implemented as a singleton `object` with a `main` method that is used to run the program. The `main` method includes ScalaFX code for the GUI and a timer to make the code run as well as handlers for the keyboard input.

Listing 16.5: Game.scala

```
 1  import scalafx.Includes._
 2  import scalafx.animation._
 3  import scalafx.application._
 4  import scalafx.scene._
 5  import scalafx.scene.canvas._
 6  import scalafx.scene.input._
 7  import scalafx.scene.paint._
 8
 9  object Game {
10    val Width = 760
11    val Height = 840
12
13    private var _lives = 3
14    private var _score = 0
15
16    def lives = _lives
17    def score = _score
18    def newLife():Unit = {
19      _lives -= 1
20    }
21
```

```scala
22    val maze = new Maze
23
24    def main(args:Array[String]):Unit = {
25      val app = new JFXApp {
26        stage = new JFXApp.PrimaryStage {
27          title = "PacMan"
28          scene = new Scene(Width, Height) {
29            val canvas = new Canvas(Width, Height)
30            val gc = canvas.graphicsContext2D
31
32            content = canvas
33
34            onKeyPressed = (e:KeyEvent) => {
35              e.code match {
36                case KeyCode.Left => maze.player.goLeft
37                case KeyCode.Right => maze.player.goRight
38                case KeyCode.Up => maze.player.goUp
39                case KeyCode.Down => maze.player.goDown
40                case _ =>
41              }
42            }
43
44            var lastTime = 0L
45            val timer:AnimationTimer = AnimationTimer(t => {
46              val delay = (t-lastTime)/1e9
47              if (lastTime > 0 && delay > 0.1) {
48                _score += maze.updateAll()
49                Renderer.render(gc, maze)
50                lastTime = t-((delay-0.1)*1000000000).toInt
51              }
52              if (lastTime == 0) lastTime = t
53              if (lives < 0 || maze.pelletCount==0) timer.stop
54            })
55            timer.start
56          }
57        }
58      }
59
60      app.main(args)
61    }
62  }
```

Lines 10-22 have basic declarations, some simple accessor methods, and a declaration of a Maze along with the instantiation of an instance of that type. In the main method on lines 24-61, we create a JFXApp and do the normal setup. Much of this should seem familiar from chapters 11 and 12. The only thing that likely seems odd is line 50. In earlier examples, we would have just said lastTime = t. However, for this example we wanted the "game ticks" to happen as close to 0.1 seconds apart as possible. In testing, the timer would fire several times before it got to 0.1 and often got to values around 0.115. The simple form would have caused the time to "slip" by 0.015 seconds in that situation. This might not seem like much, but it is a significant fraction of the 0.1 seconds that we desire. The extra math on line 50 sets lastTime to be the time at the actual 0.1 mark instead of however far we might have gone past that, so that the whole game stays close to being on time. Line 53 shows that our two conditions for stopping the timer are when the player runs out of lives or when the player clears the board of all pellets.

It is worth noting that while we are presenting one complete file at a time, this is not how it was written. Real development moves from one file to another and back again, iterating as features are added. This approach does not just help with logic, but it helps with actually getting things done. You should stop coding and run your program every so often to make sure it works. You should also debug each feature as you add it. Do not try to go on to the next feature until you have the current one working. Staying focused on a single task like that can be challenging, but it helps tremendously with the development process.

Next up is `Maze`, which is declared as a `class`. This is where a lot of the logic of the game actually takes place. The `Maze class` does not take any arguments as it always sets things up in a default configuration. Lines 4-9 declare a number of `private vars`, setting them all to default values. This can be dangerous, especially the use of `null`, except that lines 11-12 calls the `initMap` and `initContents` methods that gives them all values. Earlier iterations of the code set those values at the point of declaration, but it is useful to be able to call them later.

Listing 16.6: Maze.scala

```scala
import scalafx.scene.paint.Color

class Maze {
  private var cells:Array[Array[Int]] = null
  private var _player:PacMan = null
  private var _ghosts:List[Ghost] = Nil
  private var _fruit:Option[Fruit] = None
  private var count = 0
  private var nextFruitType = 0

  initMap()
  initContents()

  private def initMap():Unit = {
    cells = Array(Array(0,0,0,0,0,0,0,0,0,0,0,0,0,0,0,0,0,0,0),
                  Array(0,1,1,1,1,1,1,1,0,1,1,1,1,1,1,1,1,1,0),
                  Array(0,2,0,0,1,0,0,0,1,0,1,0,0,0,1,0,0,2,0),
                  Array(0,1,1,1,1,1,1,1,1,1,1,1,1,1,1,1,1,1,0),
                  Array(0,1,0,0,1,0,1,0,1,0,1,0,1,0,1,0,0,1,0),
                  Array(0,1,1,1,1,0,1,1,1,0,1,1,1,0,1,1,1,1,0),
                  Array(0,0,0,0,1,0,0,0,3,0,3,0,0,0,1,0,0,0,0),
                  Array(0,0,0,0,1,0,3,3,3,3,3,3,3,0,1,0,0,0,0),
                  Array(0,0,0,0,1,0,3,0,0,-1,0,0,3,0,1,0,0,0,0),
                  Array(3,3,3,3,1,3,3,0,3,3,3,0,3,3,1,3,3,3,3),
                  Array(0,0,0,0,1,0,3,0,0,0,0,0,3,0,1,0,0,0,0),
                  Array(0,0,0,0,1,0,3,3,3,3,3,3,3,0,1,0,0,0,0),
                  Array(0,0,0,0,1,0,3,0,0,0,0,0,3,0,1,0,0,0,0),
                  Array(0,1,1,1,1,1,1,1,1,0,1,1,1,1,1,1,1,1,0),
                  Array(0,1,0,0,1,0,0,0,1,0,1,0,0,0,1,0,0,1,0),
                  Array(0,2,1,0,1,1,1,1,1,3,1,1,1,1,1,0,1,2,0),
                  Array(0,0,1,0,1,0,1,0,0,0,0,0,1,0,1,0,1,0,0),
                  Array(0,1,1,1,1,0,1,1,1,0,1,1,1,0,1,1,1,1,0),
                  Array(0,1,0,0,0,0,0,0,1,0,1,0,0,0,0,0,0,1,0),
                  Array(0,1,1,1,1,1,1,1,1,1,1,1,1,1,1,1,1,1,0),
                  Array(0,0,0,0,0,0,0,0,0,0,0,0,0,0,0,0,0,0,0))
  }
```

```scala
37
38     private def initContents():Unit = {
39       _player = new PacMan(9, 15, -1)
40       _ghosts = List(new Ghost(9, 7, 1, 0, Color.Red),
41                      new Ghost(8, 9, 1, -50, Color.Cyan),
42                      new Ghost(9, 9, 0, -25, Color.Pink),
43                      new Ghost(10, 9, 3, -75, Color.Orange))
44       _fruit = None
45       count = 0
46       nextFruitType = 0
47     }
48
49     def player = _player
50     def ghosts = _ghosts
51     def fruit = _fruit
52
53     def rows = cells.length
54     def columns = cells(0).length
55
56     def cell(x:Int, y:Int):Int = {
57       if (y < 0 || y >= cells.length || x < 0 || x >= cells(y).length) {
58         if (y == 9) Maze.Open else Maze.Wall
59       } else cells(y)(x)
60     }
61
62     def pelletCount:Int = cells.map(_.count(c => c == Maze.Pellet || c ==
         Maze.PowerPellet)).sum
63
64     def updateAll():Int = {
65       // Update moving stuff
66       player.update(this)
67       if (player.dying) {
68         if (player.animationCount > 10) initContents()
69         0
70       } else {
71         ghosts.foreach(_.update(this))
72         count += 1
73
74         // Eat pellets
75         val pelletScore = if (player.x>0 && player.x<cells(0).length-1) {
76           val c = cells((player.y+0.5).toInt)((player.x+0.5).toInt)
77           cells((player.y+0.5).toInt)((player.x+0.5).toInt) = Maze.Open
78           if (c == Maze.PowerPellet) {
79             ghosts.foreach(_.scare())
80             100
81           } else if (c == Maze.Pellet) 10 else 0
82         } else 0
83
84         // Check ghost intersects
85         val (ghostScore, dying) = ghosts.map(g => {
86           if ((g.x-player.x).abs < 0.8 && (g.y-player.y).abs < 0.8) {
87             if (g.scared) {
88               g.eaten()
89               (200, false)
90             } else (0, true)
```

```
91        } else (0, false)
92     }).foldLeft(0, false)((acc, gt) => (acc._1+gt._1, acc._2 || gt._2))
93     if (dying) {
94       player.setDying
95     }
96
97     // Check fruit intersect
98     val fruitScore = fruit.map(f => if ((f.x-player.x).abs < 0.8 &&
99         (f.y-player.y).abs < 0.8) Fruit.Scores(f.fruitType) else 0).getOrElse(0)
99     if (fruitScore > 0) _fruit = None
100
101     // Add fruit
102     if (count%200 == 0) {
103       if (fruit == None) {
104         _fruit = Some(new Fruit(9, 11, nextFruitType))
105         if (nextFruitType < Fruit.Orange) nextFruitType += 1
106       } else {
107         _fruit = None
108       }
109     }
110
111     pelletScore+ghostScore+fruitScore
112   }
113  }
114 }
115
116 object Maze {
117   val Door = -1
118   val Wall = 0
119   val Pellet = 1
120   val PowerPellet = 2
121   val Open = 3
122
123   val Up = 0
124   val Right = 1
125   val Down = 2
126   val Left = 3
127   val XOffset = Array(0,1,0,-1)
128   val YOffset = Array(-1,0,1,0)
129 }
```

The `initMap` method sets up the `cells` member with a 2D array of integers that represent the initial configuration of the maze. The numeric values map to walls, pellets, etc. The meaning of each value is given down in the companion `object` on lines 117-121. The `initContents` method sets up the player, ghosts, and fruit as well as some counters used for adding fruit. This method is called any time the player dies to reset things back to where they started.

Lines 49-62 have a number of fairly basic methods that are used to allow other classes to get information about the maze. The `cell` method lets other code find out what value is at a particular location in the maze. It is a bit more complicated than just a look up in the array because it includes bounds checking. For squares that are outside of the bounds, it gives back the `Wall` value unless they happen to be on the row that connects from one side to the other, in which case it gives back `Open`.

The meat of the `Maze class` is the `updateAll` method that covers lines 64-114. This gets called by the timer in `Game` roughly once every 0.1 seconds. It takes no arguments, and results in an `Int` that is the number of points scored by the player in that tick. The method begins by calling `update` on the player. After this, there is a check to see if the player is in the process of dying or not. If so, it checks if that has been happening for 10 ticks yet. If it has, it calls `initContents` to reset the board. Either way, it gives back 0. The more interesting branch is when the player is not dying. Then each of the ghosts are updated, and the `count` variable is incremented. You can then see subsections that are labeled with comments for eating pellets, running into ghosts, running into fruit, and replenishing fruit. The value that is returned is the sum of points the player got from eating pellets, ghosts, and fruit.

The ghost interactions are the one subsection that likely needs more explanation, mainly because it uses a `foldLeft` operation. This code runs through the ghosts and `maps` each one to a tuple. That tuple has the number of points the player gets from that ghost in the first element and whether the ghost killed them in the second. The `if` on line 86 checks if they are close enough to interact. If they are and the ghost is scared, we tell that ghost that it has been eaten and return points and say the player was not killed. If the ghost was not scared, then there are no points, and we say the player was killed. If they are not close enough to interact then there are no points and the player is not killed.

The results of that `map` are put through a `foldLeft` operation that combines them. The first elements, which are the scores, are added up, while the second elements, whether or not the player died, is or'ed together. If any one ghost killed the player, the player is dead.

This same logic could be done with a `for` loop and two `vars` for `ghostScore` and `dying`. That code would look like the following.

```scala
var ghostScore = 0
var dying = false
for (g <- ghosts) {
  if ((g.x-player.x).abs < 0.8 && (g.y-player.y).abs < 0.8) {
    if (g.scared) {
      g.eaten()
      ghostScore += 200
    } else {
      dying = true
    }
  }
}
```

Lines 123-128 in the companion `object` declare a number of constants that are used in other `class`es to support navigation around the maze.

Up next is the `PacMan class`. This `class` is rather true to the initial design. The data from our design is split between arguments that are passed in from outside code and some values that are always set to defaults. There are also a number of accessor methods.

Listing 16.7: PacMan.scala

```scala
class PacMan(
    private var _x:Double,
    private var _y:Double,
    private var _dir:Int) {

  private var _animationCount = 0
  private var keyDir = -1
  private var _dying = false
```

```
9
10    Game.newLife()
11
12    def x = _x
13    def y = _y
14    def dir = _dir
15    def animationCount = _animationCount
16    def dying = _dying
17
18    def setDying():Unit = {
19      _dying = true
20      _animationCount = 0
21    }
22
23    def update(maze:Maze):Unit = {
24      if (dying) {
25        _animationCount += 1
26      } else {
27        if ((x == x.toInt) && y == y.toInt) {
28          val allowed = (0 to 3).filter(d => {
29            maze.cell((x+Maze.XOffset(d)).toInt, (y+Maze.YOffset(d)).toInt) > 0
30          })
31          if (allowed.contains(keyDir)) _dir = keyDir
32          if (!allowed.contains(dir)) _dir = -1
33        } else if ((keyDir+2)%4 == dir) {
34          _dir = keyDir
35        }
36        if (dir >= 0) {
37          _x += Maze.XOffset(dir)/4.0
38          _y += Maze.YOffset(dir)/4.0
39          if (x < -1) _x = maze.columns
40          else if (x >= maze.columns) _x = -1
41          _animationCount += 1
42        }
43      }
44    }
45
46    def goLeft:Unit = keyDir = Maze.Left
47    def goRight:Unit = keyDir = Maze.Right
48    def goUp:Unit = keyDir = Maze.Up
49    def goDown:Unit = keyDir = Maze.Down
50  }
```

The first thing that should stand out here is line 10, which calls `Game.newLife()`. With this line of code we make it so that creating a new `PacMan` instance automatically reduces the number of lives the player has. Lines 18-21 add a method that was not in the design which is called by the `Maze` when the player runs into a ghost that is not scared.

As with the `Maze`, the main logic of this `class` is contained in the `update` method on lines 23-44. If the player is dying, it only increments the animation count. Otherwise, it checks to see if the position of the player is a whole number in both x and y. The reason for this check is that the player can only turn at an intersection, which means that the fractional parts are zero. If they are at whole numbers, the filter on lines 28-30 checks which directions are safe for the player to turn in. This code uses the `XOffset` and `YOffset` arrays that are defined in the `Maze` companion `object`. You should spend a few minutes looking at those values to understand how they work. Up is 0, and if you look at the offsets, the values at index 0 are 0 for x and -1 for y, which is the combination you need for moving up.

After the valid directions have been determined, line 31 checks if the direction the player has pressed is allowed. If it is, that becomes the new direction for movement. This style of handling the input gives the game the feel that is typical for Pac-Man™, where you can select a direction before you get to a turn, and the character will go in that direction when it gets there.

Line 32 checks if the current direction is allowed. This line is critical because without it, the PacMan will go right through walls when it runs into them. The `else if` on lines 33-35 allow the player to turn around even if they are located at a fractional block.

Lines 36-42 check is the player is moving and if so, it moves them and updates the animation count. The movement includes checks that wrap the player from one side to the other if they should be wrapping. Lines 37-38 include one aspect that is deceptively simple yet very important, that is the division by 4. Clearly this makes it so that the PacMan moves 1/4th of a cell on each tick. What is not so clear is that you cannot just divide by any value you want. If you change this to dividing by 5, the results are very odd. The player will not be able to take certain turns, and it will even go through walls and cause an exception when it goes out of bounds. Why is that? Remember that arithmetic with Doubles is not exact. Fractional powers of two can be represented exactly. Other values tend to be infinite repeating sequences of bits. 1.0/5.0 or 0.2, does not have an exact binary representation. 1.0/4.0 does. So the decision to divide by 4.0 was not arbitrary. Due to the way that turning is only allowed on whole values, some care must be taken to use values that have exact binary representations and which add to whole numbers. So 3.0/8.0 has an exact binary representation, but you cannot add them nicely to get to 1.0. If you want to adjust the speed in ways that does not do exact binary arithmetic, you could change the checks to something like `(x-x.toInt).abs < 1e-8`, using the technique described in section 13.3.

Note that outside code cannot directly set `_x` and `_y` or any of the other values after creation. This was true for Maze as well, but that might not have seemed as interesting given how much of the game is controlled by the Maze. For PacMan, this really highlights the significance of encapsulation given to us by writing this in an object oriented method. If the player ever moves or behaves in a way that it is not supposed to, you know immediately that these 50 lines are code are where you need to look because nothing else can move or alter the direction of a PacMan instance.

Up next is the Ghost class, which is very similar to PacMan in many ways. The Ghost has the ability to be scared and to be eaten by the player. In addition, the movement has to be done completely through code, not based on input from the user.

Listing 16.8: Ghost.scala

```scala
import scalafx.scene.paint.Color

class Ghost(
    private var _x:Double,
    private var _y:Double,
    private var _dir:Int,
    private var _animationCount:Int,
    val color:Color) {

  private var _scared = false

  def x = _x
  def y = _y
```

```
14    def dir = _dir
15    def animationCount = _animationCount
16    def scared = _scared
17
18    def scare():Unit = {
19      _scared = true
20      _animationCount = 0
21    }
22
23    def eaten():Unit = {
24      _x = 9
25      _y = 9
26      _scared = false
27    }
28
29    def update(maze:Maze):Unit = {
30      if (animationCount >= 0) {
31        if ((x == x.toInt) && y == y.toInt) {
32          if (maze.cell(x.toInt, y.toInt-1) == Maze.Door) {
33            _dir = 0
34          } else {
35            val allowed = (0 to 3).filter(d => {
36              val cell = maze.cell((x+Maze.XOffset(d)).toInt,
                      (y+Maze.YOffset(d)).toInt)
37              cell > 0 && (d+2)%4 != dir
38            })
39            _dir = allowed(util.Random.nextInt(allowed.length))
40          }
41        }
42        if (dir >= 0) {
43          _x += Maze.XOffset(dir)/4.0
44          _y += Maze.YOffset(dir)/4.0
45          if (x < -1) _x = maze.columns
46          else if (x >= maze.columns) _x = -1
47        }
48      }
49      _animationCount += 1
50      if (scared && animationCount > Ghost.ScaredTime) {
51        _scared = false
52      }
53    }
54  }
55
56  object Ghost {
57    val ScaredTime = 100
58  }
```

Here again, the main logic is in the `update` method. The `Ghost` does not do anything unless the animation count is non-negative. This allows us to easily get the ghosts to leave their home base one at a time. The logic for turning is like that for `PacMan` in that they can only turn when x and y have no fractional part, and it finds the allowed directions using a `filter`. One difference is that the random selection is not allowed to send the `Ghost` back in the opposite direction it is currently heading. The `Ghost` picks randomly from those allowed directions. Lines 32-33 are a special case to get the `Ghost` to come out of their home area as soon as they move below the `Door` cell.

The end of `update` checks if the `Ghost` has reached its maximum time being scared and sets that value to false if it has reached that max. The companion `object` defines the constant for how long the ghosts remain scared.

The main thing missing from this `Ghost` implementation is that they always turn randomly, with no regard to where the player is. In reality, they should generally head toward the player when they are not scared, and they should head away from the player when they are scared. In the actual game, they also move slightly faster than the player when they are not scared and slightly slower when they are. The intelligent direction selection can be implemented with the help of the recursive shortest path algorithm introduced in chapter 15. Making these enhancements is left as an exercise for the reader.

The simplest `class` is `Fruit`. It does nothing more than store a location and a type. There is also a companion object that stores information related to Fruit.

Listing 16.9: Fruit.scala

```scala
import scalafx.scene.paint.Color

class Fruit(
    val x:Double,
    val y:Double,
    val fruitType:Int) {
}

object Fruit {
  val Cherry = 0
  val Strawberry = 1
  val Orange = 2
  val Scores = List(100, 300, 500)
  val Colors = List(Color.Red, Color.Fuchsia, Color.Orange)
}
```

The last file that is part of the Pac-Man™ implementation is the `Renderer`. Because this implementation only allows for one of these at a time, it has been declared as a singleton `object`. The fact that there are no data members in this tells us that it should probably be an `object` instead of a `class`. A `class` would allow us to create multiple instantiations, but why should we make multiple instances of a type that does not have any data in it? Generally, the answer is that we should not; so, in that situation, an `object` declaration is the better choice.

The `object` has a single method called `render` that takes a `GraphicsContext` and a `Maze`. It then draws everything in that `Maze` to the `GraphicsContext`.

Listing 16.10: Renderer.scala

```scala
import scalafx.scene.canvas._
import scalafx.scene.paint._
import scalafx.scene.shape.ArcType

object Renderer {
  def render(gc:GraphicsContext, maze:Maze):Unit = {
    gc.fill = Color.Black
    gc.fillRect(0, 0, Game.Width, Game.Height)
    val cellWidth = Game.Width/maze.columns
    val cellHeight = Game.Height/maze.rows
    for (i <- 0 until maze.rows; j <- 0 until maze.columns) {
      maze.cell(j, i) match {
```

```
13      case Maze.Wall =>
14        gc.fill = Color.Blue
15        gc.fillRect(j*cellWidth, i*cellHeight, cellWidth, cellHeight)
16      case Maze.Pellet =>
17        gc.fill = Color.White
18        gc.fillOval(j*cellWidth+3*cellWidth/8, i*cellHeight+3*cellHeight/8,
               cellWidth/4, cellHeight/4)
19      case Maze.PowerPellet =>
20        gc.fill = Color.White
21        gc.fillOval(j*cellWidth+cellWidth/4, i*cellHeight+cellHeight/4,
               cellWidth/2, cellHeight/2)
22      case Maze.Open =>
23      case Maze.Door =>
24        gc.fill = Color.LightGray
25        gc.fillRect(j*cellWidth, i*cellHeight+cellHeight/4, cellWidth,
               cellHeight/2)
26      }
27    }
28    maze.ghosts.foreach(g => drawGhost(gc, g, cellWidth, cellHeight))
29    drawPacMan(gc, maze.player, cellWidth, cellHeight)
30    maze.fruit.foreach(f => {
31      gc.fill = Fruit.Colors(f.fruitType)
32      gc.fillOval(f.x*cellWidth+cellWidth/4, f.y*cellHeight+cellHeight/4,
               cellWidth/2, cellHeight/2)
33    })
34    gc.fill = Color.White
35    gc.fillText(Game.score.toString, 10, 20)
36    for (i <- 0 until Game.lives) {
37      gc.fill = Color.Yellow
38      gc.fillArc(400+i*cellWidth, 0, cellWidth, cellHeight, 45, 270, ArcType.Round)
39    }
40  }
41
42  private def drawPacMan(gc:GraphicsContext, pm:PacMan, cellWidth:Double,
           cellHeight:Double):Unit = {
43    gc.fill = Color.Yellow
44    if (pm.dying) {
45      val centerAngle = 90
46      val openAngle = pm.animationCount*18
47      gc.fillArc(pm.x*cellWidth, pm.y*cellHeight, cellWidth, cellHeight,
               centerAngle+openAngle, 360-2*openAngle, ArcType.Round)
48    } else {
49      val openAngle =
               45*math.sin(pm.animationCount/2.0)*math.sin(pm.animationCount/2.0)
50      val centerAngle = pm.dir match {
51        case Maze.Up => 90
52        case Maze.Right => 0
53        case Maze.Down => 270
54        case Maze.Left => 180
55        case _ => 0
56      }
57      gc.fillArc(pm.x*cellWidth, pm.y*cellHeight, cellWidth, cellHeight,
               centerAngle+openAngle, 360-2*openAngle, ArcType.Round)
58    }
59  }
```

```
60
61   private def drawGhost(gc:GraphicsContext, ghost:Ghost, cellWidth:Double,
         cellHeight:Double):Unit = {
62     gc.fill = if (ghost.scared) {
63       if (ghost.animationCount < Ghost.ScaredTime-20 || ghost.animationCount%10>2)
           Color.DarkBlue else Color.AliceBlue
64     } else ghost.color
65     gc.fillArc(ghost.x*cellWidth, ghost.y*cellHeight, cellWidth, cellHeight, 0,
         180, ArcType.Round)
66     gc.fillRect(ghost.x*cellWidth, ghost.y*cellHeight+cellHeight/2, cellWidth,
         cellHeight/3)
67   }
68 }
```

Lines 11-27 draw the grid of the `Maze`, including the pellets. There are separate methods that handle the drawing of the `PacMan` and the `Ghosts`. After those are called, the `Fruit` is drawn then the score is written out as are graphical representations of however many lives the player has left.

The result produced by this looks like figure 16.2. The `Renderer` could be enhanced in a number of different ways, especially the representations of the ghosts and the fruit, but even with this fairly short section of code, the result is highly recognizable and the full code produces a game that it generally playable.

Hopefully these different examples have given you a feel for the general concepts of object-orientation and how you think about programs when using it. There is a lot more left to explore, but that is left for *Object-Orientation, Abstraction, and Data Structures Using Scala*[1].

16.6   End of Chapter Material

16.6.1   Summary of Concepts

- Objects are constructs that combine data and the functionality that operates on that data.

- The standard way of making objects in Scala is to define a `class` and instantiate objects from it. You can think of a `class` as a blueprint for making objects.

   - A `def` declaration in a class defines a method. A `val` or `var` declaration defines member data.

   - `classes` can take parameters. By default they are not members. Adding `val` or `var` in front of the name makes it member data.

   - By default, constructs declared inside of a `class` are public and can be seen by any code. A big part of the power of object-orientation comes from the ability to hide things so that other code cannot get to it. The `private` modifier makes it so no code outside of the `class` can see the declaration. There is also a `protected` modifier that makes declarations visible only to subtypes.

- Scala also allows `object` declarations which create singleton objects. These cannot take arguments and are not instantiated with `new`. A single instance just exists for the ability to use it.

FIGURE 16.2: This figure shows the window created by running the Pac-Man™ game.

 – Scripts are good for small programs, but larger programs are written and run as applications. The entry point to an application is defined by a `main` method in an `object`. Top level declarations are split to different files. Code is compiled with `scalac` and run using `scala`.

 – An `object` with the same name as a `class` can be declared in the same file, and it becomes a companion object. Companions have access to `private` declarations.

16.6.2 Exercises

1. Alter the `Ghost class` so that it makes more intelligent direction choices to head toward or away from the player. Note that they cannot always make the best choice or the game becomes impossible.

2. Alter the `Ghost class` so that they move faster than the player when they are not scared and slower when they are.

3. You can turn any script that you wrote previously into an application by embedding everything into an `object` in the `main` method. The results of this should compile with `scalac` and the run with `scala`.[4]

   The result of this simple conversion is not generally ideal. Any method or `case class` declarations should generally be pulled out of `main`. The methods likely should go inside of the `object`, but outside of `main`. If those methods are not generally useful, they should be made `private`. The `case class`es could go inside of the `object` where they might or might not be `private`, depending on how generally useful they are. However, if they truly stand on their own and have meaning, they should go outside the `object` and into a separate file bearing their name.

   The results of this modification might not compile. That will depend on the quality of the original code and whether you used variables that were declared outside of methods in the methods. If you use variables defined outside of the methods in their body, you have one of two choices. Generally your first choice should be to pass those variables in by adding extra parameters to the method. If a variable really deserves to be a data member/property of the `object`, it can be moved up a level so it too is in the `object`, but not in `main`.[5]

   Your goal for this exercise is to run through this process on a number of scripts that you wrote earlier in the book. When you do this it is recommended that you make a subdirectory for each script you are converting, then copy files into there. This way you not only preserve the original script, you make it easier to compile that single application. Doing this on a number of scripts will really help to build your feel for how this new approach differs from what we had been doing previously.

4. Another thing that you can do to help you get used to some aspects of the object-oriented model is to convert `case class`es over to normal `class`es. This is generally a less significant modification, but it will require some changes in other parts of the code such as forcing you to use `new` when instantiating an object. You will also lose the `copy` method.

---

[4]Remember that when you run an application with `scala` you give only the name of the `object`, not the name of the file. Most of the time the file name should start with the `object` name so you are just leaving off the `.scala`.

[5]Note that if a variable is set by user input, it almost certainly needs to stay in `main`. Having singleton `object`s that request users to enter input will lead to very odd and unexpected behaviors in larger programs.

While converting scripts to applications has some general use to it, keep in mind that not all **case class**es need to be changed. The **case class**es exist in Scala for a reason that goes beyond the educational purposes that have benefited us in this book. It is quite possible that a number of your **case class**es should ideally remain that way over the long term. Practice with changing them is good for you to see how that matters, but it is not always something you want to do.

16.6.3   Projects

All the projects in this chapter switch from running as scripts to running as applications. That means that you need to run **scalac** first to compile the application, then run **scala** on just the name of the **object** you put the **main** method in. This also allows you to split code up into separate files for different **class**es and **object**s. All code must be inside of a **class** or an **object** for this approach to work. If you put the files for the project in their own directory you can compile them all by running **scalac *.scala**.

1. One way to build on top of project 15.3 is to give the user the ability to plot functions. For this to work, the formula parser needs to support variables.

   In the formula parser, there were two possible cases where you did not find an operator: a number and parentheses. The variable becomes a third possibility. To keep things simple for now, the only variable allowed in a formula will be x. So if you do not find an operator and the value of the **String** you are parsing is "x", you will say it is a variable. To use variables, you need an extra argument to the evaluation, a **Double** that represents the value of the variable x.

   To fit in with the contents of this chapter, you are to make your code for this object oriented by having it run as an application and including two separate **class**es called **Plot** and **Formula**. The **Formula class** is where your earlier parsing code will go. The **class** should take one argument that is a **String** for the formula. You will put the parsing code into a method in that **class**. The method should take a single argument of **x:Double**. It does not need to accept a **String** because the **class** as a whole knows what formula is being used. You could also add things like trigonometric function support so that you can plot more interesting curves.

   The **Plot class** will keep a **private var** List that tells it all the formulas that are to be plotted. This **List** could contain instances of **Formula** or of some **case class** that groups a **Formula** with a **Color** so that different functions are drawn in different colors. The **Plot** also needs to keep bounds for the x and y axes. You should put a **draw(gc:GraphicsContext)** method into **Plot** that can be called from the application itself. That method will loop through the different **Formula** objects, evaluating them at proper points inside the range to be plotted, and connecting the points to draw a line for the function.

   You can decide exactly how to create the application. Clearly the **object** will need to have a **main** and include a **JFXApp** that has a **Canvas** and a **GraphicsContext** with code that calls **draw** on a **Plot** object. If you want an extra challenge to make the application more useful you can add the capability for the user to edit settings on the **Plot**. As with previous plotting work, it is recommended that you not try to support labels on axes. However, if the range or domain crosses zero, you might consider drawing in lines for the axes themselves.

2. If you did project 15.3 and you did gravity integrations before that, then you can consider doing this option. The basic idea is that you will add variable support to your formula as described in the previous project, and then have the user type in a force formula that is used instead of gravity.

   In the GUI for your integrator you should have a text field that lets users input a formula. You should use "x" for the distance between the bodies. You can also define standard variables that are used, like "ma" and "mb" for the masses of the first and second particles in an interaction. Using this, gravity would be input as "-ma*mb/(x*x)". A spring connection would be "-x". You can play with other options as well and see how they change the way particles move.

   To make this work with the `Formula` type described above, you will need to build a `String` where the variables like ma and mb have been replaced by the proper numeric values for the masses of the particles being considered. You might even consider keeping those formulas in an `Array[Array[String]]` so that they do not have to be rebuilt each time. Then you have the `Formula` evaluate using the proper distance between the particles.

   You could also extend the `Formula` type so that it accepts variables for x, ma, mb, and possibly others that you want to define. This second approach is less than ideal for the `Formula` type as it makes it less generic, but is probably the easiest way for you to get some flexibility is user force functions.

   Note that this version of the integrator is going to be significantly slower than what you did before because every force calculation involves `String` parsing.

3. If you did the Tron game in an earlier project, you could update it to be object oriented. For Tron you can make a class that represents a light cycle including its position, direction, and color. It could have a method called `move` that alters the position appropriately for the direction. Other methods could include turning left or right that are called by handlers for key presses or a `draw` method that puts a point at the appropriate location. You might also encapsulate whether a cycle is alive or not into the `class`.

   With this change it is now feasible to add in more players and have walls from one player disappear when that player dies. Extend the game to have three human players (or add a computer player) and make it so that when a player dies, the line for that player goes away.

4. If you worked on any other type of graphical game in earlier chapters, you should also be able to convert it to be an application with an object-oriented approach by making the different elements of the game into `class`es that store the relevant information and have appropriate methods.

5. If you implemented the Mandelbrot set calculation and display previously, this is the project for you. First, use what you have learned in this chapter to make a class called `Complex` that represents complex numbers and alter your original Mandelbrot calculations to use that. Once you have done that, you will take further advantage of it by adding the ability to display Julia sets of quadratic polynomials.

   Every point in the Mandelbrot corresponds to one of these Julia sets. The equation for a Julia set looks much like that for the Mandelbrot set, $z_{n+1} = z_n^2 + c$ with c being a fixed value and z_0 being the point in the plane you are calculating for. By contrast, the Mandelbrot set used a different c value for each point and $z_0 = 0$ so

$z_1 = c$. Again, you iterate until either the value leaves the area around the origin or until some pre-specified number of iterations.

If the user clicks on a location (use the `MouseClicked` event) then you should pop up another window (use `Stage`for this) with a different panel that will draw the Julia set associated with the point the user clicked on. You can decide what functionality you want to put into window. It should start with a bounds of $r \in [-2, 2]$ and $i \in [-2, 2]$ for the real and imaginary axes, respectively. It should not be hard to add zooming, but that is not required.

6. This project continues project 8.5, where you had to figure out how many elements of each type were on the different sides of a chemical equation. You should recall that the output of that program was something like the following.

```
C: a*1=d*1
H: a*4=c*2
O: b*2=c*1+d*2
```

We want to treat this as a system of linear equations and solve it to find the proper values of a, b, c, and d. These particular formulas give you three equations for four unknowns. If you remember systems of equations, that is typically not a good thing as the system is underdetermined so there are an infinite number of solutions. The easy way to fix that is to assume that a certain number of the coefficients are 1 so that you are left with equal numbers of equations and unknowns.

The form given above would be fine for solving this equation by hand, but for solving it on a computer we will want to move things around a bit. The idea is that we want to get it into the form $Ax = y$, where A is a matrix giving the explicit numbers in the equations, x is a vector with the variables we are solving for, and y is a vector with whatever constants wind up being on the right side of the equal sign. Let us make this more explicit by rearranging the equations above into something more like the form we want.

```
1*a-1*d=0
4*a-2*c=0
2*b-1*c-2*d=0
```

Now we have to pick a coefficient to set to 1 so we get down to equal numbers of equations and coefficients. Any will do equally fine, but programmatically you will probably find it is easiest to set the last coefficient to a constant value (so everything is still zero indexed in your code). In this case that is d. If we do that and move constants to the right side we get the following equations.

```
1*a=1
4*a-2*c=0
2*b-1*c=2
```

This can be transformed into $Ax = y$ if

$$A = \begin{bmatrix} 1 & 0 & 0 \\ 4 & 0 & -2 \\ 0 & 2 & -1 \end{bmatrix}$$

and

$$y = \begin{bmatrix} 1 \\ 0 \\ 2 \end{bmatrix}.$$

Both x and y are column vectors here so they have a single column and multiple rows. Remember that

$$x = \begin{bmatrix} a \\ b \\ c \end{bmatrix}$$

and that is what we want to solve for. The way you will do this is through a process called Gaussian elimination. It turns out that there are many methods of doing this that have different numerical properties. Gaussian elimination is not the best, but it is the simplest to describe and sufficient for our purposes. Gaussian elimination is also exactly what you would do if you were solving this problem on paper so hopefully it will make sense.

What we do in Gaussian elimination is multiply and add together rows in A and y to remove coefficients and turn A into a triangular matrix. We might also have to swap rows at certain times. In fact, we will do that generally to improve numerical stability. To begin with, we want to remove the a term from everything but the first row. We could do this with A as it is, but for numerical reasons it is best if we keep the largest coefficient. You will see other reasons for this when we remove b. So, the first thing we do is we note that the largest coefficient of a is 4, and it is in the second row; so, we swap the first and second rows. Note that we swap them in both A and y.[6] This gives the following values.

$$A = \begin{bmatrix} 4 & 0 & -2 \\ 1 & 0 & 0 \\ 0 & 2 & -1 \end{bmatrix}$$

and

$$y = \begin{bmatrix} 0 \\ 1 \\ 2 \end{bmatrix}$$

Now we eliminate the a terms in the second and third rows by multiplying their values appropriately so that when we subtract them from the top column we do not have anything left. If the a term is already zero in a row we can leave it alone. In this case we will remove the a term from the middle row by multiplying it by 4 and subtracting the top row from it. We will do nothing with the bottom row. This gives the following values.

$$A = \begin{bmatrix} 4 & 0 & -2 \\ 0 & 0 & 2 \\ 0 & 2 & -1 \end{bmatrix}$$

and

$$y = \begin{bmatrix} 0 \\ 4 \\ 2 \end{bmatrix}$$

Now the top row is set so we look at the smaller nested matrix ignoring the first row and column. We want to eliminate the b coefficients from that. Here it really matters

---

[6]The fact that value in both A and y need to be swapped at the same time means that your implementation could represent both A and y in a single `Array[Array[Double]]` where each inner array stores a row of A followed by the corresponding y value.

that we swap up the row with the largest coefficient because what is there right now has a zero coefficient and that will cause division by zero if we do not move it. We do the swap and since the last row already has a zero in the b coefficient we will not do anything else. That leaves us with the following values.

$$A = \begin{bmatrix} 4 & 0 & -2 \\ 0 & 2 & -1 \\ 0 & 0 & 2 \end{bmatrix}$$

and

$$y = \begin{bmatrix} 0 \\ 2 \\ 4 \end{bmatrix}$$

This we can solve by working our way up the matrix. The bottom row tells us that c = 2. We plug that into the next row and get 2*b-1*c = 2*b-2 = 2 and find b = 2. It is easy for us to say that, but we should probably examine how the computer will find these values. For the first one, the computer simply does c = y(2)/A(2)(2). Then as we move up we have to do some loops to subtract things off before we do the division. In this case b = (y(1)-c*A(1)(2))/A(1)(1). Note that we are always dividing by the component on the diagonal, and we subtract off all the terms to the right of the diagonal. Now we are down to our top equation which is a = 2*c/4 so a = 1. In the program that will be a = (y(0)-c*A(0)(2)-b*A(0)(1))/A(0)(0). The values of y(0) and A(0)(1) are both zero, but the math in the program will include them, and it will not matter.

Your program should take an input just like the earlier project where you type in a chemical equation. It should then print out the same output as in that assignment and follow that with a solution. In this case it would print out "a=1 b=2 c=2 d=1" as the last line of output. While this example does not show it, there might be some decimal points in there as well. If you want an extra challenge you can scale things up so they are all integers.

Make this object oriented by making an application and having a **class** that represents a chemical formula. You can add methods to access the information that is calculated using the approach described above.

7. It is time to make your map program into a true text adventure game. It does not matter how you do this, but you need to add in a goal to the game and have it inform the player if he/she wins or loses. Most of the original text adventure games had some type of puzzle solving where the items could interact with one another or with rooms in certain ways to make things happen. You probably do not want to go the RPG route and have equipment, combat, and the like though simple combat might be feasible. You must add a "help" command which prints out the objective of the game and how to play. This should read and print the content of a file so you do not have to type in the full help in print statements.

8. If you did project 15.1 for the ray tracing with recursion and the graphics projects before it, now it is time to add lighting and proper colors and make it so that you have a real image. You will also organize things into an application with **objects** and **classes**.

You already have a **case class** scene that includes spheres, planes, and lights. So far you have not made use of the lights. Each of the geometry objects and lights should have a color attached to it. You need to write a function that takes a ray and the scene and tells you what color to use. If it never hits anything the color should be black. It is recommended for this that you make your own **class** called **DColor** that uses **Doubles** for the red, green, blue, and alpha values. You can put in methods to scale and combine these in meaningful ways. This might include a + method for adding **DColors** and a * method for scaling when multiplied by a **Double**.

The reason for using your own **DColor class** is that you can represent colors with components that go outside the normal 0.0 to 1.0 range when you are doing calculations. You only have to bring things back to that range when you build a normal **Color** object. That operation makes another good method for the **DColor** type.

You should also update the existing ray tracer code so that it uses **classes** for the various components including points/vectors, spheres, planes, lights, etc. The **Point** and **Vect**[7] **classes** in particular can have methods for doing mathematical operations that will make your code much simpler and easier to read. Methods like **mag** and **normalize** could also come in handy.

To color a point you need the color of the object that was hit, and the lighting on that point. To get the lighting, send a ray from the point of contact to the light source. If there is any geometry in the way (t value between 0.0 and 1.0), then that point is shadowed and the light does not get to it. If there is not anything in the way, that light contributed to the point. The color should be the component wise product of the colors of the light and the object hit, scaled by the cosine of the angle between the normal and the direction to the light. To find the cosine, keep in mind that $\vec{a} \cdot \vec{b} = |a| \, |b| \cos \theta$. To make this easier, you should probably store the colors of lights and geometry using **DColor**. This hopefully helps to demonstrate why putting proper methods in the **DColor**, **Point**, and **Vect** will help to make the code easier to write and to read.

If the geometry that is hit is reflective, you do the recursive call to find what it hits and add the resulting colors,scaled by the appropriate fractions, together to get the color value of the incoming ray.

9. You can make the scheduling script from project 15.6 into an object-oriented application. To do this, you want to put a body on your **Course case class**. It is possible that you will leave it as an immutable **case class** and just add methods to it. The methods should probably include one that takes another **Course** and determines if the two overlap or not. There might also be methods designed to do specific types of **copy** operations that you find yourself doing frequently, such as making a copy of a course at a different time or with a different professor.

   The fact that the choice of professor often impacts how much you want to take a course, you could split this off and make a **class** called **Professor** that has the name of the **Professor** and a favorability rating. The **Course** could then hold a reference to **Professor** and have a rating that is independent of the professor. A method can be written to give you the combined favorability. This way you can easily copy course options with different professor selections and have that change automatically taken into account.

   You can also build a mutable **Schedule class** that keeps track of what the user wants in a schedule along with what has been met so far from the courses that have been put inside of the schedule. Put this all in an application and tie it together in a GUI.

---

[7]There is a standard collection type called **Vector**. We suggest the shorter name **Vect** so as not to conflict with that.

10. To upgrade the recipe project from 15.5 to be more object-oriented, you will convert `case class`es to have methods and possibly make them mutable `class`es. You will also move the main functionality of the script into an `object` with a `main` method.

    In addition to having `class`es for `Pantry`, `Recipe`, and `Item`, you might want to have a `class` that keeps track of all the known `Item` types that have been used in the program. That `class` should have methods that facilitate looking up items by names or other attributes so that users do not wind up creating duplicates of `Item` objects.

11. The music library script from project 15.9 can be converted to an object-oriented application by filling out the different `case class`es with methods or turning them into mutable `class`es.

    You can also make types for play lists or other groupings of information that would benefit from encapsulation. Put the main script into an `object` with a `main` and you have an application.

12. There are a number of features of the L-Systems script from project 15.4 that call for object-orientation. Starting at the top, an L-System should be represented by a `class` that includes multiple `Productions` as well as an initial value. The `Production` type should be a `class` that includes a `Char` for the character that is being changed and a `String` for what it should result in.

    One of the advantages of encapsulating a production in an L-System is that you can make probabilistic productions. These start with a single character, but can result in various different `String`s depending on a random value. Whether a `Production` is deterministic or not does not really matter to the whole system as all the system does is ask for what a given character will be converted to.

    Using a `String` to represent the state of the L-System at any given time works fine, but there is some benefit to creating your own `class` that is wrapped around a `String`. This `class` might also include information on what generation you are at in applying the productions. The advantage of this is that the type makes it clear this is not any random `String`; instead, it is the state of an L-System.

    You should make these changes and adjust the script to go into an `object` with a `main` to create a complete application.

Additional exercises and projects, along with data files, are available on the book's web site.

Chapter 17

Wrapping Up

We have reached the end of our tour of basic programming for problem solving with Scala. While this book has covered a good number of different topics, this is only a basic introduction to programming, which is but a stepping stone into the field of computer science. The goal of this book was to lay a foundation that you can build on going forward. Now it is your job to continue the construction process.

17.1 What You Have Learned

This book focused primarily on basic constructs used to tell a computer how to do things. You learned the use of conditional logic and looping constructs that could be put around simple sequential commands to express actions that you wanted to have the machine carry out. You learned how to break problems into pieces. Primarily the approach was to build separate functions that handled one part of the problem at a time, and ways of grouping data that made sense. At the very end you got a brief introduction to the approach of decomposing problems into objects that combine data with functionality. This learning to break problems apart should have had an impact on how you looked at many things through your life and in your studies outside of computer science. The basic constructs of conditional, functions, loops, and basic collections will be applicable, with minor modifications, to any other language that you might study in the future.

The modern world is teeming with data. Data is produced by all types of devices during all types of activities. You have learned how to give instructions to a computer to make it process data for you. There are limits on how much you can do by hand, but a computer can read and process millions or even billions of values quickly. You now know how to pull this data from files or over networks. You can deal with data in flat text or XML. You can use higher-order constructs to do the processing.

You also have the ability to write programs that display GUIs and draw custom graphics, allowing you to make programs that interact with the user in whatever manner you desire.

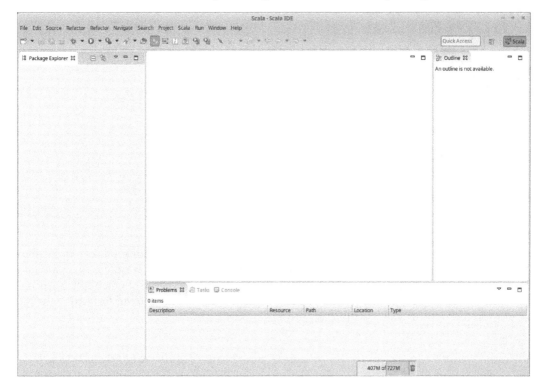

FIGURE 17.1: This is the blank workspace that you will see when you start up Eclipse for the first time.

17.2   IDEs (Eclipse)

There is still a lot more for you to learn about Scala, programming, and the field of computer science. In order to not only learn more, but to maintain the skills you have developed, you need to practice. Many students find the combination of text editor and command line tools that we have used through this book to have an unnatural feel, as it is not the way they normally interact with their computers. Most students are more used to a GUI interface than to command line. Having discussed applications and the object-oriented approach to programming in chapter 16, it is now possible to use development environments that have this style of interaction.

Programs that provide complete support for writing code, running programs, and debugging are called IDEs, short for Integrated Development Environments. The officially supported IDE for Scala is Eclipse. You can download a version of Eclipse with support for Scala already included at `http://scala-ide.org`.[1] If you download, install, and run this, when you go to the workspace you should see a window like figure 17.1.

When you want to code in Eclipse, you cannot just create a file and go. First, you have to create a project. You can do this with File > New > Scala Project or by right clicking on Package Explorer area and selecting New > Scala Project from the context menu. You

---

[1]The version of Eclipse used by the bundle often lags a bit behind the newest version of Eclipse. You can also download Eclipse for Java from `http://java.oracle.com` and add the Scala plug-in using the link at `http://scala-ide.org/download/current.html`.

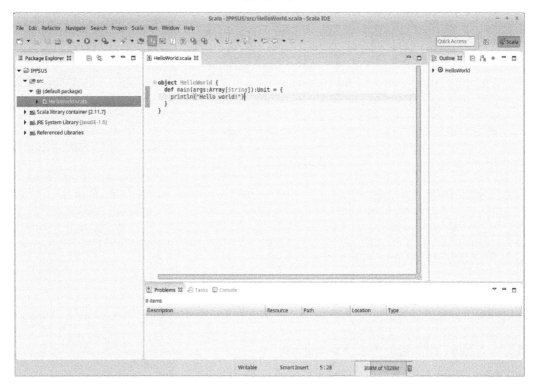

FIGURE 17.2: This is the Eclipse workspace after we have created a project and an object that runs "Hello World".

probably want to make a new project for each major program you write. You should be able to use the default options for the project after you enter the name.

Once you have a project, click on the triangle to expand it and right click on the src directory to bring up a menu that will let you create a file with code. Given what you have learned, you should select to make a "Scala Object". As we did with our scripts, you could start with a "Hello World" program, so call the object `HelloWorld`. Add a main and a print statement into the `object` that is created. At this point, your Eclipse window should look something like figure 17.2.

To run this code, right click on the text editor or on the "HelloWorld.scala" file in the Package Explorer and select Run As > Scala Application. This will pop up a console at the bottom of the workspace that shows the output. You can create more `object`s or `class`es to build whatever applications interest you.

Note that if you want to make a program with ScalaFX there is one other step that you will have to take. You have to point Eclipse to the ScalaFX JAR file that you have been using with your scripts. To do this, right click on your project and select Build Path > Add External Archives... then go select the JAR file on your file system. Once you have done that, you will be able to include ScalaFX types in your applications and run them.

17.3   Next Steps

You could redo lots of the things that you have done through this book in Eclipse, but what are your options for continuing to learn? If you want to continue to work with Scala and gain a solid foundation in key concepts of computer science, we would certainly recommend *Object-Orientation, Abstraction, and Data Structures Using Scala*[1]. In addition to the topics that are listed in the title, this book gives you an introduction to parallelism through multithreading and a coverage of networking.

Whatever route you take, you really should strive to learn more about computer science. These remarkable machines have become integrated in every aspect of our daily lives, often in ways that are remarkably non-obvious because they have become so ubiquitous. Even when you do not see them, computers and the software that they run have a dramatic impact on your life. Learning how to tell these incredible devices how to do what you want is an increasingly essential skill for modern life.

17.4   End of Chapter Material

17.4.1   Exercises

1. Back in project 1.2 you were asked to compare programming to three other activities based on an extremely limited introduction. Now that you have significantly more experience, revisit that same question.

2. Write one of the projects from chapter 16 in Eclipse.

Appendix A

Getting to Know the Tools

Everything you do on a computer involves the running of software. At the most basic level, your computer has some operating system on it that controls everything that happens, organizes data, and lets you execute other programs inside of it. What operating system and tool set you choose to use has a significant impact on how you go about doing things. This chapter walks you through some tools that you might use and might not be familiar with. It then moves on to look at some of the tools that are part of a standard Scala distribution that we will use through the rest of the book.

Your choice of tools begins with your choice of operating system. It is very likely that you are familiar with common graphical environments on a Windows PC or Macintosh environment. For that reason, we are not going to take time to introduce them. Instead, this appendix talks about how you interact with various operating systems using a text interface. There are a number of reasons why it is good to be able to use this type of interface. For our purposes, the most significant is that it is the way we will interact with some of the Scala tools that we will be using.

A.1   Unix/Linux (includes Mac OS X)

This section covers the environment you will encounter in a Linux and Unix environment. Mac OS X is built on top of BSD Unix so everything here applies to the Mac. If you are on a Mac you can open a terminal to get access to the command-line tools.

A.1.1   Command Line

Most people these days are used to the point and click approach of a Graphical User Interface, GUI. This is not how people have always interacted with computers. Indeed, a GUI takes a significant amount of processing power to run. For many years, computers had to have a simpler interface. Even today, there are times when you will need to interact with a computer without a GUI. This can happen when something goes wrong with the computer or if you need to use a computer that is located far away. There are also situations where people will choose to not run a GUI on a computer simply because they do not want to waste the processing power.

Even when a computer is running a GUI, you can choose to interact using a command line interface. Using the command line itself provides you with abilities that are not easily accomplished with a GUI. The reason why many people choose not to use a command prompt is that it does have a bit of a learning curve. You have to know some basic commands to be able to do anything. If you are going to spend much time working on computers it is worth putting in the small amount of effort to learn the basic commands for the command prompt interface on your OS. Odds are that you will find that not only does it speed things up on a regular basis, but there will be times when it allows you to do things that you simply could not do otherwise.

Figure A.1 shows a terminal window under Linux that gives a command prompt. Your command prompt might look different than this. In general, the prompt should display some useful information followed by a character that signifies the end of the prompt. In this figure the prompt shows the user name and computer name followed by the current directory and a $. It is the $ that signifies the end of the prompt. Commands that you enter appear after the $. Anything the command displays will be printed out below the prompt. After the command finishes, another prompt will be given.

A.1.1.1   Files and Directories

In a GUI you organize your files in folders. You have icons that represent different file types or folders and you can place files/folders inside of folders to organize them. These concepts came from the systems already in place on the command line. The only change is terminology. The term folder works well from a graphical standpoint. On the command line they have long been called directories.

The first set of commands we will learn about allow us to work with files and directories so that we can navigate around and do the types of things you are used to doing by clicking, double-clicking, or drag and dropping. Here is a list of commands with a brief description of each. You will notice that in general they are abbreviations for what they do.

- cat – Display the contents of one or more files.

- cd – Change directory.

- cp – Copy one file to another name or location.

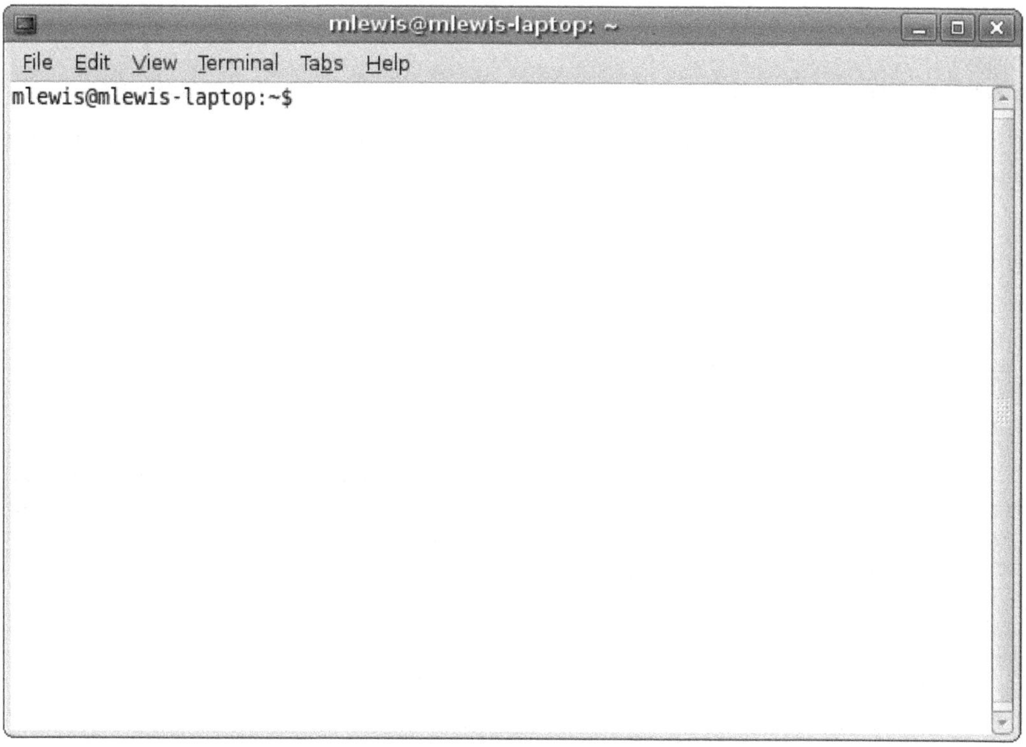

FIGURE A.1: An example of a Linux terminal with a command prompt.

- `less` – Display out the contents of one or more files with the ability to move forward and backward, search, etc. (Less is more than more.)

- `ls` – List the contents of a directory.

- `mkdir` – Make a directory.

- `more` – Display the contents of one or more files with the ability to page through or search.

- `mv` – Move a file to a new name or location.

- `pwd` – Stands for "print working directory". This prints out the directory you are currently working in.

- `rm` – Remove a file.

- `rmdir` – Remove a directory if it is empty.

The big advantage of the command line is that each command can be followed by arguments. Just entering a command is like double clicking on a program icon. It is the arguments that let you specify additional information. Most of these commands do not do anything unless you give them one or more file/directory names to work on.

To see how you go about using these commands we will begin with pwd. It is a very simple command and can come in handy if you have lost track of where you are in the directory structure. At the command prompt simply type in pwd and hit Enter. Exactly what you get will depend on many things, but it should print out one line of output showing you the current directory. It could look something like this.

```
mlewis@mlewis-laptop:~$ pwd
/home/mlewis
mlewis@mlewis-laptop:~$
```

In this case, the directory is /home/mlewis. After printing the output we get a new prompt to enter the next command.

Aside

In Unix and Linux systems, the base of the entire file system is /. That is why it appears at the beginning of the /home/mlewis directory. A directory that begins with / is an absolute directory.

The next command we want to enter is the ls command. Type in ls, followed by an enter and you will get a listing of all of the files in your current directory. This is the most basic usage of the command. You can also use it to see the contents of some other directory by following the ls command with a directory name. Here we see a listing of the contents of the root directory.

```
mlewis@mlewis-laptop:~$ ls /
bin dev initrd lib lost+found opt sbin tmp vmlinuz
boot etc initrd.img lib32 media proc srv usr vmlinuz.old
cdrom home initrd.img.old lib64 mnt root sys var
```

If you list a directory and it does not start with / then it will be a relative directory which means that the location of that directory is relative to (within) the current working directory. So, the directory that you list will be appended to the end of the current directory. For example, you might have a directory called Desktop in the current directory. Typing in ls Desktop will list the contents of the Desktop directory under the current directory.

```
mlewis@mlewis-laptop:~$ ls Desktop
AdobeReader.desktop QtCreator.desktop
```

You can also specify multiple files or directories, and they will all be listed as seen here.

```
mlewis@mlewis-laptop:~$ ls Desktop /
/:
bin dev initrd lib lost+found opt sbin tmp vmlinuz
boot etc initrd.img lib32 media proc srv usr vmlinuz.old
cdrom home initrd.img.old lib64 mnt root sys var

Desktop:
AdobeReader.desktop QtCreator.desktop
```

As you can see, the contents of the current directory are listed on the lines below /: and the contents of Desktop are listed on the line below Desktop:.

You might not think that you would normally want to list multiple things at once, but it is a feature people use all the time with wild card characters. If you put a * into a name, the system will replace it with all the files or directories that match if the * is replaced with zero or more characters. Here is an example of that usage.

```
mlewis@mlewis-laptop:~$ ls /vmlinuz*
/vmlinuz /vmlinuz.old
```

For the first file, the * is replaced with nothing. For the second it is replaced with ".old". You can also use a ? to represent any one character. These wild cards are not ever seen by the ls command of other commands. The command shell replaces them with all of the matching files, and it is the multiple file names or directory names that get passed to the command.

The ls command has a lot of other possible options as well. The options are all preceded by a hyphen. The most commonly used option is -l, which tells ls to use a long display format. The following is an example of that.

```
mlewis@mlewis-laptop:~$ ls -l Desktop
total 8
-rwxrwxrwx 1 root root 1261 2008-11-28 12:21 AdobeReader.desktop
-rw-r--r-- 1 mlewis mlewis 250 2009-12-22 11:03 QtCreator.desktop
```

The long format puts each file or directory on a separate line. In addition to the file names, it shows a number of other pieces of information. The first ten characters show permissions on the files. Permissions are discussed in section A.1.1.4. After the permissions is a number showing how many links there are to this file. After that are the user name of the owner of the file and the group that the file belongs to. The last two pieces of information are the size of the file, measured in bytes, and the date and time of the last modification of that file.

The next most commonly used option is -a. This tells ls to list all the files, including the hidden files. Hidden files in Unix and Linux are files whose names begin with a period. We see here a usage of this combined with the long option. Only one hyphen is needed when options are combined.

```
mlewis@mlewis-laptop:~$ ls -al Desktop
total 16
drwxr-xr-x 2 mlewis mlewis 4096 2010-02-27 20:02 .
drwxr-xr-x 106 mlewis mlewis 4096 2010-08-09 16:57 ..
-rwxrwxrwx 1 root root 1261 2008-11-28 12:21 AdobeReader.desktop
-rw-r--r-- 1 mlewis mlewis 250 2009-12-22 11:03 QtCreator.desktop
```

This adds two additional lines. The letter d at the far left edge tells us that these are actually directories. The . directory is the name for the current directory. The .. directory is the name of the directory in which the current directory sits. We might also refer to it as the parent directory.

By default, commands work on the things in the current directory. For this reason, one of the commands you will use the most is the command that changes the current directory, cd. The cd command is followed by a single directory name. The directory will be the new current directory. As we just saw, each directory has . and .. as directory names inside of it. If you want to move up a level in the directory structure, simply execute "cd ..". You can also enter cd without a directory at any time to change back to your home directory. The following command changes our current directory to the Desktop directory.

```
mlewis@mlewis-laptop:~$ cd Desktop
mlewis@mlewis-laptop:~/Desktop$
```

Note the prompt used in this example changes so that we can see we are in this new directory.

Other tasks that you do frequently when you are manipulating files in a GUI are to copy, move, or delete. The commands for these are cp, mv, and rm. The cp and mv commands are used in roughly the same way. One way to use them is to enter one or more file names and then end with a directory name. In that usage all of the files listed are either copied or moved to the directory listed at the end. If you include a name that has a wild card in it, it will be expanded to all the file names. We can see this in use here:

```
mlewis@mlewis-laptop:~/Desktop$ mv ../*.txt .
```

Here we move all the files from our home directory that end with .txt into the current directory. This line with mv uses both .. to refer to the directory above the current one and . to refer to the current directory. Another way to use those commands is to enter two file names, where the first file will be copied to the second name with cp or renamed to the second name with mv. Assume that one of the files in the Desktop directory was named text.txt. After the following command executes we will have copied that file into a new file named text2.txt.

```
mlewis@mlewis-laptop:~/Desktop$ cp test.txt test2.txt
```

If you want to remove a file, use the rm command. The rm command can be followed by one or more files. Again, wild cards can be used to match multiple files. Here we are getting rid of all of the files that end with .txt.

```
mlewis@mlewis-laptop:~/Desktop$ rm *.txt
```

Use caution when removing files on most Linux based systems with the command line, it is not forgiving and you cannot easily "undo" what was done. This is true whether you are using rm or if you happen to mv or cp to a file name when you already had a file of that name. While most of the GUIs that you have used probably move files you delete to some form of trash can, really just a different directory, and then allow you to clear that out later, doing rm on a file really deletes it at that moment. You cannot go digging in the trash to find it. In general when you delete a file with rm on these systems, it is gone and your only chance of getting it back is if you have a backup.[1]

This can be especially dangerous with the way that the command line deals with wild cards. Take the rm command above. If you accidentally insert a space between the asterisk and the dot, you will delete every file in the current directory. This is because the * alone matches everything.[2] So be careful when you are using rm and look over the command before you hit enter to make sure you did not mistype anything or insert any unwanted spaces.

Directories exist to keep files organized. For this reason, you should probably make directories for each different major grouping of work that you have. You might also nest these to further refine the organization scheme. You can make a new directory with the mkdir command. Simply follow the command with the names of one or more directories you want to create. For example, it is probably a good idea to make a directory to store all of the work for this book or the course you are taking. You might make sub-directories in there for different assignments/projects or the code you write in class. So that we do not clutter up our Desktop directory, here make a sub-directory called projects in which we can store all of our assignment and project files.

```
mlewis@mlewis-laptop:~/Desktop$ mkdir projects
```

---

[1]There is a -i option for rm, mv, and cp that will cause the program to prompt you before any file is removed. This is not practical if you want to remove a lot of files, but it could be considered a good default for most of what you do.

[2]There is one minor safety in place. The * alone does not match files that start with a ., so hidden files would not be deleted in that situation. That will not make you feel much better though if you have just erased five different multi-hour coding projects.

By default, rm does not remove directories. To remove a directory use the `rmdir` command. Like `mkdir`, you follow it with the name of one or more directories you want to remove. `rmdir` will only remove a directory if it is empty. So you would need to remove the contents to the directory first before you can remove it.

A.1.1.2 Aside

If you really want to delete an entire directory and all of its contents, including potentially other directories, there is an option for `rm` that will do that. The `-r` option tells `rm` or `cp` to recursively run down into a directory and either remove or copy all of its contents. This is very helpful if you want to copy a directory that has a lot of things in it. It can also be of great help if you want to remove such a directory. Remember to tread carefully though. Using `rm` with the `-r` flag has the possibility to wipe out a lot of files that you might have wanted to keep around. The command below uses `-r` with a copy to copy over an entire directory and all of its contents into the current directory.

```
mlewis@mlewis-laptop:~/Desktop$ cp -r ../Music/ .
```

The last thing you might want to do with your files is actually look at what is in them. There are many ways to do this. The commands we will consider here work best with plain text files. We will be dealing with a lot of those in this book. The most basic way to look at such a file is with the `cat` command which simply prints the contents of one or more files to standard output. A step above that is the `more` command which will display one screen at a time and let you move down one line by hitting enter or a whole screen by hitting space. You can also search for things by typing what you want to search for after pressing /. Yet another step up is the `less` command which allows you to move around freely using arrow keys or page-up and page-down as well as search as you could with `more`.

A.1.1.3 Helpful Tips

Many people who are not familiar with the command line are initially turned off a bit by the amount of typing they have to do. People who use command line all the time do not necessarily like typing more, they simply know more tricks to do things with the keyboard that do not require a lot of typing. A couple of handy tricks to help you use the command line more efficiently include tab completion and viewing your command history.[3]

Typing in complete file names can be tedious and worse, is often error prone. If you mistype something it will not work and you have to enter the whole thing again. Worse, it might work and do something you did not really want to do. Because of this, tab completion is probably the most helpful feature you will find in your command line environment. If you type in the first few letters of a file or directory then hit tab, the shell will fill in as many characters as it can. If there is only one file that starts that way, it will give you the whole file. If there is more than one, it will fill in as much as it can until it gets to a point where you have to make a choice. If you double tab it will print out the different options that fit.

You should try this by going back to your home directory (remember you can do this at any time by typing in `cd` and hitting enter without giving it a directory) and then typing `cd De` and hitting tab. Odds are good that the word `Desktop` will be completed for you

---

[3]Tricks like those discussed in this section vary depending on the exact command shell you are using. A command shell is a separate software program that provides direct communication between the user and the operating system. This book describes features of the `bash` shell, which is standard for many Linux installs.

with a / at the end because it is a directory. Use `ls` to see if there are two files/directories that start with the same first few letters. Type in `cat` followed by the first two letters and hit tab. The shell will complete as much as it can then stop. Hit tab twice quickly and it will show you the different options. It might look something like this.

```
mlewis@mlewis-laptop:~$ cd Do
Documents/ Downloads/
```

Not only does tab completion save you from typing a lot of extra characters, it never misspells a file or directory name. Use tab often and it will save you a lot of key strokes and a lot of little mistakes.

It is not uncommon to want to enter the same command or a very similar command more than once. The easiest way to do this, if it is a command you entered recently, is to use the up and down arrow keys to navigate backward and forward through the command history. When you get to a command, you can edit it if you want to make changes.

Using the arrow keys is not all that convenient if the command was something you entered a long time ago. The simplest way to repeat an old command exactly is to start the line with an exclamation point (often read "bang") and follow it by some of the command. This will go back through the history and find the most recent command that started with those characters. That command will be executed without the chance to edit it. The limitation that you cannot edit it or see exactly what the command is you will execute before you execute it means that ! is most useful for commands that start in a rather unique way.

You can also press `Ctrl-r` to get the ability to search through history. After pressing `Ctrl-r`, start typing in characters that occur consecutively in the command you want. It will bring up the most recent command that includes what you have typed in and you can edit it.

Another command that is very useful in general is the `man` command. This is short for manual and basically functions as help pages for commands or installed programs on the systems. The most basic usage is to type in `man` followed by the command that you want information on. You can also put in a `-k` option to do a search for something in the man pages.

A.1.1.4 Permissions

All modern operating systems have permissions that control who can get access to different types of files. What files you can access depends on who you are and what permissions you have. The act of logging into a machine determines who you are. You can use the command `whoami` to see who you are logged onto the machine as.

```
mlewis@mlewis-laptop:~$ whoami
mlewis
```

In this case the prompt also displays the user name, but that will not always be the case. Each user can also be a member of various groups. You can use the `groups` command to see what groups you are a member of.

```
mlewis@mlewis-laptop:~$ groups
mlewis adm dialout cdrom floppy audio dip video plugdev fuse lpadmin admin
```

The combination of who you are and the groups that you are in will determine what you have access to on a machine.

On Unix and Linux every file has read, write, and execute permissions for their owner, their group, and others. Those, along with the owner and the group of a file are displayed by `ls` when you use the `-l` option. Let us go back to the `Desktop` directory and look at the long listing of the files again.

```
mlewis@mlewis-laptop:~/Desktop$ ls -l
total 8
-rwxrwxrwx 1 root   root   1261 2008-11-28 12:21 AdobeReader.desktop
-rw-r--r-- 1 mlewis mlewis 250  2009-12-22 11:03 QtCreator.desktop
```

There are two files here. The first ten characters tell us the permissions on the file. The first one is either 'd' or '-'. The 'd' would tell us it is a directory. Neither is a directory so both of these begin with '-'. After that are three groups of `rwx` where any of those can be replaced by '-'. The letters stand for read permission, write permission, and execute permission with '-' being used when that permission is not granted. The first set of `rwx` is the permissions of the user the file belongs to. The second set is for the group the file belongs to. The third is for others. The first file listed here gives full permissions to all three. The second gives read and write to the user and only read to the group and others. Shortly after the permissions appear two names. For the first file they are both `root`. For the second they are both `mlewis`. These are the user and group owners for the file. The name `root` is the superuser on Unix and Linux machines.

To make this more interesting, let us use the `mkdir` command to make a new directory and then list the contents of the directory again.

```
mlewis@mlewis-laptop:~/Desktop$ mkdir NewDir
mlewis@mlewis-laptop:~/Desktop$ ls -l
total 12
-rwxrwxrwx 1 root   root   1261 2008-11-28 12:21 AdobeReader.desktop
drwxr-xr-x 2 mlewis mlewis 4096 2010-08-11 19:59 NewDir
-rw-r--r-- 1 mlewis mlewis 250  2009-12-22 11:03 QtCreator.desktop
```

This new directory is also owned by `mlewis` and is part of the group mlewis. Note the letter 'd' first thing on the line telling us that this is a directory. The owner, `mlewis`, has full read, write, and execute permissions. Anyone in group `mlewis` will have read and execute permissions as do other users on the machine. If you try to use a file/directory in a way you do not have permissions for, you will get an error message. To go into a directory and see what is inside of it you need both read and execute permissions.

So now that you know how to tell what permissions a file has, the next question is how do you change them. The `chmod` command is used to change permissions on a file. There are quite a few different ways to use `chmod`. We will just introduce one of them here. After `chmod` you specify how you want to change or set permissions, then give a list of the files you want to make the changes to. The simplest way to specify changes is to use character codes for which set of users and the permissions involved and separate them with a '+', '-', or '=' to say if you want to add them, remove them, or set them. The different permission sets are specified by 'u' for the user, 'g' for the group, 'o' for others, and 'a' for all. The rights are specified by the letters 'r', 'w', and 'x' as we have already seen.

Let us say that we are going to put things in this new directory that we do not want anyone but the user to have the ability to see. In that case we would want to remove the read and execute permissions from the group and others. We can see how that is done and the result of doing it here.

```
mlewis@mlewis-laptop:~/Desktop$ chmod go-rx NewDir/
mlewis@mlewis-laptop:~/Desktop$ ls -l
total 12
-rwxrwxrwx 1 root   root   1261 2008-11-28 12:21 AdobeReader.desktop
drwx------ 2 mlewis mlewis 4096 2010-08-11 19:59 NewDir
-rw-r--r-- 1 mlewis mlewis 250  2009-12-22 11:03 QtCreator.desktop
```

Had we wanted to give everyone full permissions, we could have used a+rwx as the second argument to chmod.

Less common than changing permissions is changing ownership. If you own the file you can change the user and the group of the file with the chown command. The most likely scenario for this is if you want to change the group to a shared group so that a select set of users can have access to something. If the group were called project then you would type in "chown :project files".

Aside

As with the cp and rm commands, there are times when you will want to change the permissions of a large number of files in an entire directory. Both chmod and chown accept the -R option so that they will do this.

A.1.1.5   Compression/Archiving

If you spend much time on a system, you will likely need to interact with compressed or at some point. Both of these play the role of taking many separate files and turning them into one large file. Archiving is used to collect multiple data files together into a single file for easier portability and storage. Compressing is used to reduce the size of the file to save storage space or to share more easily. A large, compressed file should be smaller than the sum of the files that went into it. Later, the file can be expanded back to its original size. Many programs that you might want to put on your computer come in the form of compressed files. If you have assignments or projects that span multiple files, your instructor might want you to combine them into a single file to turn them in. There are many different compression and archiving utilities. We will just talk about the main ones here.

- tar – Archive files to a single tarball.[4]

- gzip/gunzip – Compress one or more individual files to the gzip format.

- zip/unzip – Compress multiple files to a single zip file. The zip format is used extensively on many operating systems.

The first two are used in the Unix/Linux world. They do archiving and compression separately, but are often used together to do both. The archiving program is called tar. It was originally created for archiving files from disk to tape. It simply collects multiple files together into a single file called a tarball. It can also extract files from a tarball. As with our other programs, tar is controlled by passing it arguments. You can use the man pages to see the many different options for tar. We will only discuss the basic ones.

The three main ways you will interact with a tarball are to create one, view the contents of one, or extract files from one. These different options are given with the characters c, v, and x. Most of the time you will want to interact with a file, and for that you will give the f option. The next thing after the f option should be the file name of the tarball you are creating, viewing, or extracting. Tarball files should typically end with the extension .tar. If you are creating a tarball, that will be followed by the file and directory names of the things you want to put into the tarball. Here are some sample invocations.

---

[4]Tarball is a jargon term for an archived group of files collected together into one file. You might think of tar as a sticky substance that holds things together.

```
tar cf assign1.tar header.txt Assign1/
tar tf assign1.tar
tar xf assign1.tar
```

The first command is what you might execute to create a tarball called `assign1.tar` which contains a header file and all the contents of the directory `Assign1`. The second command might be used to verify that everything you want is in the file or to check what is there before it is extracted. The last command would extract the contents.

The counterpart to `tar` is `gzip`. This command will compress one or more files. Simply follow the `gzip` command with a set of file names to compress. That will create a new set of compressed files that end with the extension `.gz`. The `gunzip` command will unzip any file to give you back the original file. You can use the `-r` option to recursively descend into a directory.

The combination of `tar` and `gzip` is so common that you can get `tar` to automatically zip or unzip files using `gzip`. Simply put the `z` flag in the options to `tar` and either `gzip` or `gunzip` will be invoked. Files created that way are typically given the `.tgz` extension.

Both `tar` and `gzip` are rather specific to the Unix/Linux world. If you have to deal with compressed files going to or from other systems, you might want to use the more broadly used `zip` format. The `zip` command can be followed by the name of a zip file and the set of file names to zip up into the specified zip file. If you want it to zip up a whole directory structure you will need to use the `-r` option. There is an `unzip` command that will extract the contents of a zip file.

A.1.1.6 Remote

One of the great strengths of Unix/Linux is the true multi-user capabilities that they have. These systems will allow multiple people to be logged in at once. A computer could be in full use even if no one is sitting at it. In fact, this is a big part of why you should be familiar with the command line. Running graphical interfaces on machines across long distance Internet connections can be very unpleasant. However, the plain text form of a command line will be quite responsive even across slow networks.

So how do you log into one computer from another one? If you are on a Unix/Linux box and you want to log into another there are several commands that you could use to make a connection. The most basic, and likely least useful, is `telnet`.[5] Simply type in `telnet` followed by the name or IP address of the machine that you want to connect to. This opens a very simple network connection between the machines. Because `telnet` connections are not secure and are, in a way, too flexible, you will probably find this approach blocked by the security setting on most machines.

Next up the scale is `rsh`. This stands for remote shell and can be invoked in the same way as `telnet`. The `rsh` connection is also not secure and as such, is blocked on many systems. However, it might be allowed on your system if you are logged into a trusted machine inside of the local firewall.

The connection type that you probably should use by default is `ssh`. This stands for secure shell and, as the name implies, all communication across a ssh connection is encrypted using public key cryptography. As such, if a system is going to let you log in remotely, this is the method that most system administrators are likely to leave open for you.

When you connect remotely to a machine you will be asked for your password and, assuming you get it right, you will be dropped at a command prompt on that machine.

---

[5]You will have occasion to use `telnet` in the second half of the book when we start writing our own networked code. One of the project ideas from the second half of the book is a networked text-based game called a MUD that users would connect to with `telnet`.

Everything that is discussed in this section will work on the remote machine just as well as the local one, no matter how far away the machine is. When you are done entering commands and you want to come back to the current machine, type `logout` or `exit` to terminate the remote session.

Aside

If your username on the remote machine is different from that on the current machine you can use "`ssh` *username@machine*" to specify the remote username. Also, the `ssh` command has one argument that will be particularly helpful once you get past chapter 11. The `-Y` option allows the remote machine to send windows back to the current machine. Use "`ssh -Y` *machine*" to activate this.

What if you are on a Windows machine? How can you connect then? Your windows machine should likely have `telnet` on it if you bring up the command prompt. As was mentioned above, this type of connection will likely be blocked for security reasons. Thankfully, there are free `ssh` programs that you can get for Windows. One of these is called `putty` and it will allow you to `ssh` into a remote machine and use it from a Windows box. If you decide you like the Unix/Linux command prompt you can also install Cygwin under Windows and get `rsh`, `ssh`, and all the other commands we will talk about on your Windows machine.

Aside

You can get remote windows sent back to your machine through Putty using X-forwarding. This is a bit more complex so you should look at the help for Putty to see how to do it.

An even better approach if you have a Windows machine is create a dual boot so you can run Linux on the machine. There are versions of Linux such as Ubuntu® and Linux Mint® which are specifically targeted at novice users or people who want to just get things running without much difficulty. You can download these from their websites at `http://www.ubuntu.com` and `http://linuxmint.com` respectively. If you do not want to dual boot your machine, you can even run Linux in a window under Windows using free virtualization software such as VirtualBox®(`http://www.virtualbox.org`) or VMWare Player®(`http://www.vmware.com/products/player/`).

`telnet`, `rsh`, and `ssh` all give you ways to log into a remote machine and execute commands on that machine. Sometimes what you need is to move files from one machine to another. For this you should probably use `scp` or `sftp`.[6] The `scp` command stands for secure copy. You use it much like you would the normal `cp` command. The only difference is that your file names can be prepended with a machine name and a colon to say that they come from or are going to a different machine. A sample invocation is shown here.

```
scp cs.trinity.edu:Desktop/fileINeed.txt .
```

This copies the file `fileINeed.txt` from the `Desktop` directory on the machine `cs.trinity.edu` to the current directory on the current machine. If the second argument

---

[6]There are also `rcp` and `ftp` programs, but just like `rsh` and `telnet`, secure systems likely will not let these through.

has a machine name with a colon then the file will go to that remote location. As with `cp`, the `-r` option can be used to copy an entire directory structure.

The `sftp` command is a secure version of the classic `ftp` program. `ftp` stands for File Transfer Protocol. It allows you to navigate local and remote directory structures, see what files are on the remote machine, and then move them across. The full usage of `sftp` is beyond the scope of what we want to cover here, but you should be aware it exists in case you need to move files around and are not certain what you want to move or where it should be moved to or from.

A.1.1.7 Other Commands

There are many other commands that are available to you on the Unix/Linux command line. You do not have to be familiar with all of them to get things done with the command line. In fact, after this appendix you should have enough to get through the tasks that you will need for this book. There are just a few more commands you might find helpful that we will list here.

- `clear` – Clears the terminal so you get a fresh screen. This is only cosmetic, but it can be helpful at times.

- `df` – Stands for disk free. This will list all of the different volumes on the current disk system and show you information on their usage.

- `du` – Stands for disk usage. This will give you a rundown of how much disk space is being used in the current directory and its sub-directories.

- `echo` – Prints whatever follows the command back to the terminal.

- `find` – Find files that have different properties specified on the command line.

- `grep` – Searches for text inside of the specified files.

- `head` – Print the first lines of a file.

- `ps` – Show information about processes running on the machine.

- `tail` – print the last lines of a file.

- `touch` – Updates the edit time on the specified file to the current time.

- `top` – Lists the top resource consuming programs currently running on the machine.

- `w` – Tells you who is logged onto your machine, where they are coming from, and what they are running.

- `wget` – You give this command a URL and it will download the file at that URL to the current directory.

- `which` – This command should be followed by the name of an executable. It will tell you the full path to that executable if it is in the current path.

As a final tip, you can follow any command with `&` to make it run in the background.

A.1.2 I/O Redirection

So far we have been talking about the different commands we can run from command line. We have typed input into the console when it was needed, and the commands have output to the terminal screen so we could see it. The Unix/Linux command line gains a significant amount of power from the fact that you can redirect input and output. The simple forms of redirection have the output of a program go to a file or have the input come from a file. These will be of use to you in many of the projects in later chapters even if you do not see the benefit right now.

To send the output of a program to a file you put a greater than, >, followed by a file name after the command. The output will go into that file. Here is an example.

```
mlewis@mlewis-laptop:~/Desktop$ ls -l > list.txt
mlewis@mlewis-laptop:~/Desktop$ ls -l
total 16
-rwxrwxrwx 1 root root 1261 2008-11-28 12:21 AdobeReader.desktop
-rw-r--r-- 1 mlewis mlewis 259 2010-08-12 22:32 list.txt
drwx------ 2 mlewis mlewis 4096 2010-08-11 19:59 NewDir
-rw-r--r-- 1 mlewis mlewis 250 2009-12-22 11:03 QtCreator.desktop
mlewis@mlewis-laptop:~/Desktop$ cat list.txt
total 12
-rwxrwxrwx 1 root root 1261 2008-11-28 12:21 AdobeReader.desktop
-rw-r--r-- 1 mlewis mlewis 0 2010-08-12 22:32 list.txt
drwx------ 2 mlewis mlewis 4096 2010-08-11 19:59 NewDir
-rw-r--r-- 1 mlewis mlewis 250 2009-12-22 11:03 QtCreator.desktop
```

The first call to `ls` does not print to screen. Instead, it sends it to the file `list.txt` because of the greater than followed by that file name. We can then do another `ls` and let it print to screen to verify the new file is there. Indeed, we have a new file that is 259 bytes in length. We can also use `cat` to view the contents of the file, and you see the output you expect from `ls`. There is one interesting aspect to it. The file we are outputting to is there, but it has a size of zero. This is because the file is created as soon as we execute the command, but it does not get contents written to it immediately. That happens over time as parts of the output get big enough that they have to be moved from a buffer in memory to the file.

Using > for redirecting output will create a new file if none is there and will wipe out any existing file if one is there. If you want a command to add onto an existing file, use » instead. This will redirect the output so that it appends the output to an existing file.

We can also tell a command to take it's input from a file instead of from standard input using the less than, <, symbol. None of the commands we have looked at require additional input from standard input unless it requires a password which you do not want to put in an unencrypted file. Many of the programs that you write will need to have you type text into them. You will often have to run these programs multiple times, and it can get very tedious to input the same values over and over while you try to get the program to work. Having it take the input from the file can save you time as you only have to enter it into the file once, and then use the file when you are testing the program.

The real value of redirection power comes from the ability to send the output of one command into the next command as the input. This is called "piping", and it is done with the vertical bar, |, a symbol that is read as pipe. Perhaps the most common usage of the pipe is to send an output that is long into `grep` to search for something in it. This example here will find any files in the current directory that were last edited in December of 2014. For the directory listings we have done so far this type of thing would not be of much use, but it is more helpful in large directories like the one this example was executed on.

```
mlewis@mlewis-laptop:~$ ls -l | grep 2014-12
drwxr-xr-x 6 mlewis mlewis    4096 2014-12-22 11:03 qtsdk-2009.05
drwxr-xr-x 4 mlewis mlewis    4096 2014-12-23 17:33 QtWorkspace
-rwx------ 1 mlewis mlewis 2291366 2014-12-28 12:21 SwiftVis-0.3.0.jar
```

Another example is when some programs that are long running produce very long output files. Even if you use `grep` to cut the output down to just lines that have what you are interested in, that might still be many pages of material. You could pipe that output to programs like `head`, `tail`, or `less` to allow you to look at them in a manner that is easier to handle. For example, if you have a log file for a program that includes certain lines that have the word "Step" that you care about and you only want to see the last few you could do this.

```
grep Step log | tail
```

A.1.3 Text Editors (vi/vim)

Most programming languages have programs that are written in plain text. As such, you can use any simple text editor that you want to edit your programs. Word processors, such as Microsoft Word® or Open Office® are not plain text editors. You should not use them to edit your programs. Word processors store all types of additional information such as the font and format of text in addition to the text itself. Programs need only the straight text. On a Windows machine, Notepad is an example of a text editor.

You could use Notepad to edit Scala programs, but not all text editors are equally useful for programming. Due to the fact that programming is a fairly common usage for plain text editors, many text editors have features in them that are specifically aimed at programming. On Unix/Linux a primary example of this is the `vi` editor. While there are lots of different editors that one can choose from on Unix/Linux, we will work with `vi` because it is lightweight and is installed on virtually all such machines.[7] Many other editors will bring up separate windows or are not part of a default installation. If you decide you do not like `vi` or have another preference, you are more than welcome to use whatever editor you want in going through this book.

To start running `vi`, simply type in `vi` followed by the name of the file you want to edit. If the file exists, you will see its contents on your screen. If not, you will get a mostly blank screen with tildes ( ) down the left side and some extra information at the bottom.

The first thing to know about `vi` is that it has two main modes. When you first start it up, you are in `command` mode. To type things you need to be in an `editing` mode. When you are in an editing mode you can type just like you would in Notepad. What gives `vi` the extra power that is helpful for programmers is the ability to use the command mode. To get from command mode into a edit mode, type one of the following.

- `i` – Insert before the current character.

- `I` – Insert at the beginning of the current line.

- `a` – Append after the current character.

- `A` – Append at the end of the current line.

---

[7]Many Linux installs technically come with `vim` instead of `vi`. `vim` stands for "`vi` improved" and everything covered in regards to `vi` here will work for `vim`. On those systems executing `vi` will typically run `vim` by default, so, it is likely you will never notice a difference. In fact, a few features described in this section, like multiple undo and redo, only work in `vim`, not in standard `vi`.

- **R** – Start replacing characters from the current position.

Most of the time you will probably just use **i**, but there are occasions when the others can be helpful. The line at the bottom of your terminal should change to show you that you are now in an edit mode. After you get into edit mode you start typing. In **vim** you can also use the arrow keys along with **Home**, **End**, **Page Up**, and **Page Down** to move around. If you are in a true **vi** install that will not work, instead you will have to return to command mode to move the cursor around.

To get back into command mode, simply hit escape (**Esc**). The bottom line will let you know that you are back in command mode. In command mode you can move the cursor around. In **vim** the special keys will still work for this. If you are in a true **vi** install, you will need to use the following keys to move around.

- **h** – Move left.

- **j** – Move down.

- **k** – Move up.

- **l** – Move right.

- **Ctrl-d** – Page down.

- **Ctrl-u** – Page up.

You can do other edits while you are in command mode. Here are some of the other keys that do things for you in command mode.

- **x** – Delete a character.

- **dd** – Delete a line and place on the clipboard. (Precede with a number for multiple lines.)

- **yy** or **Y** – Yank a line. This copies to the clipboard. (Precede with a number for multiple lines.)

- **p** – Paste the clipboard after the current line.

- **P** – Paste the clipboard before the current line.

- **r** – Replace a single character.

- **J** – Join lines.

- **/** – Search for something. Enter the search string after the /.

- **n** – Repeat the last search.

- **cw** – Change the current word. This removes up to the next white space and goes into insert mode.

- **.** – Repeat the last command.

- **u** – Undo the last command.

- **Ctrl-r** – Redo the last undone command.

If you type in a number before any command in `vi`, that command will be repeated that number of times. This includes going into an edit mode. So if you type 100 and 'A', any text you add in will be added 100 times when you hit `Esc`. If that is not what you wanted, you will probably find that the 'u' command comes in handy.

Some commands work well in patterns. For example, if you want to do a find and replace where you look at each instance before doing the replace, you might use '/' to search for the first instance, then use 'cw' to replace the word in question and hit `Esc`. After that, you use 'n' to move to the next instance of the string and '.' to repeat the 'cw' command if you want to replace that instance. The '.' command is something that would only have minimal value in writing a term paper, but comes in handy a lot more with programming tasks.

There are also a whole set of commands that you invoke by typing a colon. For example, you might have wondered how you save files or how you get out of `vi` all together. Here are just a few of the colon commands that you might find helpful.

- `:n` – Jump to the nth line of the file where n is a number.

- `:w` – Save the file as it is.

- `:q` – Safe quit of vi. This will not work if you have made changes since the last save.

- `:q!` – Unsafe quit of vi. This will quit and throw away any unsaved changes.

- `:wq` – Save the file and quit vi.

You should spend some time getting used to these commands. If you really want to see what `vi` can do, read the `man` page. Many students who are new to `vi` fall into the pit of hitting 'i' as soon as they start up and then using it like Notepad until they are ready to save. This approach ignores all the powerful features of `vi` and, in the long run, slows you down. Try to get used to hitting `Esc` whenever you pause in your typing. If nothing else, it will allow you to also hit ':w' so you save your changes frequently. On systems with `vim`, you can run `vimtutor` to get a guided tour.

---

vi **Settings**

You can change settings in `vi` by editing a file in your home directory with the name ".exrc". In particular, you should consider adding this line.

```
set tabstop=2
```

This makes it so that tabs display as four characters. You can use 4 if you want, but the default is 8 which is a bit too large to work well with Scala.

A.2   Windows

Based on installation statistics, odds are good that the computer you use most of the time is running some form of Using Microsoft Windows®. That will not prevent you from programming or doing any of the activities described in this book. One of the advantages

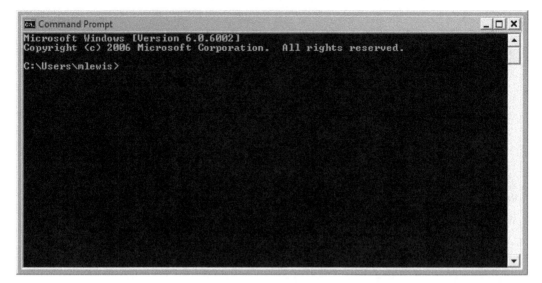

FIGURE A.2: This figure shows the Windows command prompt. This is the command-line interface for Windows.

of the Scala language is that it is platform independent and will work equally well on many operating systems, including Windows.[8]

A.2.1　Command Line

Most people interact with Windows using the GUI. This is so much the case that you might not even realize that there is a command-line interface. The `Command Prompt` is the command line interface program used to execute commands in Windows. How you access this program depends on the version of Windows you are using. In Windows 7, look under the `Accessories` option on the `Programs` menu you will see an option for `Command Prompt`. In Windows 8, you can access this program by right-clicking on the Start button or pressing the `WIN` key and `X` key together. Each of these different actions brings up a menu with `Command Prompt` as one of the menu options.[9] If you select and run this, you should see a window that looks like figure A.2. Like the terminal for Unix/Linux, this is a text interface that gives you some type of prompt with basic information. In the figure it shows you the current directory followed by a '>'. In this case we are in the directory "C:\Users\mlewis".

While the Windows command prompt looks and acts much like a Unix/Linux terminal, it has a different set of commands. The history of Windows and its command prompt are rooted in DOS (Disk Operating System) and the commands that are used in the command prompt for Windows today are largely the same as for that purely command-line based operating system. One of the differences between Unix/Linux and Windows that should be noted immediately is that unlike Unix/Linx, the Windows system is not case sensitive. Commands will be presented here in lowercase, but they will work fine in uppercase or mixed case.

---

[8]Even though you can do your coding under Windows, you might want to consider installing Cygwin or running Linux through virtualization or as a dual-boot configuration just to get experience with something new.

[9]As of the time of writing, Microsoft® had announced that the `bash` shell will be included in Windows 10, but details on how to access it are not available. Once that is available, Windows users can use the same tools as described for Unix/Linux.

A.2.1.1 Files and Directories

As with the Unix/Linux environment, the folders that you are used to in a GUI are called directories on the command line, and they are used to organize files on the computer. There are a number of different commands that can be used for working with directories and files with the Windows command prompt.

- `cd`/`chdir` – Changes the current directory.

- `copy` – Copy files from one location/name to another.

- `del`/`erase` – Deletes files. This does not move them to Trash like doing a deleter from the GUI. Once this has been executed the files cannot be recovered.

- `dir` – Lists the contents of the specified directory or the current directory if none is listed.

- `md`/`mkdir` – Make the specified directory.

- `more` – Print the contents of a file one screen at a time.

- `move` – Moves a file to a different directory.

- `path` – Let's you see the current path or change the path.

- `ren`/`rename` – Rename an existing file.

- `rd`/`rmdir` – Remove a specified directory.

- `tree` – Display the contents of a directory or drive as an ASCII based tree.

There are some significant differences between directories under Windows and those under Unix/Linux. The first is that directories are separated by a backslash, '\', instead of a forward slash, '/'. This is something you have to keep in mind, but it does not significantly impact how any commands are used. The other difference is that Windows machines have different drives, each with its own directory tree.

The prompt above showed the current directory as being "`C:\Users\mlewis`". The `C:` at the beginning of this specifies the drive. The drives on a Windows machine are specified as single capital letters. By default, `C:` is the master hard-disk of the machine. If you want to deal with locations on other drives you need to be certain to specify the drive name in the path you are using. In addition, by default the `cd` command will either change drives or move you to a different directory. To do both you have to give the '`/d`' option.

Aside

The `C:` standard for the hard drive arose early on. The first DOS machines did not have hard-drives. Instead, they typically had one or two floppy disk drives. These were named `A:` and `B:`. When hard drives were added on, they took the next letter in the alphabet. Even though it is very rare to find a computer with a floppy disk drive these days, the main disk still gets the `C:` distinction and other drives use higher letters in the alphabet.

By contrast, different drives in Unix/Linux systems are represented as directories under the main directory tree that starts with /. Removable drives are typically mounted in subdirectories of `/media`.

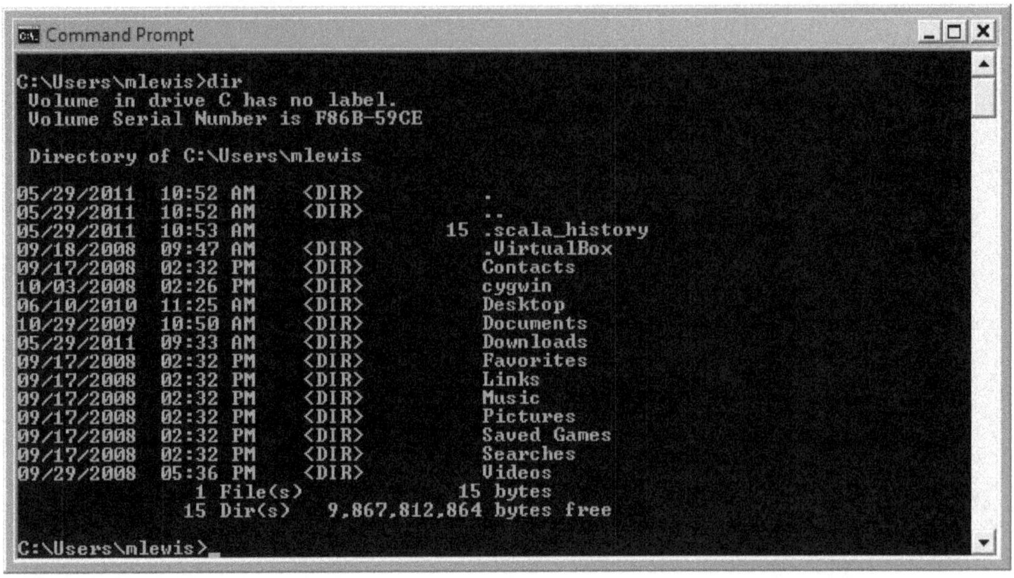

FIGURE A.3: This figure shows the command prompt from Windows after the dir command has been run.

Figure A.3 shows what the command prompt looks like after the `dir` command has been run. This shows a lot of the same information that you get from running "`ls -l`" on a Unix/Linux machine.

A.2.2   Text Editors

No matter what the operating system, programs are still generally written in plain text files so you need a plain text editor for doing that. In this section we will look at some of the options you have for this in Windows.

A.2.2.1   Edit

Working on the command line in 32-bit Windows the command "`edit`" will bring up a text based program that works much like a GUI. You can see what it looks like in figure A.4. It has a set of menus that you can access by hitting `Alt`. These let you do the normal things you would expect from an editor. The `edit` program is much more of a general purpose editor and is much more user friendly than `vi`. As a result, you can use it without too much introduction. The down side is that it lacks features that are specifically beneficial for programming. If you have a newer Windows install it is probably 64-bit and there will not be an install of `edit`. You can use a GUI based text editor instead.

A.2.2.2   Notepad

Of course, you do not have to edit your text files in the command prompt. You can feel free to bring up Notepad on a Windows machine and edit files that way. Then you can use the command prompt just for running the programs that need to use that interface. The simple GUI of Notepad has the same benefits and drawbacks of `edit`. It will work fine for small things, but it was not built for programming and will slow you down at a certain point.

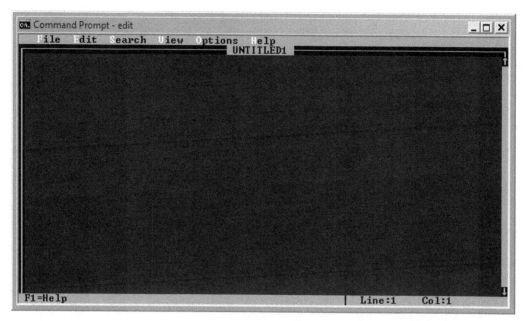

FIGURE A.4: This figure shows a command prompt with the edit program running.

A.2.2.3 Others

There are many other text editors that you could use on Windows which are more directly aimed at programming. Attempting to list all such editors and provide any type of comparison or judge their relative merits in this space would be pointless. Not only are there too many of them, they can change over time. Doing a web search for "text editor" will likely provide you with many different options. In addition, other people have done the work of compiling a list with comparisons on Wikipedia®. Simply go to `http://en.wikipedia.org/wiki/Comparison_of_text_editors` and you can see tables comparing different editors and check out other information on the ones that seem interesting.

A.2.3 Other Commands

There are a number of other commands that you might find it helpful to know when working with the Windows command prompt.

- `cls` - Clears the screen and puts you at a blank prompt. This command only impacts appearance, but can be useful at times.

- `echo` - Like the Unix/Linux command, this will print out whatever text is passed in to it.

- `exit` - This will close the command prompt.

- `find` - This is the equivalent of `grep` in Unix/Linux. It will find strings in a file for you.

- `help` - This gives you help on commands. If you enter it without specifying a command it will list the possible commands for you.

- `where` - Takes one argument which is the name of a command. It will tell you the full path of that command.

- `whoami` - Tells you the name of the user account you are logged in as.

In addition to these commands, I/O redirection with '<', '>', '»', and '|' work in the Windows command prompt just like they did in Unix/Linux. You can also use tab completion in Windows to auto-complete file names and save yourself some typing or the possibility of typos. The use of tab in Windows is particularly helpful because many files and directories have spaces in them that have to be handled properly. If you use tab completion, the system will automatically add double quotes around things so that spaces are dealt with appropriately.

A.3   End of Appendix Material

A.3.1   Summary of Concepts

- The software you use to interact with your computer comprises a tool set. The nature of these tools is extremely important to software developers.

- Most of this book is written with the expectation the reader is using command-line tools. These vary between operating systems.

- Linux and Unix based operating systems, including Mac OS X, include a powerful command-line that is often utilized.

  - There are a number of different commands that you use to give instructions to the computer using the command-line. You will likely commit these to memory over time as you use them. The real power of these commands is that they can be given arguments to make them behave in specific ways or do specific things.

  - Files are organized in directories. Some of the most important commands you will need to learn are for dealing with directories and navigating the directory structure. The name . always represents the current directory and .. represents the parent directory.

  - Input and output for commands can be redirected in different ways.

    * Use > to send the output of a program to a file, deleting anything that is there.
    * Use » to send the output of a program to a file, appending to the end of current contents.
    * Use < to have the contents of a file act as the standard input for a program.
    * Use | to have the output of one program sent to the input of another. You can chain multiple programs together with this technique called "piping".

  - Programs are typically written in plain text with a text editor. The text editor described in this appendix is vi. It is very common on Linux/Unix installs, and the capabilities of the command mode make it a good editor for programming.

- Windows also incorporates a command prompt that resembles the older DOS interface.

  - There are a different set of commands that you can use in the Windows command prompt. I/O redirection works like it does in Linux/Unix.

– The command-line text editor is called `edit`. It is not included in 64-bit Windows installs. You can use Notepad or some other GUI text editor. Microsoft Word and other word processing programs are not text editors.

A.3.2 Exercises

1. Make a directory for the work associated with this book in your user directory.

2. Enter the "Hello World" program in the directory you created. Make sure the file name that you use ends with `.scala`. Run it using the `scala` command.

3. (Linux/Unix/Mac) Use `wget` to download the file for the primary page of google. This should give you a file called `index.html`. Move this to the name `google.html`. Count how many times the word "google" appears in that file. Describe how you did this and how many times it occurred.

4. Ask your instructor if you have a web space on the machines you work on. If so, make a directory in that space called `CS1`. Put a file called `persinfo.txt` in that directory that tells your instructor a little about you and why you are taking the course. Set permissions on the directory and the file so that they are visible for the web (`a+rX` for Linux). You can test if you did this correctly by pointing a browser at your web space to make certain the files are there.

5. (Linux/Unix/Mac) The `ps` command can tell you what is running on a machine. Run "`ps -ef`" and send the output to a file called `allprocs.txt`. Look at the file to see what all is running.

6. (Linux/Unix/Mac) Do a long listing (`ls -l`) of the `/etc` directory. Pipe the output to `grep` to see how many files contain your first and last initials (e.g. `jd` for John Doe).

Additional exercises can be found on the website.

Appendix B

Glossary

Affine Transform This is a term from graphics that is defined as any transformation of the space where parallel lines are preserved. The basic affine transforms are translation, rotation, scale, and shear. Any combination of these is also an affine transform.

Argument A value that is passed into a function of method.

Array A basic collection of values that is a sequence represented by a single block of memory. Arrays have efficient direct access, but do not easily grow or shrink.

Class A construct that works as the blueprint for objects.

CPU The term Central Processing Unit is used to describe the primary computing element in a computer. This is where most of the work for the programs you are writing takes place.

Conditional An expression of the `Boolean` type that is used to determine if code is executed.

Expression A sequence of tokens that has a value and a type.

File An independent grouping of information in the static storage of a computer.

Function This is a concept from mathematics of something that maps from values in a certain domain to values in a certain range. In programming, functions can serve this same purpose, but they also more generally group statements together under a particular name to allow you to break code into pieces and give it meaningful names.

GUI Short for Graphical User Interface. This is an interface that includes graphical elements that the user interacts with using a pointing device possibly in addition to keyboard input.

Higher-Order Function This is a term used to describe a function whose inputs or outputs include function types.

if A simple conditional construct that picks between one of two options based on whether or not a Boolean expression is true. In Scala, the `if` is a valid expression, but can also be used as statements.

Instantiation The act of creating an object that is an instance of a particular type.

Iteration The act of running through steps or elements one at a time.

List An abstract data type that stores items by a numeric index and allows random access, insertion, and removal.

Logic Error This is an error in which a program compiles and runs to completion, but produces the wrong output. These are typically hard to find and fix as the computer gives you limited information about what is wrong.

Loop A construct designed to execute code repeatedly.

Multithreading The act of having a program that allows more than one thread of control to be active at a time. This effectively allows multiple instructions to execute simultaneously. This is a form of shared memory parallelism.

Parallel In the context of programming, this is when two or more things are executing at the same time. This generally requires multiple computational elements such as cores, processors, or even full computers.

Parameter This is a place holder for a value that will be passed into a function.

Recursion When a function or method calls itself. In mathematical terms, this is when a function is defined in terms of itself.

Runtime Error This is an error that causes the program to crash/terminate while running. This means that the code is valid Scala that compiles, but it has a flaw that produces throws and exception or error. When this happens, Scala prints a message and a stack trace.

Search The process of looking for a particular element or its position in a collection.

Sequence (Seq) A type of collection characterized by items having a particular order and being referred to by an integer index.

Signature The signature of a method/function includes the name along with the parameter types and the return type. From the signature you can see how a method/function should be called. The return type also lets you know what you can do with the result of the call.

Sort The act of putting the items in a sequence into the appropriate order according to some comparison function.

Stack Trace A listing of the call stack of a program. This shows what line in what function a program is at as well as what line and function called that going back many levels of calls.

Statement A set of tokens that represents a complete command to a language. In Scala, any expression can be used as a statement.

Syntax Error This is an error you are notified of during the compile stage which exists because you wrote code that violates the syntax of the language. These are typically the easiest errors to deal with because the compiler can give you an informative message with a line number. If you are working in an IDE, syntax errors will often be shown with a red underscore, much like spelling errors in a word processor.

Thread A single unit of control in a program that shares memory with other threads.

Token The smallest element of a programming language that has meaning on its own and which changes meaning if altered or broken apart with whitespace.

Type A construct that specifies a set of values and the operations that can be performed on them.

Variable A construct that associates a name to a reference to a value.

XML This is short for eXtensible Markup Language. XML is a standard, plain text format that can be used to represent various forms of data in a tree-like structure. It is called extensible because users are allowed to develop their own tags to represent data in the manner they want.

Bibliography

[1] Mark C. Lewis and Lisa L. Lacher. *Object-Orientation, Abstraction, and Data Structures Using Scala*. Chapman & Hall/CRC, 2st edition, 2016.

[2] Michael P. Marder. *Research Methods for Science*. Cambridge University Press, New York, NY, USA, 1st edition, 2011.

[3] Martin Odersky, Lex Spoon, and Bill Venners. *Programming in Scala: A Comprehensive Step-by-Step Guide, 2nd Edition*. Artima Incorporation, USA, 2nd edition, 2011.

Index